Energy and the Wealth of Nations

Charles A.S. Hall · Kent A. Klitgaard

Energy and the Wealth of Nations

Understanding the Biophysical Economy

Springer

Charles A.S. Hall
Professor of Systems Ecology
Faculty of Environmental &
Forest Biology and Graduate Program
in Environmental Science
College of Environmental Science &
Forestry, State University of New York
Syracuse, NY 13210, USA
chall@esf.edu

Kent A. Klitgaard
Professor of Economics
Patti McGill Peterson Professor
of Social Sciences, Wells College
Aurora, NY 13026, USA
kentk@wells.edu

ISBN 978-1-4419-9397-7   e-ISBN 978-1-4419-9398-4
DOI 10.1007/978-1-4419-9398-4
Springer New York Dordrecht Heidelberg London

Library of Congress Control Number: 2011938144

Printed on acid-free paper

Springer is part of Springer Science+Business Media (www.springer.com)

To Myrna, my wonderful companion on this and other journeys

– Charles A.S. Hall

To my children, Justin and Juliana Klitgaard-Ellis, in hopes that the information contained herein can make their world a little better place in which to live, and to Deb, who gives my life meaning.

– Kent A. Klitgaard

Preface

There are four books on our shelves entitled, more or less, "wealth of nations." They are Adam Smith's 1776 pioneering work, *An Inquiry into the Nature and Causes of the Wealth of Nations*, and three of recent vintage: David Landes' *Wealth and Poverty of Nations*, David Warsh's *Knowledge and the Wealth of Nations*, and Eric Beinhocker's *The Origin of Wealth*. Warsh's book is rather supportive of current approaches to economics whereas Beinhocker's is critical, but all of these titles attempt to explain, in various ways, the origin of wealth and propose how it might be increased. Curiously, none have the word "energy" or "oil" in their glossary (one trivial exception), and none even have the words "natural resources." Adam Smith might be excused given that, in 1776, there was essentially no developed science about what energy was or how it affected other things. In an age when some 70 million barrels of oil are used daily on a global basis, however, and when any time the price of oil goes up a recession follows, how can someone write a book about economics without mentioning energy? How can economists ignore what might be the most important issue in economics? In a 1982 letter to *Science* magazine, Nobel Prize economist Wassily Leontief asked, "How long will researchers working in adjoining fields . . . abstain from expressing serious concern about the splendid isolation within which academic economics now finds itself?" We think Leontief's question points to the heart of the matter. Economics as a discipline lives in a contrived world of its own, one that is connected only tangentially to what occurs in real economic systems. This book is a response to Leontief's question and builds a completely different, and we think far more defensible, approach to economics.

For the past 130 years or so economics has been treated as a social science in which economies are modeled as a circular flow of income between producers and consumers where the most important questions pertain to consumer choice. In this "perpetual motion" of interactions between firms that produce and households that consume, little or no accounting is given of the necessity for the flow of energy and materials from the environment and back again. In the standard economic model energy and matter are ignored or, at best, completely subsumed under the terms "land," or more recently "capital," without any explicit treatment other than, occasionally, their price. In our view economics is about stuff, and the supplying of services using stuff, all of which is very much of the biophysical world, the world best understood from the perspective of natural, not social sciences. But within the discipline of economics, economic activity is seemingly exempt from the need for energy and matter to make economies happen, as well as the second law of thermodynamics.

Instead we hear of "substitutes" and "technological innovation," as if there were indefinite substitutes for matter and energy. As we enter the second half of the age of oil, and as energy supplies and the social, political, and environmental impacts of energy production and consumption become increasingly the major issues on the world stage, this exemption appears illusory at best. All forms of economic production and exchange involve the transformation of materials, which in turn requires energy. When our students are exposed to this simple truth, they ask "why are economics and energy still studied and taught separately"? Indeed, why is economics construed and taught only as a social science, in as much as economies are as much, and perhaps even principally, about the transformation and movement of all manner of biophysical objects in a world governed by physical laws.

Part of the answer lies in the recent era of cheap and seemingly limitless fossil energy which, ironically, has allowed a large proportion of humans basically to ignore the biophysical world. Without significant energy or other resource constraints, economists have believed the rate-determining step in any economic transaction to be the choice of insatiable humans attempting to get maximum psychological satisfaction from the money at their disposal, using markets that have an infinite capacity to serve these needs and wants. Indeed the abundance of cheap energy allowed essentially any economic theory to "work" and economic growth to be a way of life. All we had to do was to pump more and more oil out of the ground and economic growth could happen, no matter the theory. However, as we enter a new era of "the end of cheap oil," in the words of geologists and peak oil theorists Colin Campbell and Jean Laherrère, energy has become a game changer for economics and anyone trying to balance a budget.

- Provides a fresh perspective on economics for those wondering "what's next" after the crash of 2008 and the subsequent economic malaise in much of the world
- Summarizes the most important information needed to understand energy and our potential energy futures

In summary, this is an economics text like no other, and it introduces ideas that are extremely powerful and are likely to transform how you look at economics.

Acknowledgments

We thank the Santa Barbara Family Foundation, the UK Department for International Development, and several anonymous donors for financial support, Jim Gray for excellent editing of words and ideas, Michelle Arnold for assistance in getting the book together, Rebecca Chambers and Ana Diaz for their able assistance with the data analysis, editing and graphics, and our students over the years for helping us think about these issues. The late, fantastic Howard Odum taught us about systems thinking and the importance of energy in everything, and John Hardesty, who introduced Kent to the limits to economic growth. Our colleagues Lisi Krall and John Gowdy provided valuable advice and critique for this an other projects. Their continued collaboration makes our work stronger. We also thank David Packer for believing in us and Myrna Hall and Deb York for loving support and infinite patience.

Contents

Part I   Energy and the Origins of Wealth

1   Poverty, Wealth, and Human Aspirations 3

2   Energy and Wealth Production: An Historical
    Perspective 41

3   The Petroleum Revolution 71

Part II   Energy, Economics and the Structure of Society

4   Explaining Economics from an Energy Perspective 95

5   The Limits of Conventional Economics 131

6   The Petroleum Revolution II: Concentrated Power
    and Concentrated Industries 145

7   The Postwar Economic Order, Growth,
    and the Hydrocarbon Economy 161

8   Globalization, Neoliberalism and Energy 191

9   Are There Limits to Growth? Examining the Evidence 207

Part III   Energy and Economics: The Basics

10   What Is Energy and How Is It Related to Wealth
     Production? 223

11   The Basic Science Needed to Understand the Relation
     of Energy to Economics 251

12   The Required Quantitative Skills 285

13   Economics as Science: Physical or Biophysical? 301

Part IV   The Science Behind How Real Economies Work

14   Energy Return on Investment 309

15   Peak Oil, EROI, Investments, and Our Financial Future 321

16 The Role of Models for Good and Evil.. 339

17 How to Do Biophysical Economics 351

Part V Understanding How Real-World Economies Work

18 Peak Oil, Market Crash, and the Quest for Sustainability: Economic Consequences of Declining EROI 369

19 Environmental Considerations 385

20 Living the Good Life in a Lower EROI Future 393

Index 403

Author Bios

Kent A. Klitgaard is professor of economics and the Patti McGill Peterson Professor of Social Sciences at Wells College in Aurora, New York, where he has taught since 1991. Kent received his Bachelor's degree at San Diego State University and his Masters and PhD at the University of New Hampshire. At Wells, he teaches a diverse array of courses including the History of Economic Thought, Political Economy, Ecological Economics, The Economics of Energy, Technology and the Labor Process, and Microeconomic Theory, and is a cofounder of the Environmental Studies Program. Kent is active in the International Society for Ecological Economics and is a founding member of the International Society for Biophysical Economics. Recently, his interests have turned towards the degrowth movement, and he has published multiple papers in Journals such as *Research and Degrowth* and *Ecological Economics Reviews*. He has two children, and is interested in the outdoors in general: from hiking to beach walking to the occasional round of golf (despite the high energy use of golf courses). Kent is a Californian who still surfs the frigid waters of New England when he gets a chance. This is his first book.

Charles Hall is a systems ecologist who received his PhD under Howard T. Odum at the University of North Carolina at Chapel Hill. Dr. Hall is the author or editor of seven books and more than 250 scholarly articles. He is best known for his development of the concept of EROI, or energy return on investment, which is an examination of how organisms, including humans, invest energy in obtaining additional energy to improve biotic or social fitness. He has applied these approaches to fish migrations, carbon balance, tropical land use change, and the extraction of petroleum and other fuels in both natural and human-dominated ecosystems. Presently he is developing a new field, biophysical economics, as a supplement or alternative to conventional neoclassical economics, while applying systems and EROI thinking to a broad series of resource and economic issues.

Part I

Energy and the Origins of Wealth

When first encountering the subtitle of this book, *Understanding the Biophysical Economy*, most readers probably asked, "What is a biophysical economy?" The answer is deceptively simple: the word "biophysical" refers to the material world, that which is usually, but not completely, covered by courses in physics, chemistry, geology, biology, hydrology, meteorology, and so on. This can be compared with a "social" or "anthropocentric" (i.e., human-centered) perspective. In this second perspective, which is dominant in our society, humans believe that they can make any world, or set of decisions, or economic systems, that they wish, if they can just get the policies right and enough time has passed for new technologies to come on line. The subsequent world becomes our new reality and truth.

But we must ask how do the powerful, governing physical laws, which we are all prepared to accept in physics, chemistry, and biology classes, operate outside the scientist's laboratory and the "natural" world? Scientists often think of these laws as imposing constraints on a system. Do these constraints really disappear when human ingenuity is applied to economics and markets? Most economics textbooks would lead you to this conclusion as growth is just a matter of human actions, technologies, policies, and a healthy dose of ambition. Western culture and its leading commentators (with a few exceptions such as Joseph Tainter and Jared Diamond) do have a tendency to elevate personal and social aspects of a problem, specifically, human actors and their ideas, above any biophysical considerations. Thus we learn about history as the action of great leaders. In reality, wars—if not always battles—are usually won or lost due to the biophysical resources that generals can bring to bear. Napoleon once quipped that "God usually fights on the side with the best artillery." There is little debate that the South had the better generals in the Civil War, but the North had the industrial might. The North won because of biophysical, not leadership, issues.

Most readers would not argue with the idea that we live in a world that is completely beholden to the basic laws and principles of science. These basic laws include Newton's laws of motion, the laws of thermodynamics, the law of the conservation of matter, the best first principle, the principles of evolution, and the fact that natural ecosystems tend to make soil and clean water and human-modulated systems tend to destroy both. Do economic systems operate outside these laws? Did the seemingly unconstrained technological and economic expansion of the twentieth century show that these laws were irrelevant or at least insignificant when applied to economics and the satisfaction of human needs and wants?

There is no more important question as we attempt to move beyond the recent financial trauma of the "Great Recession." Unfortunately, the biophysical laws, particularly as applied to energy, are not understood or appreciated by most people. Ironically, our focus on exploiting and investing energy in the economic process has divorced many people from the very biophysical realities that are necessary to sustain them. This includes our ways of building dwellings, living in cities, importing food, being transported and entertained, and so on. This book examines these issues through an integrated view of economics that emphasizes scientific principles and a more frequent use of the scientific method. We begin with two chapters designed to demonstrate the importance of energy in human economies through narratives about energy and the recent U.S. economy, and then more generally through an analysis of history. Chapter 3 examines in more depth how petroleum revolutionized our economies and their structures. Together they provide the beginnings of a powerful new way to think about economics.

Poverty, Wealth, and Human Aspirations

The years that ended the first decade of the new millennium were not kind to the economic situations of most people and institutions in the United States and much of the rest of the world, nor to the economic and financial theories that once explained and operated our economies so well, or so it seemed. For the majority of people it has become more difficult to meet basic obligations such as rent or mortgage payments or feeding or educating a family, and especially to do this when diminishing asset values, particularly home values, threaten future financial security. Ten to twenty percent of Americans have no job at all, a poorly paying job in the service sector, or work part time. Incomes for the middle class have been stagnant at best for decades while the size of the middle class shrinks. Many, perhaps most, new college graduates have had to greatly reduce their aspirations. The stock market and real estate have become far less reliable ways to amass wealth. Some 46 of our 50 states and many of our municipalities face crippling budget deficits, and many colleges, pension plans, charities, and other institutions are operating with diminished funds or going bankrupt. Even the U.S. government faces the prospect of seeing its credit rating diminished. "Tea Partiers" seek to cut debt and the role of government even while poll after poll shows the public does not want its health or most other benefits cut. There are many pronouncements about "waiting, or borrowing, until the economy grows again," but little evidence of that growth happening. The inflation-corrected GDP of the United States was about the same in 2010 as it was in 2004.

Such dire financial conditions have not traditionally been the stuff of the United States, where what is often called the "American dream" promised, and generally delivered, an economic situation that improved decade to decade and generation to generation. Few questioned the dominant economic paradigm, which has been called variously industrial capitalism, growth-oriented economics, or neoclassical economics. Major financial publications such as the *Wall Street Journal* [1] and even Nobel laureates in economics [2] are calling for "new economic models" because it is clear that the old ones are not serving us well. But there are no such new models forthcoming from within traditional economics, which is as befuddled as anyone. Can we do any better? We think so, but to understand how we might reform or improve our economic system, we must understand it better than is the case now. To do this we must understand our economy, and build an economics, from a biophysical or natural science perspective as well as from the presently dominant social science perspective. Why this has not happened already is a very curious subject about which we have no clear answer.

Unlimited Wants, Limited Means

Most humans have a deep-seated desire for a stable, comfortable life, indeed for affluence, but, historically at least, limited means to acquire material wealth. For thousands of years most

C.A.S. Hall and K. Klitgaard, *Energy and the Wealth of Nations: Understanding the Biophysical Economy*, DOI 10.1007/978-1-4419-9398-4_1, © Springer Science+Business Media, LLC 2012

people led lives like their parents, either hunting and gathering, farming, or undertaking some artisanal occupation. If their parents were not members of the aristocracy, or at least successful tradespeople, their own chance of seeing anything resembling affluence was close to zero. Most wealth in the past, even the very modest amount available by today's standards, depended upon ubiquitous, reliable, but diffuse energy from the sun. Owning or having access to land enabled one to capture solar energy and turn it into useful economic products such as timber, fuel wood, crops, or animal products. First sons of landowners usually inherited the land their fathers owned, second or later sons had to try something else, often the military or clergy, or, when new worlds opened up, migration. Peasants, or serfs, did not own or bequeath land, but worked on land owned by others, transforming the captured solar energy into food, fiber, and implements. For most of medieval times custom and tradition declared that the peasantry could not be displaced from the land. For the wealthy aristocracy there were normally hundreds or thousands or, in the case of kings, millions of people who worked the land and who were taxed to generate the surplus wealth that enabled these privileged few to live a much more affluent lifestyle. Even the richest kings of the past, however, did not have the affluence of a middle class person today in terms of quality and diversity of diet, transportation options, and a comfortable microclimate. From where did this tremendous new wealth come?

Economists, Human Ingenuity, and the Origin of Wealth

According to Gladwell [3] most people attribute the success of those who make a lot of money, or are otherwise successful, to their own special characteristics or efforts, that is, that they are successful because they are blessed with superior intelligence, work especially hard, have special skills, and so on. Parents everywhere preach this lesson, and certainly there is a lot of truth in it, but perhaps not as much as those fortunate individuals may wish. Gladwell undertook a

comprehensive analysis of people who by various criteria were very successful, and found that when he analyzed the success of contemporary people from Bill Gates to young Canadian hockey players, the statistically strongest predictors of success were the circumstances that they were born into, such as the time of their birth or the financial level, education, ethnicity, and so on of their parents. Bill Gates had access to computers at 13 when almost no other 13-year olds did, and came along when such skills were rare but critically useful. Professional Canadian hockey players are four times more likely to be born in the first quarter of the year than the last quarter, because their early birth made them on average larger, more mature and better players in their respective youth programs, leading progressively to higher levels of coaching and play. They were not necessarily intrinsically better hockey players, as most fans would presume, but lucky in the month in which they were born. In other words successful Canadian hockey players tended to have a *physical* advantage over those who did not make it simply because of the month in which they were born.

Economists today, in a way perhaps analogous to most Canadian hockey fans, tend not to think too much about the physical origin of wealth, but rather the importance of human efforts and ingenuity and the social question of how people maximize their happiness by spending their money on goods and services within markets. Thus economics has become almost entirely a social science today, focused on what humans think, want, and need. This is consistent with Ehrenfeld's [4] notion of "humanism," that is, a completely human-centered frame of reference.

But earlier economists did not especially think that way. In the eighteenth century, the first formal "school" of economists (called *Physiocrats* and centered in France) wrote at length about the origin of wealth and their belief that wealth came from the land (which was sometimes associated with natural resources or raw materials) and the agricultural labor needed to transform the "free gifts of nature" into useful commodities. In the first half of the nineteenth century the *classical* school of economists, including Adam Smith,

David Ricardo, and, later, Karl Marx, developed the idea that generally it was *labor* that was the principal generator of wealth, although they thought land could be important too. Land was important in the sense that it provided *use value*, that is, useful products such as food, fiber, minerals, and energy. Exchange value, the basis of price, depended solely upon the quantity of human labor needed to transform nature. Ricardo, especially, treated the products of nature as free, and concentrated primarily on commodities (now called goods and services) that could be reproduced by human labor. Critics of the classical school, such as James Maitland, concentrated on the difference between value and wealth. *Value* was a flow that resulted from the production of commodities sold on markets. That value depended primarily on the amount of human labor expended directly on production itself, as well as upon the machines that augmented human labor. *Wealth* was a stock that was derived primarily from the use values of nature, that is, ownership of land. Classical political economists all held a theory of money that declared money was a universal equivalent of commodity values. Using money as a metric could overcome the vastly different qualities of different products as well as the different processes by which they were produced. By the nineteenth century, however, wealth came to be seen as accumulations of money. A wealthy person was one who possessed large stocks of money. Thus the earlier thoughts on the origins of wealth in nature became, for the most part, lost to history. Here is where the problems with economics begin.

Later *neoclassical* economists, the school that is overwhelmingly dominant today, were even less concerned with the origin of wealth and much more on exchanges in markets, including the benefits they believed markets brought to humans. Value was something that markets assigned, that is, the price. In doing so, they denied that objective costs of production, such as the amount of human labor used in the production of a good, was the basis of value. In turn, neoclassical economists returned to the idea of use value as the source of wealth. A good was valuable to the degree it was useful to the buyer,

and use value, renamed *utility*, was an important basis of value or price. A difference with the earlier use of the term use value is that value, like beauty, was in the eye of the beholder. Consequently, price theory became essentially subjective, brought together in a strange marriage of eighteenth century utilitarian philosophy and differential calculus. This theory, which depended upon the propositions that all humans are self-interested, individualistic, and "rational" evolved intellectually over the course of the nineteenth century. By the early twentieth century the theory of production was brought within the domain of utility theory. In the relatively rare cases when the origin of wealth flowing through markets was considered, neoclassical economists often used Cobb–Douglas [5] (and similar) *production functions*. These attributed the generation of wealth to some combination of capital and labor, while occasionally considering (although not in the formal equation) land and technology. Technology was considered especially important, as it could compensate for any possible depletion of resources. This perspective was bolstered by an influential paper by Barnett and Morse [6] who found that inflation-adjusted prices for most raw materials tended not to increase over time, which they interpreted as technology compensating for any depletion. Thus with resources seemingly less important, land (and hence natural resources) was never part of the Cobb–Douglas equations. In the 1950s the important neoclassical economist Robert Solow [7] dropped even labor from the equations and said that wealth was generated mostly from capital. Technology remained important but no one knew how to measure technology directly. The most comprehensive empirical assessments of Cobb–Douglas production functions were undertaken by Denison [8] who found again and again that increases in capital and labor explained only about half the increase in economic production. Denison and others attributed the residual (i.e., the increases in wealth not explained by increases in land and labor) to human technological ingenuity. This faith in technology is shared by the majority of Americans: Scott Keeter, who directed a broad survey on the future for the *Smithsonian Magazine*'s Fortieth

Anniversary Issue in 2010 said, "If the U.S. has a national religion, the closest thing to it is faith in technology." This perception that ingenuity is of critical importance is consistent with Gladwell's finding that successful people perceive that it is human intelligence, skills, and hard work that is the basis of their financial success and wealth.

What These Economists Missed: The Role of Energy

Can these issues be explained better from an energy or biophysical perspective than by these existing narratives? The answer is clearly yes. In the mid-1700s the Physiocrats were writing that economic production was principally biological, came from land via forest, agricultural, and animal production (and sometimes mining), and occurred more or less in proportion to land area. The idea of nature, or at least heavily managed nature, as the origin of wealth had been part of French economic thought – especially of land-owners – since the late 1600s. The wealthy tended to be "country gentlemen" who owned large land holdings which intercepted large quantities of sunlight and generated a lot of valuable products through photosynthesis and the backbreaking labor of an impoverished peasantry. (The skewed distribution of this energy surplus was a major factor behind the French Revolution). Clearly the importance of land was as an interceptor and user of solar energy, essentially the only energy source of the time. The more land you owned the more sunlight you could intercept and the more economic work you could do.

Later, Adam Smith wrote during a time when there was a relatively new and greatly increasing production of wealth by concentrating human workers in centralized workshops where the workers' physical energies (sometimes assisted by water power) were used to make various manufactured items. To Adam Smith and other classical economists, labor was the principal means of making wealth because they saw it with their own eyes. Smith held that the wealth of a nation would increase as a function of (1) the number of productive laborers, and (2) the productivity of each laborer, which he believed was increased solely by organizational means (e.g., the division of labor). Smith did not give any special role to energy and, for example, mentions James Watt's "fire engines" only once, and that in an organizational context.

When Solow and other neoclassical economists wrote about the origin of wealth in the middle of the last century, capital seemed much more important than land or labor. Solow believed that capital equipment (represented by physical buildings and the machines within them or their monetary value) was the principal determinant of wealth. But physical capital equipment does not generate wealth by itself, rather it was the means of utilizing the new and increasingly large flows of fossil energy throughout society. Human labor, once so important, had by then decreased to less than one percent of the energy used to generate wealth; the rest was fossil fuels or hydro/nuclear power that flowed through Solow's capital. Thus each school of economics rightfully concentrated on the means by which wealth was generated in their time. In each case, however, what they perceived as important was related to the dominant energy flow that was generating the most wealth at their time, and because little was understood by these economists about energy or its importance in production they tended to focus on proxy values – land, labor, and capital – rather than the true causative agents. We believe that this is one of many examples by which a biophysical explanation can help us to understand the economic process by giving the actual mechanisms by which that process is occurring. The fundamental mechanisms by which all economic processes occur in all time periods require an understanding of the role of energy. Some quantitative measures of energy use are needed for our discussion, and Table 1.1 provides several useful conversions.

Energy can also explain other aspects of the economic story. When energy analyst Cutler Cleveland [9] re-examined the study of Barnett and Morse, he found that the only reason decreasing concentrations and qualities of resources

Table 1.1 Getting a feel for energy units and their conversions (J = Joule, K, M and G refer to thousand, million and billion respectively)

Useful conversions[a]	
One calorie	= 4.1868 J
One kilocalorie (cal or kcal)	= 4187 J
One BTU	= 1.055 KJ
One kWh	= 3.6 MJ
One therm	= 105.5 MJ
One liter of gasoline	= 8.45 MJ
One gallon of gasoline	= 130 MJ (million joules)
One gallon of diesel	= 140 MJ (million joules)
One gallon of ethanol	= 84 MJ (million joules)
One cord dried hardwood	= 26 GJ
One barrel of oil	= 6.118 GJ
One ton of oil	= 41.868 GJ (= 6.84 barrels)
Some basic energy costs	
One metric ton of glass	= 5.3 GJ
One metric ton of steel	= 21.3 GJ
One metric ton of aluminum	= 64.9 GJ
One metric ton of cement	= 5.1 GJ
One MT of nitrogen fertilizer	= 78.2 GJ
One MT of phosphorus fertilizer	= 17.5 GJ
One MT of potassium fertilizer	= 13.8 GJ
1 J	= Picking up a newspaper
1 million joules (1 MJ)	= A person working hard for 3 h
3 million joules (3 MJ)	= A person working hard for 1 day
11 million joules (11 MJ)	= Food energy requirement for one person for 1 day
1 billion joules (1 GJ)	= Energy in 7 gallons of gasoline
1 trillion joules (1 TJ)	= Rocket launch
100×10^{18} J (100 exaj)	= Energy used by United States in 1 year (2009)
488×10^{18} J (488 exaj)	= Energy used by world in 1 year (2005)

[a] Thanks in part to R. L. Jaffe and W. Taylor Energy info card, Physics of energy 8.21, Massachusetts Institute of Technology

were not translated into higher prices was because of the decreasing price of energy and its increasing use in the exploitation of increasingly lower grade reserves. In another example, many economists studied growth of the economy using mathematical tools such as Cobb–Douglas production functions that focused on labor and capital. They always found a large "residual," that is, about half of the increase in economic production could not be explained by the increase in labor or capital. This they attributed to technological innovation. But when physicist Reiner Kummel [10] and his colleagues examined very carefully how economic goods were produced in the United States, Germany, and Japan in recent decades they found that energy was not only important but in fact more important than either the capital or labor that had been used by economists. In other words, when Kummel added energy to the economist's Cobb–Douglas production functions he found that the unexplained residual disappeared and energy was even more powerful than capital or labor in explaining economic growth for these countries. Physicist Robert Ayers and his associates have made similar analyses focusing on energy and have come to similar conclusions: that energy and the way it is used is the most critical issue in the functioning and growth of our economy [11]. Why did the economists who studied growth using Cobb–Douglas production functions not include energy in their analyses? The present authors even sent Denison our early papers that showed the importance of energy. He replied with a nice letter indicating that we had indeed uncovered a very important relation, although his subsequent publications gave no more weight to energy than before! The explanation for his and other economists' near complete disregard for energy is probably no more complicated than our earlier statement that most economists today are social scientists who, like most humans, tend to give personal or social explanations even to biophysical processes. This is why we are trying to encourage young economists to create a new interdisciplinary economics with a biophysical basis [12].

Speaking more generally, this biophysical and energy perspective can integrate much more fully the discipline of economics with the natural sciences, and even within itself. As we noted, energy was critical to the thinking of the earliest economists, although they could not use the language we would use today because the concept of energy was not clear to them or even physical scientists at that time. Economists understood that land was important in the eighteenth century without understanding that it was because most of the energy available for economic production came from the sun. The concept of photosynthesis as energy capture was not, or barely, understood. Likewise in the time of Adam Smith, factories were becoming increasingly important. These factories employed many workers whose muscles provided much of the energy to generate the transformations of raw materials into desired products. Then, as the industrial revolution came about, it was monetary capital that allowed construction of physical capital, that is, the equipment that in turn allowed the use of coal or oil to run machinery. In all cases a biophysical analysis shows that it is the energy that does the actual work in turning raw materials into useful goods and services. Therefore, although we agree that many factors contribute to the production of wealth, the critical element is and always has been energy. Without energy there would be no economies or economics because there would be no goods or services produced or moved from place to place or through markets. The more one controlled the most important energy source of the time, the more wealth production was possible and, because wealth often buys influence, the more political power the person or people who controlled that energy had.

If you ask a physicist or agronomist how something was made, such as a car or a bushel of corn, both would probably say that you start with some raw material from the ground or the air, add energy, and start to turn it into something you want. Few products we buy (other than fresh food) closely resemble the raw materials in nature. Energy, both fossil and human labor, were required for the chemical, mechanical, or other transformations used to harvest, gather, or concentrate the materials and transform them into the desired end products. A physicist might think of energy from oil or coal and the agronomist, energy from the sun (and maybe oil for the tractor and fertilizer). Then one would add more energy to refine the stuff further into more precisely what you want: a car or corn flakes. In other words, most natural scientists would start thinking about what raw materials the product is made of and

how energy is used to upgrade it into raw material stocks and then final products. Then the social science of how the goods were distributed comes later in the process, although markets too require energy to operate.

If this description is basically accurate, and we believe it is, then why have most economists treated energy not as a critical factor of production but only as another commodity to be bought and sold? Our answer again is that economics was able to evolve almost exclusively as a social science for the past 100 years because, for most of the twentieth century, fossil energy was so powerful, so abundant, so capable of expansion, and so cheap as to be invisible and taken for granted. But what if this condition changes? From all of the discussion and debate recently about peak oil and gas, the environmental impacts of coal and the growth of "alternative energy" sources, you have probably sensed that the twenty-first century will be very different. Economics must be very different as well, and become more a *biophysical* science that reflects the actual conditions in real-world economies, one that focuses on resources and energy and not one that treats them simply as a commodity or as an externality. This book shows how the production of wealth and our economic past, present, and future can be explained and predicted much better in terms of a new biophysical economics.

A Substantive Definition of Economics

The usual definition of economics focuses on the social attributes of the field and human choice. It is "the study of the allocation of scarce resources among alternative choices." Scarce here has no relation to scarce resources as a geologist or other biophysical scientist might think, but relative to a person's purchasing power at that time and, more comprehensively, the infinite psychological wants of humans for "more."

There is a second, quite different, definition of economics coming from the great Hungarian economic anthropologist Karl Polanyi [13]. In the 1950s he wrote and edited a collection of essays entitled *Trade and Market in Early Empires*, in which he and other scholars explored the relation between the economy and broader society in ancient and medieval times. They understood that markets are not new phenomena but instead date back to antiquity. But the question was, according to Polanyi, not whether they existed (they did) but instead how important were they in peoples' day-to-day lives (not so important). To pursue this idea, Polanyi provided what he termed a *substantive* definition of economics:

> "The substantive meaning of economics derives from man's dependence for his living upon nature and his fellows. It refers to the interchange with his natural and social environment, insofar as this results in supplying him with the means of material want satisfaction."

In other words, the substantive definition of economics is how groups of humans transform nature to meet their needs. Transforming nature is hard work. In the past when this work was done mostly with one's own muscles, the amount of transformation an individual could do was physically difficult and limited in magnitude. Wealthy people of the past often did this through the hard work of others by means of social conventions such as low-wage labor, serfdom, and slavery. Think of the lovely houses and lives of ease of southern U.S. plantation owners 150 years ago, an affluent lifestyle generated on the backs of dozens to hundreds of laborers working to clear forests and plant and harvest crops. In fact slavery has been a common situation mentioned frequently in the Bible and in many ancient historical accounts. It was not a nice life (to put it mildly) and the concept became increasingly repugnant even to many of the owners of slaves. The Civil War ended slavery in the United States, but de facto slavery continued as former slaves continued to work the lands and as many poor immigrants were brought into the country from Ireland, Italy, China, and elsewhere to do hard physical work at "slave wages" or as indentured servants. People were helped in this work by the physical power of horses and by the physical work obtained from burning wood and the power of falling water. Wind was exploited by sailing ships and an occasional windmill, and increasingly coal was used for railroads and in factories. But overall most work continued to be done by human labor assisted by animals through the turn of the century. This is not to say that most people were not happy: often they were. But the production of wealth was a difficult, sweat-generating process, and most people were very poor by today's standards.

Spindletop and the Beginning of the Affluent Society

Then in 1901 something happened. The generation of wealth for entire societies (especially in the United States and also much of Europe) suddenly changed and the proportion of people with at least moderate wealth took a great upswing, as did the total quantity of wealth in the world and even the wealth per capita (Fig. 1.1). Perhaps the single most important event in a series of similar events was the development of the Spindletop oil field in Beaumont, Texas in 1901, which gave a new realization that serious wealth could be generated for the many by finding, selling, and using oil (Fig. 1.2). Before Spindletop oil certainly had been found and developed, but individual oil fields were relatively rare, small, and difficult to develop, with production of hundreds or thousands of barrels of oil per year. Spindletop alone changed all that, by producing up to 500,000 barrels per day, essentially doubling the nation's petroleum production. It was then understood that a great deal of wealth for many could be had from the oil business with relatively little work, and soon other areas were found to be nearly as productive as Spindletop. Other people looked at how relatively small investments could produce a great deal of money and, by using the ideas and technologies developed at Spindletop, oil production increased rapidly. Large additional finds were made not only in Texas and Louisiana but also in Indonesia, Persia, Romania, and many other areas. As the production of oil increased more and more every year so did the nation's wealth, far more rapidly than ever before. Oil's original use was for kerosene but soon a waste product, gasoline, found an

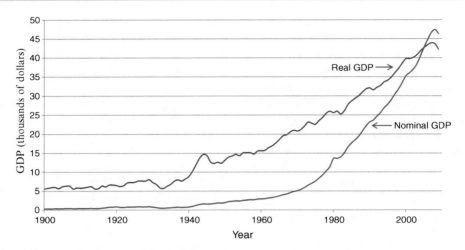

Fig. 1.1 United States Nominal and Real GDP (in 2005 dollars) from 1900 to 2008 (*Source*: U.S. Department of Commerce)

Fig. 1.2 Spindletop, Beaumont, Texas, 1901 (*Source*: Texas Energy Museum)

important new use as automobile fuel. Oil and oil-driven vehicles began to be applied to all economic areas, such as growing food and transporting it long distances; catching fish; cutting, moving and milling lumber; running all kinds of factories; and most other economic processes. Although the few continued to get most of the direct wealth, its use spread affluence to the many. The gross domestic product (GDP), an index of the total production of income by the country, began to grow exponentially (i.e., as compound interest), decade after decade (although interrupted by periodic depressions in 1921, and of course 1929), something almost unheard of before. Thus began the age of affluence for the many, or what can be called mega-affluence. That oil-based growth spread increasingly around the world and has continued for many until now.

This pattern of exponential growth of oil (and energy more generally) use and wealth for the United States, at least up to the "oil crises" of the 1970s (Fig. 1.1), fits in well with our more general energy perspective, for it focuses on the raw materials needed and the energy required to do any process, including economic production. Quite simply, it is the enormous increase in energy that has allowed our economy to undertake the transformations that extract and process those materials into the economic products and services we desire. Other things are needed, of course, such as the technology to get and use energy and a supportive political and economic environment, but the driver of wealth production is the energy to do the work of economic production. To make this clearer we examine in more detail what has probably been the largest generation of wealth to have ever occurred: the production of the vast amount of wealth represented by "the American dream."

The Creation and Spread of the "American Dream"

Traditionally the United States has been considered the world's richest nation and, perhaps of greater importance, as a place where someone with a lot of skill and effort can make a great deal of money if he or she works hard enough, regardless of the circumstances of his or her birth. For many working-class Americans the dream was not affluence, but stability: a steady job, a house of one's own, the ability to pay one's bills, take a vacation, have a cushion for old age, or have a better life for one's kids. This economic success is usually attributed to the characteristics of the people who live within its borders, to their genes, to their hard work, the beneficence of God, or

some other such factor. Education is usually considered important, and the United States has traditionally led the world in the quantity and quality of its higher education, especially at the postgraduate level. Many of the readers of this book in the United States may be taking economics or business classes in order to learn the skills necessary to become more affluent. The idea that possessing more money makes you better off is central to the economic theory of consumer behavior, which in turn is an underpinning of modern economic thought. We explain this idea in more detail in Chap. 4, including the fact that there is no clearly convincing evidence that it is true. The idea that a better education will lead to more affluence is also deeply engrained in the American psyche, and from the first days of the Republic, as per the Northwest Ordinance, land was to be set aside for schools largely for this reason.

The ability to achieve wealth in the United States is in large part a consequence of the incredible resource base once found on the North American continent. These include initial endowments of huge forests, immense energy and other geological resources, fish, grass, and, perhaps of greatest importance, rich deep soils where rain falls during the growing season. Although many other regions of the world also have, or had, a similarly huge resource base, the United States has several other somewhat unique important attributes: the fact that these resources have been exploited intensely for only a few hundred years (versus many thousand as in Europe or Asia), the presence of large oceans that separate us from others who might want our resources; resources per capita that are relatively large, an extremely low human population density in the past and even now. These meant that the resources per capita are still relatively high (Table 1.2). There are a number of reasons that the population density is low. Probably time is most important. To the best of our knowledge, humans have been in North America, at least on any substantial scale, for only 10 or 15,000 years versus 50,000 for Europe and much longer for Africa and Asia. Second is the vast depopulating of the original native population that occurred after 1492. The

Table 1.2 Population numbers and density of the United States and other countries in 2009–2010

	Total population (thousands)	Density (people/km^2)
World (land)	6,828,134	46
Bangladesh	162,221	1,127
Palestinian territories	4,013	667
South Korea	46,456	487
Puerto Rico	3,982	449
Netherlands	16,618	400
Haiti	10,033	362
India	1,182,328	360
United Kingdom	62,041	255
Jamaica	2,719	247
Germany	82,689	229
Pakistan	169,792	211
China	1,338,153	139
Nigeria	154,729	168
France	62,793	113
United States	309,535	32
Argentina	40,134	14
Russia	141,927	8
Greenland	57,000	0.026

Source: Wikipedia

third is the slowing of population growth rates commonly observed as humans become more affluent, which occurred in the United States.

Waves of Colonists to America: First Asians and then Europeans

Most scientific analysis supports the idea that people first came to the Americas during the low ocean levels that occurred 10,000–20,000 years ago when huge amounts of water were tied up in glaciers during the most recent ice age. (One should respect, however, the view of many Native Americans, including some Native American scientists, that "they have been here indefinitely"). When Native Americans arrived on this continent they found few other humans, amazing natural ecosystems, and enormous wildlife resources (their principal resource base). Because these people were skilled hunters and had very effective tools (spears and bows and arrows, as well as highly evolved social systems for hunting and, subsequently, agriculture) they had a tremendous

economic boom, increasing in numbers to perhaps 50 million people in the Americas. But there was a cost to this tremendous economic growth: the extinction of many of the species that had originally been very important in their diet. For example, we know that 10,000 years ago there were two species of elephants, 10-foot-tall beavers, and giant sloths in what is today the United States. These, and many other large species (known collectively as *megafauna*, meaning simply "large animals") disappeared soon after humans came. Scientists debate the degree to which climate change versus human hunting did in these animals, however, there is no question that everywhere that humans went on the planet the large animals disappeared soon after [14, 15]. Meanwhile other humans in the Americas were overexploiting soils in many regions, leading to collapse. That is a radical and sudden decrease in the magnitude and degree of complexity of entire societies such as happened to the Mayas of the Yucatan and present-day Guatemala [16, 17]. Whether such a collapse will occur with present-day European-Americans has been discussed by these and many other authors, most of whom consider it a distinct possibility.

The second wave of humans who entered the Americas came from Europe starting in 1492. They brought with them a whole new suite of plants, animals, and technologies [18]. From our present perspective the basic result of this was that the overwhelming majority of the people who were in the Americas in 1492 were killed directly by Europeans or by the diseases they brought, as described in *Guns, Germs and Steel* [19]. It is not a pretty story and would be called genocide today [20]. Thus the total population again was maintained at a very low level as the new people arriving from Europe were more or less no more than compensating for the net reduction of the original human inhabitants. From the perspective of the next three centuries of economics, this meant that there were still tremendous resources on a per capita basis for each European immigrant and for their children. America was a land of opportunity indeed, for there were enormous untapped resources and not too many people with which to share it. From roughly 1700 to about 1890 there was always an "empty frontier" to the west with land open for the taking and many opportunities for the ambitious and industrious. Of course European Americans rarely considered that these "empty" lands were already heavily populated with Native Americans whose sometimes settled but frequently nomadic, nonindustrial lifestyle was, in fact, very well equipped for a sustainable existence based on mostly renewable resources. The economy of the entire continent went from one relatively sustainable to one clearly not. The greatest cause for the War of Independence was basically resource scarcity: the cutting off of the trans-Allegheny frontier, first in 1763 by means of the Proclamation Act, and then later in 1775, with greater enforcement, with the Quebec Act. Open rebellion soon followed [21].

The technologies that the Europeans brought (such as mining, metallurgy, the moldboard plow, deepwater fishing, and so on), plus the development of a series of self-serving myths (e.g., "Rain follows the plow " and "Manifest Destiny") led to a massive exploitation of both renewable (e.g., soil, trees, fish, bison) and nonrenewable (i.e., gold, silver, coal) resources. This brought seemingly limitless wealth for many, although hardly all, people in a way very consistent with Polanyi's definition of economics. The inventiveness of Americans certainly added to their ability to utilize resources, and important new products and processes were developed that included, among other things, the light bulb, intercontinental railroads, steamboats, mass-produced automobiles, and the telegraph. As we said earlier, most people who thought about why the United States had become so wealthy attributed the affluence to the particular industry of the existing or immigrant European populations or to the blessing of God. Probably far fewer thought about the fact that the United States had such a huge, largely untapped resource base and very low population density compared to, for example, Europe. In other words, the United States enjoyed large resources per capita. Alfred Crosby (1988) [18] believes that Europeans were especially good at colonizing the rest of the world and exploiting the resources where they colonized because of their unique and

self-serving aggressiveness. As evidence he cites that essentially all people in the world today are where they were in 1000 AD (or CE) except for Europeans (who have colonized North and South America, South Africa, Australia, and New Zealand: i.e., all regions with temperate climates) or those who have been moved by Europeans (African slaves and their descendants, Chinese workers to the western United States). This view of the essential aggressiveness of Europeans and their ability to successfully exploit others and their resources is the essence of Jared Diamond's highly successful book *Guns, Germs and Steel*, which also focuses on certain geographical advantages that Europeans had. Europeans were not necessarily good inventors, but they were extremely good adapters: of gunpowder, agriculture, animal husbandry, metallurgy, communication through the written word, and so on. All of this was transferred to the United States, where it was applied with great gusto to a continent rich in unexploited resources, including, as we have said, timber and grass for fuel, good soils with summer rains, rich mineral deposits, and so on. Thus immigrants from Europe found that they could own what was, by ordinary European standards, a massive amount of fertile land whose fertility depended basically upon their own initiative and energy. This was the beginning of the "American dream," the ability to exploit large quantities of solar energy by massive numbers of ordinary individuals.

Industrialization, Isolationism

By the late eighteenth century new sources of energy were being developed in, especially, New England where the abundant water power potential allowed enormous (by the standards of the time) new factories to be built making textiles, shoes, chemicals, and all manner of iron tools and equipment. This allowed the development of great concentrations of workers in such towns as Manchester, New Hampshire, Lowell, Massachusetts, Boston, and New York City. Water-powered machines greatly increased the amount of goods a laborer could generate in an hour (i.e., labor productivity) and the subsequent wealth of at least some in New England. Meanwhile, as forests were cleared for agricultural land and for homesteads in New England, Europeans spread to the Southeast and then westward to virtually the entire Midwest, where enormous amounts of wood fuel were available for all manner of local industries [22]. Fish and other aquatic life were abundant too, and the world's vast numbers of whales were greatly decreased by Massachusetts seafarers in order to get whale oil, the principal source of lighting.

At the start of the nineteenth century, England and Germany had begun their great industrial transformation using the concentrated solar energy found in coal to generate enormous new amounts of high-temperature heat that allowed far more work to be done than was the case with water power, wood, or charcoal. This technology was transferred to the United States which had very rich coal reserves. In 1859 Colonel Edwin Drake drilled the nation's first oil well (actually the first in the world was in Oil Springs, Ontario the previous year), and kerosene began to replace whale oil as the lighting source of choice. The enormous wealth generated by the new industrialization allowed the "captains of industry" to become enormously rich by world standards. This, along with the great disparity in wealth between them and their workers, generated the phrase "the gilded age" for the 1890s. But it was not a smooth pattern of growth as periodic depressions caused a serious loss of wealth for many people, rich and poor. Most people continued to be poor, or at least far from affluent, making barely enough to survive and support a family. Still, in America, despite the disparities in income, the wealth distribution was quite equitable compared to Europe and most of the rest of the world, in part due to the ability of many to have access to land and its solar energy (once the Native Americans were displaced) through farming or with an axe. Large dams, built with the help of massive oil- and coal-powered machines, brought irrigation water and electricity to many in rural areas, resulting in huge additions to the availability of biological and physical energy for each American.

Then came Spindletop and many wells like it. At the beginning of the twentieth century

the United States, dominated by European Americans, was becoming the world's emerging agricultural and industrial giant. In 1900 the United States ran principally on coal, wood, and animal power, but oil became increasingly important with the new oil wells and the development of automobiles, trucks, and tractors that could run on what had formerly been a waste product of the kerosene industry: gasoline. For the first time a very large proportion of the population of an entire country was becoming fairly affluent, and some were becoming extraordinarily so. This enormous affluence was associated with, and clearly dependent upon, an increasing use of energy that expanded at almost exactly the same rate as the increase in wealth, and that made each worker much more productive (Fig. 1.4).

Curiously, even as the United States became more and more dependent upon fossil fuels for basic transportation (meaning mostly railroads), people became even more dependent upon horses for transportation of people and goods at either end of the journey [23]. Because coal-fired railroads generated a great deal of noise and were very smelly, and especially because they threw out sparks that often set houses on fire, they tended to be banned from city centers. Thus until the dominance of the internal combustion engine after about 1920, freight and passengers tended to be delivered from the railheads to the center of the city by solar (i.e., grass) -powered horsedrawn vehicles!

Two World Wars Separated by the Great Depression

By the early 1900s the spirit of isolationism was strong among the citizens of the United States who were deeply suspicious of Europe and its entrenched rivalries and frequent wars. The United States had mostly isolated itself by choice from Europe and indeed most of the rest of the world. After a long delay, the United States entered the First World War, greatly accelerating our involvement with the rest of the world even as antiwar sentiment at home was especially strong. Indeed, incumbent president Woodrow Wilson based his re-election campaign on the slogan, "He kept us out of war." The military value of oil and petroleum-based transportation was first realized by Winston Churchill who had begun the transformation of the British fleet from coal to oil just before the war. England, however, had no oil. Parliament passed the final piece of the conversion, a guaranteed contract for the Anglo-Persian Oil Company (now BP) nine days after the outbreak of hostilities. Thus began the long and often contentious association of the increasingly oil-dependent Western world and the oil-rich Middle East. The value of oil was shown clearly when the French, faced with a potential large military defeat during the battle of the Marne in 1914, rushed 6,000 French soldiers from Paris to the battlefield in taxicabs, where they helped to achieve a great victory. Petroleum was also used for the first time for airplanes and primitive tanks. The war, begun with coal-powered ships and railroads and millions of horses, ended as an increasingly petroleum-based conflict. Thus the ability of petroleum to enhance all things, including mass murder of and by armies, was tremendously enhanced.

After the war, except for the Depression of 1921, the United States had a decade-long period of peace and greatly increasing affluence, fueled in large part by the ever-increasing production of oil. In retrospect it is clear that much of that affluence, however, was wealth only on paper or speculation. In contemporary terms the increase in oil prices became an asset bubble. *Speculation* refers to people purchasing land or other resources not for their own use but in anticipation of being able to sell it later to someone else at a higher price. To do this, banks in the 1920s loaned out far more money than they actually had as assets (i.e., "money in the vault" or ownership of houses) to cover the loans. Simplistically one can think of banks as the place that people put their excess money, saving "for a rainy day," and other people can borrow that money to buy a home, for example. Because most homeowners want to keep their home and will try hard to make their payments, this is normally considered a fairly safe way to loan money, at least if the bankers have done their homework and determined that the borrowers have the means to do so.

Since the early days of capitalism (which some attribute to the rise of the Medici family in Florence, Italy) banks have also loaned out some portion of this money for others to use as investment capital, that is, money to start or expand a business, to buy equipment, build buildings, and so on, in anticipation of using them to make additional money. Nobel Laureate Paul Samuelson wrote that this process, called fractional reserve banking, probably had its origins with ancient goldsmiths who gave receipts or notes for the storage of gold. Eventually, the notes began to circulate as money when the smiths realized not all depositors were likely to return for their gold at the same time. Both processes have allowed banks to pay interest to those who put their money in the bank. Traditionally the prudence of the bank owners and directors, or sometimes government regulators, led bankers to keep a significant portion of the bank's money in the actual bank vaults, so that the people who own the money can withdraw it if they want. All banks, however, live in fear of a "run on the bank," that is, a time when too many people want to get their money out of the bank at the same time. Some speculation has always been with us, but it became much larger toward the end of the 1920s. This was because in the expanding economy the price of land and securities had been pushed up to be far higher than their real worth by people paying higher and higher prices in anticipation of even higher prices in the future.

Reality caught up with the speculators on October 29th, 1929, a day called "Black Tuesday" because of the enormous loss of wealth, and remembered today as a time when at least according to legend a number of investors committed suicide by jumping off their Wall Street buildings. On that and ensuing days, speculators and other investors lost $100 billion, a huge sum at the time. Although no more than two percent of Americans owned stock at that time, the impact of the Wall Street collapse filtered downward to local banks, who loaned out far less money to protect themselves, and thus to local economies. Speculators had borrowed money from their stockbrokers who, in turn, borrowed from banks. The spectacular losses in asset values left investors unable to repay their brokers, who then defaulted upon their own loans. Runs on the banks ensued and insolvencies rose to more than 5,000 by 1931. Before long nearly 20% of Americans had lost their jobs.

This began the period we now know as "the Great Depression" when the country slipped into a long period of little or negative economic growth, high unemployment, and the general financial difficulties of the 1930s. President Herbert Hoover, who had previously shown great skill in combating postwar starvation in Europe, attributed the primary cause of the Great Depression to the "War of 1914–1918" and the economic consequences of the peace treaty that ended the war. This attitude encouraged American isolation and individualism, which was made even stronger by the press, especially in the Midwestern states. The publisher of the influential *Chicago Tribune* carried on an enthusiastic campaign to stop the country from any international entanglements, such as aiding Britain in the days before the United States joined the Second World War. He even considered Hoover's mild reforms to try to deal with the early days of the Depression, and Hoover's tepid contact with international leaders, to be dangerous, going so far as to call the president "the greatest state socialist in history." This is pretty ironic as today Hoover is usually considered as one of our most conservative presidents. Hoover believed that the economy would correct itself given time, and used an unemployed man selling apples on a street corner as an example of someone working individually towards a recovery for all.

In fact the economy got worse, and in the next election the country rejected Hoover and turned to Franklin Roosevelt. Roosevelt ran as a fiscal conservative and believed in a balanced budget. This belief led him to raise taxes to pay for social programs. Consequently his "New Deal" did not provide a great fiscal stimulus. Yet Roosevelt had also long believed in the idea that the government should strive to improve the life of its people, especially in hard times. This belief took many forms, which ranged from job creation programs such as the Civilian Conservation Corps and Works Progress Administration, to Social Security and the reform of labor relations. Most economists,

including liberal and Keynesian economists, agree that this approach actually did not generate enough deficit spending to add a great deal to economic recovery. That took the huge increase in public spending associated with World War II, during which time the economy had tremendous growth fueled by massive increases in government spending and government debt. The commitment to a balanced budget disappeared during the war. The use of deficit spending to stimulate the economy, along with the social structure that the war helped create, led to a long period of rapid economic growth. What was not so well understood was that all of this economic expansion required cheap oil, which established our long-term structural dependence upon petroleum. The combination of increased government spending and the rekindling of the moribund industrial power of the nation had been a primary factor that clearly worked for winning the war and maintaining an ever-increasing standard of living and thus the American dream.

Nevertheless there are many to whom Roosevelt's (and later presidents') intervention in the economy was anathema, for they believed that government should stay out of what they consider peoples' own private business. But their voices were few and far between at the time. The era of the "New Economists," who based their principles on the work of John Maynard Keynes, but emphasized economic growth over all other goals, was about to begin. It was the era in which economists believed they had "conquered the business cycle." The New Economists believed that with wise application of prudent policies regarding taxing, spending, money, and interest, they could enhance the efficiency of markets and relegate depressions to the past. This confidence would not last beyond the 1970s, however, a period characterized by high unemployment and inflation. The questions of the effectiveness of government regulation are with us again in the 2010s, at a level that few economists of the 1950s and 1960s could have possibly imagined.

What is especially interesting from our energy perspective is that the depression was a time of tremendous energy availability in the United States. The East Texas field, the nation's largest ever except for Prudhoe in Alaska, was discovered in 1930, the first full year of the Depression.

Oil was cheap, but there was virtually no market for it. But when the U.S. economy finally began to recover, especially in the 1940s, there was a great deal of energy to power that expansion. An important question for today is whether there is sufficient cheap energy to power whatever recovery may take place after the recession of 2008–2011.

Meanwhile Japan, a small country without a large resource base and which had formerly looked inward for centuries, increasingly became industrialized and, of necessity, looked outward for the resources it needed. Buoyed by their success against a giant Russian fleet at the battle of the Tsushima Straits in 1905, the Japanese built a huge modern fleet. As much as half of the gross national product of Japan went to building up their military machine, and this expansion took up such a large portion of the resources available to them that, for example, Japanese families were encouraged to feed their rice to make their boys, the future soldiers, strong while the girls got to eat only the water in which the rice was boiled. Japan invaded China and Korea for coal and iron, and began to expand outward into the Pacific Ocean, for example, into Okinawa. The United States had worked to contain the imperial ambitions of the Japanese in the 1930s by both negotiated treaties and a limited military build-up in the Pacific. The Japanese realized that the expansion of their economy depended upon reliable access to oil. That oil was to be found in the Dutch East Indies (now called Indonesia). The United States, in 1941, in a largely overlooked overt act of war, blockaded Japan's access to that oil using warships. The most militant voices in the Japanese military were convinced that the only way to protect their oil resources was to deliver a knockout blow to the U.S. Pacific Fleet. Thus the desired and partly successful isolation of the United States from the rest of the world came to a screaming halt December 7, 1941 when the Japanese attacked U.S. naval bases on the island of Oahu in the Hawaiian Islands. The day after the attack President Franklin Delano Roosevelt asked Congress for a Declaration of War. Germany and Italy subsequently declared war on the United States. The Second World War that began in Europe in 1939 had begun for the United States. In many ways it was the world's first war

based upon oil, and in many ways it greatly accelerated the industrialization of the world. The role of oil in the Second World War has been especially well told by Daniel Yergin [24] in *The Prize*, his very comprehensive book about oil.

Our entry into the shooting war began with the Japanese bombing of the United States fleet in 1941, although as noted this was not the first act of war in the Pacific. The war ended in Europe with the military defeat of the Italian and German militaries and the surrender of the Fascist and Nazi governments. Again, the availability or lack thereof of fossil fuels played a key role. Toward the end of the war Germany, having lost access to the petroleum supplies of Africa and the Middle East, produced limited amounts of gasoline from coal, pioneering the same technologies (called Fischer–Tropsch) currently being considered for making liquid fuels from coal. Their production facilities, however, were destroyed by Allied bombing once the Allies gained air superiority. Air superiority was itself enabled by the fact that U.S. companies invented and then produced 100-octane aviation fuel which helped the British win the battle of Britain and the Allies to eventually gain general air superiority. The Germans were so depleted of liquid petroleum by late in the war that they had to bring the first ballistic missiles (the V-II rocket) to the launching pad with mules. In the Pacific theater, the Japanese too had run so short of oil that they initially had to leave the world's largest battleship in port for lack of fuel, and then sent it out to a last battle with only a one-way supply of oil. They used turpentine as fuel to fly some of the kamikaze (suicide) airplanes that were attempting to sink the ship that the father of one of this book's authors (Hall) was on in Okinawa. Hall's friend and colleague Tsutomu Nakgatsugowa remembers clearly as a child that all of the pine trees in his Japanese village were uprooted to make turpentine for fuel. The war ended in 1945 after the first use of atomic weapons during wartime, representing again an enormous increase in the human use of energy, both in the nuclear explosions themselves but also in the huge amount of fossil and hydroelectric energy that had been used to separate the isotopes of uranium. It was only a matter of time and technology until America's vast industrial strength prevailed.

Perhaps it was more accurately put by Pulitzer Prize winning historian David Kennedy who said that the war was won with Russian lives and American machines. And, we add, the petroleum to run them.

The Rise of Affluence for Many

On the home front, something unique occurred. The standard of living rose for a people engaged in war, as the war effort rekindled the U.S. economy that had been devastated by the Great Depression. Unemployment, which stood at more than 17% of the labor force in 1939, fell to less than 1.2% in 1944. The value of economic output more than doubled in a mere six years. Large social changes occurred during the war years too. Women entered the paid labor force in unprecedented numbers, often earning high wages in both clerical and production jobs. There was little to spend one's money on, and savings as a percentage of income rose to the highest levels in history, providing massive investment monies. People patched their clothes, recycled their metals and, encouraged by gasoline rationing, stopped driving to aid the war effort. African-Americans found relatively high-paying jobs in the labor-scarce factories, and began the slow and painful process of integrating into white society. The conflict between labor and management that so characterized the Depression era declined as the major industrial unions signed a no-strike pledge for the duration of the war while seeing both corporate profits and their wages and benefits increase.

Even larger changes were to come with the end of the war, changes that dramatically affected the drive towards affluence. A new social contract among workers, employers, and the government was in the process of creation, and this social contract provided a newly powerful nation with the "pillars of postwar prosperity" [25]. These vehicles to maintain prosperity and social stability were based on domestic economic growth and enormous international power (military and economic). Specifically:

1. Basic accord between capital and labor, at least, after a period of intense strike activity following the war, especially between the largest multinational corporations and the largest manufacturing unions by giving labor a share of productivity gains in the form of higher wages.

2. Pax Americana. The United States became the dominant military and economic power after the Second World War, with most of the world's nuclear weapons and gold, as well as being the largest exporter of oil. In addition, the international monetary system was reworked with the U.S. dollar as the key currency and the fractional reserve banking system was internationalized to allow the expansion of the money supply to accommodate growth.

3. Accord between capital and citizens. Large-scale oligopolies, the government, and the average citizen united around three basic premises: economic growth would replace redistribution as the means of improving well-being, government policy should be focused on the availability of cheap nuclear and other energy, and anticommunism.

4. The containment of intercapitalist rivalry. The tight oligopolies constructed from the 1890s onward controlled destructive price competition and allowed large corporations to control their rivalries by means of mechanisms such as price leadership, market division, and use of advertising. Initially the United States was the dominant producer worldwide, having the only viable industrial economy at the end of the war. Stable oligopolies competed on the basis of market share, not price.

A critical component of these patterns was the large increase in labor productivity during that time. This allowed both industry owners and labor, especially of the largest corporations, to do better and better. What was less emphasized, but enormously clear in retrospect, was that to allow these four pillars to operate and expand it was possible to massively increase production from oil, gas, and coal fields, some new, and some old, but barely tapped previously. Once the economic engine was started there was a great deal of high-quality energy available even though the war itself had consumed some seven billion barrels of oil (about the same as recent annual consumption by the United States). The United States began using many times as much energy per person as had been the case relatively few decades before.

In addition, the nation was left with the enormous munitions facilities built at taxpayer expense at, for example, Muscle Shoals, Alabama. These facilities used the Haber–Bosch process, invented in Germany just before the First World War, to make ammonia [26]. This chemical process for the first time allowed humans to access directly the enormous amount of nitrogen in the atmosphere, which was extremely valuable for the munitions, agricultural, and chemical industries. Before Haber and Bosch perfected their chemical synthesis the primary sources of nitrates were manure, the large deposits of bird guano found off the South American coast, and the sodium nitrate deposits in the Atacama desert. Peru and Chile had fought the Guano Wars over access to the bird droppings. But eventually guano mining exceeded replenishment and the resource vanished. Another source needed to be found. Nearly 80% of the atmosphere is nitrogen (N_2) but this nitrogen is very difficult to access because of the triple bonds in the di-nitrogen molecule (i.e., N_2). Until 1909 only the tremendous energy of lightning or some very special algae and bacteria could break these bonds. Gunpowder and fertilizer depended upon the exploitation of rare deposits of nitrates concentrated by birds over millennia. Fritz Haber, in one of the most important scientific discoveries ever made, found that by heating and compressing air mixed with natural gas, that is, by adding hydrogen and large amounts of energy to the nitrogen in the air, and with the right catalyst, the N_2 molecule could be split and turned into ammonia (NH_3). This in turn could be combined with nitrate (itself created by oxidizing ammonia) to generate ammonium nitrate which is the basis for both gunpowder and the most important fertilizer.

When in 1946 there was no further need for massive amounts of explosives the U.S. federal government asked whether there might be any other use for these factories. The answer came back from the agricultural colleges: yes, we can

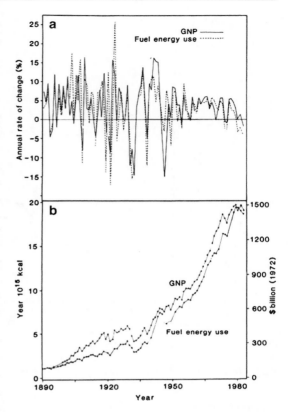

Fig. 1.3 Total wealth plotted along with total energy use for the U.S. economy 1905–1984. The top graph shows the rate of change for each (Source: Hall et al. 1986)

use it to greatly increase agricultural yield, and this is what happened. This "industrialization of agriculture" freed food production from its former dependence upon manure and, encouraged by the concurrent development of machinery, far fewer Americans were needed to grow our food. This increased the exodus to the growing number of urban industrial jobs, the increased use of oil, gas, and coal, and the massive generation of wealth. Over the course of the twentieth century America continued to change from a relatively poor, largely agricultural, rural country into an increasingly industrialized and urban country while becoming vastly more wealthy, by most accounts, in the process. Meanwhile the energy required to do all this economic work was increasing exponentially (Fig. 1.3). New economic theories were launched to explain the

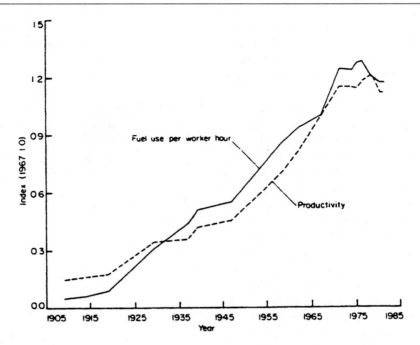

Fig. 1.4 Mean U.S. labor productivity per worker hour (in constant dollars) and energy used per worker hour, 1905–1984, when data acquisition was stopped. The graph is scaled so that each equals 1.0 in 1965 (*Source*: Cleveland et al. 1983 and Hall et al. 1986)

enormous increase in wealth with, however, essentially no mention of the energy enabling and facilitating the expansion by those chronicling the process.

Europe and Japan had been decimated by the fighting. Every warring nation except the United States saw their industry and infrastructure in ruins, and the Allies, especially Britain, were deeply in debt to the United States. The new peace was to be an American-dominated peace, with the terms dictated by Americans. The American-led Marshall Plan helped rebuild the war-devastated economies of Europe. After some 15 years of the Depression and war the international monetary system was in need of serious rebuilding. The gold standard, which had served as the foundation of international trade since the mercantile days of the 1600s, was a casualty of the Depression. In 1944 an International Monetary Conference was convened at a ski resort in New Hampshire called Bretton Woods. Under the auspices of the new system, known as the Bretton Woods Accord, the U.S. dollar replaced gold as the basis for international trade and investment.

Only the dollar was stated in terms of gold, the value of all other currencies was expressed in dollar terms. In essence, the rest of the world was willing to give the U.S. interest-free loans in their own currencies just to hold our dollars. The United States reaped several benefits from the new configuration on the world level. The value of U.S. investments abroad grew at nearly nine percent per year from 1948 to 1966. The terms of trade, or the ratio of export prices to import prices grew by 24% over the same period. People of this country bought in a buyer's market (i.e., in conditions favorable to the buyer) and U.S. corporations sold in a seller's market. Finally, U.S. business gained access to crucial raw materials and additional cheap energy, despite the fact that the United States was the world's leading exporter of oil at the time. America's industrial might and monetary control formed an important foundation for growing affluence. America became extremely powerful both economically and in terms of energy use.

The Depression era had witnessed a considerable amount of strife between labor and capital.

By the late 1930s strong industrial unions organized to win recognition, higher wages, and better working conditions. After a flurry of strike activity immediately following the war, relations between large businesses and their employees stabilized. In 1948 an epoch-making contract was signed between General Motors and the United Auto Workers. In this contract the UAW gave up their claim to joint management of the company and control over the trajectory of technology. In return they received a larger share of the company's profits in the form of higher wages and benefits. The contract linked increases in wages to increases in productivity, or output per worker. In this climate of American peace, labor stability and productivity (i.e., value added per hour of worker input) grew at a brisk pace and the after-tax earnings of American manufacturing workers grew by more than 50% from 1948 to 1979. This was responsible for spreading some of the wealth earned by business to the pockets of the American worker. More than most factors, this new social contract, based on shared gains from increased productivity, helped establish the American dream.

One specific example of the link between energy and economic prosperity rarely understood by most economists is that of the role of energy in the dollar value of the products generated by a worker working for 1 hour. Increased labor productivity allowed the employer to pay his or her worker more even while making a larger profit. This increased productivity is normally assigned to technological progress. What is less understood is that labor productivity increased in direct proportion to the amount of energy used per worker hour (Fig. 1.4). At that time labor productivity in the United States was two or three times that of a European worker, not because the worker worked harder or was more clever, as commonly assumed, but because he had big machines using two or three times more energy helping him do the job! Again what is often attributed exclusively to technology was in fact equally based on increasing the availability and use of cheap energy, which was much cheaper in the United States than in most other nations.

The Increasing Role of Government

The idea that government participation in the economy should be minimal, which had been around at least since the time of the Physiocrats and Adam Smith, went by the wayside starting with the Great Depression and continuing into the postwar years. The strategy for ending the Depression, the New Deal, created not only the alphabet soup of government agencies, but also an attempt to involve the federal government in economic planning. This planning was augmented and extended in the Second World War, undoubtedly the greatest public works program in the history of the United States. After the war Congress passed a law entitled The Employment Act of 1946. This law mandated the government to pursue taxing and spending policies that would result in reasonably full employment, stable prices, and economic growth. In this era of "New (Keynesian) Economics" budget deficits were sometimes purposefully created. They became an important tool of economic policy rather than a dangerous aberration that must be avoided at all costs.

The increased spending, which was often financed by debt rather than taxes, injected increased purchasing power into the economy to help maintain postwar affluence. The government created new programs to subsidize home mortgages and home ownership, an important component of the expanding realization of the American dream. Spending on social programs also increased. In 1968 a state-supported health initiative for the elderly called Medicare was passed into law to supplement the retirement insurance program (Social Security) created during the Great Depression. For the first time, being old no longer meant being poor for the majority of American workers. This act represented the culmination of a whole series of social spending programs during the 1960s. Spending for income maintenance programs and education increased during the presidency of Lyndon Johnson, who envisioned a "Great Society." But spending also rose for military purposes as the United States became more deeply involved in a prolonged war in Vietnam. Although this expansion of spending eventually helped to initiate the end of the American dream, more than two decades of prosperity and increasing affluence for a growing number of Americans ensued. The United States was affluent enough to spend more on health care and education and create more opportunities for those formerly left out of the general economic expansion. General affluence increased even while waging war, at least initially. Wages and profits continued rising at least for a large proportion of the population.

The engine that held this increasing prosperity together was economic growth, that is, the increase in the material economy expressed in the dollar value of the goods and services we produced in a year (this is called gross domestic product or GDP). The fuels for that were a social structure that prompted growth, expanding international markets, and exponentially increasing use of oil and coal and gas. All through this period energy use increased in almost direct proportion to the economy; the fossil energy was there to do the actual work of an expanding economy. GDP more than doubled from 1945 to 1973, increasing from about $1.8 trillion to over $4.3 trillion in inflation-corrected (e.g., year 2000) dollars. Energy was readily available, very cheap, and the incentives to use it abounded as "the good life" was increasingly sold using advertising.

The "Oil Crises" of the 1970s: Hints at Limits to Economic Growth

As the 1970s approached all four pillars of the American success story began to fracture. Europe and Japan caught up and surpassed the United States in terms of technology and productivity growth. New technologies, a more restrictive regulatory climate, and a new type of merger (conglomerates) destabilized the tight oligopoly control of manufacturing. This would further destabilize corporate structure in the 1980s and 1990s. The rise of state-owned oil companies presented another threat to the control of intercapitalist rivalry. Bretton Woods was abandoned,

and U.S. oil production peaked in 1970 and became subject to "supply shocks." In 1973 the United States experienced the first of several "oil shocks" that seemed, for the first time, to inject a harsh note of vulnerability into the united chorus of the American dream for all. Before the 1970s nearly all segments of American society – including labor, capital, government, and civil rights groups – were united behind the agenda of continuous economic growth. The idea that growth could be limited by resource or environmental constraints, or, more specifically, that we could run short of energy-providing fossil fuels was simply not part of the understanding or dialogue of most of this country's citizens. But this was to change in the 1970s.

In the popular phrase of economists, the economy began to "overheat." Consumer spending had more than doubled from $1.1 trillion in 1945 to nearly $2.5 trillion in 1970 (in 2000 dollars) as workers spent the dividends from the social contract from 25 years earlier on the many goods they had been deprived of in the Depression and the war and as general affluence increased. As the U.S. economy retooled in the postwar era, investment spending likewise rose from about $230 billion to $427 billion in 2000 dollars, aided by steadily increasing numbers of people, consumer credit, and corporate profits. Government spending, driven by the expansion of social programs during the time of President John Kennedy's "New Frontier" and President Lyndon Johnson's "Great Society," and the costs of fighting the Vietnam War, in constant 2005 dollars, increased from $405 billion in 1950 to more than $1 trillion during the same 20-year period. Unemployment fell at a relatively steady pace, dropping from about 6.5% of the labor force in 1958 to only four percent in 1969. Hourly earnings of manufacturing workers after taxes rose from about $2.75 per hour in 1948 to about $4.50 in 1970 when both were expressed in 1977 dollars. As spending increased faster than the ability to produce goods (given the relatively modest levels of unemployment) prices began to rise. The specter of "creeping inflation" began to enter the lexicon of economists and citizens alike.

In 1973 the United States (and much of the world) experienced the first "energy crisis." Crude oil, selling for $2.90 per barrel in September, soared to $11.65 by December. The price of gasoline shot up suddenly from 30 to 65 cents a gallon in a few weeks and the available supplies declined. Americans became subject to gasoline lines, large increases in the prices of other energy sources, and double-digit inflation. Home heating oil became much more expensive, as did electricity, food, and even coal! Few people understood that the production of oil in the United States had reached a peak in 1970, and had begun to decline. Although the specific initiation of the price increase began with a bulldozer that in 1970 ruptured a pipeline carrying oil from the Persian Gulf to the Mediterranean, the peak of oil production in the United States, the U.S. resupplying the Israeli military in their war with Egypt, the long history of Western arrogance in the Middle East all set up the circumstances in which a minor event could generate an enormous impact.

In 1979 the world experienced another oil shock. According to the Energy Information Agency, the current dollar price of domestic crude oil rose from $14.95 in 1978 to $34 per barrel in 1980. This would amount to nearly $100 per barrel in 2010 prices. Consequently, the 1980 price of gasoline increased again to an average of $1.36, equal to $3.19 in 2009 prices. The increases were directly in response to the withdrawal of supply by the new Islamic Republic after the collapse of the U.S.-backed government of Reza Pahlavi in Iran, but again the inability of the United States to supply its own consumption underlay all. Many of the economic ills of 1974, such as the highest rates of unemployment since the Great Depression and rising prices, were repeated in the late 1970s and early 1980s when oil once again became less available and more expensive due to restrictions in supply brought about by the Organization of Petroleum Exporting Countries (OPEC, including many oil-rich countries in the Persian Gulf, Venezuela, and Indonesia). Americans became used to energy as a topic that was in the newspaper every day, and, especially in the colder northern tier of the United

States, conversation was often about wood as a fuel to heat one's house, or the fuel efficiency of the then-new Japanese imported cars versus the familiar Fords and Chevrolets.

The American economy, used to being overwhelmingly the strongest in the world, suffered as businesses in the countries American aid helped restore after the Second World War now became effective competitors. This was partly because energy prices, which were once much cheaper in the United States, became effectively the same around the world. Higher-priced American labor was no longer compensated for by cheaper American energy. Real wages began to fall in the United States. By the end of the 1970s Japanese autoworkers were earning more per hour than their American counterparts. The unemployment rate increased to nearly 10% in 1982, a number unheard of since the Great Depression of the 1930s, and prices of everything increased at nearly 10% per year. Unemployment and inflation were supposed to be inverse to each other according to the well-established economist's Phillips curve, but here they were simultaneously increasing, something called "stagflation." Labor productivity ceased to increase, also something unheard of formerly. The news was so bad that the Reagan administration stopped gathering data on this important economic parameter. For many, it seemed like the world was falling apart.

Stagflation, which was difficult to explain by means of standard Keynesian theory, is easy to explain from an energy perspective: as energy prices increased and supplies declined, the dollars circulating in the U.S. economy were increasing more rapidly than new energy was added to do economic work. As a result each dollar bought fewer goods and services. In addition, the monopolized corporate structure allowed business to pass on increased costs of production in the form of higher prices. As more of society's output was required to get the energy necessary to run the economy, costs of everything from food to packaging were pressured upwards; this resulted in an increase in joblessness as there was less money available for purchases. In fact adding the energy and historical perspective provides a ready explanation for stagflation: as energy use was increasingly restricted (by supply and higher prices) the economy contracted. Inasmuch as the energy supply contracted more than the dollar supply, there was also inflation. This explanation shows the power of energy analysis and the inadequacy of pure economic models that exclude the fundamental role of energy. In systems language, the economic models focused almost entirely on the internal dynamics of the system but were insensitive to changes in forcing functions because they had not been included in the model structure.

The Limits to Growth

At about this time a series of quite pessimistic reports about the future came out, with the most important being the "Club of Rome's" *Limits to Growth* [27] and *The Population Bomb* [28] by Paul Ehrlich. These reports added to the concerns based on the predictions by Shell Oil geologist M. King Hubbert [29] of the inability of both the United States and the world to keep increasing petroleum production. These reports implied in various ways that the human population appeared to be becoming very large relative to the resource base needed to support it – especially at a relatively high level of affluence – and that it appeared that some rather severe "crashes" of populations and civilizations might be in store. Meanwhile many new reports appeared in scientific journals about all sorts of environmental problems including acid rain, global warming, pollution of many kinds, loss of biodiversity, and the depletion of the earth's protective ozone layer. The oil shortages, the gasoline lines, and some electricity shortages in the 1970s and early 1980s all seemed to give credibility to the point of view that our population and our economy had in many ways exceeded the world's "carrying capacity" for humans, that is, the ability of the world to support it.

Universities hired many new people in the previously obscure disciplines of ecology and environmental sciences, and there was a great surge of interest by students in issues of resources and the environment. Although courses in environmental economics were added to many college catalogues, economists generally ignored these issues or, if anything, modeled nature as part of the economy, and added in environmental factors to the list of things that would be regulated by rational individuals responding to price incentives. The notion of biophysical limits to growth, based on biophysical constraints, got a chilly reception from the community of mainstream economists, although the idea of an economy limited by nature began to develop a following among political economists in the early 1970s [30–32]. Although economists have written about the *internal* limits to growth since the eighteenth century, these new works raised a new possibility: our futures would be limited by nature as well. Historically, humans have been able to transcend nature's limits by employing increasing amounts of energy to the problems at hand. But were we nearing those limits? If so, the age of convenience and growth would be replaced by living within our means or even degrowth. The message was not popular. President Jimmy Carter discussed on television the need for Americans to conserve, and even installed solar collectors on the White House roof. He said that the American people should view the energy crisis as "the moral equivalent of war." For many people it did seem like humans had reached the limit of the abilities of the Earth to support our species.

Most economists did not accept the absolute scarcity of resources. The return to growth, they said, was just a matter of implementing a series of proper incentives and market-based reforms, as well as dispensing with the dangerous ideas of absolute limits. A series of scathing reports appeared directed at those scientists who wrote articles with that perspective, such as Passell et al. [33]. They argued that economies had built-in market-related mechanisms to deal with short-term (relative) scarcities. Technical innovations and resource substitutions, driven by market incentives, would solve the longer-term issues. Critics of the early antinuclear movement belittled the idea that using less electricity or generating it from less dangerous sources was remotely viable. For them it was generate more nuclear power or "freeze in the dark."

Crumbling Pillars of Prosperity

In retrospect, we can now say that the pillars of postwar prosperity began to erode in the 1970s and early 1980s, and that changes in the social sphere also began to complicate and add to the biophysical changes derived from the decline in the availability of cheap oil. Even though the oil market had stabilized and cheap energy returned to the United States in the late 1980s, the changes in the structure of the economy were long lasting. The economy ceased growing exponentially, although it continued to grow linearly but at a decreasing rate, from 4.4% per year in the 1960s to 3.3, 3.0, 3.2 to 2.4% in the following decades. Many formerly "American" companies became international and moved production facilities overseas where labor was cheaper and oil, no longer cheaper in the United States compared to elsewhere, was the same price, although cheap enough to pay for the additional transport required. The decrease in labor costs when production facilities were moved to other countries outweighed the costs and the process of globalization accelerated. Productivity growth (formerly strongly linked to increasing energy used per worker hour) in manufacturing industries began to slow, falling from 3.3% per year in the 1966–1973 period to 1.5% from 1973 to 1979 to essentially zero in the early 1980s. Reductions in the rate of growth in the energy-intensive sectors of utilities and transportation were even greater, whereas construction and mining showed actual declines in output per worker hour. As productivity growth slowed so did the growth in workers' hourly income, from a substantial 2.2% per year from 1948 to 1966 (which would lead to a doubling of incomes in 32 years) to 1.5% in 1973 to 0.1% in 1979. Corporate profits also decreased from nearly 10% in the mid-1960s to a little more than 4% by 1974. Things seemed bad for both capital and labor.

Mainstream economists seemed at a loss to explain this phenomenon. Their statistical models, which relied on the amount of equipment per worker, education levels, and workforce experience left more factors unexplained than explained.

Even the profession's productivity guru, Edward Denison, had to admit that the 17 best models explained only a fraction of the problem. Fortunately two other approaches yielded far better explanations. Economists associated with the "Social Structure of Accumulation" approach [25] developed a statistical model that explained 89% of the decline, and attributed most (84%) of the slowdown in productivity growth to decreases in work intensity. Under the social contract of the postwar era, unions were able to limit speedup by a series of work rules that limited how hard workers could be driven. Despite increases in the numbers of supervisors, businesses (especially manufacturing firms) could not increase the amount of output per worker at will, especially without increasing wages. The biophysical approach also yielded promising results. Howard Odum had been writing about the importance of energy in the economy for a decade, as had others [34]. In a 1984 article in the prestigious journal *Science*, Cutler Cleveland, Charles Hall, Robert Costanza, and Robert Kaufmann [35] found that they could explain 98% of the decline in output growth by the decline in fuel energy after the oil crises of the 1970s. The two concepts may be linked because the increase in fuel-intensive machinery is one factor in how intensive work can be made [36, 37].

Things looked bad for the United States in the international arena as well. The United States had rebuilt Europe and Japan with the latest technology soon after World War II, and by the 1970s these former "second-rate trade partners" turned into fierce competitors. The commitment to energy efficiency in Europe and Japan far surpassed that of the United States. Moreover, labor relations in other countries were far less contentious than they were at home. Terms of trade, or the ratio of export prices to import prices, fell from about 1.35 in the early 1960s to only 1.15 by 1979. Adding to the difficulties faced by the United States, the world monetary system came unglued by the early 1970s. The system, developed in Bretton Woods, New Hampshire, depended upon the United States being the world's most productive economy, and upon its willingness to let other countries redeem their dollar holdings in gold. However, when declines in productivity and the terms of trade and the mounting costs of the Vietnam War came home to roost, the value of the dollar relative to other currencies plummeted. President Richard Nixon suspended the convertibility of dollars to gold. The international trade system was now a free-for-all, and the new and more chaotic system contributed to a fall in corporate profits. Presidents Nixon, Ford, and Carter were unable to break the political stalemate of rising labor costs caused by union power and a commitment to low rates of unemployment in spite of their best efforts. Something had to give.

In 1979 the editors of *Business Week* opined that to restore the nation's affluence labor would have to learn to accept less. The *Wall Street Journal* was calling for "supply side economics," an approach associated with increasing the rate of exploitation of natural resources by decreasing government environmental and other regulations. In the same year, on the steps of the Statehouse in Concord, New Hampshire, former actor and California Governor Ronald Reagan, then a presidential candidate, declared that "for the country to get richer, the rich have to get richer." Reagan won the 1980 presidential election and instituted what the Social Structure of Accumulationists termed "A Program for Business Ascendancy" or what the *Wall Street Journal* praised as "Supply-Side Economics." This constituted a sharp turn to the right in American politics. The Reagan administration focused far more on inflation than on the restriction in growth, and immediately confronted unions, and further disciplined workers by moving to create a sharp recession by means of policies that raised interest rates and hence severely restricted the amount of money in the economy and, consequently, jobs.

By the mid-1980s home mortgages carried 20% interest rates, and business loans were nearly as expensive. In order to increase America's power in the world they instituted an aggressive program of military build-up and returned to what former President Theodore Roosevelt termed "Big Stick Diplomacy." Inflation rates subsided and corporate profits rose, but these victories came at a cost. Unemployment rose to almost 10%, inequality increased as the percentage of Americans living in poverty jumped from about 11% to a little more than 13%, and the number of rich households (who earned more than nine times the poverty level) went from less than 4% in 1979 to nearly 7% in 1989. Compared to earlier times most Americans thought that the economy was a mess. Few blamed it on energy, but in retrospect we can say that the pillars of postwar prosperity were eroded in the 1970s and early 1980s because there was no longer unlimited supplies of cheap energy, which caused changes in the economic and social sphere that had begun to affect prosperity.

The Twenty-Year Energy Breather

By the mid-1980s the price of gasoline had dropped again as the inflation-adjusted (2010) price of crude oil fell from $98.52 per barrel in 1980 to $15.84 in 1998. The new Prudhoe Bay field in Alaska, the largest ever found in America, added to our oil production and helped mitigate, to some degree, the decrease in production of other domestic oil. Around the world many earlier discoveries had become worth developing in the 1970s, and cheap foreign oil flooded the market. As a result, energy as a topic faded away from the media and so in the perception of most people. For most people who thought about it at all, the reason that the energy crisis was "solved" was that the market was allowed to operate by generating incentives from the higher prices. In fact this was largely true, for although domestic production continued to fall year by year (Fig. 1.5) foreign-derived oil was increasingly imported to the United States from other countries and we shifted the production of electricity away from oil to coal (a generally dirtier but more abundant form of energy), to natural gas (generally a cleaner form), and to nuclear energy. So it indeed did look like the economy, through price signals and substitutions, had in fact responded to the "invisible hand" of market forces. Conservative economists felt vindicated, and the resource pessimists beat a retreat, although the economic

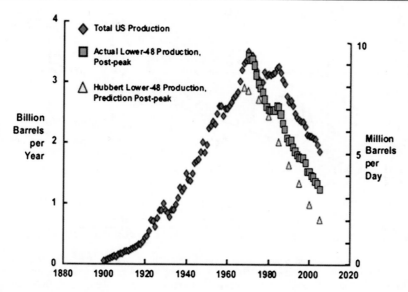

Fig. 1.5 Production of oil in the United States (with and without Alaska) compared to Hubbert's 1969 prediction for the lower 48 (*Source*: Cambridge Energy Research Associates)

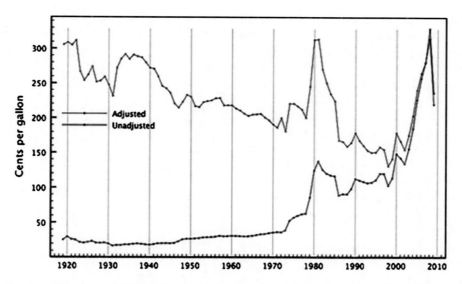

Fig. 1.6 Gasoline price corrected (2005 dollars) and not corrected for inflation in the United States (*Source*: USDOE)

stagnation of the 1970s as indicated by declining rates of GDP growth, continued until the present day in the world's mature economies.

By the early 1990s inflation had subsided and the world economy grew at about three percent a year. Inflation-corrected gasoline prices, the most important barometer of energy scarcity for most people, stabilized and even decreased substantially from $3.41 per gallon in March 1980 to $1.25 in December 1998, in response to an influx of the foreign oil (Fig. 1.6). Much of this new wealth was generated not through working for wages but by owning stocks. Wages fell and assets surged, but, as in earlier times in history, stock

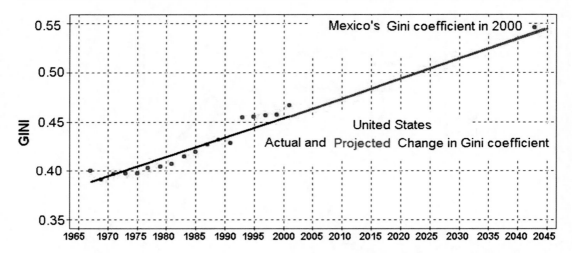

Fig. 1.7 Gini coefficient for the United States, earned by which is, approximately, the ratio of income earned by the top 20% compared to the interest earned by the bottom 20%. This graph shows that since about 1970 there has been increasing inequality of wealth in the Unites States, with the wealthiest 20% gaining an increasingly large proportion of the economic pie and the poorest 20% getting a smaller and smaller portion (*Source*: SustainableMiddleClass.com)

ownership was not spread evenly throughout the economy. The majority of stock market gains accrued to the top one percent of the income distribution. Increasingly many landscapes were filled with very large houses that were far larger than the basic needs of a family, and purchased primarily as luxury items, for the perceived status or on speculation: to sell for a higher price a few years in the future. This process was driven by market forces, as housing represents both investment and shelter for most Americans. One's house is generally a person's greatest asset or repository of wealth. However, large houses, especially those filled with myriad electronic appliances, are also extravagant energy users. So declining real energy prices combined with market forces produced a growing stock of larger houses that used more energy even though many appliances had become much more efficient. Discussions of energy or resource scarcity largely disappeared from public discourse, or were displaced by new concerns and courses about environmental impacts on tropical forests and biodiversity. Income inequality between the rich and poor, as measured by the Gini index, increased greatly, both absolutely and in comparison with other industrialized nations (Fig. 1.7; Table 1.3). Indeed it seemed that some

Table 1.3 Recent Gini indexes for a select group of nations: the lower the number the more equitable the distribution of wealth

Japan	24.9
Sweden	25.0
Germany	28.3
France	32.7
Pakistan	33.0
Canada	33.1
Switzerland	33.1
United Kingdom	36.0
Iran	43.0
United States	46.6
Argentina	52.2
Mexico	54.6
South Africa	57.8
Namibia	70.7

Source: Sustainable middleclass.com

100 years after the first Gilded Age, America had entered a new one.

In the United States conservatives led by President Ronald Reagan were successful in convincing many formerly apolitical or even labor union people that their own personal conservatism in issues such as family, society, religion, and gun ownership could be best met through conservative economic and political groups whose agendas were historically opposed to the interests of the working people. These groups and their representatives in government were very much opposed to government in general and any interference with individual "freedom," especially intervention in the market. Thus they opposed, for example, government programs to generate energy alternatives (such as solar power or synthetic substitutes for oil), believing that market forces were superior for guiding investments into energy and everything else. They also tended to be opposed to restrictions on economic activity based on environmental considerations and even mounted campaigns to discredit scientific investigation into environmental issues such as global warming. (However, it is important to point out that many conservative people are extremely interested in the conservation of nature.) One specific thing that President Reagan did was to remove the solar collectors installed by President Carter on the roof of the White House even though they were working fine.

These new conservative forces tended to be opposed to government policies that restricted such freedoms (i.e., gas mileage standards and speed limits). Both liberals and conservatives tended to support free trade and hence contributed to the movement of many American companies or their production facilities overseas where labor was cheaper and pollution standards often less strict. One effect was probably a substantial contribution to the improved efficiency of the economy (GDP per unit of energy used) as polluting and expensive heavy industries were moved overseas. For example, strong federal programs to improve solar collectors and the like were often eliminated as government interferences. By 2000 the country seemingly had recovered from the stagnant 1970s and the recessions of the 1980s and early 1990s, although prosperity was based on a growing level of debt, just as it was in the mid-1980s. Stock values began to increase steadily and the general economic well-being of many Americans led to a general sense of satisfaction in market mechanisms. The collapse of the Soviet Union and the end of its influence in Eastern Europe effectively brought the Cold War to an end, and the free market approach

to economics came to dominate the economics profession. The ideas of John Maynard Keynes, emphasizing government intervention and considered to be orthodox in the golden age of postwar prosperity, fell into disrepute in many of the nation's leading graduate schools. The apparent success of England in the late 1980s under conservative Margaret Thatcher led to additional impetus that the conservative free-market approach to economics worked. The presidential administrations of George H. W. Bush and Bill Clinton alike pressed a free trade agenda and reduced spending on social programs. As markets became "liberalized," prices of basic commodities from coffee to cotton to oil declined by more than 100%. The terms of trade greatly improved for the United States, but poverty rates and debt soared in Africa and other developing regions where coffee growers, for example, had to compete with each other for the limited markets in the rich countries.

Our energy perspective has a different view, of course. First of all, much of the economic expansion of Presidents Reagan and George H. W. Bush was paid for with debt, so that the administrations of these supposedly fiscally conservative presidents (and Congresses at the time) actually generated far more debt, even when corrected for inflation and increased GDP, than even the supposedly "free-spending liberal" Franklin Roosevelt did for domestic programs in earlier times. It is important to understand that although the United States and Great Britain, for example, appeared to be doing much better economically under "conservative" administrations, both countries happened to enjoy low prices for oil and for energy in general while there were conservative leaders. (Less conservative Bill Clinton and Tony Blair benefited too.)

In the United States there was a decline in the proportion of GDP needed to pay for energy, from a maximum of 14% in 1981 to about six percent in 2000. This effectively gave the U.S. people some 6–8% additional discretionary income (that not required for basic food, shelter, and clothing), which could be spent on big houses and stocks. In addition, declining oil prices, an input to most basic commodities, reduced general inflation. In England Margaret Thatcher received a great deal of credit for her nation's economic recovery, but few attributed her success to the simple fact that the vast North Sea oil field came on line during her administration, greatly reducing former costs for imports. A large part of that oil was sold abroad, thus very large revenues for the government were generated, allowing the reduction of other taxes. Clearly conservatism alone could not fully explain England's success, as nominally socialist Netherlands was also doing very well economically at that time fueled by the vast Groningen gas field, whose profits allowed for social benefits to be extended to everyone. Energy analyst Doug Reynolds [38] generates a strong case that the collapse of the Soviet Union, often attributed to strong actions by the United States, was actually mostly a consequence of the partial collapse of Soviet oil production over the previous three years, greatly reducing the revenues that went to the central government and leading to many problems such as the inability to pay military pensions. So, again, these historical data about energy help explain what is normally attributed solely to political or economic leadership. A more difficult question, of course, is how to govern well when the abundant resource "rug" is pulled out from under the economy, a question of great importance as we write this book.

Political and Economic Response to Oil Price Increases Since 2000

A rather comfortable economic situation in America became subject to some disquiet as oil prices once again increased in the early 2000s and the overpriced stock market fell by nearly 20% as the technology bubble burst. Those who particularly benefited from the extensive surplus wealth of the 1990s often shifted their money into the housing market, as real estate was perceived to be a safer investment than technology. Government programs initiated by the Clinton administration and encouraged through the Bush administration, were designed to put more people into their own homes for political and social reasons. Oil prices relaxed a bit through about 2006

but then increased rapidly in 2007 and enormously more in the first half of 2008. For those who read widely there was a new set of economic predictions emanating from various oil industry analysts. Followers of M. King Hubbert, including Colin Campbell and Jean Laherrere [39], warned that the "peak" in oil production would soon be upon us and that the end of cheap oil would almost certainly follow, and with it significant economic consequences. The new Bush administration, apparently with its own inside information on declining oil production prospects, called for the drilling for oil in the Alaskan Wildlife Preserve and enhanced oil and gas development. Something that was barely noticed was that global oil production stopped growing in 2004. Colin Campbell had predicted at the Association for the Study of Peak Oil (ASPO) meeting in Lisbon that we were likely to see an *undulating plateau*, rather than a steep peak, for global oil. He reasoned that initial shortfalls in oil would lead to price increases, which would lead to economic recession, which would lead to a reduction in demand and lower oil prices, which would lead to economic recovery, which would lead to a new cycle. This basic pattern seems to have been exactly what has happened from 2004 to at least mid-2011.

The stock market continued to be sensitive to oil price changes, and the value of the Dow Jones kept struggling to increase beyond its inflation-adjusted peak in 1998 when corrected for overall price increases (Fig. 1.8). Although in some senses (high employment and increasing wealth of the more affluent) the economy was doing quite well in the first seven years of the 2000s, many questioned how much of the apparent affluence was real and how much was based on debt, as both real estate speculation and debt soared. From 1997 to 2005 the financial sector debt grew from 66% to more than 100% of GDP. Household debt rose accordingly, from 67% to 92% of GDP. Many private and public pension systems were based on the assumption that stocks would continue to grow at historical rates of eight percent or more, as had been seen in the "good times" (but speculative) period during the late 1990s.

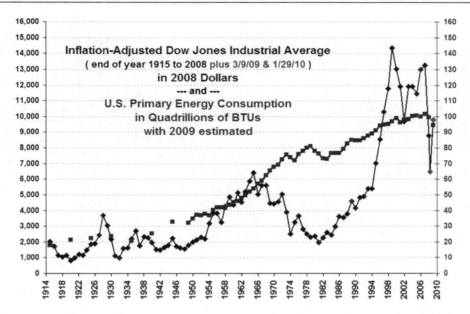

Fig. 1.8 Inflation-corrected Dow Jones Industrial Average (2008 dollars) scaled to plot in the same graph space as total energy used by the U.S. economy. Over long periods of time the slopes are very similar but the Dow Jones snakes around the total U.S. energy use, with the deviations from that line presumably reflecting psychological aspects (Figure courtesy of William Tamblyn)

When the stock bubble disappeared in 2000, many large companies were found to have placed not nearly enough money into their pension funds. Many workers who had worked hard all their lives with the expectation that they would have a good solid pension found they had little or nothing. Some were fortunate in having the federal government bail them out, but there is not enough money in that fund to cover even a fraction of the people who will have lost their pensions. Public entities, which are required by law to meet their pension obligations, fell about $500 billion in the hole. All forms of debt, including that of the federal government, increased faster than did the economy as a whole, as measured by the growth of gross domestic product. The federal government took in about $55 billion more in taxes than it spent in the last year of President Clinton's administration. By 2003, powered by tax cuts at the top end of the income distribution and increased military spending, the debt soared to an annual deficit of more than $500 billion by 2006.

The federal government, attempting to avoid inflation, did not "print" more money but became increasingly dependent on loans from Asia, especially China, to pay its bills. These loans, and those of the 1980s, will be a tremendous financial burden on young people who are reading this as undergraduate or graduate students, yet our government seems unwilling to raise taxes or reduce total spending. As of 2006 wars in Iraq and Afghanistan cost more than $8 billion *per month!* The situation only got worse following the financial meltdown of 2008, and Presidents Bush and Obama set new deficit records. Various healthcare initiatives imply enormous future federal spending requirements. In addition, individuals had been living far beyond their means by borrowing heavily on credit cards (Fig. 1.9). There is another unseen debt as well, that of delayed maintenance of society's infrastructure such as bridges, roads, levees, schools, and so on, not to mention degradation of the natural infrastructure of clean water, soil, and biodiversity.

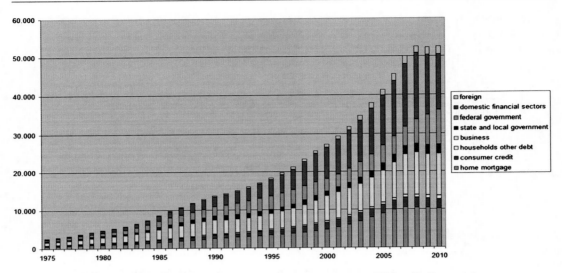

Fig. 1.9 Cumulative debts for the United States in trillions of dollars (*Source*: Wikimedia Commons)

What does this debt mean in energy terms? In 2008 the United States owed various entities such as banks and pension funds in Japan and China and other countries some $8,000 billion. If we were to pay off this debt all at once, and those who received these dollars (say retired Japanese Toyota workers) chose to spend it on beef, fish, rice, or Ford automobiles from America, it would take at the average energy use per unit of economic activity of the U.S. economy (about 8 MJ/dollar then) an estimated 64 exajoules worth of energy to make those goods, equal to 10 billion barrels of oil or half of U.S. known oil reserves remaining then to make the products those foreign people purchased. Meeting the interest payments transfers income to the holders of debt, and has nearly the same effect. In other words with our massive foreign debts people in other countries have a huge lien (i.e., obligation to repay a debt) on our remaining energy reserves or whatever their replacements might be. If this debt becomes too burdensome one way out of this problem is hyperinflation.

After the Treaty of Versailles in 1918 Germany was obligated to pay some $30 billion in "reparations" to France and England. They paid the international debt in "hard currency," mostly borrowed from U.S. banks, and they paid off their own domestic debts in deflated marks. The impact

was to enormously devalue their currency (Fig. 1.10). Prices rose by nearly 21% per day and doubled every 3.7 days. In 1918 it took 1 mark to purchase about 0.4 grams of gold. By November 1923 it took 100 billion reichmarks to buy the same quantity. This greatly undermined the entire financial system and helped lead to the rise of the Nazi party. Moreover, when the crises of the early 1930s crippled the U.S. banking system, American banks were either unwilling or unable to continue the loans to Germany. The Germans defaulted on their reparations, and Britain and France suspended the repayment of their war debts to the United States. The ensuing collapse of international trade and the gold standard was a primary factor in the depth and length of the Great Depression. The East African nation of Zimbabwe recently experienced just such hyperinflation with an inflation rate so large that prices doubled each day. Thus the United States continues to have enormous wealth and potential for creating wealth, although it may be increasingly constrained by the new shortage of cheap energy. Our debt to other nations and to nature also make our financial future potentially precarious. We need a careful systems analysis using both conventional and biophysical accounting to determine what is real wealth production and what is not, whether there is, as in the past, much

Fig. 1.10 German children playing with "bricks" of paper money, which had completely lost their value

potential for future growth to pay off this debt, and what might be the effects of future increases in energy costs.

The financial "crises" that occurred in the second half of 2008 add another dimension to our analyses. Many financial firms, highly respected for decades, collapsed or were accused of excessively risky and even quite shady financial undertakings. The government was asked to bail out all kinds of financial entities, and many people lost from one-third to one-half of their savings as housing and Wall Street prices collapsed. As of this writing it is far too early to tell whether this is just a "correction" to excessive speculation and "irrational exuberance" or, as seems increasingly likely, a genuine new direction for Wall Street. We suspect that if Wall Street is to grow again in the future beyond where it was in 2007, huge new energy supplies, or an unprecedented and unlikely increase in efficiency will need to be found. Barring that, many Americans will have to readjust greatly and permanently their perspective on wealth production through the stock market and probably in regard to economic growth in general. That this transition would be difficult, financially, intellectually, and emotionally for many, is an understatement.

We leave an examination of whether the vast increase in oil prices in the first half of 2008 were directly responsible for the economic meltdown of the second half of 2008 for a more comprehensive analysis in Chap. 18. In the meantime note that the total increase in energy use in the United States began to flatten out considerably starting in about 2000 as it had for domestic sources in 1995 (Fig. 1.8). Thus if the production of real wealth is as dependent upon the use of energy, as we believe, then we have left a long period of increasing energy and wealth and entered a period where it may no longer be possible to produce much more, or perhaps even as much, of either.

Why Does the Energy Issue Keep Emerging?

Given the very large jumps in the price of crude oil and gasoline, both down and mostly up, that occurred in the first eight years of the new millennium, many people have started to think about energy again. Why do "oil crises" keep reoccurring? Despite conservative claims that market processes and technology make considerations of any "limits to growth" and physical restrictions on energy resource supplies obsolete, world shortages and price fluctuations continue. Why have the stock market and real estate been failing so frequently to increase in value as they had in the past?

In the long term, markets and technologies have been a means of enabling humans to increase their wealth and material well-being. But wealth does not come out of thin air but only from the use of energy and the exploitation of physical resources. Thus an associated and necessary aspect of this increase in wealth is that the same factors, markets and technologies, that have enabled and encouraged us to become wealthy have also enabled and encouraged us to run through the world's resources more rapidly. It is quite possible that we are beginning to reach the limits of the Earth's ability to provide cheaply and easily the resources we have taken for granted. The periodic oil price increases are small reminders that, eventually, the piper must be paid, for as we like to say, Mother Nature holds the high cards. Humans are indeed industrious and ingenious, but that industriousness and ingenuity still require the Earth itself to provide the raw materials and fuels that are the basis for most wealth production and the capacity to absorb our wastes. Humans appear to be increasing energy and economic costs to the economy through indirect effects of industrialization, for example, increased damage by hurricanes from a warmer ocean, increases in sea level, possible crop production declines, tropical soil drying, increased rates of flooding and tornadoes, and so on. The *Stern Report* [40] says that the price for mitigating future environmental effects from global warming might be 20 times more than the cost of acting now to reduce our impact on the planet. This is one of many reasons why we must include more natural science in our economics, which we do throughout this book.

Debt, Inequality, and Who Gets What

Whatever the future of the total production of wealth in the United States, there are several clear and unsettling trends that will affect the future of energy supplies. The first is the enormous increase in debt in recent years (Fig. 1.9). Thus, much of the apparent prosperity of the recent past was based on debt and the ability to run an economy when debt expands more rapidly than income and wealth became highly suspect. The limits to debt constitute a limit to growth. The United States generated huge debts (at the time) as the administration of Franklin Roosevelt, especially during World War II, spent far more money than it took in. Since then the debt economy has escalated even further. The standard answer of mainstream economists is that economic growth allows us to carry the debt without peril to the rest of the economy. The crucial question for the world economy is whether there is the energy available today to facilitate or even allow the growth that might make the debts of today and the future fundable. If the traditional internal limits to growth, such as demand and productivity, coincide with the biophysical limits, every economic problem will be rendered more difficult. The era of peak oil is likely to be the era of degrowth.

The long history of the United States is based upon the strength of the middle class. Since about the 1960s, however, capital and wealth more generally have been concentrated increasingly in the hands of the wealthy (Fig. 1.7). The postwar history of the United States was based upon the spreading of income among workers and the poor. This provided the income to purchase the tremendous increase in the world's output after the Second World War. But increasingly in recent years the wealth of the United States has been concentrated in the hands of the rich, mostly as a consequence of expanding financial markets and tax policies that increasingly favor them, at least relative to earlier tax policies that would (in World War II) tax up to 94% of a wealthy person's income. As we describe in greater detail in Chap. 7, progressive taxes were curtailed in the 1980s, and the corporate tax burden has been falling since the early 1950s [25]. Historically, increases in inequality have faced limits as well. When income became too concentrated at the top, as it did in the 1870s, 1890s, and 1920s, a depression followed as citizens lacked the purchasing power to buy the products of industry.

Excess capacity increased, investment declined, and unemployment soared. By many measures this is the situation the United States faces today.

Are We Seeing the End of the American Dream?

A central question that we continue to explore in this book is whether this American dream (or European or Chinese or whoever else's dream) is sustainable, and what we might do to maintain it over the long term. Sustainability is a relatively new issue in economics, but one that is increasingly on many economists' agendas. What sustainability is, of course, is highly dependent upon the perspective of who is asking the question. To an anthropologist or developmental economist a sustainable economy might mean one that persists in time in the face of competition or aggression from other cultures or entities. To a conservation biologist sustainable economy might mean one that does not degrade biodiversity, and to resource-oriented persons (like ourselves) it might mean not "living beyond the planet's biophysical means of supporting one's culture." We prefer a concise biophysical definition of sustainability. To be sustainable, an economy must live indefinitely within nature's limits. In other words an economy must persist over the long haul without excessive depletion or degradation of the energy and material flows – and the physical milieu – of the biophysical system that contains and supports economic activity. A sustainable economy must be able to provide not only jobs but, ideally, also meaningful work and meaningful lives for those human beings who make up "the economy." By this definition we are very, very far from sustainable. To us it is dishonest and unethical to declare as sustainable so many "green" entities, as we see daily in the media, that in fact require the use of fossil fuels and nonrenewable fuels or other depletable resources. The fact that a product or process is marginally better or greener in these respects than their competition (or can be made to look so) does not make them sustainable.

Historically, especially in the post-WWII era, the vehicles to maintain prosperity and social stability were economic growth domestically and enormous power (military, monetary, productivity) internationally. The productivity increases and cost containment that were the basis for these pillars were dependent upon cheap oil and ignoring many environmental issues such as CO_2 release. As we can no longer do this, the inherent tendency towards stagnation that characterizes mature market economies is exacerbated by biophysical limits. As these pillars of prosperity have weakened, the prospects for a dream instead of a nightmare decline as well [41, 42].

What is the basis for our perspective? What does it mean to live within nature's limits? Ultimately it comes down to maintaining the per capita resource stocks and flows required for human existence (at what level of well-being?), and the degree to which the atmosphere and oceans can handle the wastes of the human economy. The number of people in the United States and elsewhere continues to increase greatly (Fig. 1.11). For example, when Hall was born (1943) there were about 137 million Americans, and a little more than two billion people in the world. There are now more than 310 million people living in the United States and seven billion people in the world. So the resources that form the basic inputs into our national and global economies have to be divided by roughly three times more people, and this is in only one person's (incomplete) lifespan. Global populations may well double or at least increase by another 50% in the reader's lifetime. Our most important mined resource is oil, and although it is not clear yet whether global oil production has peaked for all time, it is clear that per capita oil use (or oil use per person) peaked in about 1978 (Fig. 1.12). In other words, a growing amount of oil (until recently) has been used by an even more rapidly increasing world population. The traditional economist argues that this is not crucially important as various technologies have allowed humans to generate more resources, or more wealth from the resources that we do use. We do not argue with the idea that technological innovation is

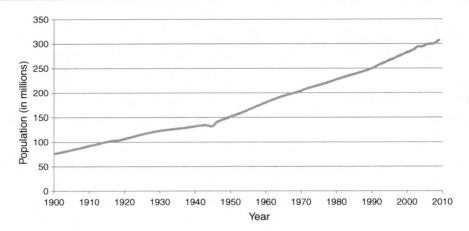

Fig. 1.11 U.S. population

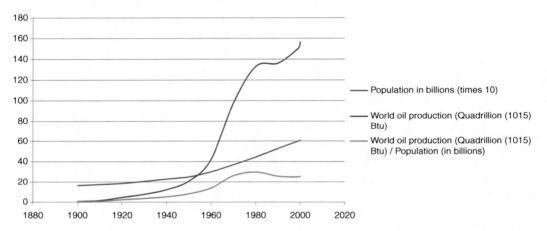

Fig. 1.12 Total and per capita oil production for the world. The energy units are in "Quads," or quadrillion BTUs, which is approximately equal to exajoules. Growth in energy has nearly stopped since 2008

important, however, we also show in later chapters how this concept is extremely misleading and cannot form the only solution for our future and that of our children.

We can begin by considering petroleum, perhaps our most important resource beyond sunshine, clean water, and soil. Most everything we do is based on cheap oil [34, 35]. Where we live relative to where we work, what we do for our work, how much leisure time we have and how we spend it, the price of our food, most of our purchases, and how much education we can afford, to name but a few, are largely dependent on adequate supplies of cheap oil. For example, it takes the energy of about a gallon of oil a day to

feed each American, about 80 barrels of oil to provide an undergraduate education at one of our colleges, and the energy equivalent of about ten gallons per day to keep us supplied with all the goods and services that we demand through our economic activity. In earlier days this level of affluence was available only to a tiny elite of society, and was usually provided by slave labor or indentured servitude. The net effect is that each of us today has some 60–80 "energy slaves" doing our bidding, effectively "hewing our wood and hauling our water."

The incredible thing about oil and gas is the almost complete absence of an understanding of its importance to the average American, and their

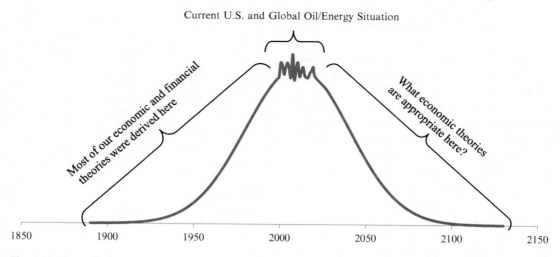

Fig. 1.13 Conceptual view of relation of our economic concepts and the Hubbert curve for global total oil use. Most of our economic concepts were derived during a period of increasing energy use. They are having trouble explaining economic events during the present period of peak oil. How will they do during the decline in energy availability?

failure to understand how critical it is to our economy. At a meeting of ASPO–USA (the Association for the Study of Peak Oil) in 2006 Denver Mayor Hickenlooper, who understood the importance of oil and its restrictions, said, "This land was originally settled by the Sioux. Everything that the Sioux depended upon, their food, clothing, shelter, implements and so on, came from the bison. They had many ceremonies giving thanks and appreciation to the Bison. We today are as dependent upon oil as the Sioux were on the bison, but not only do we not acknowledge that, but most people do not have a clue." The second critical thing about oil is peak oil, and that as of 2010 global oil production clearly is no longer increasing and may indeed be decreasing. Almost certainly it will decrease substantially in the future as we enter, in the words of geologist Colin Campbell, "the second half of the age of oil."

We conclude by saying yes, the American dream was the product of industrious and clever people working hard within a relatively benign political system that encouraged business in various ways, but that all of these things also required a large resource base relative to the number of people using it. A key issue was the abundance of oil and gas in the United States, which was the world's largest producer in 1970. But in 1970 (and 1973 for gas, although there may be a second peak) there was a clear peak in U.S. oil production, and although the continued increase in oil production worldwide buffered the United States (and other countries) from the local peak, it seems clear by 2010 that global oil production has reached its own peak whereas demand from around the world continues to grow. This mismatch between supply and demand resulted in a sharp increase in the price of oil and many economic problems that we believe it caused, at least in part, including the stock market declines, the subprime real estate bust, the failure of many financial corporations, the fact that some 40 plus of 50 states are officially broke, and that there is a substantial decrease in discretionary income for many average Americans. As developed later in Chap. 18 we believe that all of these economic problems are a direct consequence of the beginning of real shortages of petroleum in a petroleum-dependant society (Fig. 1.13).

The Future of the American Dream?

"Macondo (The Gulf of Mexico oil spill site) has eventually gripped the media and political eye. It is time for sober reflection on the global energy predicament and not for knee jerk reactions. How important is primary energy production and consumption for the OECD way of life? It links to economic growth, tax receipts and all that these pay for, pensions, manufacturing, food production, defense, leisure, comfort and security." These words, presented by energy analyst Euan Mearns on *The Oil Drum* (Europe) June 16, 2010 sum up our dilemma. As the country wrestles with the terrible environmental and economic consequences of the oil spill, does the disaster signal the end of finding new oil to support the fishing and recreation industries that are bemoaning the impact? Will this single event be the turning point in our hope of maintaining our national affluence? [41, 42] Have we finally caught up with living beyond our means through debt and seeing the beginning of the end of the continual hope that technology, such as deepsea drilling, will extend the American dream forever?

Franklin Roosevelt ran up huge debts to reconstruct the economy after the Depression and during World War II. Except for Bill Clinton, all presidents since (and including Ronald Reagan) increased debt (both corrected for inflation and the size of the economy) more than did Franklin Roosevelt during the New Deal (Fig. 1.9). Will the economy continue to grow so that we can pay off those debts if there is no longer cheap energy? What kind of jobs will be available if Americans have less and less discretionary income? Do we still want more labor productivity, which has usually meant subsidizing each laborer with more and more energy? Or do we want less labor productivity, that is, more energy productivity, by subsidizing each increasingly valuable unit of energy with more and more labor, to keep our people employed? How will we maintain and enhance the value of pension funds if the stock market no long grows in real terms? What about our inner cities? Can we find ways to employ those desperately in need of a job? Do population increases enhance our economic well-being or simply divide up our remaining untapped resources among more and more people? Do we need a completely new approach to economics during times when energy is declining? What indeed will the American dream mean in the future? Can we generate a new American dream with fewer material goods and more leisure? Will these issues limit our ability to support and educate our children? These are new economic questions that require a new way of thinking about economics. Much of the rest of this book tries to provide some of the information to answer these questions.

Questions

1. What does the word biophysical mean? How does it relate to economics? With what word(s) would you contrast it?
2. What factors are likely to influence your own economic success in life?
3. Although energy is barely discussed in Physiocratic, classical, or neoclassical economics, explain how each of these schools of economic thought focuses on the dominant energy flows of their time.
4. Why was Spindletop an event of great economic importance for the United States?

References

1. Whitehouse, M. 2010. Economists' Grail: A Post-Crash Model. Wall Street Journal, November 30.
2. Stiglitz, J. 2010. Needed: A new economic paradigm. Financial Times, Aug 20, 2010.
3. Gladwell, M. Outlier: The Story of Success. New York: Little, Brown.
4. Ehrenfeld, P. 1978. The Arrogance of Humanism. Oxford: Oxford University Press.
5. Cobb, C.W. and. Douglass, P.H. 1928. A theory of production. American Economic Review 18 Supplement 139–156.
6. Barnett, H. and Morse, C. 1963. Scarcity and Growth: The Economics of Natural Resource Availability. Baltimore, MD: Johns Hopkins Press.
7. Solow, R. 1956. A contribution to the theory of economic growth. Quarterly Journal of Economics. 70: 65–94.

8. Denison, E.F. 1989. Estimates of Productivity Change by Industry, an Evaluation and an Alternative. Washington, DC: The Brookings Institution.

9. Cleveland, C.J. 1991. Natural resource scarcity and economic growth revisited: Economic and biophysical perspectives. In Ecological Economics: The Science and Management of Sustainability, pp. 289–317.

10. Kummel, R. 1989. Energy as a factor of production and entropy as a pollution indicator in macroeconomic modeling. Ecological Economics. 1: 161–180.

11. Ayres, R. and Warr, D., 2005. Accounting for growth: The role of physical work. Change and Economic Dynamics 16: 211–220.

12. Hall, C.A.S. and K. Klitgaard. 2006. The Need for a New, Biophysical-Based Paradigm in Economics for the Second Half of the Age of Oil. Journal of Transdisciplinary Research Volume. 1, Issue 1, 4–22.

13. Polanyi, K., C. M. Arensberg and H. W. Pearson eds. 1965 Trade and Market in the Early Empires Economies in History and Theory. Free Press. New York.

14. Hanson, D. M. and M. Galeti. 2009. Forgotten Megafauna. Science. Volume. 324, 42–43.

15. Martin, P.S. 1973. The Discovery of America. Science. Volume. 179: 969–974.

16. Tainter, J. 1988 The Collapse of Complex Societies. Cambridge University Press. Cambridge, England.

17. Diamond, J. 2004 Collapse: How Societies Choose to Fail or Survive. Viking Press. New York.

18. Crosby, Alfred. 1986 Ecological Imperialism: The Biological Expansion of Europe, 900–1900. Cambridge University Press. Cambridge, England.

19. Diamond, J. 1997 Guns, Germs and Steel: The Fates of Human Societies. New York: Norton.

20. Zinn, Howard 1980. A People's History of the United States. New York: Harper & Row

21. The two greatest causes for war among the perhaps one third of the colonial population who supported independence was basically resource scarcity – the cutting off of the trans-Allegheny frontier, first in 1763 by means of the Proclamation Act, and then later in 1775, with greater enforcement, with the Quebec act. Open rebellion soon followed.

22. Perlin, J. 1989 A Forest Journey: The Role of Wood and Civilization. Harvard University Press. Norton, N.Y.

23. Greene, A. N. 2008 Horses at work: Harnessing Power in Industrial America. Harvard University Press. Cambridge.

24. Yergin, Daniel 1991 The Prize: The Epic Quest for Oil, Money, and Power. New York: Simon & Schuster.

25. Bowles, S., D. Gordon, T. Weisskop 1990 After the Wasteland. ME Sharpe. Armonk, NY.

26. Smil, V. 2001. Enriching the Earth. Fritz Haber, Car Bosch and the Transformation of Western Food Production. MIT Press. Cambridge.

27. Meadows, D., D. Meadows and J. Randers. 2004. Limits to Growth: The 30-Year Update. Chelsea Green Publishers. White River, V.T.

28. Ehrlich, P. 1968 The Population Bomb. Ballantine Books. New York.

29. Hubbert, M.K. 1969. Energy Resources. In the National Academy of Sciences–National Research Council, Committee on Resources and Man: A Study and Recommendations. W. H. Freeman. San Francisco.

30. Hardesty, John, Clement, Norris C. and Jencks, Clinton E. 1971. The political economy of environmental destruction. Review of Radical Political Economics 3(4): 82–102.

31. England, Richard and Bluestone, Barry. 1971. Ecology and class conflict. Review of Radical Political Economics 3(4) 31–55.

32. Daly, Herman. Towards a Steady-State Economy. 1973. London: W.H. Freeman and Company, Ltd. Tables.

33. Passell, P , M. Roberts, and L. Ross. 2 April 1972. Review of Limits to Growth. New York Times Book Review.

34. Odum, H. T. 1973 Environment, Power and Society. New York: Wiley Interscience.

35. Cleveland, C., R. Costanza, C. Hall and R. Kaufmann. 1984. Energy and the U.S. Economy. A Biophysical Perspective. Science. Volume. 225, 890–897.

36. Jorgenson, D. W., and Z. Grilliches. 1967. The Explanation of Productivity Change. Review of Economic Studies. 249–283.

37. Maddala, G. S. 1965. Productivity and Technical Change in the Bituminous Coal Industry. Journal of Political Economy. 352–265.

38. Reynolds, D. 2000. Soviet Economic Decline: Did an Oil Crisis Cause the Transition in the Soviet Union? Journal of Energy and Development. Volume. 24, 65–82.

39. Campbell, C., and J. Laherrere. 1998. The End of Cheap Oil. Scientific American. March: 78–83.

40. Stern, N. 2007 The Economics of Climate Change. The Stern Review. Cambridge University Press. Cambridge.

41. Luce, E. 2010 "Goodbye, American Dream. The Crisis of Middle-Class America". The Financial Times. 30 July 2010.

42. Brinkbaumer, K., Hujer, M.. Muller P. Schulz, T. 2010. Is the American Dream Over? Das Spiegel 2010.

Energy and Wealth Production: An Historical Perspective

2

The History of Formal Thought on Surplus Energy

There are many scientists from different disciplines who have thought deeply about the long-term relation of humans and wealth production. Most have concluded that the best general way to think about how different societies evolved over time is from the perspective of surplus energy. To chemists Frederick Soddy and William Ostwald, anthropologist Leslie White, archeologist and historian Joseph Tainter, sociologist Fred Cottrell, historian John Perlin, systems ecologist Howard T. Odum, sociologist and economist Nicolas Georgescu-Roegen, energy scientist Vaclav Smil, and a number of others in these and other disciplines, human history, including contemporary events, is essentially about exploiting energy and the technologies to do so. This is not the perspective taught in our schools and the role of energy is essentially missing from our dominant books and teaching about history. Instead human history usually is seen in terms of generals, politicians, and other personalities.

This chapter develops the alternative perspective that the fates of past civilizations and other events of the past can be better understood from the perspective of the importance of energy, and in particular surplus energy. *Energy surplus* (or net energy) is defined broadly as the amount of energy left over after accounting for the costs of obtaining the energy. The energy literature is quite rich with papers and books that emphasize the importance of energy surplus as a necessary criterion for allowing for the survival and growth of many species, including humans, as well as human endeavors, including the development of science, art, culture, and indeed civilization itself. Each acknowledges that other issues such as human inventiveness, nutrient cycling, and entropy (among many others) can be important, however, each is of the opinion that it is energy itself, and especially surplus energy, which is key. Survival, military efficacy, wealth, art, and civilization are believed by all of the above investigators to be a product of surplus energy. For these authors the issue is not simply whether there is surplus energy but how much, what kind (quality), and at what rate it is or was delivered. The interplay of those three factors determines net energy and hence the ability of a given society to divert attention from life-sustaining needs such as growing sufficient food or the attainment of water towards trade, warfare, or luxuries, including art and scholarship. Indeed, humans could not possibly have made it this far through evolutionary time, or even from one generation to the next, without there being some kind of net positive energy, and they could not have constructed such comprehensive cities and civilizations, or wasted so much in war without there being substantial surplus energy in the past.

C.A.S. Hall and K. Klitgaard, *Energy and the Wealth of Nations: Understanding the Biophysical Economy*, DOI 10.1007/978-1-4419-9398-4_2, © Springer Science+Business Media, LLC 2012

The Prehistory of Human Society: Living on Nature's Terms

Human populations must first feed themselves, and after that generate sufficient net energy to survive, reproduce, and adapt to changing conditions. Although people in most industrial societies today hardly worry about getting enough to eat, for much of the world and much of humanity's history getting enough food was the most important task. For at least 98% of the 2 or so million years that we have been recognizably human, the principal technology by which we as humans have fed ourselves has been that of hunting and gathering. Contemporary hunter-gatherers – such as the !Kung of the Kalahari desert in southern Africa – probably live as close to the lifestyle of our long-ago ancestors as we will be able to understand (Fig. 2.1). Most hunter-gatherer humans were probably little different from cheetahs or trout in that their principal economic focus was on obtaining enough food and hence for getting their requirement for surplus energy directly from their environment. Studies by anthropologists such as Lee and Rapaport

confirmed that indeed present-day (or at least recent) hunter-gatherers and shifting cultivators acted in ways that appeared to maximize their own energy return on investment.

Richard Lee, in particular, studied the energetics of the !Kung while they were relatively unaffected by modern civilization [1]. A charming although romanticized view of their culture is readily accessible in the movie, *The Gods Must Be Crazy*. Life for a hunter-gatherer is basically about taking nature as it is found and finding ways to survive on those resources. Because most early human hunter-gatherers lived in tropical environments, the key challenge was gaining the needed food energy. For the !Kung, this was undertaken by women gathering mongongo nuts and men hunting anelope and other animals. Mongongo nuts are the most abundant resource that provides the largest part of the energy and protein consumed by the !Kung, although game is very much appreciated and gives needed additional protein to the diet. Life is good for the !Kung, at least it was before their major contact with civilization. According to Lee's studies, the !Kung spent far fewer hours working each day than do people

Fig. 2.1 !Kung people, modern day hunter-gatherers, probably represent how all of our ancestors lived their lives for far more time than even the time since the start of agriculture (*Source: Science*)

Fig. 2.2 Map of the various waterholes in the Kalahari desert that the !Kung migrate from and to over the seasons. The exhaustion of easy food in the region of one wateringhole necessitates movement to another (*Source*: Lee, 1973)

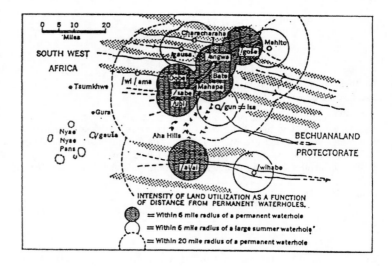

living in industrial societies; a lot of their time was spent in leisure activities. Young women tended to be sexually active (which was considered normal) from as early as age 9, but tended not to get pregnant until about age 18, when they had sequestered enough body fat so that the pregnancy was possible (i.e., it appears that the human body protects young women from pregnancy when they do not have enough energy surplus to carry a fetus) [2]. Life for the !Kung was not always simple, however, for they lived in a desert and were constrained by their need for water and food. In their homelands of Botswana there are only relatively few waterholes, and it is essential to set up camp near one of these waterholes (Fig. 2.2). Mongongo trees are spread more or less randomly around this part of the Kalahari desert so initially the !Kung can derive all the food they need from relatively short excursions from their camp. As time goes on they deplete the nuts (and game) within easy reach, so that each day they have to make a longer and longer trip to gather enough mongongo nuts to feed their families. At some point they have gathered all the mongongo nuts within a day's hike. Then they have to make much longer, overnight trips to get them. They eat a lot of food both going and coming back, therefore they consume a substantial portion of the food they went out to get! This

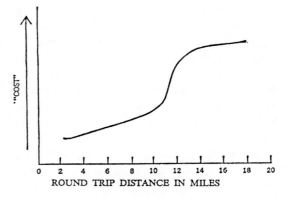

ROUND TRIP DISTANCE IN MILES

Fig. 2.3 Determinants of !Kung EROI. At a distance of about 11 miles energy cost increases greatly because an additional day is needed. When the !Kung have exhausted the mongongo nuts within 1 day's walk they have to make a substantially greater investment to walk 2 days to get a new supply of nuts (*Source*: Lee, 1973)

greatly increases their energy investment and lowers what we call their energy return on investment, or EROI (Fig. 2.3; see Chap. 13). This makes it desirable at some point to make the additional investment of moving to a new water hole.

According to Lee, the !Kung lifestyle, under normal circumstances generates a quite positive energy return on investment (i.e., generates a large surplus) from their desert environment, perhaps an average of some 10 kcal returned per one

Table 2.1 Megafaunal extinctions

	Extinct	Living	Total	% Extinct	Landmass (km²)
Africa	7	42	49	14.3	30.2×10^6
Europe	15	9	24	60	10.4×10^6
North America	33	12	45	73.3	23.7×10^6
South America	46	12	58	79.6	17.8×10^6
Australia	19	3	22	86.4	7.7×10^6

Late Quaternary (last 100,000 years) extinct and living genera of terrestrial megafauna (>44 kg adult body weight) of five continents. Adapted after Martin (1984). Data for extinct and living European megafauna from Martin (1984). For Australia it may be that as many as eight genera were already extinct before human arrival (Roberts et al., 2001). If so, this reduces both the number and percentage of megafaunal extinctions that could conceivably be attributed to human activity

Source: Wroe et al. [32]

of their own kcal invested in hunting and gathering. In normal times these cultures had plenty to eat, and the people tended to use the surplus time made available from their relatively high EROI lifestyles in socializing, child-care, and storytelling. The downside was that there were periodic tough times, such as droughts, during which starvation was a possibility. It is probable that our ancestors had a fairly positive EROI for much of the time, although periodic droughts, diseases, and wars must have occasionally, or perhaps routinely, taken a large toll. Thus even though the !Kung, and by implication other hunter-gatherers, had a relatively high EROI, perhaps 10:1, human populations tended to be relatively stable over a very long time, barely growing year to year from millions of years ago until about 1900. Thus even this relatively high energy return was not enough to generate much in the way of net population growth over time.

It is increasingly clear that our stone-age hunter-gatherer ancestors, as hunter-gatherers today, tended to be quite good hunters. This hunting prowess resulted in an enormous environmental impact on the large birds and mammals of the earlier world. As humans spread about the world they encountered in each new place large, naive herbivorous animals of the sort we do not see anywhere on Earth today. For example, the new arrivals in North America found giant beaver, rhinoceros, two species of elephants, camels, and so on. Human arrivals in

Australia found giant flightless birds, whereas the first humans in Italy found large turtles (no longer extant) and so on. None of these large animals are there today, and except in Africa, there few animal species larger than 100–200 kg left. These large animals were abundant prior to human arrival (Table 2.1). (Of course bison, bears, moose, and elk are large and still with us, although in greatly reduced ranges.)

What caused their extinction? There are two competing hypotheses. First, because the climate was warming rapidly 10,000 years ago it is possible they succumbed to some effect of climate change. The second hypothesis is that humans hunted these animals to extinction. These large animals had no previous reason to be afraid of anything as small and puny as a human being and humans could simply walk up to these animals and stick a spear into their sides. Africa still has many, many very large herbivorous species, probably because the animals coevolved with humans as they slowly became more proficient hunters with better weapons. All around the world where humans came later most or all the larger animals disappeared within 2,000 years of human arrival. This certainly supports the idea that it was humans who did them in [3]. The fact that these same animal species had survived many previous climate changes lends support to the human-caused extinction idea. Thus, significant environmental impact from humans is hardly new.

African Origin and Human Migrations

All available evidence suggests that humans and their predecessors evolved in Africa, which is the only place we have found human fossils or evidence dating to roughly 1.7–1.8 million years ago [4]. Take a mental time trip to East Africa about 2 or 2.5 million years ago. You would be in the cradle of the evolution and development of all that makes us human. Remarkably you would find not one, but perhaps half a dozen types of early humans (or hominids), each group as distinct from another as chimpanzees from gorillas. Most of these protohominids were found in small migratory bands more or less at the transition of forests to drier savannas. We continue to learn more about our ancestors. The finding in the 1990s of the fossils of what appears to be the ancestor of humans who lived some 4–6 million years ago is cause for great excitement among those who are determining our lineage. This creature, named *Ardipithecus ramidus* (Ardi for short), walked more or less upright but still spent much, perhaps the majority, of its time in trees (Fig. 2.4).

Recent research has found that a human uses only about one quarter as much energy to walk 100 meters as a chimpanzee, so there obviously has been a tradeoff favoring more energy-efficient walking over the ability to both walk and climb trees well, as chimpazees can. Probably most of the Ardis made, or at least used, tools of some sort, for we understand now that even chimpanzees have a rather astonishing ability to make many different types of tools, including stone anvils. Most of their tools were made from organic materials and hence are not well preserved, so we know little about the past tool making of either chimps or protohominoids. By about 2.5 million years ago our ancestors had developed quite sophisticated methods for making stone knives and spear points by striking or stroking one rock on another in repeated and often sophisticated patterns. There are even a number of ancient "industrial complexes" in, for example, Kenya's Olduvi Gorge, a rich hunting ground for information about our ancestors (Fig. 2.5). Spear points and knife blades are actually energy technologies: energy- (force-) concentrating devices that allow the strength of a human arm to be multiplied many times when concentrated on a line or point (Fig. 2.6). This allowed humans to exploit many new animal resources and eventually the colonization of cooler lands. Our ancestors were using stone tools for roughly two and a half million years, which is equivalent to about 100,000 human generations.

These stone spear points and knife blades were more or less the first in a long series of

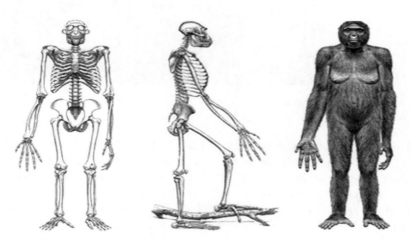

Fig. 2.4 Ardi, *Ardipithecus ramidus* is a new-found fossil that is neither man nor ape but probably represents our human ancestors some 4,000,000 years ago (*Source*: *Science*, Jay Matternes)

Fig. 2.5 Olduvai Gorge (from Shunya website). Many very early human remains have been found here as well as early "industrial" sites, where stone tools were manufactured

Fig. 2.6 Spear heads

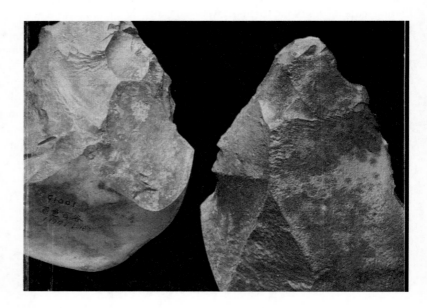

technological advances that helped increase the flow of energy to humans, thus greatly expanding the ability of humans to exploit the energy available in the various plant and animal resources in their environment. It also greatly increased the climates in which they could live because of their ability to kill large animals and use their skins for clothes (Fig. 2.7). Another important new energy technology was fire, which allowed people to stay warm in cooler climates, but more important, increased the variability and utility of plant foods, as cooking broke down the tough cell walls that plants (but not animal) cells have. Many humans left the relatively benign climate

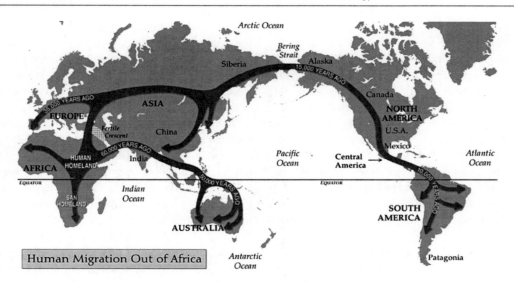

Fig. 2.7 Human migration patterns. All humans originated in Africa but then took various routes to establish new groups of people

of Africa probably a little less than two million years ago. The remains of both humans and their tools of that era have been found in the present-day Middle East, Georgia, and probably Indonesia [5]. By a million years ago human remains were common all through Asia, but curiously humans did not appear to colonize Europe until roughly 500,000–800,000 years ago. The first humanoid colonists of Europe do not appear to be our direct ancestors, for morphologically modern humans (popularly known as Cro-Magnons as distinct from the earlier Neanderthal stocks) appear to have left Africa in a separate migration only about 100,000 years ago. There are very strong debates in the anthropological literature as to whether all of these groups of people are our ancestors or just the Cro-Magnon variety of a large suite of early humans. Modern DNA analysis seems to favor the separate stock concept. For whatever reason, perhaps interracial warfare, climate change, or some indirect result of competition, the Neanderthal stocks as well as the many other protohominid variations were eliminated from Europe by 35–40,000 years ago, leaving, it seems as of 2010, a few of their genes mixed with those of Cro-Magnon stock.

One of the many changes that took place as humans moved out of Africa was that humans tended to lose their melanin, a protective pigment that helped people living in Africa avoid various skin diseases such as skin cancer. When humans were exposed to much less sun for long winter periods, while in the meantime covering their skins with animal hides, they did not get the benefit of the sun producing vitamin D within human skin. This made humans much more susceptible to rickets, a debilitating vitamin deficiency disease that results in broken bones, obviously a great problem for hunter-gatherers. Inasmuch as the dark pigment melanin protects skin, but also decreases its ability to make Vitamin D, darker skin is less advantageous in areas with less year-round intense sun. Hence skin color, something of enormous and often egregiously misplaced cultural importance, is simply a reasonable evolutionary response to humans leaving or not leaving the tropics.

The Dawn of Agriculture: Increasing the Displacement of Natural Flows of Energy

Sometime about 10,000 years ago, in the vicinity of the Tigris and Euphrates valleys of present day Iraq, a momentous thing happened [6, 7]. Humans previously had been completely constrained by their limited ability to exploit entirely natural food chains, due to the low abundance of edible plants. They found that they could increase the flow of food energy to themselves and their families enormously by investing some seeds into more food for the future. How this happened is lost to antiquity, but as described by Jared Diamond in *Guns, Germs and Steel* [8], it probably happened as people observed that their own kitchen middens (garbage areas) produced new crop plants from the seeds that had been deliberately or inadvertently discarded.

The implications for humans were enormous. The first, seemingly counterintuitive, is that human nutrition, on the average, declined. One of the best studies to document this was by Larry Angel, who studied the bones of people buried over the past 10,000 or so years in Anatolia, roughly the border region of modern-day Turkey and Greece [9]. Angel was able to date the bones he found in ancient burial grounds, and could learn many things about the people who once lived there from the bones themselves. For example, their height and general physical condition, as well as functions of the quality of nutrition, could be determined by the length and strength of the bones. Bones could also show the number of children a woman had by the scars on the pubis, whether that person had malaria by the appearance of the bone marrow-producing regions of the bone, and so on. The data indicate that the people actually became shorter and smaller with the advent of agriculture, indicating a decrease in nutritional quality. In fact the people of that region did not regain the stature of their hunter-gatherer ancestors until about the 1950s. Thus although agriculture may have given the first agronomists an advantage in terms of their own energy budgets, that surplus energy was translated relatively quickly into more people with only an adequate level of nutrition as human populations expanded. Or perhaps, as outlined below, more of the farmers' net yield was diverted to artisans, priests, political leaders, and war, leaving less for the farmers themselves. One of the clear consequences of agriculture was that people could settle in one place, so that the previous normal pattern of human nomadism was no longer

the norm. As humans occupied the same place for longer periods of time it began to make sense to invest their own energy into relatively permanent dwellings, often made of stone and wood, which left more durable artifacts for today's archeologists.

A second major consequence of agriculture was an enormous increase in social stratification as economic specialization became more and more important. For example, if one individual was particularly skilled at making agricultural implements or understood the logic and mathematics (i.e., best planting dates) of successful farming, it made sense for the farmers of the village to trade some of their grain for his implements or knowledge, initiating, or at least formalizing, the existence of markets. From an energy perspective, relatively low-quality (because so many people had the necessary skills) agricultural labor was being traded for the high-quality labor of the specialist. The work of the specialist can be considered of higher quality in terms of its ability to generate greater agricultural yield per hour of labor. Considerable energy had to be invested in training that individual through schooling and apprenticeship. The apprentice had to be fed while he was relatively unproductive, anticipating greater returns in the future. Thus we can say that the energy return on investment of the artisan was higher than that of the farmer, even if less direct, and often his pay and status as well.

Eventually, the concept of agriculture spread around Eurasia and Africa (Fig. 2.8). A new phenomenon appeared with the development of agriculture, the large net profits from the farmers and the permanent settlement of certain regions: cities and other manifestations of urbanization. The first place this occurred appears to be in the Tigris–Euphrates valleys; and one of the first cities was known as Ur, from which we derive the word urban. Today we call that ancient civilization Sumeria and the people Sumerians. There were many great cities of that time (roughly 4,700 years ago) and region, including Girsu, Lagash, Larsa, Mari, Terqa, Ur, and Uruk. These cities grew up in what had been at first a heavily forested region, as can be understood from the

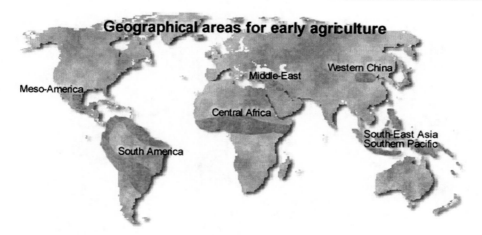

Fig. 2.8 Origins of early agriculture

massive timbers in remaining ruins, although today there are essentially no trees and no cities in that region. In fact the forests were gone by 2400 B.C., the harbors and irrigation systems silted in or required enormous energy to maintain, the soil became depleted and salinized, barley yield dropped from about 2.5 tons per hectare to less than one, and by 2000 B.C. the Sumerian civilization was no longer extant. The world's first great urban civilization, in fact its first great civilization, used up and destroyed its resource base and just disappeared over a span of 1,300 years. These stories are well understood and told in fascinating detail in many places including by Perlin, Michener, and Tainter [10–12].

The interaction of people with cultivars (plants that humans cultivate) also greatly changed the plants themselves. All plants are in constant danger of being consumed by herbivores, from bacteria to insects to large grazing or browsing mammals or, formerly, herbivorous dinosaurs. The evolutionary response of plants to this grazing pressure was to derive various defenses, including physical protection (such as spines, especially abundant in desert plants) and, more commonly, chemical protection in the form of alkaloids, turpenes, tannins, and so on. These compounds, usually derived at an energy cost to the plant, place a heavy burden on herbivores or potential herbivores by discouraging consumption or by extracting a high energy cost on those

specialized herbivores that can eat them, for the energy cost of detoxifying poisonous compounds is very high [13, 14]. Humans do not like these frequently bitter poisonous compounds either, and for thousands of years have been saving and planting the seeds from plants that taste better or have other characteristics that humans like. Partial exceptions are, for example, mustards, coffee, tea, cannabis, and other plants whose bitter alkaloids are poisonous if that were all we ate but an interesting dietary supplement in small doses. Consequently our cultivars are, in general, quite poorly defended against insects and have required us to invent and use external pesticides, with complex consequences. Many of our cultivars would not survive in the wild now, and have coevolved with humans into systems of mutual dependency. A visitor from outer space might conclude that the humans have been captured by the corn plants who use us for their slaves to make their lives as comfortable and productive as possible! Meanwhile all kinds of pests were themselves adapting to the concentration of humans and their growing and stored food, often with disastrous impacts on humanity [15].

At roughly the same time that agriculture was spreading around the world humans made another extremely important discovery: metallurgy. Prior to the advent of metallurgy essentially all tools used by humans were derived directly from nature: stone, going back perhaps 50,000 years

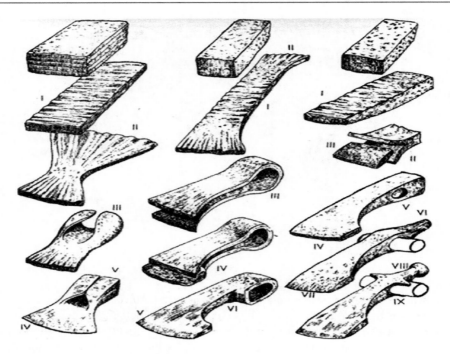

Fig. 2.9 Early metallurgy (*Source*: *National Geographic*)

(Fig. 2.5) fashioned with increasing sophistication, wood, bones, antlers, and so on. According to Ponting [16], the first evidence of the smelting of copper was found in Anatolia from about 6000 B.C., although the near contemporaneous existence of residuals of smelting from all continents at only slightly later in time implies that probably many groups of people had roughly the same idea by about 5000 B.C. (Fig. 2.9). Eventually very specialized furnaces were developed, as is indicated by archeological digs from 5,000 to even 10,000 years ago in Africa, Europe, South America, and Asia. Early copper and bronze tools were replaced over time with iron as people learned to make hotter fires. We have been using metal tools for roughly 8,000 years, or about 400 generations. So most of our history as a species is without metal tools. An important component of the transition is that the stone tools could be made with only a very small energy investment, essentially all as human muscle power, whereas the metal tools required a much larger investment in terms of cutting trees, making charcoal, and of

course the energy of the wood itself. Early smelting was probably technically inefficient but had the advantage, at least initially, of the availability of very high grades of ore.

Smelted metals had a number of advantages compared to materials derived directly from nature: metals were harder and could take a sharper edge, increasing the cutting work that could be done by human muscles, and the sharper knife blades and spear points concentrated energy onto a smaller surface and enhanced the process of humans exploiting nature, for example, by accelerating the rate that people could cut trees (and of course each other) with bronze versus stone axes. Perlman [10] has chronicled the tremendous increase in human cutting of forests in a wonderful book, *A Forest Journey*. He makes the point in this book that massive deforestation is an old phenomenon, and that India, China, and most of the Mediterranean were pretty thoroughly deforested by the time of Christ. In most cases the most severe deforestation was to get fuel for metallurgy.

The scenario often went something like this (with Crete as a good example). A group of people would find and develop a rich ore deposit of, for example, copper. This metal would be very valuable in trade and the people would become prosperous. Cutting of trees for smelting also cleared land for agriculture, and the wealth and well-being of the people increased not only from the trade in metals but also from the large increase in the area under agriculture in the rich forest soils where the trees had been cut. Things would tend to go very well for roughly a century. But once rich forest soils were exposed to agriculture and rain they would tend to erode, and the agricultural yields would decline. That civilization would decline as ore deposits and soils wore out, until they collapsed: meaning that the number of people being supported decreased dramatically. According to Perlin (and many others [10, 16, 17]) this process has occurred again and again and again throughout history. India and Greece have had three separate major deforestations, with the forests growing back each time human populations became lower. The great works of literature, for example, Thucydides' *The Peloponnesian Wars*, were written about events enormously affected by large resource and environmental events (i.e., the exhaustion of sufficient forests for Athenians to smelt silver or make ships) although such resource issues were rarely considered by historians until recently [10].

Other important energy-related events were occurring in these prehistoric times. Perhaps most important was the domestication of useful animals, some of which predated agriculture, and some occurred more or less simultaneously. The domestication of animals and the increased sophistication of animal husbandry was important in increasing energy resources for humans in at least two ways. First, because these animals ate plant material that humans did not, this greatly increased the amount of energy that humans could harvest from nature, especially in grasslands. Second, oxen and especially horses as draft animals greatly increased the power output of a human (Table 2.2). This power was useful for transport, for agricultural preparation (which came later), and for war.

Table 2.2 Evolution of power outputs of machines available to humans

Machine	Horsepower
Man pushing a lever	0.05
Ox pulling a load	0.5
Water wheels	0.5–5
Versailles water works (1600)	75
Newcomen steam engine	5.5
Watt's steam engine	40
Marine steam engine (1850)	1,000
Marine steam engine (1900)	8,000
Steam turbine (1940s)	300,000
Coal or nuclear power plant (1970s)	1,500,000

Source: Cook [33]

The story of how the use of animal technology was passed throughout Eurasia has been developed elegantly by Diamond, so we say little except to repeat his main conclusion that geography was critical. Most of the important domestic animals came from Eurasia and could be passed east and west much more easily than north to south. Our most important animals, the sheep, cow, horse, pig, and chicken were "corralled" in Eurasia and developed into today's domestic animals. The increasing familiarity with beasts of burden and the development of roads and caravan technology in turn allowed for the development of long distance trade. Meanwhile sailing and navigational skills were developed and passed on, and Cottrell writes well about the importance of using wind power in ships to enormously enhance the amount of work (carrying goods) that one person could do. Trade between cultures enriched the knowledge and the biotic resources of many human groups.

As agriculture, settlement, and commerce expanded, there became a greater need for maintaining records, and some time about 3000 B.C. formal writing was developed, apparently simultaneously in Egypt, Mesopotamia, and India (and perhaps other places). Writing allowed technologies to be maintained from one generation to another and transferred among cultures. Cumulatively all of these new technologies increased the energy flow to the human population, which slowly but relentlessly increased. These old records have allowed us to estimate

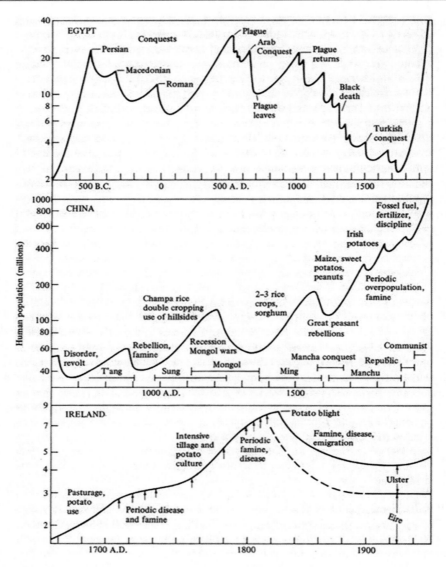

Fig. 2.10 Human population changes in Egypt, China, and Ireland, regions that had relatively well-developed bureaucracies and hence good data (*Source*: Lieth and Whittaker, 1975)

some earlier patterns of human population changes (Fig. 2.10). They suggest that the human population record is hardly one of continuous regular growth, but rather one of periodic growth and decline. Sometimes this is manifest as catastrophic decline and the virtual or absolute cessation of that population or, more commonly, the political structure that once held them together.

Edward Deevey [18] has suggested that there were three main increases in human populations associated with first the corralling of animals, then the development of agriculture, and then the industrial revolution. We are still experiencing the latter as global human population growth continues strongly, although at a somewhat lower rate than a few decades ago.

Human Cultural Evolution as Energy Evolution

As we keep pointing out, most of the major changes that occurred in the ability of humans to exploit more and more of the resources around them were either directly about, or clearly associated with, increased use of energy. Spear points and knives are energy-concentrating devices, fire allows greater availability of plant energy to humans, agriculture greatly increases the productivity of land for human food, and so on. These evolutions of the ability of humans to control more and more energy, for example, the evolution of wind and water power, is probably best told in Fred Cottrell's wonderful book, *Energy and Society*, published more than half a century ago [19]. Cottrell was a railroad man for most of his life and then a college professor near the end. Always impressed, like us, with the energy that undergirds all that humans do, Cottrell's focus was on the development of what he called "converters," that is, specific technologies for exploiting new energy resources.

Cottrell's early chapters focus on herding and agriculture as a means of exploiting biotic energy and then on water and wind power. He shows the historical importance of a city being located relatively downstream on a river, so that the natural flow of the water would allow that city to exploit all upstream resources easily, such as timber, agricultural products, game, and ores. Of course there was always a problem with this: barges had a one-way trip and so had to be built anew at the top of the watershed for each new trip. Also crews had to walk or otherwise get themselves back upstream. Nevertheless a barge could carry an enormous load compared to a person alone, who can carry only about 25 kg at a maximum for any serious distance, or a pack animal such as a horse that can carry about 100 kg. Thus the use of a barge carrying, say, 10 tons of goods and with a crew of four increased the efficacy of each person by a factor of 25–100.

The development of a sailing ship likewise increased the energy that subsidized a human porter enormously. According to Cottrell's calculations, an early sailing ship such as used by the Phoenicians (more or less the equivalent of modern-day Lebanon) increased the load that a human could carry by some factors of 10, and by late Roman times as much as a factor of 100. The Romans needed to import large quantities of grain from Egypt because, in part, they had depleted their own soil. But, according to Cottrell, the Romans were not the only ones who had an eye on this grain, and initially the Romans lost a lot of grain to pirates. This required the Romans to transport the grain in heavily guarded narrow warships, and a significant part of the grain was consumed by the soldiers on board. Thus one further energy investment had to be made by the Romans: clearing the Mediterranean of pirates. Once this was done, proper wide-beamed sailing merchant vessels could be used and Egypt finally became a large net energy source for the Romans. Cottrell gives many other examples of the increasing use of energy by humans over time, including very interesting chapters on the growth of railroads in England, steam power, and industrialized agriculture.

The Possibility, Development, and Destruction of Empire

Agriculture and its greatly increased yield brought with it the possibility of the concentration and storage of food, specialization, and through greater populations, military–political power. These concepts are again ably reviewed in Diamond, Tainter, Ponting, and others. From our energy perspective, agriculture allowed for huge energy surpluses as a result of high return ratios from large energy investments. Thus agriculture allowed a massive increase in the ability of people to generate culture and cultural artifacts. We have bare glimpses of these in the remaining artifacts of ancient cultures such as the main building at Ur (Fig. 2.11), temple complexes and the great wall of China. What we see of these ancient civilizations today are more or less beautifully shaped and carefully put together stones, and increasingly over time, more sophisticated

Fig. 2.11 Ruins of ancient
city of Ur

ornamentation, pottery, and metal household
implements. By digging a little deeper we can
find other enormously impressive artifacts of past
civilizations: irrigation systems to bring water
over large distances and large pyramids of stacked
stones. These artifacts imply huge energy surpluses relative to hunter-gatherers.

In hunter-gatherer cultures there was normally
relatively little differentiation in what different
people did, except for divisions by sex and age.
Agricultural surpluses allowed a greater differentiation of labor and with it a greater difference in
wages, status, and social power. This differentiation led in time to extreme differences in political
power. This power was enhanced as professional
military men became increasingly common,
exemplified in the ancient Assyrian cultures.
Most people had very little status or wealth and
tilled the soil or took care of domestic matters.
Only a very small proportion – merchants, tech-
nocrats, and political leaders – lived lives of
increasing affluence and luxury. Over time the
difference between rich and poor increased
enormously.

As the concentration of wealth and power
increased, as central granaries became more
important, and as military power and war became
increasingly institutionalized, there were increasing opportunities for the development of empire.

An empire is defined as large geographic areas
under the rule of a central place and chief, and
maintained through what we might call civil servants or bureaucrats (although "lieutenants" is
probably more accurate). Tainter and others have
developed the concept of a pattern that they
believe has occurred again and again through history. One city or local culture becomes very successful through effective agriculture, mining, or
trade and the resultant growth in population and
economy. Often it becomes increasingly wealthy,
allowing it the surplus energy to support soldiers
and expropriate larger and larger areas of land
around its periphery while exploiting the subjugated people's energy surplus. Because war is
expensive, it becomes increasingly important for
the central city to impress others with their
wealth, a sign of surplus energy available to be
used, potentially, against others. Therefore huge
public investments are made in public structures,
temples, administrative centers, markets, roads,
food, storage facilities, and so on. If they are successful, outsiders decide it makes sense to become
aligned with this most powerful culture, even at
the expense of tribute in the form of agricultural
products, precious metals, or other materials.
Thus, the culture expands, often many times.

At some point the culture, through its growth,
begins to exhaust the initial resources that made

it rich. Another problem is that as cultures increased in linear dimensions, the energy cost of moving resources (e.g., taxation grains) to the central city became greater and greater. If the provinces sense difficulties in the central city they might become a bit more restless, requiring increasing investments in military forces or status symbols in the central city. According to Tainter, eventually the citizens of both the central city and the provinces become tired of paying high taxes for what is essentially "maintenance metabolism" (see p. 234). Due to diminished revenues, the physical and social infrastructure is not maintained, leading to the collapse of the empire. Tainter, an archeologist, ecologist, and historian, says that this has occurred repeatedly (he gives more than 20 examples in his first chapter) through prehistory and history. Ponting develops a similar scenario in many detailed examples and with a bit more emphasis on resource depletion, as does Charles Redman.

Mediterranean Cultures

There are some quite detailed assessments of the rise and collapse of earlier civilizations from the perspective of the energy and other resources required for development and maintenance. Mediterranean cultures are a good place to start thinking about these questions for a number of reasons. First, many of the most important ideas for the contemporary world, including democracy as a form of government, mathematics as we know it, and concepts in art and culture originated in this region. Second, the Mediterranean world offers a well-documented, well-studied suite of examples for us to explore and to understand the importance of energy and other resources in helping to shape the events that many of us recognize from traditional historical accounts. Third, this region remains today a vibrant and sometimes contentious region with many issues going way back in time. Many of the readers of this book will have been educated in the history of the region, which allows us an opportunity to examine familiar territory through our different lens of energy-based analysis.

Greece

Contemporary western democracies usually trace their ancestry back to ancient Athens. 2500 years ago these were vibrant, dynamic, frequently wealthy cities with some truly remarkable accomplishments, including defeating enormously larger Persian forces and producing some of humanity's still greatest architecture, sculpture, literature, and ideas about government. Athens and its sister city-states were also venal, domineering, and frequently squabbling cultures, squandering remarkable opportunities for "the good life" on pointless wars. The most important city-states were Athens and Sparta. Today we remember Athens as an incredible cauldron of art, ideas, and famous men, and Sparta as a culture completely dominated by preparing its young men for war (hence "Spartan conditions" is a term used today for harsh, uncomfortable, and arduous conditions). Athens too was a militaristic and imperialist culture and excelled in maritime combat. Athens and Sparta lived for many years in an uneasy truce that eventually ended in distrust and shifting alliances. From 431 to 404 BC these states and their allies initiated more than 25 years of intense combat that has been elegantly told by Thucydides [20]. Thucydides was once one of Athens' generals, but the price of losing a battle in Athens, which had happened to him, was dismissal from the army. This gave him the time to write a comprehensive history (*The Peloponnesian Wars*, a classic of history) of what ensued during this war, which was a stalemate for decades.

One interesting energy-related analysis of the Peloponnesian Wars, from which the following is borrowed, is found in Perlin's book, *A Forest Journey*. Perlin surveys the Peloponnesian Wars from the perspective of the forests and forest-derived energy required for the military activity and generation of the wealth required to finance the war. Anyone visiting Greece today is impressed by the nearly total absence of extensive and robust forests, so that it is quite curious to think of Greece and its southern part, the Peloponnesian peninsula, as heavily forested.

Plato, as late as the sixth century B.C., remarked that not long before his own time the hills surrounding Athens provided the huge building timbers he could still see in the buildings of Athens, and that these hills even contained forest-dwelling wolves that were a threat to livestock. Perlin believes that these abundant forests probably saved Greece from Persian domination as they provided timber to construct the Athenian fleet that defeated the Persian monarch Xerxes at Salamis. This was followed by the construction of an even larger 200-ship navy so that ambitious Athens could become the mightiest marine force in Greece. The Athenians were running into timber shortages, however, because of intense demand for fuel and construction wood in the city (including for immense wooden cranes to build the Parthenon) and because an immense vein of galena ore that had been discovered in the nearby town of Laurion. The ore could be smelted using charcoal as an energy source to produce silver, which was then spent on the new fleet, public works such as the Parthenon, and personal luxuries. This immense ore deposit made Athens extremely wealthy and powerful, however, it was at the expense of many of the forests of the region. This became a large problem because the Persians still controlled the timber supply regions to the east and north, including especially the Strymon valley. Ten thousand Athenians sent to colonize the mouth of that river, to ensure a timber supply for Athens, were slaughtered by the locals. A second invasion was somewhat more successful resulting in the capture, at least for several years, of the port city of Amphipolis. When that city was lost later (this was the battle Thucydides lost) Athens struggled for timber throughout the ensuing decades and centuries.

The Peloponnesian War that followed was principally between Athens and Sparta, but also included other Greek city-states. It was ruinous to all the participants. Due to wood being used for all instruments and means of war, it depleted the remaining forests of southern Greece, and the soil eroded as a consequence. The war spread even to Sicily, which the Athenians attacked unsuccessfully in a vain attempt to seize the forests to build a giant armada. In the meantime Sparta had seized the forest reserves on the peninsula that belonged to Athens and other states. Sparta then turned to Macedonia, made a new alliance, and built a new fleet. Meanwhile plague had entered Athens, greatly decreasing their number of soldiers. The Spartans made an alliance with their former Persian enemies and constructed a new fleet from Persian forests. They caught the Athenian fleet on shore with their crews foraging for dinner, and Athens was finally and permanently defeated, leaving the city destitute and without fuel or too much in the way of food. Thus although we learn of the war in terms of battles, generals, and so on, much of the background was about energy (to smelt silver to pay the armies and for obtaining timber and metal for weapons and armor) and other resources (e.g., wood for ships), the depletion of which contributed to the eventual outcome. The golden age of Athens was over, as was the city's contribution to our present culture.

Rome

Rome, founded in about 750 B.C. (according to myth by the abandoned twins Romulus and Remus, who were supposedly nurtured by a female wolf), was initially a group of neighboring hill towns that increasingly become incorporated into a city. Rome kept expanding through trade and military conquest, until it comprised much of the world known to Romans. The Romans learned early on that wealth could be gained much more easily through conquest and subsequent taxation than through other means, and thus kept expanding their boundaries. Subjugation and taxation were of course not especially popular among those subjugated, but the "*Pax Romana*" (Roman peace) imposed by the strong Roman military force actually decreased local conflict for many. The city was ruled by a series of kings until about 400 B.C., when it was changed to a republic ruled principally by a senate of patricians.

The Roman Empire lasted 500 years from roughly 44 B.C., when Julius Caesar appointed himself emperor, to 476 A.D., although the

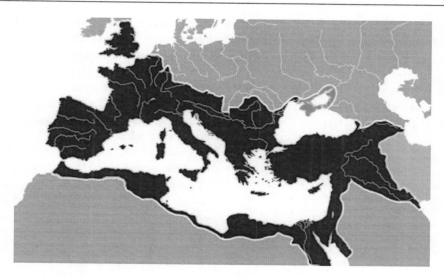

Fig. 2.12 Maximum extent of Roman empire (*Source*: Ronald F. Tylecote, 1987)

eastern portion at Constantinople lasted for 1,000 years more. The Empire reached its maximum extent about 117 A.D., when it encompassed essentially all areas around the Mediterranean, including all or most of the present countries of Italy, France, Spain, England, Greece, and Egypt, as well as the North African coast, Syria, the Middle East, and the regions around the Black Sea (Fig. 2.12). Rome had at its height about 1,000,000 people (of whom only about 10% were citizens) and the entire Empire contained as many as 70,000,000 people. This empire was carved out, maintained, and governed essentially by human energy: by citizen soldiers on foot who traveled on campaigns each year, utilizing wonderfully engineered stone roads that spread throughout the Empire (hence "all roads lead to Rome"), although ships were used over the Mediterranean itself. Imperial Rome was probably the most populous city in the world until the eighteenth century. The task of feeding roughly 1,000,000 people was enormous, especially following the passage of a law that guaranteed free grain to Roman citizens. The Roman invasion and subjugation of Egypt was not simply about Caesar's lust for Cleopatra, but also about shoring up Roman food supplies after the soils of Italy had been depleted by Roman farmers. Fortunately for the Egyptians, and for the

Romans, the annual flooding of the Nile replenished Egyptian soils. This occurred every year until the closing of the Aswan Dam in the 1960s. The concentration of artisans in Rome, and the *Pax Romana* that existed within the Empire, brought unprecedented economic prosperity to many, many people, and Roman engineering and architecture (borrowed heavily from the Greeks and others) generated massive and often wonderful public works throughout the Empire. Swamps were drained, creating new agricultural land and ending malaria. Although Rome is mostly thought about as a militaristic imperial force, and it certainly was, day-to-day life and influence were probably more a function of extensive and very effective trade, engineering, and agriculture.

Although Roman emperors were often venal, cruel, and corrupt, the best of them espoused very noble ideas about civilization and citizenship. There was a succession of good and bad emperors and other leaders, often representing different classes of people. For example, Julius Caesar although an aristocrat by birth, represented especially the interests of the common citizen class, although those who killed him also claimed to represent more the interests of the general Roman citizen. Either way, as was the period when Athens was at its height, this was a remarkable period for civilization. Some of the leaders, including

Marcus Aurelius, appear in history's lens as quite enlightened. Edward Gibbon, the eighteenth century historian who wrote *Decline and Fall of the Roman Empire*, described the period best or at least most eloquently [21]. Gibbon believed that Rome in the second century might have been the greatest time of all for humanity.

> In the second century of the Christian Era, the empire of Rome comprehended the fairest part of the earth, and the most civilized portion of mankind. The frontiers of that extensive monarchy were guarded by ancient renown and disciplined valor. The gentle but powerful influence of laws and manners had gradually cemented the union of the provinces. Their peaceful inhabitants enjoyed and abused the advantages of wealth and luxury. The image of a free constitution was preserved with decent reverence: the Roman senate appeared to possess the sovereign authority, and devolved on the emperors all the executive powers of government. During a happy period of more than fourscore years, the public administration was conducted by the virtue and abilities of Nerva, Trajan, Hadrian, and the two Antonines.
>
> If a man were called to fix the period in the history of the world, during which the condition of the human race was most happy and prosperous, he would, without hesitation, name that which elapsed from the death of Domitian to the accession of Commodus. The vast extent of the Roman empire was governed by absolute power, under the guidance of virtue and wisdom. The armies were restrained by the firm but gentle hand of four successive emperors, whose characters and authority commanded involuntary respect. The forms of the civil administration were carefully preserved by Nerva, Trajan, Hadrian, and the Antonines, who delighted in the image of liberty, and were pleased with considering themselves as the accountable ministers of the laws. Such princes deserved the honor of restoring the republic, had the Romans of their days been capable of enjoying a rational freedom.

Nevertheless there were always economic troubles, generally related to natural resources, including grain and wood, and the failure to maintain the solar-based systems that generated them. The general consumption of the Romans always exceeded the revenues. Common and necessary raw materials, such as wood, became more and more difficult to obtain as forests increasingly far from Rome were cut and turned to agricultural land, whose productivity tended to decrease over time. To meet its expenses, the government increasingly debased the gold and silver currency, causing extreme inflation, a fascinating story told in detail by Walker [22]. The Roman denarius was adulterated from being 98% silver in 63 A.D. to 0% (i.e., all copper or other such metals) by 270 A.D. as the main silver mines at, for example, Rio Tinto were depleted. As the denarius was adulterated, its purchasing power decreased proportionally.

Lead may have had an impact too, as the bones of ancient Romans have very high levels of lead, probably reflecting its use in pipes and in wine making. Nevertheless it is quite remarkable what humans can do based on essentially solar energy plus their own (or slave) muscle power alone. Perhaps it is better to conclude that the energy that built and maintained Rome was hardly the muscle power of Romans and the agriculture of Italy, but rather that of the millions of subjugated people in the provinces (and their land) who grew the necessary grain and cut the necessary wood to maintain the level of concentrated wealth in Rome. Perlin calculates that to run the baths at Caracalla for 1 year, 114,000,000 tons of wood were required, a truly prodigious quantity that had to be transported from tens to hundreds of miles by human or horse power.

Over time the Romans "became soft," hiring or forcing others to do their military service and grow their food. Vast expenditures went into public buildings and sports (if that word can be used) complexes, the most important of which is the Coliseum, where thousands of exotic animals were brought in and put into combat with slaves. They even staged naval battles in the Coliseum by flooding the interior with water. Clearly Hollywood has had its precedents. But by 200 A.D. the Empire began to be nibbled away by soil erosion, plagues, crop failures, and the Germans and Asians who desired the wealth that was within. Ultimately the city itself was successfully stormed by the Goths, Visigoths, and Vandals, with the full fall generally agreed to be in 476 A.D. Of course Rome the city is still there, with many artifacts from earlier times, although it is hardly the center of an empire.

The most interesting and from our perspective, insightful, analysis of the decline and fall of Rome (other than Gibbon's monumental books) is that of Joseph Tainter [12], who examined the entire process from the perspectives of the energy cost and gain of each activity. The main way that the ancients gained wealth was through conquest. Whatever wealth had accumulated in a region was the result of the slow accumulation of solar energy. This included mineral wealth, for the metals had to be mined by solar-powered human activity and then smelted using wood for fuel. Obviously this was hard work, and many preferred the much easier (although possibly fatal) path of conquest. As the Roman Empire become larger and more powerful it also became more complex to maintain and defend the provinces and eventually Rome itself. According to Tainter, increasing complexity is usually how problems are solved. But there is a high energy cost to complexity that makes its use eventually counterproductive. Tainter develops in a very compelling narrative how complexity, for example, through the maintenance of distant governmental administration and bureaucracies, garrisons, communications, and so on, and the importing of grain from ever more distant provinces, imposed an ever-increasing energy drain on the empire and how this led eventually to its susceptibility to decay and invasions. Basically the necessary investments in maintaining centralized administrative and military control become increasingly expensive and counterproductive, especially as the limits of an empire are pushed farther and farther from the centralized control, necessitating increasing energy costs for transport and to maintain the compliance of other people. Combining the language of Tainter and that of economists we might consider this decreasing marginal returns to complexity, which Tainter shows us occurred again and again and eventually led to the collapse of most empires.

We know less about the next 500 years in what had been the Roman empire, partly because few historians have given us as comprehensive assessment of the subsequent events as we have for the years of the Roman Empire. These years are often called the Dark Ages or the Middle Ages and are left at that. It is important to remember that life went on, Romans or Italians or whatever we wish to call them continued to live in Italy (as the French did in Gaul and so on), solar energy was used through agriculture and forestry to maintain people as they had been for millennia, and people lived, loved, fought, and died, while populations grew and sometimes declined from plague. Sometimes they left stone or occasionally literary artifacts, but more usually leaving behind only more depleted soils and forests. What was left of knowledge and culture and civilization tended to be kept alive in monasteries and in civilizations farther to the east.

The Rise of Islam

The prophet Mohammed, originally a merchant but eventually a political and religious leader, united the Arabian peninsula in the seventh century A.D. [23]. His followers expanded the empire under his influence so that within 100 years after his death they controlled a large area stretching from Central Asia through the Middle East and along North Africa to Spain. The empire expanded again in about 1200 to become what was probably the largest land empire ever. Although the ethnic diversity of this empire was enormous and the political administration diverse and far from centralized, the people were united in their devotion (or subjugation) to Islam and in their use of the Arabic language, in which the Muslim holy book, the Koran, was written. Known in the west for their fierceness, once subjugation occurred the Muslim leaders tended to be relatively tolerant and left others within their administrative units (including Christians and Jews) to their own devices as long as they paid their taxes. At that time most of the economies of the Muslim empire were either agricultural or grazing-animal based. Likewise, conquest was generally through foot soldiers or cavalry, so that we can assume both the economy and expansion were nearly completely based on solar and biomass base for energy.

The Muslim world was increasingly focused in Cairo following the Arabic conquest of Egypt in the seventh century A.D. Originally Muslims eschewed naval warfare and even sea-based trade, focusing on land-based expansion by trade or voluntary conversion or sometimes conquest. Day to day the main events were much more likely to be about trade than conquest. For example, Muslims had regular overland trade to China along the very lengthy "silk road." Eventually they became seafarers, focusing initially on the Sea of Arabia and then the coasts of India and Africa. Their long presence in Africa is reflected in the principal language in Kenya remaining Arabic, and also in the name Swahili, which means coast. Arab traders brought coffee, originating in Ethiopia, to the rest of the world, and this is reflected in the scientific name for the best coffee, *Coffea arabica*.

Increasingly the Byzantines, as the residents of the eastern Roman Empire were called by the early Middle Ages, attacked Egypt and other Arabic possessions using ships and caused great destruction. Again the use of solar energy to make timbers for ships and wind energy to move large quantities of people and goods by ships gave enormous power to those who were able to exploit it. In response, the great Arab leader Calif-Abd-al-Melik in the late seventh century initiated a great program of shipbuilding. This program was based in Egypt, but Egypt had few trees and none of a size to allow the construction of strong ships. Large cedars, many 170 ft in length, were imported from Lebanon, although that was very expensive. Consequently the shipbuilding had to be moved to what is now Tunisia, which at that time was heavily forested. A very strong fleet was constructed which captured Sicily (with its huge forests) and established a beachhead in Spain. In time the Mediterranean became essentially an Arab lake, as it had been previously a Roman one. The only ones to challenge this were the Venetians, who had access to the forests of the Po and Adige river basins. Thus the exploitation of wind energy allowed the Muslims to conquer and hold on to huge new land holdings, and to generate great wealth through trade. They were more or less the masters of the Mediterranean world for nearly a 1000 years (or more, considering that today most of North Africa is Muslim).

Among the many who accepted Islam as their religion were the Turkic peoples of Central Eurasia, who established a very strong empire beginning near present-day Constantinople, and eventually spreading under the Ottoman group influence through much of the Islamic world. They also spread into the West and were finally stopped at Vienna in 1683. According to Cameron, although this was not a tightly integrated empire, it persisted and spread for a very long time because it did not subjugate those it conquered but only asked for taxes which were not excessive. This approach to empire seems to be a relatively successful one compared to brutal repression. Arabic influence spread through European culture, leaving, for example, its imprint in the English language with words such as "arsenal" (construction house, originally), "algebra," and "algorithm," reflecting the great advances made in mathematics within the Muslim world during what we now call the Dark Ages and more recently IDRISI (an important GIS tool named after the great twelfth century Arabic–Sicilian geographer of that name.)

The Muslin world, Ottomans in particular, often found themselves in direct competition with the Christian world. Several specific events stand out. The Christian invasions of Muslim-controlled Jerusalem known as the great Crusades (1095–1099, 1147–1149, 1188–1192, 1202–1204, 1217–1221, 1228–1229, 1248–1250) reflected the growing wealth, power, and some would say arrogance of Europe. It represented not only a chance for the faithful and adventurous to attempt to wrest the Holy Land from the "infidels" but also opportunities for plunder, rape, trade, and extension of commercial influence. The first Crusade caught the inhabitants of Jerusalem by surprise, and an enormous bloodbath of mostly Muslims (but also Christians) by Christians followed as the city was wrested from "infidels." None of the subsequent Crusades were as successful militarily. Some of the related events were especially pernicious. On the fourth Crusade the European knights and their camp followers (tinkers, blacksmiths, prostitutes, and so on) tired of walking and riding horses, stopped in Venice to attempt to purchase passage by ship to the Holy Land. The Venetians, crafty businessmen and politicians, took their gold for passage, loaded the heavily armed men onto ships and set off for what they said was the Holy Land. The Venetians had some old scores to settle with the inhabitants of Constantinople, then a Christian remnant of the old Holy Roman Empire. On the way they took a left, and the unsuspecting knights were deposited before the city of Constantinople which the Venetians said was Jerusalem. When they asked their Italian ship captains why the city was adorned with crosses they were told that this was a Muslim trick. So they attacked the city, eventually subduing the inhabitants, and looted, raped, and pillaged for several months. The Venetians received not only payment for ship passage, but ensured that Constantinople would no longer be a threat to their commercial interests in the Aegean and Adriatic Seas, for example, for wood in the region, at least for a while. In the long term the plan perhaps backfired as the weakened Christian city of Constantinople fell later to Islamic invaders from the east, and the importance of the Venetian empire and Christianity in that region faded. Those who wish might say that indeed God works in mysterious ways.

Thus the enmity of much of Islam today towards the West and for the exploitation of the region's oil resources is hardly new, and lives on today as great distrust by many Muslim cultures for the motives of the West. It is hardly surprising that as the West has become so dependent upon oil from the Muslim world there are many who view the relation with great suspicion.

The Lasting Legacies of Ferdinand and Isabella

Another place that Muslims and Christians clashed was in Spain. Muslims came to Spain from the south across the Mediterranean, and from the ninth to the thirteenth century controlled most of the Iberian peninsula. While there they developed very sophisticated agricultural and horticultural systems and essentially tolerated diverse other cultures. Christian influence filtered in from the North beginning about the tenth century, culminating in the expulsion of both Moors and Jews by King Ferdinand and Queen Isabella, whose names

are familiar to most Americans because they also supported Columbus, both in 1492. The result was disastrous for the Spanish economy because the Moors were much more sophisticated agriculturists than the Christians, at least for the southern part of Iberia, and because many skilled Jewish people were forced to leave. Many of these Moors and Jews probably went as colonists into the Americas, feeling no longer welcome in Spain. The wealth of Spain, originally based on sophisticated agriculture and trade, was partially restored only by the brutal exploitation by Spain of the inhabitants of the New World as they extracted gold, silver, and other minerals with the aid of slaves, wood fuel, and wind power for their sailing ships. The food production system exported to the New World by the Spanish was one based on cattle raising, as this was the system favored by the Christian Spanish. The often sophisticated agricultural systems (e.g., extensive terracing) in place in Central and South America were displaced, even destroyed, by the very crude cattle-based latifunida system brought from Spain. In both southern Spain and Central America the cattle were turned out to graze in the much more productive original gardens that were often highly terraced, representing generations of careful investments of human energy. Because cattle return much less food per hectare per year than crops, the overall productivity of these systems for food energy was greatly lowered. Thus in a sense the actions of Ferdinand and Isabella destroyed two great agronomical systems and replaced them with unsophisticated grazing systems with perhaps one tenth or one twentieth the capacity to produce usable food energy for humans.

Entire forests, such as in southern Bolivia, were cut to supply timber for mines and provide energy for smelting. Much of the Tarija region of southern Bolivia, for example, was deforested to support the silver mines in Potosi, and the timbers were transferred nearly a 1000 km horizontally and thousands of meters vertically on the backs of mules and slaves [24]. The deforestation resulted in some of the most extensive erosion found on the face of this earth, which covers nearly 5,000,000 ha. Spain grew rich on the imported gold, but a curious phenomenon happened. The Spanish efforts in the New World doubled the quantity of gold in the old, but it decreased its value to less than half! What had happened was no different from what happens when a modern country prints too much money: inflation. Gold has little utilitarian value, but is rather a medium of exchange. The real wealth of Europe came from the fields, forests, fisheries, and artisans, that is, the investments of solar and human (and occasionally wind and water) energy into the process of turning raw materials into real wealth: food, clothes, shelter, tools, utensils, and so on. Much of that gold ended up eventually in the great cathedrals of Europe. We cannot visit these beautiful cathedrals without thinking of the death, erosion, and human misery occasioned by the procurement of that gold.

Other Regions of the Earth

While Europe was living in the Dark Ages, independent and often very sophisticated cultures were developing in China, India, and the Americas, each of which had much greater and often more sophisticated human populations than did Europe. Again these were solar-powered agrarian cultures for the most part, and depended year after year on intensive human labor and of course the sun as a source of energy. Several grass-based nomadic civilizations, including that in Mongolia led by Ghangis Khan, also established very extensive empires that in his case reached nearly to Europe. In the Americas very extensive city-states developed, flourished, and eventually collapsed. For example, the Olmecs and Maya of present-day Mexico and the Inca in Peru followed such fates. But, as we said at the beginning of the Mediterranean section, these cultures are not our focus here.

The Energetics of Preindustrial "Modern" Societies: Sweden, The Netherlands

There have been several especially comprehensive analyses of preindustrial solar-powered economies in the Netherlands and Sweden by De

Zeeuw [25] and Sundberg [26]. These analyses indicate that it was possible to generate a very significant energy-based economic machine on plant material alone. The longer view, however, is that eventually these "renewable" systems tend to become depleted.

In the period 1640–1740 the Dutch had created a very profitable ceramics industry in the vicinity of the city of Delft, near Rotterdam. Even today it is possible to purchase very fine china by the name of Delft. Making pottery is energy intensive, as the raw material (basically clay with metal decorations; in the case of Delft, characteristically blue) has to be heated to high temperatures. The fuel for this in the Netherlands was originally peat, partially decomposed sphagnum moss, which was abundant in the low-lying areas of the Netherlands. To this day large rectangular holes, called polders, remain where the peat was extracted four centuries ago.

A particularly thorough energy analysis of the economy of that time has been undertaken for Sweden by Sundberg. In 1550 Sweden was overwhelmingly rural and very poor. Most of Sweden is too cold for much agriculture, which was concentrated in the south of the country. Most of the citizens lived scattered throughout the vast forests where they cut trees for charcoal, which was used for a variety of purposes, most importantly for smelting the abundant silver, copper, and especially iron ore. Thus Sweden had at that time two particular assets in terms of natural resources: vast forest areas and rich iron ore. In order to make iron, high temperatures (above 1,000°C) must be used. This is not possible from timber alone, but can be done with charcoal, which is basically wood heated in the absence of oxygen so that it is nearly pure carbon. Charcoal is made by taking trees and piling them into a large earth-covered structure containing dozens to hundreds of trees. Then the pile is fired and allowed to smolder for days.

In 1600 approximately 15–20% of all Swedes lived in small family groups scattered throughout the forest. Their houses were quite small and simple. Over time more and more of the Swedish forest was cut and burned, and because trees grow slowly in the cold climate, eventually the vast Swedish forests were destroyed almost in their entirety. The Swedes faced an enormous energy crisis, and many froze in the winter because they had insufficient fuel and insufficient food. Starting in about 1850 vast numbers of Swedes moved to the United States, especially to the northern Midwest, where they felt right at home among the snow and the pine forests.

In Sweden in the 1600s the resulting charcoal was taken to regional metal processing centers and the iron and copper ore turned into metallic implements. The principal products of Swedish iron factories in the time period 1600–1800 were very good cannon. The Dutch were the first to take full advantage of these cannon, and mounted them in warships that made them rulers of the European seas for about 100 years, until the English became better at the game. The Dutch invested in cannon because they allowed them, essentially, to steal whatever they wished from other nations. This was considered fair game, at least by the rules of the newly emerging mercantile capitalist economy (although not by the conquered and colonized).

Crosby has commented upon the particular aggression and greed of Europeans compared to others about the world. By 1641 the Dutch trade and military empire extended as far away as Malaysia, where a Dutch fort and windmill can still be found in Malacca. If other nations wanted to trade in waters where the Dutch ruled they had to either pay tribute to the Dutch or suffer the loss of some of their ships and ports. The Dutch got very rich as a consequence. Thus the energy of photosynthesis of Swedish forests was translated into dominance of the seas by the Dutch using wind-driven ships to carry far more Swedish cannon than land armies could muster. These energies also generated a very high level of comfort for Dutch burghers, and the leisure to generate some of the world's greatest art. Then as now affluence had a source somewhere in extensive use of energy. But that affluence for the Dutch did not last either, for it was the British defeat of the Dutch at the Straits of Malacca that catapulted the British into prominence as a mercantile power. Then the bulk of the eighteenth century was spent in British conflict with the French.

Finally at the end of the Seven Years War and the great British naval victory at Trafalgar, British hegemony was established over the world's seas and the long period of *Pax Britannica* began.

Throughout world history, however, most people remained very poor. Societies often adjusted to these mean circumstances by generating limited social expectations and mechanisms that allowed people to be comfortable with only these very limited economic circumstances and opportunities. One's rewards would be found (they said) after death, or in serving God modestly, or in leisure (in many societies men hardly worked but spend much of the day in cafés or smoking or drinking coffee while the women tended the fields or shops as well as the children). Fortunately death rates were high and the population did not expand greatly beyond the means of the land to support the people who were there. People may have been as happy as, or even happier than, today, we don't know, but the economic circumstances for most were barely above what it took to remain alive and to have and raise children. Some very few adventurous souls would join armies going to faraway places to exploit new resources and peoples (i.e., the rampant European colonialism of one, two, three, and four centuries ago and the crusades long before that). When the Americas opened up, massive numbers of Europeans were ready to move to the new "empty" continents to try to better their fortunes, sometimes paying little respect to the fact that the continents were already heavily peopled with Native Americans. In other words once material opportunities opened, there were plenty of Europeans ready to give it a try to improve their own personal financial situations. Even so for almost all individuals it was extremely hard to make a living. This was normally accomplished through hard physical labor to chop down trees, or to farm or work a mine or in a factory. Records of colonial Americans, for example, show that people spent almost all of their time and money just surviving, although they may have done that in reasonable comfort. The concept of spending money for recreation simply did not exist for most, as there was relatively little surplus wealth or surplus energy in these solar-based societies.

Throughout history in many societies it was deemed just fine to attack another city or nation and simply steal whatever wealth they had accumulated. Although this may sound offensive to us in fact it was highly regarded by many in antiquity. Great writers of past times, chronicled approvingly again and again the stories of a leader of one state who plunders another state, bringing glory and treasure to himself and his own state. Vikings, living in northern lands of very low productivity sought wealth in raiding parties that terrorized much of Europe for 1,000 years. Wooden Viking ships with charcoal-derived iron nails and weapons and woolen sails were constructed and equipped entirely using solar energy. Europeans stole entire continents from Native Americans on solar power (again winds and charcoal plus genocide and settlement), with God as well as gunpowder and European germs on their side [8, 27]. Today this process continues through the economic principle of "globalization," which is viewed by many principally as a means by which the more-developed world legitimizes its extraction of resources and cheap labor from the less-developed world. Others believe that trade benefits all (see Chap. 7).

We stop our history here for the history of industrial society is treated in the next chapter as well as in parts of many other chapters, including the end of Chap. 10.

A Somewhat Cynical View of Human History

It is very impressive to examine from today's perspective the views of the ancients with relation to war. Plutarch's *Lives* [28] is a book about famous ancient Greeks and Romans, written several thousand years ago by a distinguished Roman historian. One of your authors (CH) tackled this book with vigor, wanting to better himself because his classical education, once the signature of a well-educated person, was limited to two undistinguished high school years of Latin

under the fierce eye of Miss Meservey. He was also interested in what might be the characteristics of leaders whose reputation had lasted thousands of years. He was quite surprised by what he found: the largest group of the people singled out for praise by Plutarch made their mark by plundering other culture's cities. Plutarch recounted with favor and apparently without irony how these people brought fame and riches to their own cities or regions. These great leaders of the past appeared to be simply robbers with plunderers of accumulated solar-based profits. Human history has been in large part about mustering armies to rape and plunder, and about the efforts of others to counter these robbers. Modern Italy, Scotland, and many other European landscapes are full of ancient stone fortifications that must have taken an enormous portion of the time and energy reserves of the ancient citizens to construct. The evolution of more powerful cannon reduced the effectiveness of these fortifications until they were reconstructed to stronger specifications.

America too is constructed on conquest and plunder, from the obvious example of early English and Spaniards stealing the lands of Native Americans to a United States military expedition taking what is now California and the rest of the U.S. Southwest from the Mexicans in the 1830s by force. Empires seem to have more or less gone out of style during the twentieth century as nationalism and ethnic foci have taken the main stage and as most of the less militarily powerful world has filled up with hungry armed people who are difficult to manage.

Occasionally we can get a quantitative glimpse of the enormous input required to fuel the expansion of empires, and also the misery suffered by the common person during both the times of the expansion and the collapse of empires. Little was known about energy during most of history but we can get some glimpses and make some rough calculations. Napoleon was famous for his "cannon park" of 366 cannon, each capable of hurling a 6- to 12-pound iron ball. He took this formidable machine with him to Russia, an incredible and ultimately disastrous campaign that resulted in the death of most of his army. The Russian army under Kuznetsov chose not to stand up to Napoleon's well-oiled military machine but instead retreated before him, stopping only briefly at Borodino to give some serious resistance before melting away, leaving Napoleon to be defeated later by "General Winter." Military historian John Keegan has calculated the energy requirements to feed that cannon park. The 300 plus cannon required 5,000 horses to pull them along plus soldiers and teamsters to handle the horses and man the cannons. The men required about 12 tons of food a day and the horses 50 tons of hay, so many additional horses were required to bring along the fuel for men and horses pulling the cannon. One of Keegan's main points is that Nelson's fleet at Trafalgar carried six times the firepower at one fifth the logistic cost by exploiting wind energy. This indicates the importance of being able to exploit a relatively large energy resource, in this case the wind.

In three successive summers one of your authors (CH) happened to read three historical books on European history and some important military invasions in search of empire: the first Peter Massey's on Peter the Great and his attack south into the Crimea in 1696, the second Phillipe De Segur's (a nobleman in Napoleon's army) record of Napoleon's Russian campaign in 1804, and the third Anthony Beevor's *Stalingrad*, the story of the farthest point that Nazi Germany had penetrated into Russia in 1942–1943. Each of these books is a masterful summary of enormous military campaigns. But it came as a shock to me when I looked at the first map in the third book, for I had been looking at essentially the same map in each of the two previous books, centered on the region between the Baltic, the Black Sea, Moscow, and the Caspian region. Each of the books tells of initial tremendous success and enthusiasm for the "glory" of conquest, and in each of them the invading armies were humbled eventually by the peasant armies, climate, and lack of enough fuel within the devastated invaded

regions to support horses, tanks, and soldiers. The suffering of the soldiers, officers, and the commoners caught in the middle in each was immense, and in each the tales of massacre and barbarous behavior on all sides was appalling. No new territory was gained by any of these campaigns, despite the enormous expenditure of resources. Educated German officers in 1942 knew well of Napoleon's appalling retreat in Russia, and watched as day by day General Winter imposed the same horrible fate on their own army. At the end it all seemed so stupid. Except for the massacre and displacement of Native Americans (and other aborigines) by Europeans it seems that since 1800 (and probably long before) most land has remained in the hands of those who were there first. But that certainly has not stopped many invasions as some attempt to conquer others' land.

Thus much of history can be seen as times of very limited abilities to do much more than survive on one's own resources, and that the main path to personal or national wealth was through exploiting others through warfare. Much of history can be viewed as a series of attempts by one group to exploit or dominate others, either by directly stealing their wealth (represented as the long-term gradual accumulation of net solar energy in precious metals, jewels, and edifices) or by gaining access to their resources. We end our brief historical review at a point before the fossil fuel era gave a tremendous boost to our ability both to generate wealth at home and to inflict carnage and misery upon each other [23]. We do note an optimistic pattern: the long age of arrogant European colonization, empire by conquest, and continuous international conflict appears to be behind us following the end of World War II. With the rise of industrialization and the enormous ability to increase wealth that fossil fuels and their technologies allowed, plus a growing appreciation of the cultures of others and the costs of war, the concept of empires and subjugation of others seems to have largely stopped. But war and its misery continue for all kinds of other reasons.

The Repeated Collapse of Empires

There are several dicta of history that are important here. The first is that "history is written by the winners" and the second is that most human endeavors of the past are barely or not at all recorded. The scholars who think the most about this are archaeologists, and the archaeologist (and anthropologist and historian and energy analyst) who has the most to say about this issue is Joseph Tainter. Tainter's magnum opus is *The Collapse of Complex Systems* (although we have found his 1992 paper "Evolutionary Consequences of War" to be equally pungent). Both are incredibly good reading. Tainter lists a minimum of 36 once-great civilizations that exist today only as a series of rocks and other hard materials, often under desert sands. The list goes on and on. One has only to visit the great museums of anthropology in, for example, Mexico City or Jalapa, to get a perspective on what incredible civilizations there were in the past, and how so many have crumbled.

Why do most military invasions fail, and how did it come to pass that so many once proud and powerful civilizations fell apart so completely and, often, so quickly? There are probably many reasons but we believe that the energy-based mechanisms put forth by Tainter offer the best clue. The pattern that Tainter has developed seems so very powerful: that as a civilization generates some successful means of generating wealth (i.e., surplus energy) and is able to feed its people and keep its enemies at bay, the power of the central city and of the chief can increase dramatically. Wealth and resource flows to the center increase dramatically with early successful invasions of neighbors. But the very success of the expansion/subjugation eventually leads to the collapse of many of these civilizations because of the increasing and eventually unsustainable energy costs of the necessary increase in complexity, that is, the energy cost of maintaining the required food production and distribution systems for the increasingly populous central city from increasingly distant granaries, and the energy cost of armies necessary to enforce discipline on larger and larger subjugated people. This eventually exhausts the treasuries and the real resources of the central authority, and the lands revert to the original inhabitants.

Summary

All of life, including human life in all of its manifestations, runs principally on contemporary sunlight that enters the top of our atmosphere at approximately 1.4 kilowatts per square meter (5.04 MJ per square meter per hour). Roughly half that amount reaches the Earth's surface. This sunlight does the enormous amount of work that is necessary for all life. The principal work that this sunlight does on the Earth's surface is to evaporate water from that surface (evaporation) or from plant tissues (transpiration) which in turn generates elevated water that falls back on the Earth's surface as rain, especially at higher elevations. The rain in turn generates rivers, lakes, and estuaries and provides water that nurtures plants, animals, and civilizations. Differential heating of the Earth's surface generates winds that cycle the evaporated water around the world, and sunlight of course maintains habitable temperatures and is the basis for photosynthesis in both natural and human-dominated ecosystems. These basic resources have barely changed since the evolution of humans (except for the impacts of the ice ages) so that preindustrial humans were essentially dependent upon a constant although limited resource base. Over time humans increased their ability to exploit larger parts of that natural solar energy flow through technology, initially with spear points, knives, and axes that could concentrate human muscular energy, and then with agriculture and dams, and now with fossil fuels.

The development of agriculture allowed the redirection of photosynthetic energy captured on the land from the many diverse species in a natural ecosystem to the few species of plants (called cultivars) that humans can and wish to eat, or to the grazing animals that humans controlled. Curiously the massive increase in food production per unit of land brought on by agriculture did not, over the long run, increase average human nutrition but mostly just increased the numbers of people. Of course it also allowed the development of cities, bureaucracies, hierarchies, the arts, more potent warfare, and so on, that is, all that we call civilization, as nicely developed by Jared Diamond in *Guns, Germs and Steel*. For most of humanity's existence most of the energy used was animate – people or draft animals – and derived from recent solar energy. Generally humans themselves did most of the work, often as slaves but more generally as physical laborers which, in one way or another, most humans were. For thousands of years, from the period of the beginning of empires 5,000 or more years ago until the widespread use of coal for steam power in about 1850, the principal source of energy for any large-scale agriculture or public works was masses of human power, principally but not always as slaves or near slaves (i.e., serfs). By one account the Cheops pyramid represents essentially the entire energy surplus of the Nile civilization of about 3,000,000 people at that time, and required the labor of 100,000 people over 20 years. A second very important source of solar energy was from wood, which has been recounted in fascinating detail in books by Perlin, Ponting, and Smil. Massive areas of the Earth's surface (Peloponnesia, India, parts of England and many other locations) have been deforested three or more times as civilizations have cut down the trees for fuel or materials, prospered from the newly cleared agricultural land, and then collapsed as fuel and soil became depleted. Archaeologist Joseph Tainter recounts the general tendency of humans to build up civilizations of increasing reach and infrastructure that eventually exceeded the energy available to that society.

Both the natural biological systems subject to natural selection and the preindustrial civilizations that preceded our own were highly dependent upon maintaining not just a bare energy surplus from organic sources but rather a substantial energy surplus, or large net energy, that allowed for the support of the entire system in question, whether of an evolving natural population or a civilization (29–31). Most of the earlier civilizations that left artifacts that we now visit and marvel at, including pyramids, ancient cities, monuments, and so on, had to have had a huge energy surplus for this to happen, although we can hardly calculate what that was. An important question for today is to what degree does the past critical importance of surplus energy apply to contemporary civilization with its massive although possibly threatened energy surpluses.

Surplus Energy and Contemporary Industrial Society

Contemporary industrial civilizations are dependent on fossil fuels in addition to sunlight. Today fossil fuels are mined around the world, refined, and sent to centers of consumption thousands of miles away. These fuels have allowed for accelerated exploitation of solar energy and for the huge increase in food production, water transport, and sanitation that has allowed the human population to grow enormously over the past 100–200 years. For many industrial countries, the original sources of fossil fuels were from their own domestic resources. The United States, United Kingdom, Mexico, and Canada are good examples. Many of these initial industrial nations, however, have been in the energy extraction business for a long time so they tend to have both the most sophisticated technology and the most depleted fuel resources, at least relative to many countries with more recently developed fuel resources. For example, as of 2010 the United States, originally endowed with one of the world's largest oil provinces, was producing only about 45% of the oil that it was in the peak year of 1970, Canada had begun a serious decline in the production of conventional oil, and Mexico recently was startled to find that its giant Cantarell field, once the world's second largest, had begun a steep decline in production at least a decade ahead of schedule. Meanwhile the global human population continues its upward course, although at a decreasing rate (Fig. 2.13). The next chapter examines the role of oil in our society in much greater detail.

Fig. 2.13 Global human population (*Source*: United Nations)

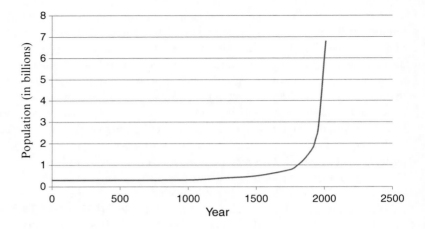

Questions

1. Discuss several examples of how preagricultural humans exploited solar energy and the relation of the energy they obtained to their own personal energy investments.

2. How are spear points related to energy?

3. How does agriculture concentrate energy for humans? How does this process support a larger population?

4. The human use of fire assisted in opening up a huge new food resources of agriculture for humans. Can you explain what the connection might be?

5. What was the relation of agricultural surplus to human specialization?

6. Many former dominant human cultures have collapsed. Can you give an example and the reasons thought likely for that happening?

7. Name at least two important legacies of the reign of Ferdinand and Isabella.

8. What does surplus energy mean to civilizations?

References

1. Lee, R. (1969). Kung bushman subsistence: An input-output analysis. Environment and Cultural Behavior. pp. 47–79.

2. Shostak, M. (2000) Nisa: The life and words of a !Kung Woman. Harvard University Press. Cambridge, Mass.

3. Martin, P.S. (1973). The discovery of America. Science. Volume. 179, 969–974.

4. This analysis is based mostly on a series of papers in Science magazine of March 2, 2001.

5. Culotta, E., A. Sugden and B. Hanson. (2001). Humans on the move. Science. Volume 291. 1721-1753.

6. De Candolle, A. (1959) Origin of cultivated plants. Hafner Publishing Co., London, England.

7. Sauer, C.O. (1952). Agricultural origins and dispersal. American Geographical Society. New York, New York.

8. Diamond, J. (1999) Guns, germs and steel: The Fates of Human Societies. W.W & Company, Inc. Norton, New York.

9. Angel, J.L. (1975). Paleoecology, paleodemography and health in. S. Polgar (ed). Population ecology and social evolution. Mouton Press: 667–679.

10. Perlin, J. (1989) A forest journey: The role of wood in the development of civilization. W.W. Norton & company Inc., New York.

11. Michener, James A., (1963). Caravans. Ballantine Books. United States.

12. Tainter, J. (1988) The collapse of complex societies. Cambridge University Press. Cambridge, England.

13. Abrahamson, W.G. and T.N. Taylor. (1990). Plant-animal interactions. McGraw-Hill Encyclopedia of Science & Technology.

14. Feeny, Paul. (1970). Seasonal changes in oak leaf tannins and nutrients as a cause of spring feeding by winter moth caterpillars. Ecology 51: 565–581.

15. McNeill, W. H. (1976) Plagues and peoples. Anchor Press/Doubleday. Garden City, New York

16. Ponting, C. (1992) A green history of the world: The environment and the collapse of great civilizations. Penguin Books. Saint Martens, NewYork.

17. Carter, V. G. and T. Dale. (1974) Topsoil and Civilization. University of Oklahoma Press. Norman

18. Deevey Jr., Edward S. (1960). The human population. Scientific American. Volume. 203: 194–204.
19. Cottrell, F. (1955) Energy and society. New York: McGraw-Hill.
20. Thucydides. 1954 (Translation by Warner). History of the Ppeloponnesian wars. Penguin Books. New York.
21. Gibbon, E. (1776) et Seq. The decline and fall of the Roman Empire: VOLUME I. Random House
22. Walker, D. R. (1974) et Seq. The metrology of the roman silver coinage. Part I: From Augustus to Domitian. Part II. From Verva to Commodus. Part III. From Pertinax to Uranius Antonius. British Archaeology Report. Oxford. Supplementary Series 5, 22, 40.
23. Adas, M. (1993) Islamic and European expansion. Temple University Press. Philadelphia.
24. Hall. C. A. S. (2006). Integrating concepts and models from development economics with land use change in the tropics. Environment, development and sustainability. Volume. 8: 19-53
25. Zeeuw, Jan W. (1978). Peat and the dutch golden Age. The historical meaning of energy-attainability, A.A.G. Bijdragen, 21
26. Sundberg, U. (1992). Ecological economics of the swedish baltic Empire: An Essay on Energy and Power, 1560–1720. Ecological economics. Volume. 5: 51–72.
27. Crosby, Alfred. (1986) Ecological imperialism: The biological expansion of europe, 900-1900. Cambridge University Press. London
28. McFareland, John W. (1972) Lives from plutarch. Random House. New York
29. Thomas, D.W., J. Blondel, P. Perret, M.M. Lambrechts, J.R. Speakman. (2001). Energetic and fitness costs of mismatching resource supply and demand in seasonally breeding birds. Science. Volume. 291: 2598–2600.
30. Odum, H.T. and R. Pinkerton. (1955). Time's Speed Regulator. American Naturalist.
31. Hall, C. A. S. (2004) The Continuing importance of maximum power. pp. 107–113. in Brown, M. and C. A. S. Hall. The H. T. Odum primer: an annotated introduction to the publications of Howard Odum. Ecological Modeling, Vol. 78.
32. Wroe, S., Field, J., Fullagar, R. and Jermin L.S. (2004). Megafaunal extinction in the late quaternary and the global overkill hypothesis. Alcheringa 28, 291–331.
33. Cook, E. 1976. Man, Energy, Society, W. H. Freeman, San Francisco.

The Petroleum Revolution

3

The First Half of the Age of Oil

This chapter focuses on the importance of fossil fuels (coal, gas, and oil) and especially petroleum (meaning natural gas and oil, or sometimes just oil). First we want to ask why petroleum, and especially oil? Why has petroleum been so important, and why is it so hard to unhook ourselves from it? To do that we need to look more broadly for a moment at the energy situation that has faced, and that faces, humanity. Solar energy, either directly or as captured by plants, was and is the principal energy available to run the world or the human economy. It is enormous in quantity but diffuse in quality. As we have developed in the previous chapter, the history of human culture can be viewed as the progressive development of new ways to exploit that solar energy using various conversion technologies, from spear points to fire to agriculture to, now, the concentrated ancient energy of fossil fuels. Until the past few 100 years human activity was greatly limited by the diffuse nature of sunlight and its immediate products, and because that energy was hard to capture and hard to store. Now fossil fuels are cheap and abundant, and they have increased the comfort, longevity, and affluence of most humans, as well as their population numbers [1].

But there is a downside, for fossil fuels are made principally of carbon. The use of carbon-based fuels generates a gaseous by-product, carbon dioxide (CO_2) that appears quite undesirable. Now we are constantly bombarded with recommendations of our need to "decarbonize" our economy because of the environmental impacts, such as climate change and ocean acidification that the increases in carbon dioxide appear to be causing. These impacts are likely to become much more important in the future. Consequently there have been considerable efforts to come up with fuels or energy sources not based on carbon. To date that effort has failed completely, for, according to the data compiled by the U.S. Energy Information Agency, the amount of CO_2 produced each year continues to increase at about 3 percent a year (unless there is a recession). With so many apparent options how come we cannot unhook ourselves from carbon? Why is it that most of our energy technologies continue to rely on the chemical bonds of carbon (most usually combined with hydrogen as hydrocarbons)?

The answer lies in basic chemistry: the only effective and large-scale technology that so far has been "invented" for capturing and storing that energy is photosynthesis. Humans use the products of photosynthesis for all or most all of our fuels simply because there is no alternative on the scale we need. This is because nature, the source of our fuels, has favored the storage of solar energy in the hydrocarbon bonds of plants and animals. The reasons are that these elements are abundant and "cheap" to a plant, and most important, capable of forming *reduced*, or low oxygen, energy-containing chemical compounds. Hydrogen and carbon, which essentially do not exist at the Earth's surface, are so important that plants have evolved the technology to split water

C.A.S. Hall and K. Klitgaard, *Energy and the Wealth of Nations: Understanding the Biophysical Economy*,
DOI 10.1007/978-1-4419-9398-4_3, © Springer Science+Business Media, LLC 2012

and atmospheric carbon dioxide to get hydrogen and carbon, which they combine to form energy-rich *hydrocarbons* and, with a little oxygen, *carbohydrates*. There simply are not other elements in the periodic table that are sufficiently abundant and capable of such ready reduction. Nitrogen, for example, is abundant as N_2 but much more expensive energetically to split, and sulfur is less available. In addition, carbon has four valence electrons, capable of forming four bonds with other atoms and hence the very complex structures of biology. Bonds with hydrogen greatly increase the capacity to store energy in a molecule. Thus plants and animals are carbon- and hydrogen-based because nature had no choice. Human cultural evolution has exploited this hydrocarbon energy profitably mostly because they had no choice but to use the products of photosynthesis. Now we are stuck with the carbon dioxide while we try to figure out if there possibly can be an alternative that is energetically feasible.

The Industrial Revolution

Beginning on a small scale about 1750 then increasingly rapidly to about 1850 when there was a rather remarkable change in the hydrocarbons that humans used from the recently captured solar energy of wood and muscle power to the enormously more powerful fossil fuels. This was the beginning of the "industrial revolution," although perhaps a more proper name would be the "hydrocarbon revolution." Humans had begun to understand how to use the much more concentrated energy found in fossil (meaning old) fuels. Why did they do this? The answer is simple. People want to do more work because to do so is profitable. They want more of some raw material transformed into something useful that they can eat, trade, or sell. Fossil hydrocarbons have greater energy density than the carbohydrates such as food and wood, and as a consequence they can do much more work: heat things faster and to a higher temperature or operate machines that are faster and more powerful (Table 3.1). The first fossil hydrocarbon used was coal, first used on a large scale in the eighteenth century, then in

the twentieth century, oil, and now increasingly, natural gas. The global use of hydrocarbons for fuel increased nearly 800-fold since 1750 and about 12-fold in the twentieth century alone, and this has enabled our enormous economic growth (Fig. 3.1).

Economists usually call rapid increases in economic activity development. Hydrocarbon-based energy is important for three main areas of human development: economic, social, and environmental [2]. Hydrocarbons have generated an enormous increase in the ability of humans to do all kinds of economic work, greatly enhancing what they might be able to do with their own muscles or with those of work animals by using fossil-fueled machines such as trucks and tractors (Table 2.2). Perhaps most important, this work includes very large increase in the production of food.

The industrial revolution started in England in roughly 1750 but by about 1960 the world was using more petroleum than coal, and oil continues to be our most important energy source [2]. Now we live in, overwhelmingly, the age of oil. Some have said that we now live in an information age, or a postindustrial age. Both are only partly true. Overwhelmingly we live in a petroleum age. Just look around. All transportation, all food production, all plastics, most of our jobs and leisure, much of our electricity, and all of our electronic devices are dependent upon gaseous and especially liquid petroleum. This has been, and continues to be, the age of oil, and of hydrocarbons more generally. Perhaps the industrial revolution should be renamed the "hydrocarbon revolution" because that is what happened: humans moved from using various carbohydrates as their principal means of doing economic work to using hydrocarbons.

One reason that this is the age of oil, and hydrocarbons more generally, is that there continues to be a strong connection between energy use and economic activity for most industrialized [4] and developing economies [5] (Fig. 3.1). Some have argued that through technology and markets we are becoming more efficient in our use of energy. But the evidence for that is ambiguous at best. As yet unpublished top-down macroeconomic analysis (i.e., simply dividing inflation-corrected GDP

Table 3.1 Energy density of oil and other fossil fuels

Fuel Type	MJ/l	MJ/kg	BTU/Imp Gal	BTU/US Gal	Research octanenumber (RON)
Regular gasoline/petrol	34.8	~47	150,100	125,000	Min. 91
Premium gasoline/petrol		~46			Min. 95
Autogas (LPG) (60% propane and 40% butane)	25.5–28.7	~51			108–110
Ethanol	23.5	31.1	101,600	84,600	129
Methanol	17.9	19.9	77,600	64,600	123
Gasohol (10% ethanol and 90% gasoline)	33.7	~45	145,200	121,000	93/94
E85 (85% ethanol and 15% gasoline)	33.1	44	142,750	118,950	100–105
Diesel	38.6	~48	166,600	138,700	N/A (see cetane)
BioDiesel	35.1	39.9	151,600	126,200	N/A (see cetane)
Vegetable oil (using 9.00 kcal/g)	34.3	37.7	147,894	123,143	
Aviation gasoline	33.5	46.8	144,400	120,200	80–145
Jet fuel, naphtha	35.5	46.6	153,100	127,500	N/A to turbine engines
Jet fuel, kerosene	37.6	~47	162,100	135,000	N/A to turbine engines
Liquefied natural gas	25.3	~55	109,000	90,800	
Liquid hydrogen	9.3	~130	40,467	33,696	

Neither the gross heat of combustion nor the net heat of combustion gives the theoretical amount of mechanical energy (work) that can be obtained from the reaction. (This is given by the change in Gibbs free energy, and is around 45.7 MJ/kg for gasoline.) The actual amount of mechanical work obtained from fuel (the inverse of the specific fuel consumption) depends on the engine. A figure of 17.6 MJ/kg is possible with a gasoline engine, and 19.1 MJ/kg for a diesel engine

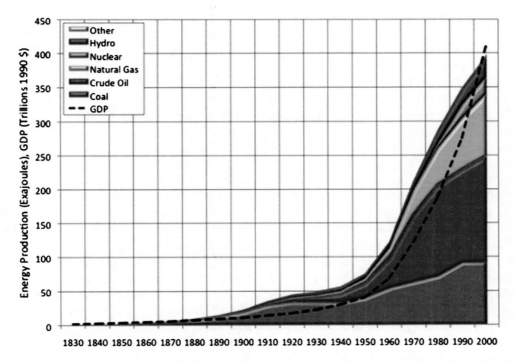

Fig. 3.1 The global use of hydrocarbons for fuel by humans has increased nearly 800-fold since 1750 and about 12-fold in the twentieth century. The most general result has been an enormous increase in the ability of humans to do all kinds of economic work, greatly enhancing what they might be able to do by their own muscles or with those of draft animals (*Source*: Ref [33])

by total energy used) undertaken by graduate student Ajay Gupta indicates that for most countries of the world there remains a very strong link between energy use and economic activity as measured by inflation-corrected GDP. Fundamentally there is no general trend of countries becoming more or less efficient in turning energy into GDP. One apparent exception is the United States, where there is an apparent decline in the ratio of energy used per unit of gross domestic product. Energy analyst Robert Kaufmann suggests that although there have been some real improvements in fuel efficiency (driven by higher fossil fuel prices) the increases in efficiency are due principally to a shift to higher-quality fuels, and especially to structural changes in national economies as richer nations move their heavy industries overseas to reduce pollution or find cheaper labor [6]. There may be another reason as well that the United States, but few other nations, appears to be becoming more efficient in our use of energy. According to the organization Shadow statistics, the United States has been engaged in a systematic "cooking of the books" on the official measure of inflation, that is, a deliberate official underestimate of inflation since 1985 to make governments look good. Correcting for any or all of these actions would greatly decrease the perceived improvements of efficiency in the U.S. economy. In addition it is clear from Gupta's data that the main way that countries develop (i.e., get richer) is through using more energy to do more economical work [5].

Energy prices have an important effect on almost every major aspect of macroeconomic performance because energy is used directly and indirectly in the production of all goods and services. Both theoretical models and empirical analyses of economic growth suggest that a decrease in the rate of energy availability will have serious impacts on the economy [7]. For example, most U.S. recessions after the second World War were preceded by rising oil prices, and there tends to be a negative correlation between oil price changes and both stock prices and their returns in countries that are net importers of oil and gas [8]. Energy prices have also been key determinants of inflation and unemployment.

There is a strong correlation between per capita energy use and social indicators such as the U.N.'s Human Development Index, although that relation is much more important at low incomes than high; in other words, increasing energy use is far more important at improving quality of life for the poor than for the rich. By contrast, the use of hydrocarbons to meet economic and social needs is a major driver of our most important environmental changes, including global climate change, acid deposition, urban smog, and the release of many toxic materials. Increased access to energy provided the means to deplete or destroy once-rich resource bases, from megafaunal extinctions associated with each new invasion of spear-equipped humans, to the destruction of natural ecosystems and soils through, for example, overfishing and intensive agriculture, and other types of development. Harvard biologist E. O. Wilson has attributed the current mass extinction to what he calls HIPPO effects: Habitat destruction, Invasive species, Pollution, Population (human), and Overgrazing. All these activities are energy intensive. Such problems are exacerbated by the increase in human populations that each new technology has allowed, as well as the overdependence of societies on previously abundant resources. Energy is a double-edged sword.

Peak Oil: How Long Can We Depend on Oil?

The critical issue with oil is not when do we run out, but when can we no longer increase or even maintain its production and use. We believe that "peak oil," the time when humans can no longer count on increasing oil production no matter what their effort, is more or less now, and that this will become the most important issue facing humanity. This critical issue can be understood on two levels: first as a simple fact, less, not more, oil over time, and second by a more thorough understanding of the properties and attributes of oil, which we do next. Although the exact timing of peak oil for the world remains somewhat debatable it is clear that it must be soon because each year we use two to four times more oil than we find. What is even more obvious is that our old rate of increase of 3 or 4% a year has declined since 2004 to from 0% to 1%, and that oil availability per capita is declining

At present, oil supplies about 40% (and natural gas about 20%) of the world's non-direct solar energy, and most future assessments indicate that the demand for oil will increase substantially if that is geologically, economically, and politically possible. What do we know about the future availability of oil? Predictions of impending oil shortages are as old as the industry itself, and the literature is full of arguments between optimists and pessimists about how much oil there is and what other resources might be available. There are four principal issues that we need to understand in order to assess the availability of oil and, by extension other hydrocarbons, for the future. We need to know: the quality of the reserves, the quantity of the reserves, the likely patterns of exploitation of the resource over time, and who gets and who benefits from the oil. All of these factors ultimately affect the economics of oil production and use.

Quality of Petroleum

Oil is a fantastic fuel, relatively easy to transport and use for many applications, very energy dense, and extractable with relatively low energy cost and (usually) low environmental impact (Table 3.1). What we call oil is actually a large family of diverse hydrocarbons whose physical and chemical qualities reflect the different origins and, especially, different degrees of natural processing of these hydrocarbons. Basically oil is phytoplankton kept from oxidation in deep anaerobic marine or freshwater basins, covered by sediments and then pressure-cooked for 100,000,000 years [9]. In general, humans have exploited the large reservoirs of shorter-chain "light" oil resources first because larger reservoirs are easier to find and exploit, and lighter oils require less energy to extract and refine [10]. The depletion of this "easy oil" has required the exploitation of increasingly small, deep, offshore,

and heavy resources. Oil must first be found, then the field developed, and then the oil extracted carefully over a cycle that typically takes decades. Oil in the ground is rarely what we are familiar with in an oil can. It is more like an oil-soaked brick, where the oil must be pushed slowly by pressure to a collecting well. The rate at which oil can flow through these "aquifers" depends principally upon the physical properties of the oil itself and of the geological substrate, but also upon the pressure behind the oil that is provided initially by the gas in the well.

Progressive depletion also means that oil in older fields that once came to the surface through natural drive mechanisms, such as gas pressure, must now be extracted using energy-intensive secondary and enhanced technologies. As the field matures, the pressure necessary to force the oil through the substrate to the collecting wells is supplied increasingly by pumping more gas or water into the structure. EOR or enhanced oil recovery is a series of processes by which detergents, CO_2, and steam have been used—since the 1920s—to increase yields. Too-rapid extraction can cause compaction of the aquifer or fragmentation of flows which reduce yields. So our physical capacity to produce oil depends upon our ability to keep finding large oil fields in regions that we can reasonably access, our willingness to invest in exploration and development, and our willingness not to produce too quickly. Thus, technological progress is in a race with the depletion of higher-quality resources.

Another aspect of the quality of an oil resource is that oil reserves are normally defined by their degree of certainty and their ease of extraction, classed as "proven," "probable," "possible," or "speculative." In addition, there are unconventional resources such as heavy oil, deep-water oil, oil sands, and shale oils that are very energy-intensive to exploit. Thus although there are large quantities of oil left in the world, the quality of the actual fields is decreasing as we find and deplete the best ones. Now it takes more and more energy to find the next field and, as they tend to be of poorer quality, more and more energy to extract and refine the oil to something we can use (Table 3.2).

Table 3.2 How reliable are official energy statistics? (From Lewis L. Smith)

OPEC	Cum Prod End 2003	% Depleted	Indicated total	Remaining reserves G				BP estimates interpreted
				PFC	ASPO	Salameh	BP	
Iraq	28	22	127	99	62	62	115	Total discovered
UAE	19	31	61	42	49	37	98	Total discovered
Kuwait	32	35	91	59	60	71	97	Total discovered
Libya	23	39	59	36	29	26	36	
Saudi	97	42	231	134	144	182	263	Total discovered
Algeria	13	50	26	13	14	11	11	
Nigeria	23	50	46	23	25	20	34	?High estimate
Iran	56	51	110	54	60	64	131	Total discovered
Venezuela	47	58	81	34	35	31	78	Total discovered
Qatar	6.8	62	11	4.2	4.1	4.6	15	Total discovered
Indonesia	20	75	27	6.7	9.4	12	4.4	
Total	365		870	506	492	520	882	

Source: Iran's reserves less than half. OPEC's reserves overstated by 80%. From mushalik@tpg.com.au and http://www.energiekrise.de/e/aspo_news/aspo/newsletter046.pdf

Statistics for the oil industry are not as bad as those for the wine industry but still, they are pretty bad! This is especially true for reserves, the amounts of oil that engineers and geologists estimate could be extracted in the future from active reservoirs or promising geological formations, given present prices and technology. The three most important compilers of statistics for the oil industry are the BP, *Oil and Gas Journal*, and the U.S. DOE's Energy Information Administration. And that is all they are, compilers. They do not audit, check, or question the information supplied to them by their diverse sources. One reason is rumored to be that they are afraid of being "cut off" by any source to which they pose embarrassing questions! Just out of curiosity, I (LLS) checked the table, "Worldwide look at reserves and production," in the December 21 issue of the *Oil and Gas Journal*, pp. 20–21. Of the 200 or so political jurisdictions that merit statistical recognition by the United Nations, 107 got a line in the table, because they have "proven" oil reserves, gas reserves, or both.

There are five good reasons why an estimate of reserves for a nation should change (up or down) every year. Indeed it is almost impossible for them to remain unchanged, if the engineers and geologists have done their work correctly. These five reasons include new findings, revisions in old estimates, and, clearly, production. However, I note that in the referenced table only 29 countries (27% of the total) report no oil reserves or changed their estimate from last year. The other 78 (73%) reported exactly the same figure for this year as last year. This includes one country widely believed to be exaggerating its "official" estimate by more than 100%! Some of the "no changers" include Indonesia, Iraq, Kuwait, Norway, Russia, and Venezuela. Ironically Norway is one of the few countries that publishes good production data by oil field. You may draw your own conclusions! I gather that the situation for natural gas is a little better, but not enough to trust the data for all important producers

Quantity of Petroleum

Most estimates of the quantity of conventional oil resources remaining are based on "expert opinion," which is the carefully considered opinion of geologists and others familiar with a particular region (Table 3.2). The ultimate recoverable resource (URR, often written as EUR) is the total quantity of oil that will ever be produced from a field, nation, or the world, including the 1.1 trillion barrels extracted to date. URR will determine the shape of the future oil production curve. Recent estimates of URR for the world have tended to fall into two camps. There is a great deal of controversy, or rather range of opinion, about how much oil remains. Lower estimates come from several high-profile analysts, many of them retired petroleum geologists, with long histories in the oil industry who suggest that the URR is no greater than about 2.3 trillion barrels (in other words the 1.1 we have used and another 1.2 we will extract in the future), and may be even less [10]. The USGS estimates that this number may be about 2.4 trillion barrels, half from new discoveries and half from reserve growth, that is, increased estimates of oil available from existing fields. A "middle" estimate is three trillion barrels and the highest credible estimate is four trillion barrels. These latter three values are from the most recent study by the U.S. Geological Survey

Table 3.3 How much oil remains in the world is highly uncertain. "Reserves" are inflated with >300 B bbls of "resources"

1.1 Trillion	Depleted	Statistical reliability	Production outlook	Technical Basis
Actual Reserves: 0.9 Trillion	Proven >90%	Proven oil in place – high confidence Developed – clear recovery factor Undeveloped – good. recov. est.	Growth through actual reservoir mgmt. and performance	Improved oil recovery through existing technology
	Probable >50%	Probable oil in place – confident Developed – prelim. recovery factor Undeveloped – est. fair recovery	Growth through delineation, testing and development	Clear opportunity with existing technology
	Potential >5%	Potential oil in place – low confiden. Drilled – very low recovery factor Undrilled – recovery likely poor	Growth through pricing, delineation or IOR/EOR technology	Indicative data and potential opportunity
Contingent Resources: 1.1 Trillion	Resource: Uneconomic volume and commerciality	Likely presence but undelineated OIP or GIP	Profitability, or Technology currently inadequate	Available access but lacks good reservoir and fluids data
Prospective and Speculative Resources: 2.0 Trillion	Oil, Gas, Shale, EHC and to be discovered resources (speculative outlook)	Technically present but physically inaccessible hydrocarbons Conceptually Possible Hydrocarbons, incl. EHC's	Future resolution through exploration and relevant technology	General geological, seismic and/or physical indications

Source: From mushalik@tpg.com.au

in 2000, which if nothing else tends to cover the range of other estimates (USGS [11, 12]). Even in that study the lower values tend to be from their staff of geologists and the larger ones reflect increasingly the opinion of USGS economists who believe that price signals will allow lower grades of oil to be exploited through technical improvements and there will be corrections of earlier conservative estimates (Table 3.3).

This relatively new addition to the USGS methodology is based on experience in the United States and a few other well-documented regions. The new totals assume, essentially, that petroleum reserves everywhere in the world will be developed with the same level of technology, economic incentives, and efficacy as in the United States. Although time will tell the extent to which these assumptions are realized, the last 10 years of data have shown that the majority of countries are experiencing patterns of production that are far more consistent with the low rather than medium or higher URR estimates [13, 14]. Increasingly other estimates by, for example, U.S. and European energy agencies (EIA and

IEA) are coming in on the low side. An assessment by oil experts (the best in our opinion) Colin Campbell and Jean Laherrère shows that we are now producing and consuming two to four barrels for each barrel we find (Fig. 3.2). One would think that the best way to find and produce more oil would be to drill more, but in fact the finding of oil and gas is almost independent of drilling rate, at least at the levels we have been used to undertaking, because time is needed to determine where the next good place to drill is (Fig. 3.3).

Pattern of Use over Time

The best-known model of oil production was derived by Marion King Hubbert, who proposed that the discovery and production of petroleum over time would follow a single-peaked, more or less symmetric, bell-shaped curve (Fig. 3.4). A peak in production would occur when 50% of the URR had been extracted (he later opined that there may be more than one peak). This hypothesis seems to have been based principally on

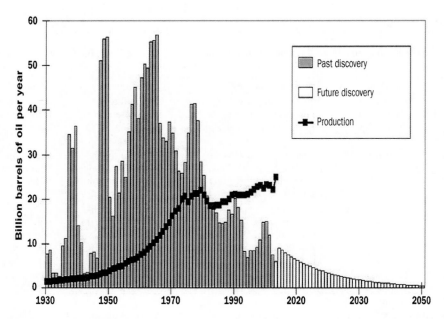

Fig. 3.2 Rates of finding and rate of production for conventional oil globally where field updates have been updated to the year that the initial strike was found (*Source*: ASPO). Note: There is another way of graphing these data by attributing "revisions and extensions" to the year of revision, not the year of initial strike. This exaggerates the finding rate of more recent years

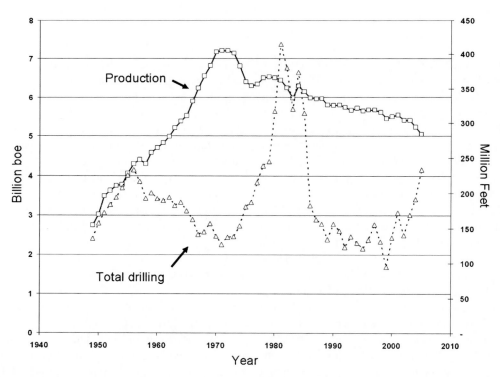

Fig. 3.3 Oil and gas production appears to be independent of drilling effort. There has been essentially no correlation between drilling effort and the rate of finding and (here given) production of oil and gas in the United States except in the first years plotted. Production increased until 1970, then peaked, then declined steadily despite enormous increase, and subsequent decrease, in drilling effort. Drilling rates tend to increase when prices are high and the converse. (*Source*: Nate Gagnon.)

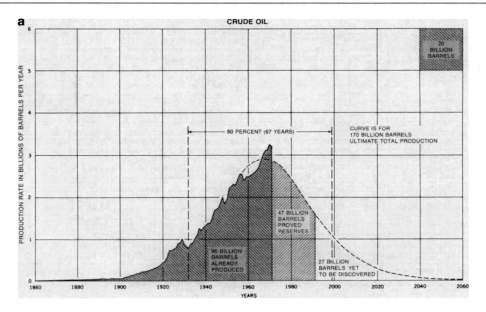

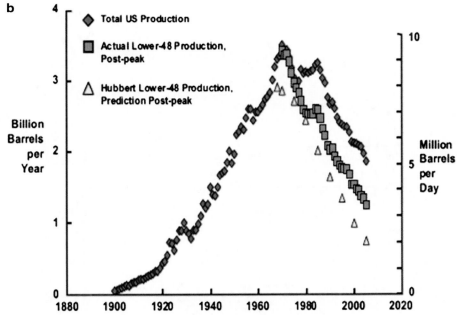

Fig. 3.4 Hubbert curves: (**a**) Original for United States. (*Source*: Hubbert, 1968); (**b**) Present for United States. (*Source*: Cambridge Energy Research Associates, 2006); (**c**) Original for world. (*Source*: Hubbert, 1968); (**d**) Present world data. (*Source*: TheOilDrum.com); (**e**) American whaling industry. Depletion; from 90% to 99% of many whale species were killed. (*Source*: Ugo Bardi.)

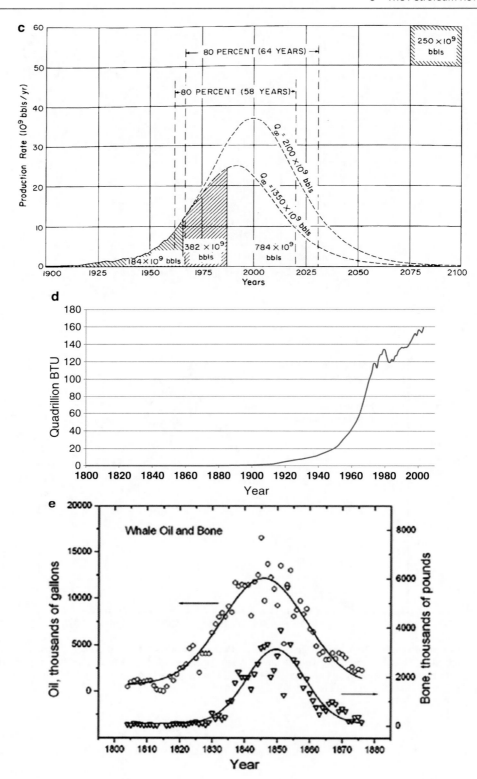

Fig. 3.4 (continued)

Hubbert's intuition and his tremendous experience examining the patterns of many, many oil fields. It was not a bad guess, as he famously predicted in 1956 that U.S. oil production would peak in 1970, which in fact it did [13]. Hubbert also predicted that the U.S. production of natural gas would peak in about 1975, which it did, although it has since shown signs of recovery and there is a second smaller peak in 2010 based on "unconventional" and "shale" gas. He also predicted that world oil production would peak in about 2000. In fact, oil production continued to increase until 2004, after which it appears to have entered an oscillation or "undulating plateau," as predicted earlier by geologist Colin Campbell.

In the past decade, a number of "neohubbertarians" have made predictions about the timing of peak global production ("peak oil") using several variations of Hubbert's approach [14–20]. These forecasts of the timing of global peak have ranged from one predicted for 1989 (made in 1989) to many predicted for 2005–2015 to one as late as 2030 [15]. Most of these studies assumed world URR volumes of roughly two trillion barrels and that oil production would peak when 50% of the ultimate resource had been extracted. The predictions of a later peak begin with an assumption of a large volume of ultimately recoverable oil. How much oil will we actually recover? The USGS study quoted above gives a low estimate (which they state has a 95% probability of being exceeded) of 2.3 trillion barrels and a "best" estimate of three trillion barrels. One analysis fit the left-hand side of Hubbert type curves to data on actual production while constraining the total quantity under the curve to two, thtree, and four trillion barrels for world URR. The resultant peaks were predicted to occur from 2004 to 2030 [17]. Brandt [18] shows that the Hubbert curve is a good prediction for most postpeak nations, which includes the great majority of all oil-producing nations. Other recent and sophisticated Hubbert-type analyses by Kaufmann and Shiers [19] and Nashawi and colleagues [20] suggest peaks in about 2013–2014, consistent with the low URR estimates of, for example, Campbell and Laherrère, at least as long as there is not much more recoverable oil than seems likely at this time

Table 3.4 Published estimates of world oil ultimate recovery (From Hall et al., 2003)

Source	Volume (trillions of barrels)
USGS, 2000 (high)	3.9
USGS, 2000 (mean)	3.0
USGS, 2000 (low)	2.25
Campbell, 1995	1.85
Masters, 1994	2.3
Campbell, 1992	1.7
Bookout, 1989	2.0
Masters, 1987	1.8
Martin, 1984	1.7
Nehring, 1982	2.9
Halbouty, 1981	2.25
Meyerhoff, 1979	2.2
Nehring, 1978	2.0
Nelson, 1977	2.0
Folinsbee, 1976	1.85
Adam and Kirby, 1975	2.0
Linden, 1973	2.9
Moody, 1972	1.9
Moody, 1970	1.85
Shell, 1968	1.85
Weeks, 1959	2.0
MacNaughton, 1953	1.0
Weeks, 1948	0.6
Pratt, 1942	0.6

Source: Volume (trillions of barrels)

[14]. If that is the case the peak may be displaced for one or two decades. An important issue that most of these studies do not consider is that most of the oil left in the ground will take an increasing amount of energy to extract (Table 3.4).

Most recent results of curve-fitting methods showed a consistent tendency to predict a peak within a few years, then a decline, no matter when the predictions were made. This is consistent with the fact that we are using at least twice as much oil as we are finding. Other forecasts for world oil production do not rely on either assumptions about URR or the use of "curve-fitting" or "extrapolating" techniques but simply draw straight lines into the future based on past increases. According to one recent forecast by the U.S. Energy Information Agency [15], world oil supply in 2025 will exceed the 2001 level by 53% [15]. The EIA reviewed five other world oil models and found that all of them predict that production will increase in the next two decades to around 100 million barrels per day, substantially more than the 77 million barrels per day produced in 2001. Several of these models rely on the 2000 USGS higher estimates of URR for oil. It should be noted that the majority of oil-supply forecasts that we examined (with the possible exception of postpeak Hubbert analyses) had a poor track record, regardless of method. It is now a well-established fact that economic and institutional factors, as well as geology, were responsible for the U.S. peak in production in 1970 [21, 22], forces that are explicitly excluded from the curve-fitting models. Thus, the ability (or the luck) of Hubbert's model (and its variants) to forecast production in the lower 48 states accurately should not necessarily be extrapolated to other regions. It is too early to tell. On the other hand, the actual data on global conventional oil production certainly show at least an undulating plateau at the time of this writing and perhaps even a production peak in 2005 (Fig. 3.4c). Certainly the old growth rate of 3–4% per year has slowed way down. This is astonishing given the continuous growth in production year after year since at least the 1980s, and it happened during times of greatly increasing oil prices. Clearly Hubbert-type peaks have occurred for oil for many nations [18] and for other resources, such as whale oil and phosphorus (Fig. 3.4e).

So, why is global oil production decreasing or at least no longer increasing? The principal reason is that most oil production comes from very large oil fields (called "elephants") and we have found very few elephants since the 1960s. Now these large oil fields are aging, and the production in many of these fields is declining by 2–10% a year. Thus although it is true that we are finding additional new oil supplies, these new fields are equal in volume to only about one-fifth of the existing ones, hence the expected decline [10, 20] (Fig. 3.6). According to Chris Skrebowski, editor of *Petroleum Review*, at least one quarter of the 400 largest oil fields in the world are in decline, and it appears impossible that new oil discoveries, most of which are not large, can possibly make up for the decline in the elephants.

Economic forecasts have not fared well in explaining U.S. oil production. In the period after the Second World War, oil production often increased as oil prices decreased, and vice versa (Fig. 3.3), a behavior that is exactly the opposite of predictions of conventional economic theory. Economic theory also assumes that oil prices will follow an "optimal" path towards the choke price: the price that is sufficiently high to cause demand for oil to begin to fall to zero. Thereafter, at least in theory, the market signals a seamless transition to substitutes. In fact, even if such a path exists, prices may not increase smoothly because empirical evidence indicates that producers respond differently to price increases than they do to price decreases [22]. In the presidential campaign of 2008, one often heard in response to the increased price of oil, "Drill, drill, drill!" In fact there is little evidence that there is any relation at all between oil and gas drilling and oil and gas production, with the exception of the early 1950s (Fig. 3.3). One way to think about this is that "Mother Nature holds the high cards." In other words, oil production will be determined much more by what is geologically possible than by human efforts or economics [23]. Significant deviation from basic economic theory undermines the de facto policy for managing the depletion of conventional oil supplies, a belief that the competitive market will generate a smooth transition from oil. We see little evidence of this happening thus far.

Whatever the exact details or the dates of peak oil it is clear that we are, in the words of Colin Campbell, in transition from the first half of the age of oil to the second half. Each half is and will be equally oil dependent, but the difference will be that between an increasing quantity being used each year to a flat and then decreasing quantity.

Net Energy from Oil

Our view is that the question is not how much recoverable oil is left in the Earth. We agree that there is a great deal, perhaps near the high end of the estimates. But what is missing from the debate

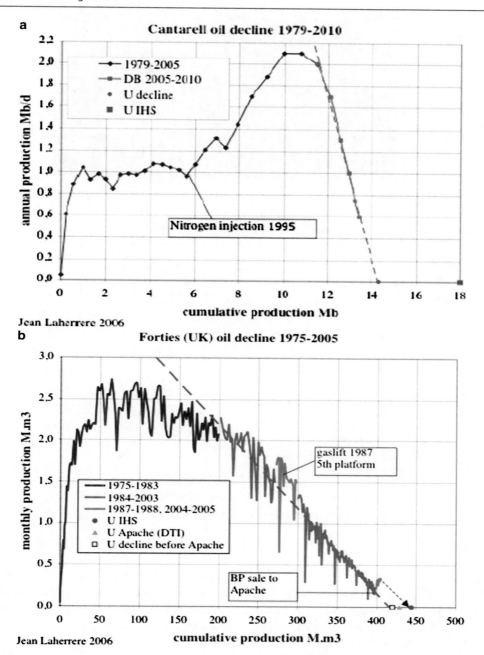

Fig. 3.5 Decline in the production of a number of important "elephants." (*Source*: Jean Laherrère). (**a**) Canterell, Mexico, once the world's second largest field; (**b**) The Brent field in the North Sea; (**c**) Prudhoe Field, the largest in the United States; (**d**) East Texas, the second largest field in the United States

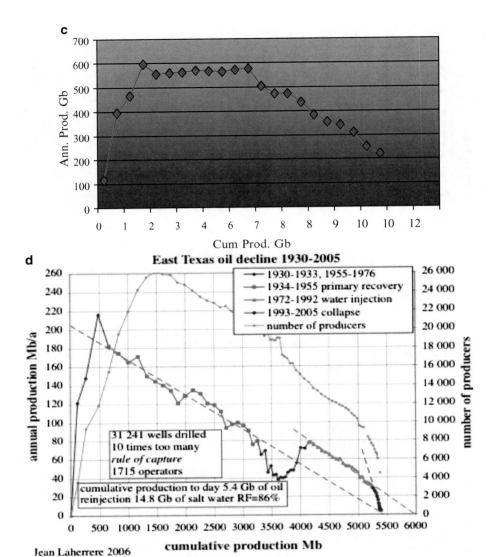

c

d **East Texas oil decline 1930-2005**

- 1930-1933, 1955-1976
- 1934-1955 primary recovery
- 1972-1992 water injection
- 1993-2005 collapse
- number of producers

31 241 wells drilled
10 times too many
rule of capture
1715 operators

cumulative production to day 5.4 Gb of oil
reinjection 14.8 Gb of salt water RF=86%

Jean Laherrere 2006   **cumulative production Mb**

Fig. 3.5 (continued)

is how much of that oil can be recovered with a significant, or perhaps any, net energy gain. These are old arguments about peak oil [24] but the new assessments are made in the absence of net energy costs. If we extrapolate essentially any time series analysis of the net energy returned from oil all of them show (if present trends continue) a breakeven point within decades. Thus we think we will reach the energy breakeven point long before we are able to exploit the larger estimates of reserves given by, for example, the USGS [25] (Fig. 3.6). In other words the total amount of oil in the ground is not a relevant number. Rather we need to know how much of that can be extracted with

a significant net energy profit. This important issue of the energy cost of getting additional quantities of oil, and how that might influence URR, is given in Chap. 14.

Geography of Oil

Oil is used by all of the nearly 200 nations of the world, but significant amounts are produced by only about 42 countries, 38 of which export important amounts. This number is declining because of the depletion of the once-vast resources of North and South America, the North Sea,

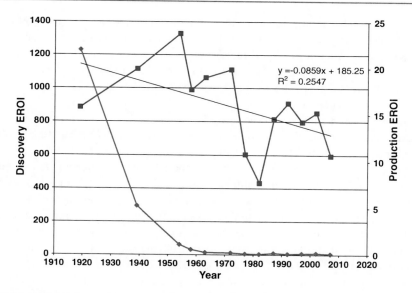

Fig. 3.6 EROI for finding U.S. oil and gas (*diamonds*) and for producing that oil and gas. (*squares*). (*Source*: Megan C. Guilford et al., Sustainability in press.)

Indonesia, and many other regions, and owing to the increasing domestic use of oil by many of the exporters. The number of exporters outside the Middle East and the former Soviet Union will drop in the coming decades, perhaps sharply, which in turn will greatly reduce the supply diversity to the 160 or so importing nations [23]. Such an increase in reliance on West African, former Soviet Union, and especially Persian Gulf oil has many strategic, economic, and political implications. Much of the world's reserves are found in nations not known to be friendly to the United States or the West more generally, in part because of the West's long history of "boots on the ground" expropriation of oil or interference with the governments of producing countries. The enormously increasing demand for oil from China and its large reserves of money are also likely to have a large impact because the Chinese should have little trouble paying for their oil even as prices rise.

Energy and Political Costs of Getting Oil

The future of oil supplies is normally analyzed in economic terms, but economic costs are likely to be dependent on other costs. In our earlier work

[1, 24] we have summarized the energy costs of obtaining U.S. oil and other energy resources and found, in general, that the energy returned on energy invested (see Chap. 14) tended to decline over time for oil and most other energy resources examined. This includes the energy cost of obtaining oil by trading (energy-requiring) goods and services for energy itself [24]. For example, the EROI for discovering oil and gas in the United States has decreased from a value of more than 1,000 to 1 in 1919 to 5 to 1 in the 2010s, and for production from about 30 to 1 in the 1970s to less than 10 to 1 today (Fig. 3.6). Likewise the EROI for the production of oil and gas globally has declined from about 36 to 1 in the 1990s to about 19 to 1 in 2006 [25]. In other words with all of our super technology we can continue to get oil and gas, but the energy cost per barrel continues to increase as we deplete the best resources. We are not aware of such estimates for other parts of the world, although we do know that both heavy oil in Venezuela and tar sands in Alberta require a very large part of the energy produced, as well as substantial supplies of hydrogen from natural gas, to make the oil fluid. The very low economic cost of finding or producing new oil supplies in the Arabian peninsula implies that it has a very

high EROI value, which in turn supports the probability that productivity will be concentrated there in future decades. Alternative liquid fuels, such as ethanol from corn, have a very low EROI. An EROI of much greater than one to one is needed to run a society, because energy is also required to make the machines that use the energy, feed, house, train, and provide healthcare for necessary workers and so on (Chap. 14).

No one who watches the news can fail to be aware of the importance of cultural and political differences between those nations that have the most oil and those that import it. How these factors will play out over the next few decades is extremely important, but also impossible to predict. Most of the remaining oil reserves are in Southern Russia, the Middle East, and North and West Africa, countries or regions with either Muslim governments or significant Muslim populations. For a long period, frustration and resentment has been building up among Muslim populations, not least because of their perception that the main Western powers have failed to generate even-handed policies to address the conflicts in the Middle East over the past half-century. Iranians still have vivid memories of the role the Central Intelligence Agency played in the overthrow of their democratically elected Prime Minister, Dr. Mohammed Mossedeq, on behalf of the Anglo-Iranian Oil Company (now BP). Another factor is that the huge revenues earned by the oil-exporting nations have been very unevenly distributed among their respective populations, adding to internal and external pressure to adopt a more equitable approach to human development. The "Arab Spring" of 2011, with its new pressures for governmental reform has greatly increased the instability of many Middle Eastern oil-producing nations, and oil prices. Much of the unrest stems in part from the failure, and some would say impossibility, of these economies to produce sufficient jobs and even food for their growing populations. Suffice it to say that there will continue to be high risks of international and national terrorism, overthrow of existing governments, and deliberate supply disruption in the years ahead. In addition, exporting

nations may wish to keep their oil in the ground to maintain their target price range. Thus, there are considerable political and social uncertainties that could result in less oil being available than existing models predict.

Deep Water and Extreme Environment

Although considerable uncertainty remains about how much oil we will extract eventually one thing is clear: oil is getting harder and harder to find [26, 27]. This can be seen by the increasing dollar, energy, and environmental cost of getting oil, and by the fact that we are undertaking major exploration and development in areas (such as very deep ocean) that were thought too difficult and expensive just a decade ago, so that half of new U.S. drilling effort now takes place far offshore. There have been amazing developments in technology that have allowed this new exploration: drilling ships unanchored to the bottom kept in place by GPS systems and huge thrusters, drill strings that go down through 2,000 m of ocean and then 5,000 m or more of rock and so on. The Deepwater Horizon oil spill of 2010 in the Gulf of Mexico has brought all these operations to the attention of the public and one of the first questions asked was: why are we working in such a difficult and potentially dangerous environment? The answer is that the oil fields that have been discovered at these depths appear to be the only large fields left that have not yet been exploited; in other words we went after the easy stuff first and left the most difficult until later. So if we are to continue to have oil we need to undertake these expensive and risky operations. The most interesting analysis of this issue is by Tainter and Patzek [28], where the authors ask whether we have expanded the complexity of our American "empire" to the point that the energy cost of getting energy itself to the center of the empire exceeds the gain from that energy. They point out that this may be analogous to other ancient empires (such as Rome) which expanded until they reached the limits of managing the complexity necessary for maintaining the society [29].

A similar analysis might be made of our large efforts in militarization in support of maintaining oil flows.

A final important issue relating to the development of new oil or its possible substitutes has been put forth by Robert Hirsch and his colleagues in several extremely insightful papers [30, 31]. Their basic point is that a critical element in finding a substitute for petroleum (if indeed a substitute exists) is time; that is, even if a workable substitute can be found (and they examine, e.g., shale oil, biomass fuels, and even greatly increasing the gas mileage of our vehicles), and assuming that government (or private) programs can be developed and money is no object, that it would take decades simply to scale up the approach. In other words if we could maintain liquid fuel use at the level of the peak of oil (perhaps about what we had in 2005–2010) that it would take decades to construct the needed infrastructure. It is a very sobering perspective.

How About Natural Gas?

Petroleum usually means liquid and gaseous hydrocarbons, and includes oil, natural gas liquids, and natural gas. Thus a chapter on oil is incomplete without some consideration of natural gas. Natural gas is often found associated with oil, although it has other possible sources, including coal beds and organic-rich shale. Oil is a natural hydrocarbon where the original plant material, often composed of 100–1,000 of carbons linked together, has been cracked or broken by geological energies to a length of (ideally) eight carbons (octane). If the cracking continues, in the extreme the carbon bonds are broken completely to a length of one carbon, usually surrounded by four hydrogen molecules, a gas called methane. This makes gas an ideal fuel because oxidizing hydrogen releases more energy and releases less carbon dioxide than oxidizing carbon. Methane is much more easily obtained, stored, and moved than is hydrogen, partly because the much smaller hydrogen molecule leaks more easily. When natural gas is held in a tank some heavier fractions fall out as natural gas

liquids, and these materials can be used, essentially, either directly or as inputs to refineries. Although natural gas was once considered an undesirable and dangerous by-product of oil production, and it was flared into the atmosphere, with time its commercial value was recognized and a complex pipeline system evolved. Now natural gas is more or less tied with coal as the second most important fuel in the United States and the world. An important question is: if oil falters can natural gas take over its role? It can even be used to propel vehicles with minimal changes to the engine and it has essentially displaced the role of oil in electricity production. It is not as energy dense or transportable as oil but it comes close, and because it is clean it has many special uses such as for baking and as a feedstock for plastics and nitrogen fertilizer.

Beginning in 2010 there was a great deal of excitement and debate about whether "unconventional" natural gas from, for example, the Marcellus shale can provide an energy renaissance for the United States. It has been known that considerable gas exists in association with certain shales, however; it was too difficult to get it out because the shale formations were too thin and a conventional vertical well simply passed through the formation without intercepting much gas. New technologies, including horizontal drilling and shattering or "fracking" the rocks with very high-pressure water have allowed considerable amounts of gas to be produced. But the environmental impacts are barely known and possibly large, and tens of thousands of wells are needed to get a significant amount of gas, thus there is great deal of controversy about the degree to which these wells should be drilled. Something less well known is that most of the gas in those areas we know best (e.g., the Barnett Shale in Texas) comes from a relatively few "sweet spots," and that the total regional production may go through most of a full Hubbert cycle in only 15 years. Meanwhile conventional gas production has peaked and dropped off to less than half the peak, so that so far the unconventional gas of all kinds is simply compensating for the drop off of conventional gas (Fig. 3.7). Thus natural gas is likely to be very important as oil production and availability

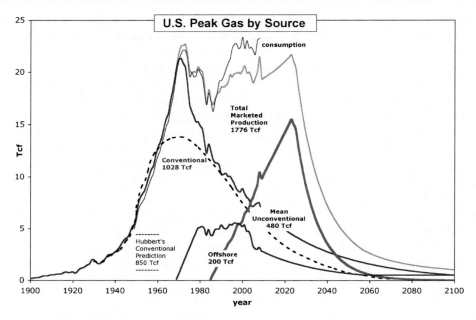

Fig. 3.7 Patterns of past and projected possible production of conventional, deep water, and unconventional (e.g., shale gas) U.S. natural gas. (*Source*: Bryan Sell, personal communication). Note: estimates for shale gas are very uncertain at this point

decline, and it will extend the petroleum age by a few decades. But then that too will be gone, and the United States will be left with little domestic production of oil or gas [32].

The Future: Other Technologies

The world is not about to run out of hydrocarbons, and perhaps it is not going to run out of oil from unconventional sources any time soon. What will be scarce is cheap petroleum, the kind that allowed industrial and economic growth. What is left is an enormous amount of low-grade hydrocarbons, which are likely to be much more expensive financially, energetically, politically, and especially environmentally. As conventional oil becomes less available, society has a great opportunity to make investments in different sources of energy, perhaps freeing us for the first time from our dependence on hydrocarbons. There is a wide range of options, and an equally wide range of opinions, on the feasibility and desirability of each. Nuclear power faces formidable obstacles. Experience of the past several decades has shown

that electricity from nuclear power plants can be a reliable and mostly safe source of electricity, although an expensive form of power, when all public and private costs are considered. The earthquake tsunami-induced accident at Fukushima Daiichi may make continued expansion unlikely in many nations. Other unresolved issues include that nuclear power generates high-level radioactive wastes that remain hazardous for 1,000s of years, possible nuclear weapons proliferation, and whether there is enough uranium to allow a significant contribution to global energy supplies. These are high costs to impose on future generations. Even with improved reactor design, the safety of nuclear plants remains an important concern. Can these technological, economic, environmental, and public safety problems be overcome? Can new reactors using thorium fuel be created that decrease the problem of dangerous by-products generated from uranium while expanding the fuel supplies? These questions remain unanswered while we increase our use of fossil fuels essentially every year.

Renewable energies present a mixed bag of opportunities. Some argue that they have clear advantages over hydrocarbons in terms of economic viability, reliability, equitable access, and especially environmental benefits. But nearly all suffer from very low energy return on investments compared to conventional fossil fuels. In favorable locations, wind power has a high EROI (perhaps 18:1). The cost of photovoltaic (solar electric) power has come down sharply, making it a viable alternative in areas without access to electricity grids, but the EROI remains relatively low, perhaps only 4:1 or less, when considered on a systems level [33]. But both of these solar energies require very expensive backups or transmission systems to compensate for intermittent production, as they are available only 20–30% of the time. With proper attention to environmental concerns, biomass-based energy generation is competitive in some cases relative to conventional hydrocarbon-based energy generation. At present liquid-fuel production from grain has a relatively low EROI [34, 35]. Hydrogen, advocated by many, is an energy carrier, not an energy source, thus requires some kind of fuel to be used to split water or run some other process to generate the hydrogen. In addition there are many problems to overcome because the small molecules leak easily and are hard to store. Hydrogen generated from renewable energy sources or electricity-driven hydrolysis is currently too expensive for most applications, but it merits further research and development.

A disquieting aspect of all these alternatives, however, is that as energy delivery systems (i.e., including backups, transmission, etc.) they all have a much lower EROI than the fossil fuels we would like them to replace, and this is a major reason for their relatively low economic feasibility in most applications [32]. This may be a very tough nut to crack. Subsidies and externalities, social as well as environmental, add difficulties to this evaluation but are poorly understood or summarized. This presents a clear case for public policy intervention that would encourage a better understanding of the strengths and weaknesses of renewable forms of energy. Policy intervention, in concert with ongoing private investment and also markets, may be necessary to speed up the process of sorting the wheat from the chaff in the portfolio of renewable energy technologies, necessary if for no other reason than to protect our atmosphere.

The Social Importance of These Supply Uncertainties

Many once-proud ancient cultures have collapsed, in part, because of their inability to maintain energy resources and societal complexity [26]. Our own civilization has become heavily dependent on enormous flows of cheap hydrocarbons, partly to compensate for other depleted resources (e.g., through fertilizers and long-range fishing boats), so it seems important to assess our main energy alternatives. Oil is quantitatively and qualitatively most important. Investments in oil have continued to increase, but supply remains flat and is likely to decrease. Some of the most promising new oil fields have turned out to be very disappointing [27, 28]. If indeed we are approaching the oil scarcity that some predict, it is barely reflected in oil prices and few investments in alternatives are being made at anything like the scale required to replace fossil hydrocarbons, if indeed that is possible. Unfortunately the majority of decision makers hold on to the fantasy that the market has resolved this issue before and will do so again. Furthermore, an increasing number of U.S. citizens believe that government programs are too ineffective to resolve any problem, including energy problems. We view this as a recipe for disaster. It is enhanced by the failure of science to be used as fully, effectively, or objectively as it should be. Failures in proper government funding for good energy analysis have led to the dominance of "science" in the media and decision making whose role is basically to support the predetermined position of those who support it. In 2011 the state of oil-supply modeling is in some ways no different than it was in Hubbert's time: a wide range of opinion exists and there is little or no objective and reliable overview.

This issue is critical at this point in time because if our civilization is to survive the next 50 years enormous new investments are necessary in whatever we will need to replace existing flows of conventional oil and gas and even coal. Energy costs now are only for the costs to extract fuels from existing reserves, not to come up with replacements once those fuels are gone. As energy prices increase citizens are probably not going to be too excited to pay even more for a program to develop the research and infrastructure to generate replacement fuels, even if we knew what they should be. According to one of our best energy analysts, Vaclav Smil, at this time there seems to be few really good options except to decrease our appetites for energy [35].

What can science do to help resolve this uncertainty? Our principal conclusion is that these critical issues could and should be the province of open scientific analysis in visible meetings where all sides attend and argue, and where resources are provided to reduce uncertainty and understand different assumptions. This analysis should be informed by professionalism, the peer review process, statistical analysis, hypothesis-generation and testing, and so on, rather than simply by the opinions of the experts one chooses or the quips on the blogosphere. These issues should be the basis of open competitive government grant programs, graduate seminars, and even undergraduate courses in universities, and our courses in economics should become at least as much about real biophysical resources, such as hydrocarbon reserves, as about market mechanisms. Also, we need to think much harder about the alternatives, including their liabilities or the need to develop a lower energy-using society. None of this appears to be part of the plans of any government or governmental agency.

Questions

1. What is meant by the phrase "The first half of the age of oil"?
2. What are energy-dense materials?
3. We have been told that we live in an "information age". Argue for or against that statement.
4. Is peak oil a fact or a concept? Defend your view.
5. What does EUR mean? How is it related to peak oil?

6. What were Hubbert's basic ideas?
7. If there are huge amounts of oil left in the Earth, does this imply adequate supplies for the foreseeable future? Why or why not?
8. How is natural gas related to oil?
9. What does "cheap oil" mean relative to the remaining oil we might be able to extract from the Earth?
10. With many alternatives, why do you think that we have continued to rely so much on oil and other hydrocarbons?

Acknowledgments We thank S. Ulgiati, R. Kaufmann, and C. Levitan for discussions.

Literature

1. Derived, with substantial modifications from Hall, C., P. Tharakan, J. Hallock, C. Cleveland and M. Jefferson. 2003. Hydrocarbons and the evolution of human culture. Nature 426: 318–322. Updates on EROI are available in a special issue of the Journal Sustainability (in press).
2. Munasinghe, M. 2002. The sustainomics transdisciplinary meta-framework for making development more sustainable: applications to energy issues. Int. J. Sustain. Dev. 5, 125–182.
3. Interlaboratory Working Group. 2000. Scenarios for a Clean Energy Future <http://www.ornl.gov/ORNL/Energy_Eff/CEF.htm> (Lawrence Berkeley National Laboratory LBNL-44029, Berkeley, California).
4. Hall, C. A. S., Lindenberger, D., Kummel, R., Kroeger, T. and Eichhorn, W. 2001. The need to reintegrate the natural sciences with economics. BioScience 51, 663–673.
5. Tharakan, P. J., Kroeger, T. and Hall, C. A. S. 2001. Twenty-five years of industrial development: a study of resource use rates and macro-efficiency indicators for five Asian countries. Environ. Sci. Policy 4, 319–332.
6. Kaufmann, R. K. 2004. The mechanisms for autonomous increases in energy efficiency: a cointegration analysis of the US energy/GDP ratio. Energy J. 25:63–86.
7. Smulders, S. and de Nooij, M. 2003. The impact of energy conservation on technology and economic growth. Resource Energy Econ. 25, 59–79
8. Sadorsky, P. Oil price shocks and stock market activity. 1999 Energy Econ. 21, 449–469. See also Hall C. and Groat, A. 2010. The Corporate Examiner 37: 19–26.
9. Tissot, B. P. and Welt, D. H. 1978. Petroleum Formation and Occurrence. Springer-Verlag, New York.
10. Campbell, C. J. and Laherrère, J. H. 1998. The end of cheap oil. Sci. Am. 278, 78–83.
11. United States Geological Survey (USGS) 2003. The World Petroleum Assessment 2000 <www.usgs.gov>

12. United States Geological Survey (USGS) 2000. United States Department of Energy Long Term World Oil Supply <http://www.eia.doe.gov/pub/oil_gas/petroleum/presentations/2000/long_term_supply/index.htm>

13. Hubbert, M. K. Energy Resources (Report to the Committee on Natural Resources) (National Academy of Sciences, Washington DC, 1962).

14. Hallock, J., Tharakan, P., Hall, C., Jefferson, M. and Wu, W. 2004. Forecasting the availability and diversity of the geography of oil supplies. Energy 30:2017–201. John Hallock, personal communication.

15. Energy Information Administration, US Department of Energy. 2003. International Outlook 2003. Report No. DOE/EIA-0484(2003), Table 16 at <http://www.eia.doe.gov/oiaf/ieo/oil.html>

16. Lynch, M. C. 2002. Forecasting oil supply: theory and practice. Q. Rev. Econ. Finance 42, 373–389.

17. Bartlett, A. 2000. An Analysis of U.S. and World Oil Production Patterns Using Hubbert-Style Curves. Mathematical Geology 32: 1–17.

18. Brandt, A. R. 2007. Testing Hubbert. Energy policy 35: 3074–3088.

19. Kaufmann, R. K. and Shiers, L. D. 2008. Alternatives to conventional crude oil: When, how quickly, and market driven? Ecological Economics, 67: 405–411.

20. Nashawi, I. S., A. Malallah and M. Al-Bisharah. 2010. Forecasting world crude oil production using multicyclic hubbert model. Energy fuels 24: 1788–1800.

21. Kaufmann, R. K. and Cleveland, C. J. 2001. Oil Production in the lower 48 states: economic, geological and institutional determinants. Energy J. 22, 27–49.

22. Kaufmann, R. K. 1991. Oil production in the lower 48 states: Reconciling curve fitting and econometric models. Res. Energy 13, 111–127.

23. Hallock, J., Tharakan, P., Hall, C., Jefferson, M. and Wu, W. 2004. Forecasting the availability and diver-

sity of the geography of oil supplies. Energy 30: 2017–201.

24. Cleveland, C. J., Costanza, R., Hall, C. A. S. and Kaufmann, R. 1984. Energy and the United States economy: a biophysical perspective. Science 225, 890–897.

25. Gagnon, N., C. A.S. Hall, and L. Brinker. 2009. A Preliminary Investigation of Energy Return on Energy Investment for Global Oil and Gas Production. Energies 2(3), 490–503.

26. Cooper, C. and Pope, H.1998. Dry wells belie hope for big Caspian reserves. Wall Street J. 12 October.

27. Hakes. J. 2000. Long term world oil supply: a Presentation Made to the American Association of PetroleumGeochemists, New Orleans, Louisiana <http:www.eia.doe.gov/pub/oil_gas/petroleum/presentations/2000/long_term_supply/index.htm>

28. Tainter, J. and Patzek, T. in press. The Gulf oil debacle and our energy future. Springer

29. Tainter, J. 1988. The Collapse of Complex Systems. Cambridge Univ. Press, Cambridge.

30. Hirsch, R., Bezdec, R. & Wending, R. (2005). Peaking of world oil production: impacts, mitigation and risk management. U.S. Department of Energy. National Energy Technology Laboratory. Unpublished Report.

31. Hirsch, R. 2008. Mitigation of maximum world oil production: Shortage scenarios. Energy policy: 36: 881–889.

32. Hall, C. A. S. (ed). Special issue of journal Sustainability on EROI (in press).

33. Prieto, P. and C. A. S.Hall. In preparation. EROI of Spain's solar electricity system.

34. Murphy, D., C. A.S. Hall and R. Powers. 2010. New perspectives on energy return on (energy) invested (EROI) of corn ethanol. Environment, development and sustainability. 13:179–202.

35. Giampietro, M. and K. Mayumi. 2009. The biofuel delusion. Earthscan, London.

36. Smil, Vaclav. 2011. Global energy: The latest infatuations. American Scientist. 99:212–219.

Part II

Energy, Economics, and the Structure of Society

Economists have long sought to understand how the economy works, and the forces that lead to economic success and economic failure. What is wealth? Where did it come from? How might we make more? How does the pursuit of wealth lead to the ways that individuals and businesses are organized in a society? Economists have also asked many questions about who received the benefits and who paid the costs of economic activity. We think it important to review the previous great ideas in economics, and to examine some of earlier economists' concepts and conclusions before we present our own perspective about that and about how energy plays out in economics. Chapter 4 considers how economies are actually about energy and how our contemporary economy is a logical extension of the role of energy in society; Chapter 5 discusses some severe flaws with the economics that we do use. The ways in which political structures are influenced by energy and the converse are discussed in Chapter 6. Chapter 7 describes the relation of energy to globalization, and Chapter 8 examines how energy might lead to limits to economic growth.

Explaining Economics from an Energy Perspective

4

Introduction

This book is written by an ecologist and an economist, and part of our objective is to assess where insights and principles from these two disciplines can be combined to understand economies better. Although the two disciplines may appear very different we believe instead that the phenomena they study are very similar in many ways. From a biophysical perspective the economies of cities, regions, and nations can be viewed as ecosystems, with their own structures and functions, their own flows of materials and of energy, with more or less diversity and stability and so on: in short with all the characteristics of natural systems, with, generally, much greater energy intensity and dominance by one species. From the perspective of individual organisms there are also important similarities between natural and economic systems. Going back to Polanyi's definition of economics from Chap. 1 ("The substantive meaning of economics derives from man's dependence for his living upon nature. . . ."), we must understand that every single species is likewise dependent upon exploiting nature: lions exploit gazelles, trout exploit insects, and plants exploit nutrients in soils and space in which to intercept sunlight. Each individual finds itself in a relentless situation where it has been selected to increase its energy gains and decrease its energy costs, for its ability to pass on its genes is possible only if it has managed to acquire a large net energy balance (e.g., Hall et al. 1986, 1993). That

most humans have strong desires to exploit and accumulate resources comes as no surprise; there have been strong selective pressures in our evolutionary past to do just that. The collective activities of humans—as represented by economies—are about humans exploiting nature also should come as no surprise, for all other species must do that too. A deer walking through the woods that eats a plant that had spent 15 years getting to that stage, killing the plant is of no concern to the hungry deer, it's just the next bite. Even if that particular plant was the last of its species and the action of the deer consigns it to extinction and this robs the deer of future food that's just how it is, an unintended but very real effect.

Humans have enormously increased their ability to exploit resources through the use of fossil fuel. Although we have been trained from birth to think about the economy as something run by money, from our perspective money is just our means of keeping track, and directing, the energy flows and energy investments that do the real economic work. The fossil fuel-based economy has given each of us the equivalent of 60–80 "energy servants" and the more money you have, the more energy servants you can have. One way to think about this is that each time you spend a dollar, roughly a coffee cup's worth of oil (or some other energy) has to be pulled out of the ground, refined, transported, and burned to provide the energy for that economic activity. For example, if you buy a bagel for a dollar, natural gas is used to make fertilizer, diesel is used to drive a tractor to plant and harvest the wheat, electricity is used to grind the

C.A.S. Hall and K. Klitgaard, *Energy and the Wealth of Nations: Understanding the Biophysical Economy*, DOI 10.1007/978-1-4419-9398-4_4, © Springer Science+Business Media, LLC 2012

wheat, and more diesel is used to ship the flour from Kansas to wherever the bagel will be made, using, of course, more energy during baking. Food eaten in the United States, on average, requires about 10 times more calories of fossil fuel for its production than is found in the food itself.

Early industrialists were little different from that deer "exploiting" plants: if the river they dammed to make lumber or grind grain killed off the salmon or the use of whale oil sent the great whales to near extinction, these effects were of little interest to most people of the time. Either they did not know about them or there was little interest in these indirect costs. Herman Melville, in *Moby Dick*, understood the very large human cost of each gallon of whale oil and is reported to have said that there was a drop of human blood in every liter of whale oil. But other than some concern for the killing of such a magnificent beast he seems to have had little to say about the depletion of whale stocks, although it is likely that he knew of the depletion of whales near Massachusetts and the North Atlantic. Even within his lifetime the world had experienced a "peak oil" with a major energy resource of that time: whale oil in 1850. Today the situation is very different from Melville's times, for with the tremendous growth of environmental and resource sciences we know a great deal now about the externalities associated with our basic economic actions.

The production and accumulation of wealth has been a central issue of economics since its earliest days, but the concept that energy is a critical factor in that production simply does not exist within that discipline. A very few economists, such as William Stanley Jevons and Karl Marx, did address energy explicitly but their contributions were infrequent and are not given much importance in their writings. Most economists have either ignored questions of energy completely or dealt with them peripherally, treating energy as a "mere" commodity, no different from other commodities such as pickles or steel and with indefinite substitutes. They focused their analyses on the production, substitution, and consumption of commodities and their exchange in markets, on economic growth, and the identification and transcendence of any possible limits to

economic growth. We know now, however, that energy is central to all of these issues and is likely to have serious effects upon, and even limit, the economist's usual goal of economic growth.

The concept of limits to growth at the micro or even macro level can be found in a variety of places habituated by economists over the past two centuries, from the political climate and laws to protect resources to the intricacies of investments. The perception or analyses of these limits in economic theory, however, are only about limits internal to the economy itself. They include factors such as imbalances of supply and demand, lack of money, and the effects of government action on the economy. The concept of external or biophysical limits simply did not exist in economic theory. Most economic theory was constructed during a time when we could transform abundant nature to meet our needs with the aid of cheap and increasingly available hydrocarbons. But as energy-dense fossil fuels peak and begin their decline, and the atmospheric consequences of using them, leads to potentially serious consequences, a new generation of economic theories must reflect these newly recognized limits. To begin this quest, this chapter assesses earlier economic theories, from an energy perspective where that is possible. We also wish to make the case that although economics has not dealt with energy very explicitly, the discipline has addressed many other important issues that help us today to understand just how energy operates within economies. Moreover it has provided some effective analytical tools with which to assess the new energy realities.

Early economic writers such as the Physiocrats understood that they lived in a time when external biophysical limits were ever present even while the Earth still held abundant resources. A problem for them was that the resources were to extract or otherwise exploit. Until the discovery and adoption of fossil hydrocarbons humans lived principally from solar flows, often from photosynthesis of nutrient-starved plants on long-depleted soils. Not surprisingly most economics then was directed toward understanding and explaining the primacy of land in overall production. But once humans discovered coal, and later oil, we were able to increasingly exploit terrestrial hydrocarbons and transcend the limits of limited solar flow. The energy density found in these new resources led to the rapid transformation of the human condition. Population, which had reached one billion only in the early 1800s, soared and reached seven billion by 2010, and meanwhile a very large economic infrastructure of factories, refineries, bridges, automobiles, suburban homes, and shopping malls was constructed. Humans could now exert a greater degree of control over nature than at any time in the past. Through the use of cheap energy we could transform nature to meet not only our basic needs but also our wildest dreams of material well-being and even avarice. Many people became convinced that mass consumption was the basic road to happiness and that the desire for more, rooted deeply in human nature, could lead to some kind of social virtue.

The Economic Irony of Industrialization

As the use of cheap hydrocarbon energy spread in the Western world, economists were in effect freed from concern over the limitations of solar flow (i.e., external limits), so they turned their efforts to transcending internal limits. How did the economy operate? Could it be left to its own devices, or was the action of a government needed to maintain balance? What is the basis of value and what conditions are needed to accumulate value to grow and meet human needs? Such questions could rise to prominence because the basic dilemma of acquiring wealth had been "solved" in the new age of hydrocarbons. Effectively taking this new source of wealth for granted, economists of the industrial era could, ironically, ignore energy and instead turn to social explanations of economic problems. Eventually their theoretical foci were turned towards the role of capital in production, without an understanding that it was the use of concentrated energy by capital equipment that allowed vastly augmented production. Other crucial economic questions addressed the prospects of finding markets for the additional output as well as

measuring technical progress and determining how to distribute the rewards of the hydrocarbon-powered economy.

Our question is whether both our life of relative affluence and the ignoring of the critical role of energy in economics can continue. Our beloved affluence is predicated upon the continued availability of not just energy, but an increasing amount of low-cost energy. Today we face a new situation: essentially all of the highest quality traditional fuel stocks appear to have been discovered and many of the best are totally or mostly depleted. How close are we to the limits imposed by energy and how quickly are we approaching these limits? As previously noted according to energy analysts Colin Campbell and Jean Laherrère [1] we have come to the end of the age of cheap oil and are entering "the second half of the age of oil," one where supplies are static or declining, rather than increasing year by year as we have come to expect. Resources of our other most important fuels, conventional natural gas and even coal, show increasing signs of such limits now or within decades. If the limits of energy are upon us and binding, then we will soon see a fundamental change in human history. We will be entering an era when we can no longer tame nature and transform it easily to meet our needs and desires, but that instead nature increasingly will dictate the terms to us.

A second aspect of this new question is that we are increasingly coming to understand that the global environment can no longer assimilate the wastes that the mass production–mass consumption economy produces. In a recent study 97% of the members of the U.S. National Academy of Sciences expressed the belief that climate change was enhanced by the human activity of burning fossil fuels. The Chinese Academy of Sciences recently released a study estimating that the glaciers feeding the major rivers of China are melting at a rate of 7% per year. At this rate the glaciers will be half their present size in a mere 10 years and 100s of millions, if not billions, of people may be deprived of water during the summer growing season. Unless some unforeseen technological advance opens the possibility of an abundant noncarbon energy source to replace oil and coal we will have to adjust human society to the duel binding constraints of energy scarcity and climate instability.

What can economics say about these issues? We next examine the way economists have viewed the interaction of internal and external limits from an historical perspective. Keep in mind that the availability of energy was implicit in their analyses whether they knew that or not, for few in the world understood the nature of energy until the twentieth century. Sometimes the external constraints of energy are direct and explicit. Sometimes they are subtle and indirect. But they are always there!

Surplus and Scarcity

Economists through the ages usually have commenced their discussion from two fundamentally different starting points: *scarcity* and *economic surplus*. Before the age of fossil fuels economic theory was based on the premise that nature limited the flow of resources, that is, there was an *absolute scarcity* of economic goods and services. After the 1870s the physical limits became much less important because of the enormous power of fossil fuels. The concept of what were the biophysical means by which the wealth was generated simply fell off the radar screen of economists. The focus of analysis shifted instead to that of *relative scarcity*, that is, of psychological choices and individual command over money. A fundamental aspect depended upon the assumption that individual human beings were "acquisitive" and "rational" whose desire for more material goods as the source of happiness could never be satisfied, and for whom the desires and preferences of others were irrelevant. No level of output, no matter how abundant, could ever satisfy fully these unlimited wants. It is a psychological, not a physical, problem. From this perspective the clash between limited means and unlimited wants is *the* economic problem. This view of scarcity as the starting point of economics underlies the usual definition of economics that we gave in Chap. 1.

The discussion of *economic surplus* begins with the premise that society can produce more than it needs for subsistence by organizational and technological means. Stated simply, an economic surplus is the difference between society's output and the cost of producing it. The *surplus approach* relates to the substantive definition of economics we introduced in Chap. 1: how human beings transform nature to meet their needs. Nature is abundant; all we have to do is exploit it. Economists of the pre-fossil fuel age relied primarily on the economic surplus approach. But by the 1870s came the dawn of the serious fossil fuel era, the industrial revolution, and the consumer society. For economists the basic starting point for thinking about economics could be reformulated from producing an economic surplus to that of relative scarcity, in essence how to distribute the new largesse of society without thinking much about how it came into being. The goal of economics became one of figuring out the optimal allocation of resources to best meet human psychological desires. In other words, economic theory was transformed from focusing on obtaining more from nature into an exercise to figure out who gets the goods and services, and how goods and services best enhance human welfare. According to these then new neoclassical economists the answers were to be found in the magic of self-regulating markets where individual pursuit of self-interest led to social harmony. Although this concept was derived from the earlier writings of Adam Smith, it was augmented by mathematical "proofs" appropriated, or better misappropriated, from energy physics. Meanwhile new research in behavioral economics shows that there is little empirical evidence to indicate that human beings actually behave in this "self-regarding" way, that was not obvious in 1900.

Economic Surplus as Energy Surplus

Economists of the seventeenth through the nineteenth centuries did not, in fact could not, focus explicitly upon energy as a source of surpluses because the concept of energy did not yet exist.

Nevertheless, the ability to extract an energy surplus from solar flow or terrestrial stocks in fact forms the basis of economic production and surplus. Contemporary energy analyst Richard Heinberg provides a framework within which to assess the economic roles of such energy surpluses [2]. He argues that throughout history humans have engaged in five strategies to expropriate energy: takeover, tool use, specialization, scope enlargement, and drawdown. Takeover was the primary method of early humans, as we appropriated more of the solar energy flow for ourselves by diverting a portion of the earth's biomass from supporting other creatures to supporting humankind. Our ancestors took over land to grow crops, first as horticulture and later as agriculture, the growing of field crops at the expense of other species. Agriculture, in essence, turned a complex ecosystem into a simple one. Plants that grew where they were not useful to humans were weeds. Animals that competed for the food were pests. As humans migrated from Africa to the far corners of the world they took over more and more biocapacity, often disrupting the natural balance. Everywhere humans have gone large mammals have disappeared. The process of acquiring energy surpluses was aided by the rapid release of chemical energy known as fire. In addition humans enhanced their abilities to exploit the solar flow concentrated by the land and photosynthetic plants by domesticating certain animals that could provide more motive power than their required feed.

Heinberg's second strategy was that of tool use. Humans have long used tools, for tools can augment the takeover of energy from other species and other societies to expropriate ever-increasing amounts of energy from the biophysical system. Specialized tools called weapons aided our ability to concentrate energy in spear points and hunt more effectively, as well as expropriate energy from other societies. Tools have evolved from those that required only human energy for their manufacture and use, such as spear points, to those that use large amounts of energy and exotic materials from external sources for their manufacture and use, such as the internal combustion engine. As the energy surplus rose to a sufficient level so that not all members of society had to work constantly simply to provide adequate food, humans could begin to specialize in activities such as tool making or soldiering. All hierarchical societies that support people who are not immediate producers of crops depend upon this. Increased agricultural productivity could now support classes of artisans, aristocrats, and intellectuals who could make better tools and social organization designed to capture even greater amounts of energy. All classical political economy, from the French Physiocrats to Adam Smith, acknowledged the role that specialization played in the origin of wealth. Howard Odum talks of all kinds of natural and human-dominated systems "self-organizing" to generate "maximum power". From this perspective humans are not doing anything that other organisms don't do; they are just "good" at it because of their technologies which are now supplemented with the "large muscles" of fossil fuels.

Another strategy of energy appropriation was that of scope enlargement, or the transcendence of limits. Justus von Liebig found that the limiting factor in the carrying capacity of any biophysical system, especially agriculture, was the factor or input least available relative to the needs of the growing plants or other ecological units. This limit could be pushed back by appropriating the biocapacity of other regions through conquest or trade. Mercantile doctrine rested de facto upon the foundation of acquiring the solar energy surpluses of other regions, and the practical aims of traders was later codified by David Ricardo into the doctrine of comparative advantage. In essence, the benefits of trade result from enlarging the scope of the energy exploited by commercial society. Industrial society depended upon the ability of urban industrial centers to appropriate the biomass of rural areas in terms of food and wood for fuel. Unfortunately, many of the nutrients that would have been returned to the soil in the countryside built up as waste in the city. Von Liebig himself referred to this system of commercial agriculture as "robbery" [3]. Scope enlargement also entailed stealing solar surpluses from others through war, exploitation, and colonization largely to provide for the increasing human population.

The last and most successful strategy for increasing carrying capacity that Heinberg describes is that of drawdown, which is, perhaps, most appropriate to the second half of the age of oil. Drawdown began occurring when we were able to change from living on steady solar flows to tap nonrenewable stocks of fossil fuels, particularly those of coal, oil, and natural gas. Drawdown was enabled by the development of sophisticated tools and greatly enhanced previous strategies. With drawdown the ability to exploit nature increased sufficiently to support a much higher population at a greater standard of living. At the beginning of the age of fossil fuels, around 1800, the world's population stood at approximately one billion. Within 200 years the world supports nearly 7 times that number (Fig. 2.13). Half of that increase came in the past 50 years following the "Green Revolution," when plant breeders combined hybrid grains with energy-intensive input packages of fertilizers, other agrochemicals, irrigation, and cultivation. Moreover, the benefits of increased yields were extended to a broader segment of the world's population.

Heinberg also points out three dangers of the drawdown strategy. First, drawdown of fossil fuels creates pollution. This can take the form of pollutants such as sulfur dioxide and nitrogen oxides that foul the air and acidify the soils and water supplies. Runoff from lands treated with nitrogen and phosphorus fertilizers creates "dead zones" in areas such as rivers, lakes, and the mouth of the Mississippi River in the Gulf of Mexico. Second, the pollution can take the form of carbon dioxide emissions, whose increasing atmospheric concentration is seen by the broad consensus of scientists as the primary driving force of climate change. Finally, terrestrial stocks of fossil fuels are finite. At the beginning of the twenty-first century we are at or near the global peak of these fuels, especially oil. As they become less available and more expensive societies dependent upon them will undergo dramatic transformation with potentially grave economic as well as social consequences [3].

As we approach the limits to the drawdown of nonrenewable stocks of nature the idea that we can satisfy human needs and economic priorities by producing and consuming ever-greater quantities of material goods should become a matter of inquiry rather than of blind faith. In the post-peak years we will need to confront the distinct possibilities of absolute scarcity and the diminished capacity to appropriate surplus energy. We need to revisit and re-examine the questions economists have asked for centuries.

What Are The Main Economic Questions?

People have been writing about economic phenomena since ancient times. Economic rules and guidelines can be found in Aristotle and the holy books of most major religions. St. Thomas Aquinas melded the New Testament with Aristotle to arrive at the Canon Law which specified, among many other things what the "just price" of any good should be, who would inherit land, and (initially at least) a ban on lending money at interest. However, economic theory would have to wait until the emergence of a society in which markets were the central agencies for getting goods and money from one set of people to another. Consequently economic thinking began to distance itself from other intellectual areas such as theology and moral philosophy beginning in the 1700s. Throughout all these historical times economists have addressed basically the same set of questions:

1. What are the origins of wealth and value?
2. How should wealth and income be distributed among economic agents?
3. What are the mechanisms that balance (or sometimes do not balance) supply and demand?
4. What are the determinants of capital accumulation and economic growth?
5. What is the proper role of the government in the economy?

As we explore the origins and development of what we call economics today, we return to these questions again and again. We also introduce the concept that we see each of these questions as being in large part about energy. Even though the questions asked by economists tended to remain the same over time the theoretical emphases, methods of inquiry, and analytical vision were so fundamentally different from one time to another that economic theory must be divided into four distinct periods and "schools of thought."

Economic Schools of Thought

It is important to think about these questions from the background of the various dominant economic "schools" of thought as they evolved over the history of economics. The first identifiable school of economic thought was known as *mercantilism*, which was grounded in the economics of long distance trade. Mercantile doctrine took the form of pamphlets written primarily by practical businessmen for the purpose of justifying the expansion of trade. Although their aims and purposes were practical, mercantilist writers did make advances in questions such as the origins of wealth and value and the accumulation of capital. In various ways mercantilism was primarily about takeover and scope enlargement.

By the end of the eighteenth century mercantilism would give way to *classical political economy*. This era began around 1759 when a French school of natural philosophers called the Physiocrats developed a theory of value that tied the origins of wealth to the photosynthetic capabilities of the land and the agricultural labor that appropriated it. Perhaps the most important event ever in economics was the development in 1776 of a general theory of economics which occurred when Scottish moral philosopher Adam Smith published *An Inquiry into the Nature and Causes of the Wealth of Nations*. Smith's book led to the great debates over distribution, population, and, in time, the concept of diminishing returns of Thomas Malthus and David Ricardo, the utilitarianism of John Stuart Mill, and call for revolution in production and distribution by Karl Marx. This 100 years generated an enormously rich and thoughtful discussion about what the proper focus and moral obligation of economics was and should be.

This period of classical economics lasted through the early 1870s. Then the discipline underwent a profound transformation in questions of value, production, and distribution. This shift in emphasis and analysis soon led to the emergence of *neoclassical* economics, based on the concept, or perhaps faith, that mechanical details of the market economies are based on the "invisible hand" of Adam Smith, and furthermore that markets are self-regulating by means of competition and flexible prices, and that these could be well represented by analytical models borrowed from physics. The originators of this idea came from the French-Swiss Léon Walras, Englishman Stanley Jevons, and Austrian Karl Menger, and focused much less on production and much more on "marginal value," that is, that the additional value of something became less the more of it you had. Neoclassical thought derived from this "marginal revolution" was fully synthesized by the early years of the twentieth century and remained the primary mode of thought until the Great Depression of the 1930s. Then systemwide economic collapse rendered the prevailing orthodoxy incapable of understanding the depth of economic decline, or formulating policies to improve it. In this climate of dislocation the theory advanced by the British economist John Maynard Keynes provided an alternative that soon dominated the profession.

The beginnings of Keynesian economics date to 1936 with the publication of *General Theory of Employment, Interest, and Money*. In this work Keynes was mostly interested in how uncertainty led to declines in capital investment and an imbalance with aggregate savings. He concluded that periodically the overall level of economic activity would fall as a consequence of investment declines, leading to an overall decline in the level of (aggregate, or total national) demand for goods and services. The economy could come to rest at an equilibrium point that was characterized by high levels of unemployment unless the economy were stimulated by an outside force. Keynes attributed the Depression to a market economy's inability to sustain sufficient demand for goods and services over the long period, as well as the misguided policies of neoclassical economics that reduced demand as they reduced business cost. Keynes believed in a mild redistribution of income from rich to poor, primarily by means of job creation, and government stimulation of demand during recessions. Keynes was somewhat of an advocate of economic planning and restricted trade.

A more "business-friendly" although perhaps somewhat sanitized Keynesian economics was synthesized, primarily in the United States, in the 1950s. Most students of economics learn that Keynes was mostly about the government's use of its power to tax and spend (known as *fiscal policy*) and its control over the price and quantity of money (*monetary policy*) to keep the economy on an even keel. For decades it appeared to many that Keynesian economics was the longed-for antidote to periodic business downturns until it itself fell victim to the prolonged economic stagnation following the peak of U.S. oil production in the 1970s and the subsequent "energy crises." This was because it was unable to "deliver the goods" of economic growth with stability. Neoclassical economics made a strong comeback from the 1980s until the global financial collapse of 2008 and the subsequent recession. Recently Keynesian economics has seen somewhat of a revitalization, but also a great deal of resistance to Keynesian measures that exists in the circles of economic policy as well as in economic theory. As of 2011 there is no clear agreement of what kind of economics works and what kind does not.

Economists then and now rarely understood or even thought much about energy. Properly understood, however, the focus of each of the main schools of economic thought were in fact on the main energy drivers of the economies of their times: for mercantilists the use of wind and animal energy (including slaves) to expropriate the diffuse solar production of other regions, for Physiocrats land as a collector of solar energy, for classical economists labor as a means of production in factories, and for neoclassical economists capital (which is the means of applying the increasingly important fossil fuels to economic production). All the theories that dominate economic thought today were developed on the upslope of the Hubbert curve, during a time characterized by the enormously increasing availability, and declining cost of obtaining, energy (Fig. 1.13). To some degree they all "worked,"

for each year more energy was made available to do more economic work no matter what the policy in vogue.

In general the entire discipline of economics has paid astonishingly little attention to energy even though energy was the basis of economic activity and growth. Rather economics has treated energy as it treats any other material resource: as a commodity, useful but ultimately substitutable by other commodities. The only entities that were deemed worthy of particular attention were labor, capital, and, occasionally, land. Perhaps the principal reasons that few economists of earlier eras dealt explicitly with the substantive issues of energy was because energy as an entity was not understood by anyone for much of that time. In that environment, economics became a social science even while, in fact, it was very much about biophysical things. Energy issues lay not far beneath the surface of economic concepts such as surplus and scarcity. Before the era of classical political economy English manufacturers had learned to smelt iron by the use of metallurgical coal and coke, and by the 1800s a fossil-fuel-based industrial revolution was in full swing. The writers of the later classical period, such as Marx, acknowledged the role played by fossil fuels and mass-production in industry and agriculture as driving forces behind increasing the output of an economy. Marx also recognized the contradictions and the needs of their own decline through depletion. But there is no excuse for economists today ignoring the role of energy in economies.

We will need a new set of economic theories for the second half of the age of oil: theories that do not treat Nature's bounty as a free gift. Moreover we must contend with the problem of limits to growth. In the past humans were able to transcend the boundaries and limits imposed by nature largely by the application of increasing quantities of cheap fossil fuels. But the era we are entering will almost certainly be the end of this. As high-quality fossil fuels increasingly run short, and the use of all carbonaceous fuels compromises our atmosphere and other natural systems, the specter of living within our means while protecting our homes becomes more and more difficult. This is likely to mean the end of the growth economy and will cause us to reconsider the meaning of technological change. As we do this, and as we begin to develop new economic theories appropriate to a new age we need to consider that many important questions and insights exist in the writings of the economists of the past. Thus we examine next the most important ideas of earlier economists.

Purposes and Visions of the Main Schools of Economic Thought

The different schools of thought, which we summarized above, often asked similar questions but had very different visions of how the economy worked. They directed their writings towards different purposes, and used very different analytical methods. We now ask how each approached the main questions of economics.

The Mercantilists

Mercantilist writers were most often practical business people, not academics. All defined the purpose of the economic endeavor as the accumulation of treasure in the coffers of the nation state. However, the transformation of nations such as England from a less-developed country that exported raw materials and imported finished goods into nations that imported raw materials and exported manufactured products brought forth a change in mercantilist theory and policy. The primary theoretical difference was an argument about whether restricted or expanded trade would best serve the accumulation of treasure. (Sound familiar?) Early mercantilists, sometimes known as bullionists, took the position that trade was a pump for wringing gold from a domestic economy. This argument made some sense when a nation exported raw materials, based on the appropriation of solar flow and for which there were many substitutes, and imported finished goods – based on the harnessing of human energy supplemented by the power of wind and water – for which there were few. The terms of trade, or ratio of export prices to import prices, were

against the raw material exporter, and they suffered from declining terms of trade. In this case the accumulation of wealth is served well by the restriction of trade.

By the end of the sixteenth century, however, England had become a manufacturing nation and was exporting its products to Europe and to the world. Mercantile thought then turned to crafting an argument that justified the expansion of trade as the primary mechanism to augment a nation's stock of precious metals. The most widely recognized tract of high mercantilism was *England's Treasure by Forraign Trade,* written in 1630 by Thomas Mun and published, after his death, in 1664. Mun's primary purpose was to persuade legislators to abolish the ban on exporting gold. He argued that the export of gold could facilitate the accumulation of treasure if that export led to a positive balance of trade, or the excess of exports over imports. To accomplish this goal Mun and his followers advocated state policies of the regulation of trade. The mercantilists stood for the expansion of trade, however, they were not advocates of free trade.

At that time the ability to extract an energy surplus was limited by the lack of concentrated energy sources. The ability to extract solar flow and turn it into products with economic value could be enhanced only by organizational change, primarily in the form of plantation agriculture and slave labor. Mercantile doctrine contained no insights as to how to reduce the costs of production. They simply had no energy or other basis to do so. Rather they focused on the gains of the trade itself and the accumulation of treasure. Mercantile doctrine was a matter of scope enlargement by means of expanded trade. As much money was to be made in transportation as was to be made in the initial appropriation of the embodied energy in crops and precious metals. But expanded and speedy transportation was limited by energy availability. Ships were constructed from wood (biomass) and powered by the solar flow of the winds, which may or may not have blown in the desired direction at the desired speed. Trade, in the mercantile era, was a dangerous and slow endeavor, albeit often a profitable one.

Mercantilists, not surprisingly, took the position that the origin of value, or price, lay in the process of exchange, and they meant to control the terms of that exchange. Their primary mechanisms were colonization, commercial treaty, and war. For most of the sixteenth century the British battled the Spaniards for control of New World colonies. The seventeenth century was spent engaged in rivalries with the Dutch for control of colonies in the East Indies as well as the Caribbean, and the eighteenth and early nineteenth centuries saw the prolonged conflict between the British and the French. Mercantilists demanded the aid of their governments in determining the terms of trade. The British Parliament passed a series of restrictions (the Navigation and Trade Acts) to assure the positive balance of trade at the expense of their mercantile rivals and the colonies themselves. In Cromwell's time (1651) Britain also instigated the Corn Laws, which regulated and restricted the importation of food from the continent. By the time that Adam Smith penned the original *Wealth of Nations* British supremacy was in sight. With the triumphant end of the last mercantile war against Napoleon the world settled down to a long peace, but a peace on British terms: *Pax Britannica.*

Classical Political Economy

Classical political economists had an entirely different set of purposes. Both the Physiocrats and the first important classical political economist, Adam Smith, desired to overturn the mercantilist doctrines of regulated trade. The Physiocrats, who gave us the term *laissez-faire* ("leave us alone"), sought a change from small-scale peasant crop production to large-scale commercial agriculture. One can reasonably assert that Smith's 1776 *Wealth of Nations* was the greatest antimercantilist tract ever written. Not only did he believe that state regulation inhibited commerce, but also that mercantilist doctrine retarded domestic production. Smith pursued and developed the idea that markets could lead to the expansion of well-being, guided as if by an unseen hand, rather than by the heavy and visible hand of state regulation. Half a century later, David Ricardo would refine the doctrine of mutual benefit from unregulated trade.

The classical political economists, taken as a school, desired to build an economic science and to uncover the origins of wealth. They did this largely through a substantive, and historically specific, study of economic surplus. Their method was essentially a narrative, supplemented by abstract propositions and the occasional recourse to numerical tables. All classical political economists were policy oriented. Adam Smith advocated not only the end of mercantile restrictions, but increased expenditures for public education and a high wage economy, Thomas Malthus and David Ricardo debated the perpetuation or abolition of the Corn Laws limiting the import of food from continental Europe, J. S. Mill argued in favor of reforms to diminish the gap between those living in wealth and poverty as well as for the emancipation of women, and Karl Marx used his analysis of the contradictions of capitalism to argue for the replacement of the current economic order by one that placed the control of economic surplus in the collective hands of workers.

These political economists grounded their analyses of the origins of wealth and value in the process of production, rather than in the process of buying and selling, or exchange, as did the mercantilists. Moreover, all used social class as their unit of analysis. The familiar "factors of production" of land, labor, and capital had their origins in the actual, and historically specific, social structure of their days. The primary questions of interest for the classical economists were those regarding the production, accumulation, and distribution of economic surpluses. Their theories of capital were historically specific and related to those of accumulation and value. "Capital accumulation is regarded as a necessity prior to production and production as necessity prior to the exchange of commodities" [4]. Price formation, which has come to dominate modern microeconomics, was of minor concern to them.

Neoclassical Economics

These visions and methods are in sharp contrast with neoclassical economics, which was enunciated in the 1870s and continually refined until the present day. The neoclassicals were interested in the development of universally applicable theory, modeled after physics and independent of its historical context. Nobel Prize-winning economist Robert Solow stated this clearly, if somewhat tongue-in-cheek.

> My impression is that the best and brightest in the profession proceed as if economics is the physics of society. There is a single universal model of the world. It needs only to be applied. You could drop a modern economist from a time machine—a helicopter maybe, like the one that drops money—at any time in any place, along with his or her personal computer; he or she could set up business without even bothering to ask what time and what place [5].

British economist G. L. S. Shackle stated that the principle around which neoclassical economics was organized, the principle that served as the equivalent of gravity in celestial mechanics, was self-interest [6]. But neoclassical economists focused not on the pursuit of self-interest, as did Smith and the classical school, but upon the maximization of personal self-interest through the mechanism of people buying what they want to in markets. Their approach was mathematical and abstract, and based upon relative scarcity as a universal principle. In short neoclassical economics was the marriage of differential calculus with utilitarian philosophy. The classical focus on social class as the unit of analysis was replaced with that of the individual, and the role played by accumulation gave way to a stress upon static equilibrium, and allocative efficiency. A neoclassical analysis of growth was not to appear until the 1950s when it was enunciated by the aforementioned Robert Solow.

Perhaps the greatest break with classical political economy came in the area of *value theory*. Classical political economists all commenced their analyses from the viewpoint that value and wealth were created in the process of production and that value could be calculated objectively from the costs of production. Neoclassical economics was, and continues to be, grounded in the proposition that value, like beauty, is in the eye of its beholder, and a matter of subjective well-being or utility. Their overall objective was not to pursue the origins of wealth as much as to show, under ideal theoretical conditions, that market economies are self-regulating

by means of small, or marginal, fluctuations in prices driven by competition on the individual level. The result of voluntary trades, based solely on the maximization of self-interest, leads us to a situation of Pareto efficiency (named after its originator, Vilfredo Pareto) where no one individual can be made better off without making another worse off. Government intervention could do no good, and much harm, as it would distort the signals of the market, which is seen as a perfect carrier of information [7].

Keynesian Economics

Economies in general, and capitalist economies in particular, suffer from strong cycles of expansion and recession. These cycles, and in particular the recessions, tend to bring enormous hardships to people as workplaces close and fewer people are employed. John Maynard Keynes had, unlike his neoclassical predecessors, developed a theory that these cycles were caused by internal conflicts. The market as a system was not self-regulating. In his 1936 work, *The General Theory of Employment, Interest, and Money,* Keynes showed that a mature market economic system could reach equilibrium at considerably less than a full employment level. Consequently the market could not be left to its own devices to restore balance, especially if were already "balanced" generated by high levels of unemployment. Keynes considered himself a "moderate conservative" and was primarily interested in saving the market economy from its own worst feature of periodic depressions accompanied by high rates of unemployment. Instead of believing that market forces of competition and flexible prices would correct the ills of depression Keynes thought that the imbalance of savings and investment led to a deficiency of aggregate demand, that is, for goods and services. Rather than wishing to replace capitalism with another form of organization and governance, Keynes believed that judicious use of government policy could boost the overall level of demand and reduce it during recessionary times. In the 1950s a new generation of economists calling themselves Keynesians would attempt to "fine tune" the economy by spending more when the economy was contracting and less when it was expanding so rapidly as to make prices rise. These actions would, they thought, tend to smooth out economic fluctuations over time. One can argue that in fact it worked, as the proportional fluctuations in the U.S. economy decreased to much less than before the general acceptance of Keynes's ideas. We explore this period in our chapter on the postwar economic order.

The Role of Biophysical Economics

All of the economic schools mentioned so far were growth oriented to greater or lesser degrees. The main disagreement then, as now, was how would growth be best achieved? Classical political and neoclassical economists tended to focus upon market processes in achieving accumulation and growth. Karl Marx explored the internal contradictions that inhibited the accumulation process. Keynesian economics relied on the role of the government to provide growth stimulus when private economy could not. In the absence of growth employment would stagnate and human well-being would decline. In the early classical era growth could be achieved principally by organizational means; the capacity to increase material output by means of technological change barely existed. It was only in the later stages of classical political economy, neoclassical and Keynesian economics, that the ability to increase output dramatically by means of harnessing energy-dense fossil fuels was possible.

What, then, should be the purpose of biophysical economics, the approach we are advocating in this book? Clearly it must deal with a world that is increasingly dependent upon stocks of fossil fuels, the depletion of those stocks, and the increasing difficulty of achieving growth as depletion occurs. Unlike the utilitarians, biophysical economics considers and encourages the possibility that humans are capable of achieving happiness by means other than the acquisition of ever-increasing quantities of material goods, goods that cannot be produced with declining resources. As such it calls back to the center stage

the question of distribution: for generations that question has been suppressed for if the pie has been getting larger then everyone can get a larger piece. But if the size of the pie is not growing, who should get how large a piece?

Biophysical economics serves as a wakeup call to the impending and inevitable end of the economy based on high-quality fossil fuels, and with it the end of growth economics. It also provides important caveats as to which of the many alternatives proffered has a good chance of succeeding by providing guidelines for the assessment of alternative sources of energy. How we can live well within nature's limits is a question we can no longer afford to postpone or subsume to a series of equations unconstrained by reality. But to answer this whole new set of questions we must first assess how economists have addressed the age-old ones, for these questions remain as relevant for these new conditions as they were for the circumstances when they were asked. In other words for a relatively few decades—a century and a half at most—in the most favorable situations has a year-by-year increase in general affluence been the normal condition. It was not true back when early economists were writing and it appears no longer true. So we must pay attention once again to their questions, but we need to do that while including an energy perspective.

The Main Questions of Economics: #1. What Are the Origins of Wealth and Value?

We begin our discussion of the main questions of economics by distinguishing between income and wealth; throughout the ages the distinction has not always been clear. Wealth has long been seen as an abundance of goods that are available to a society or to an individual. In preindustrial societies wealth was the stocks of what nature bequeathed us. But as the economy began to grow and develop, wealth began to be defined as the sum of what humans produced, in other words an accumulation of the flows of value extracted from nature. The question as to whether wealth is a stock or a flow has been debated ever since economic theory developed and the resolution has never been conclusive. The distinction is also complicated by the level of analysis. Most individuals see wealth as a stock of assets that produce a flow called income. On the level of society as a whole, wealth is measured by average per capita income [8]. Since the rise of the neoclassical era wealth has been seen as a stock called *capital*, whereas "capital" has been extended to describe all factors of production. Ecological economists regularly refer to the stocks of nature as *natural capital*. Mainstream labor economists see their discipline as the study of *human capital*. In the end questions of capital and income resolve to a discussion of wealth and value.

For the scholastics who shaped the ideas of the medieval days of feudalism, the origins of wealth lie in the land, specifically with the ownership of land. Those who owned and controlled the photosynthetic capability of the land were wealthy. Those who did not own land were not. The nobility and the church owned the land, and the elaborate principles of medieval law served to concentrate land ownership. Primogenitor demanded that all land was given to the first-born son. Daughters of landowners were expected to marry sons of other landowners. The ban on usury prohibited merchants from acquiring wealth through the charging of interest, and profits by means of trade were limited by the "just price" which covered only the costs of production, transportation, and the return necessary to keep one in "his station in life." Social mobility was seen as a mortal sin. The peasantry, known as serfs, labored primarily in the fields of their feudal lords having but 1 or 2 days per week to till the Commons for their own subsistence needs. All paid taxes to the nobles and tithes to the church.

After what historian Barbara Tuchman calls the "calamitous" fourteenth century the 1000 year stability of feudalism began to fracture, and by the beginning of the sixteenth century the merchants, much reviled by the nobility and the church, began gaining control of society. Wealth was in new hands that found new uses for it. Art and music prospered and proliferated as did commerce. The age of exploration ushered in the age of long-distance trade as well as the Renaissance. The forests and mines of the New World augmented the long-depleted stocks of the old. The writers of the new mercantile period began to redefine the meaning of wealth, from control over land and its biomass to accumulation of "treasure," or stocks of precious metals. This was the essence of mercantilist "economics". By the middle of the seventeenth century thought on how best to accumulate wealth changed from the treasure itself to the gains made by trade. Treasure, and therefore wealth, would flow to those nations that achieved a positive balance of trade. As much money could be made in control of shipping and customs as could be made mining and refining the treasure itself.

How Classical Economists Approached This Question: The Development of Political Economy

For classical economists, who called themselves "political economists," wealth, a stock, and value, a flow, originated in the process of production, rather than that of exchange, as the mercantilists believed. Furthermore, the idea that united the diverse classical political economists was that value could be determined objectively by adding up the costs of production. They believed that human labor, assisted by tools, land, and organization of the labor process, was the source of value. The first classical political economists, the Physiocrats, asserted that value originated in the land and the agricultural labor that appropriated the earth's biomass by planting, harvesting, and transporting food. Only nature created a net product (or *produît net*). Manufacturers were considered sterile in that they only transformed the value created by the land. From their perspective they added no net product.

In the English-speaking world, in contrast, economic theory extended the creation of value to manufacturing as well as agriculture. The generally acknowledged founder of British political economy was a Scot, Adam Smith. Smith is most often recognized for his belief that the "invisible hand" of the market would transform individual self-interest into social harmony. He actually began his 1776 opus, *The Wealth of Nations*, by raising the question of value. Smith diverged from both the mercantilists and the Physiocrats. He asserted that the origin of value could be found not in the bounty of nature and agricultural labor, but labor in general, specifically in the productivity of labor and the number of productive laborers. Wealth could be increased only by increasing labor productivity. Wealth was the accumulation of values generated by producing goods and services for sale on the market. He was writing in the era before fossil fuels were applied widely to manufacturing, and his theory reflected his time.

Smith's observations, the most famous being that of a pin factory, led him to believe that the primary method of augmenting the wealth of a nation was to implement the division of labor, where the production process would be subdivided into separate and more productive tasks. Smith, who was a professor of moral philosophy, then had to connect the division of labor to an overall "system of perfect liberty" found in the unencumbered operation of free markets. He did so with a surprisingly simple statement: "The division of labor is limited by the extent of the market" [9]. In order to reap the benefits of the division of labor a manufactory must have access to a sufficiently wide market to sell the products the division of labor made possible. An important constraint on that perspective, however, barely understood by Smith, was that the market itself was limited by the reliance on solar flow and animal power to transport products of the division of labor.

Smith also deals with the origins of the division of labor. Partly he attributes it to human nature. We all have an ingrained propensity to "truck, barter, and exchange," in addition to possessing a desire to increase the number of necessaries, conveniences, and amusements available to us. Always the historian, Smith addresses the question of how much any particular commodity (known today as a good or a service) was worth in earlier times as well as in his own day. He argues that in the "rude and early stage" of soci-

ety, before the development of tools and private property, the value of any commodity consisted of the amount of human labor embodied in production (meaning the hours of labor that had been used to make something). Workers could generally fashion their own tools. A distinct tool-manufacturing sector would have to wait for the application of more concentrated energy. "Labor was the first price, the original purchase money that was paid for everything. It was not by gold or silver, but by labor that the wealth of the world was originally purchased... If among a nation of hunters, for example, it usually costs twice as much labor to kill a beaver which it does to kill a deer, one beaver should naturally exchange for or be worth two deer" [10]. In this stage of development the whole product of labor belonged to the producer. But in eighteenth century society, characterized by the division of labor, this situation would not hold. At that time "modern" society enhanced the production of each worker through various kinds of equipment, and the owners of capital stock, who provided the equipment and advanced the wages before the crops were harvested, demanded a share of the output. So, too, do the owners of the land. Smith argued that the "natural price" or value can be obtained by adding up the natural prices of land, labor, and capital. Smith was not particularly clear about this, and had to devote pages upon pages to determining the natural rates of wages, rents, and profits. Three decades later David Ricardo would criticize this approach, and replace it with a defense of a pure labor theory of value. Smith links production to exchange by arguing that natural price forms a center of gravity around which market price, determined by the short-term interaction of supply and demand, is derived. Smith then goes on to explain the original accumulation of stock by the virtuous behavior of those frugal individuals who save. "Capitals are increased by parsimony and diminished by prodigality or misconduct." When the frugal abstain from immediate consumption they add to their capital. They use this capital to set to work industrious persons, and as capital accumulates the potential productivity embodied in the division of labor rises too. In the end for Smith the source of the increase of wealth can be found primarily in the increased labor productivity of an increasing population and the virtuous behavior of frugal savers.

The next great English-speaking political economist was David Ricardo, whose 1817 *Principles of Political Economy* represents the definitive statement of classical political economy. Although Ricardo had little to say about the origins of wealth, he made significant contributions to the theory of value. Ricardo was the premier advocate of a pure labor theory of value. He believed Smith to be incorrect when he separated labor *embodied*, the amount of human labor time used in production, and labor *commanded*, or what that labor is worth in terms of purchasing alternative commodities. Ricardo reconciled the two when he declared that capital was simply "dated labor". Most capital at the time was known as circulating capital, or the money advanced to purchase labor. Inasmuch as capital can be reduced to labor, the value of any commodity, or good produced for sale rather than use, was determined solely by the amount of human labor embodied in production.

The problem of dealing theoretically with long-lived fixed capital is an old one, indeed. Ricardo believed that market processes would equalize profit rates. But if one commodity were produced in a more capital-intensive process, problems emerged. If the amount of total capital were the same for two producers, then an equal profit rate meant selling the goods for the same prices, as the market also equalized price. But if, for example, wages increased, it would have a much greater impact on the more labor-intensive commodity. Two goods with unequal amounts of labor would have different prices according to the labor theory of value. But competition in markets would yield the same price. It seemed mechanization was incompatible with the labor theory of value. Ricardo was never able to solve this problem because his theory did not reflect reality—the less efficient, more costly production would simply be less profitable—as Marx discussed. In fact he died at his desk working on it. Ricardo never dealt directly with energy. Nonetheless, he

provided two theoretical tools that critically inform energy analysis to this day: the "best first principle" and "diminishing marginal returns." We deal with these principles in the next section on income distribution.

The problem of reconciling the labor theory of value with mechanization fell to another classical economist, the German philosopher turned political economist, Karl Marx. Marx was the first prominent political economist to write in the industrial era, which fundamentally altered his view of how the economy worked. Marx was both fascinated by and admiring of the increased output made possible by the application of fossil fuels to production. "The bourgeoisie, during its rule of scarce 100 years, has created more massive and more colossal productive forces than have all preceding generations together" [11]. According to Adam Smith, ten men in his time, using the system of the division of labor, make 48,000 sewing needles every day. A single needle-making machine, however, makes 145,000 needles every hour. One woman or one girl superintends four such machines, and so produces nearly 600,000 needles in a hour, or over 3,000,000 in a week! [12]. Marx thought that this was a marvelous means of making labor more productive, and he clearly understood, but did not dwell upon, the role of energy in this process.

Marx's contribution to value theory was the *theory of surplus value*. He argued that the origin of profit lies in the ability of capitalists to increase labor productivity and appropriate the result as capital, rather than to distribute it to workers. The ability to work, which Marx called labor power, was a commodity bought and sold on a labor market. As a student of Ricardo's labor theory of value, Marx believed that the value of labor power was the cost of its reproduction, in other words a subsistence wage. The application of the division of labor and fossil-fuel driven machinery could reduce the cost of production below the social value or prevailing price. Although machinery did not add surplus value in and of itself, it so revolutionized labor productivity that the value of society's output would rise. The difference between cost and revenue, or profit, could be reinvested in even more productive technologies.

Those companies that did not invest were driven from the marketplace. Price, or value, could be determined objectively by adding up the value expended on machinery (which Marx called constant capital), the money advanced to purchase labor power (which Marx called variable capital), and a share of the social surplus value. Those who survived invested in productive capital (or means of production) that, with the application of fossil fuels, would drive down the unit cost of production and allow the most innovative capitalists to undersell the market. "The battle of competition is fought by the cheapening of commodities," and mass production required the application of energy-dense fossil fuels. Arguments using these ideas are commonly heard today by those who believe that free markets generate cheaper products.

In the first volume of his 1867 opus *Capital* Marx turned to the accumulation that occurred prior to the emergence of industrial capitalism [12]. His chapters on "the so-called primitive accumulation" chronicle the process by which former artisan producers and independent farmers (even before the evolution of industrial capitalism) were forcibly "stripped of the means of production" by those with more financial or political power and left with only their labor power to sell. Furthermore, Marx analyzes the effects of mercantile strategies where fortunes were built on colonization, slave labor, and war. Unlike Smith, who attributes the origins of wealth and capital to the virtuous behavior of the frugal saver, Marx declares "If money… comes into the world with a blood-stain on its cheek, capital comes dripping from head to toe, from every pore, with blood and dirt" [13]. Thus Marx added, or continued to add, a moral dimension to how economies worked under different systems.

Neoclassical Economics and the Theory of Value

An important problem facing economists in 1870 was what is often called the "water versus diamonds" paradox. Water was, and still remains, essential for human life. But because it was abun-

dant, and often available for the taking in rural areas, it did not command a high price. In the parlance of classical political economy water had great use value, but little exchange value. Diamonds, on the other hand, had little use value, except as ornaments, but a very high exchange value. Classical political economists would attribute this to the great amount of human labor that had to be expended in mining the stones, cutting them, and polishing them for the market. Water, on the other hand, took little labor to harvest from the ground.

The newly evolving neoclassical economists saw this as a "paradox". But from our perspective the reason was not some fundamental problem with the classical view, but was because the neoclassicists did not separate use value from exchange value. Unlike classical economists, who saw exchange value as independent from use value, the early neoclassical economists viewed use value, now called *utility*, as the source of exchange value. Thus the relative prices of water and diamonds now became a paradox to them because how could something so useful be so cheap, whereas something with little use, such as diamonds, command such a high price? Their resolution was to make exchange value subjective. Diamonds were costly because people liked them, they were not especially abundant, and people were willing to pay a lot of money for them. Water was mundane but abundant. Scarce commodities carried a higher price.

This subjective approach became the jumping-off point for neoclassical economics, whose practitioners began a "new" approach to economics in the early 1870s, and still dominate the profession today. They differed from the classical economists in not being particularly interested in the origin of wealth, other than to agree with Smith that the origins of wealth could be traced to the virtuous behavior of individuals. By and large they accepted the idea that wealth was a stock. From the beginning Swiss economist Léon Walras, one of the originators of neoclassical economics, saw the study of economics as the transformation of stocks of natural resources into human-satisfying utilities, with production relegated to a rather irrelevant intermediate position [14].

Thus neoclassical economists changed the focus of the discussion from an objective theory grounded in economic surplus and the (labor) costs of production to a subjective utility grounded in psychological scarcity which ultimately was translated into willingness to pay. To create the core of neoclassical economics this idea was married to utilitarian philosophy, based on the propositions that individuals rationally endeavor to increase their happiness, and to differential calculus. If a commodity provided utility (greater happiness) more of that commodity would provide more total utility.

The focus of early neoclassical thought was also upon marginal utility, or the extra utility received from consuming one more unit of the good. Neoclassical economists believed that it was marginal utility, also known as the final degree of utility, or rareté, that determined value or price. Marginal utility declines as more of a commodity is consumed. Thus the first liter of water that might be consumed would have a nearly infinite value, and each subsequent liter was less valuable to the subjective tastes of the consumer. Because water was abundant, it was not worth too much. Theoretical "rational consumers" were thought to continue to trade with each other until the marginal utilities of the two traders equalized. At that point neither party would benefit from additional trading. No individual consumer can be made better off by trading without making another worse off. This is the genesis of what is called *Pareto efficiency*. The reader should note the irony that although the neoclassical concept of value is based on "economic scarcity" this is only relative, not absolute scarcity. Even though industrialization made possible an abundance of goods neoclassical economists were talking about scarcity only from the perspective of an individual's infinite wants.

The theory of neoclassical economics assumes that in a money economy consumers will continue to purchase a "set" of two or more commodity "bundles" even though they have less and less additional value to them. Therefore she experiences diminishing marginal utility. The consumer will cease buying when the ratio of marginal utilities equals the ratio of prices, resulting in "consumer equilibrium." In other words, when a consumer trades off good x for good y at the same rate that the market trades them off she or he will be in the best possible position. As prices change so too, does the equilibrium position, with lower prices generally resulting in higher quantities purchased. Although the initial assumptions require that interpersonal utilities cannot be compared, they can be aggregated mathematically. The standard "rite of passage" for every student of intermediate microeconomics is to decompose these changes into income and price effects and derive a downward sloping demand curve, despite the complete unreality of the assumption.

The change that neoclassical economics brought to economics was a change in the conception of value. Classical economics believed that humans generated value objectively by transforming the products of nature into things humans wanted through the actions of labor. Neoclassical economists, on the other hand, thought that the origin of value was subjective. Value was determined by human preferences, and these preferences were revealed by what humans chose to purchase in the marketplace. This was, at least in theory, a very democratic process in that any consumer is as important as any other in that his purchases will "send a market signal" to the whole economy about what that economy should be producing.

In order for a consumer-based price theory to replace a value theory based on costs of production and social classes, let alone come to dominate economic thinking, a reasonably large cohort of consumers must exist. This consumer class was created by the application of fossil fuels to economic production. The industrialization of agriculture began to drive food prices down by the early 1830s, and the increase in productivity made possible by the application of coal to machinery drove down the price of wage goods. Moreover, mechanization was accompanied by an increase in the ranks of supervisory employees who enlarged a nascent middle class whose incomes allowed the expansion of consumption and the expansion of the market [15].

By the late years of the nineteenth century neo-classical economists expanded their early marginalist roots by extending the marginal utility approach to the analysis of production. They believed that production functions mirrored utility functions. Factor price ratios (such as the ratio of wages to profits) were substituted in their equations for the price ratios of utility theory, and ratios of marginal productivities, or the change in output with respect to the addition of one more factor, took the place of ratios of marginal utilities. Producer equilibrium occurs when the two ratios are equal. Moreover, the theoretical distinction between production and distribution found in classical political economy simply vanished. The theory of production and the theory of distribution are one and the same in neoclassical economics.

The neoclassical theory of production does not deal explicitly with energy. The typical production function is simplified to include only capital and labor as the independent variables that produce output. This is true despite the fact that one of the founders of neoclassical theory, William Stanley Jevons, had written on the critical importance of energy directly in *The Coal Question*, published 12 years before his path-breaking marginalist manifesto, *Theory of Political Economy*, in 1871. We explore Jevons' premarginalist classic in our final section on accumulation and growth.

Keynes and the Taming of Economic Cycles

John Maynard Keynes, who influenced the application of economic theory to day-to-day economics more than nearly anyone else since Adam Smith, had little to say about wealth and value, or price formation. He accepted, on face value, utility theory and marginal productivity theory, and was relatively uninterested in price formation. He did base his critique of the labor market on the proposition that wages were "sticky" and did not fall as workers attempted to protect their standards of living. This, however, was not original to Keynes, as Keynes' neoclassical mentor Arthur Cecil Pigou had worked on this topic.

The Main Questions of Economics: #2. How Are Wealth and Value Distributed?

Some schools of thought find the question of distribution of the rewards of production to be fairly uninteresting. Some find it the focal point of their analyses. In general, classical political economists found questions of production and questions of distribution to be interrelated but analytically separable. Neoclassical economists, however, found them analytically identical. The neoclassical theory of production, known as marginal productivity, is also the neoclassical theory of distribution. Marginal productivity theory stated that each "factor of production" would receive exactly its additional contribution to production. John Maynard Keynes, for the most part, accepted the marginal productivity theory of distribution, with a few, but important reservations. The theories of distribution are but peripherally related to energy, however, they are sufficiently important to economics to deserve specific treatment.

Classical Political Economy and the Unequal Distribution of Wealth

The unequal distribution of wealth was the fundamental problem that had been addressed by the Physiocrats. French agriculture yielded little surplus product, as production was on a small-scale subsistence basis with basic wooden (biomass) implements and little application of fertilizer. What little surplus existed was appropriated to support the lavish court in Versailles, and to subsidize a set of pampered workshops dedicated to the hand production of luxuries. The Physiocratic program advocated instead the reinvestment of agricultural surpluses on the farm and the creation of large-scale commercial agriculture on the English model. The first economic model ever, the *Tableau Economique*, was designed to illustrate the problem of unequal distribution of wealth. Its modest reforms, however, ran afoul of Louis XVI, and were ultimately doomed to failure. The Physiocrats ultimate success was the influence they had upon later theorists such as Adam Smith and Karl Marx.

Neither the mercantilists nor Smith treated the problem of income distribution very seriously. Mercantilists, focusing on trade and exchange as the source of wealth, had little to say about the internal order of the domestic economy. This is hardly surprising as the ability to fundamentally transform the process of production by utilizing fossil energy had yet to be developed. Their main focus was the distribution of subsidies. Mercantile doctrine held that a trader was worth several artisans and artisans are worth many farmers. Therefore subsidies should flow towards those engaged in international trade. Profits were to be made and hence encouraged in the carrying trade and in the exploitation of colonial resources, not by means of reducing the cost of production at home or elsewhere.

Smith, too, wrote relatively little about income distribution, which is surprising given that he was a professor of "moral philosophy." Smith did believe that some degree of inequality was natural and that it provided incentives for increased productivity. "Wherever there is great prosperity there is great inequality. For every rich man there must be at least 500 poor, and the affluence of the few presupposes the indigence of the many." Yet at the same time he believed: "No society can surely be flourishing and happy of which the far greater part of its members are poor and miserable" [16]. Smith truly believed that accumulation of capital would raise living standards for all in the long term, although inequality would persist. In the final book of The Wealth of Nations Smith held out that a commitment to education would also raise the status of the working poor, a position commonly held by many in society today. In his chapter on wages Smith also wrote at length on the factors contributing to the differences in wages, including the difficulty of learning the trade, constancy of employment, the degree of responsibility, and the uncertainty of success. Smith held a special distaste for the landed aristocracy who loved to reap what they had not shown. He considered rents to be primarily a monopoly extraction on the part of proprietors who did not labor productively. To this day, the term "rent seeker" is one of the most powerfully negative epithets leveled by conservative economists at those who do not obtain their incomes by labor or investment.

The next prominent English-speaking political economists writing in the period following the death of Adam Smith in 1790 were the Right Reverend Thomas Robert Malthus and stockbroker-turned-landowner David Ricardo. Surprisingly, neither was particularly interested in the origin of wealth. In his 1798 *First Essay on the Principle of Population* Malthus provided a narrative history of the transition from savagery (known today as hunting and gathering) to modern societies. Like Smith he favored the (supposedly) virtuous behavior of the parsimonious wealthy classes over that of the prodigal poor. Unlike Smith, he seldom addressed the issues of capital accumulation. Malthus directed his analysis as to why populations remained stable in early societies and not to why capital accumulated.

David Ricardo subordinated the question of wealth creation to secondary status. For him the real question was one of distribution, and distribution changed according to the specific historical period. Like Malthus he accepted the division of society into classes of landlords, capitalists, and laborers as natural and inevitable. Ricardo believed that the proportions of the whole produce of the earth which will be allotted to each of these classes, under the names of rent, profit, and wages, will be essentially different in different stages of society, depending mainly on the actual fertility of the soil, on the accumulation of capital and population, and on the skill ingenuity and instruments employed in agriculture. He said, "To determine the laws which regulate this distribution, is the principle problem in Political Economy" [17].

The Origin of the Concepts of Diminishing Marginal Return and Comparative Advantage

Ricardo and Malthus were writing during the late eighteenth and early nineteenth centuries when there was a great rivalry between landowners and emerging capitalists for control of the British economy and society. English law limited the import of cheaper grains (corn) from Continental Europe. This benefited the landed classes by extending the margin of cultivation to poorer quality lands, most of which they owned. Simultaneously the law increased rents and raised wages, because wages were determined by subsistence, and ultimately the costs of extracting an energy surplus from poor land. This limited the power and income of the rival capitalists as most of the wealth of society had to go for the necessary food and hence to landowners. David Ricardo and Thomas Malthus undertook great debates concerning the Corn Laws, which limited food imports. This debate was the genesis of two of the most sacred principles of modern economics: *diminishing marginal returns* and *mutual gains from trade*, technically known as *comparative advantage*. David Ricardo devoted his life to the pursuit of political economy and the repeal of the Corn Laws by crafting myriad arguments in support of the interests of the emerging class of capitalists. His primary aim was to change the distribution of income and wealth from the less productive landed classes to the more productive capitalists, although he himself was a landowner. Malthus argued for just the opposite, the redistribution of income and wealth towards the landowners.

Ricardo enunciated a theory of rent based on the principle of diminishing marginal returns because the price of food (or "provisions") depended upon the costs of production (primarily labor costs) at the no-rent margin (or the land of lowest fertility). The owners of more fertile lands received a rent, so that food grown on more fertile, and less costly, land would sell at the same price as food that was more costly to produce. Ricardo's theory also depended upon the *best first principle*. Farmers, being no fools, would tend to utilize the most fertile, and most accessible, land first, and poorer lands second. In other words returns diminished at the margin of cultivation, that is, the poorest land that was still put into production to meet total food needs. As we show in later chapters this principle is also useful for explaining peak oil and the falling energy return on investment over time. But in the pre-fossil-fuel age the only thing that stood in the

way of the redistribution of incomes towards productive commercial farmers and manufacturers was the cumbersome Corn Laws limiting the import of cheap grains. If these laws were repealed the cultivation of poorer quality lands could be postponed or eliminated.

Ricardo crafted his arguments in the context of benefits to the nation rather than in terms of benefits to a particular class. He reasoned that free trade among nations in finished commodities would result in more goods for a cheaper price than if each nation produced all that it needed on a self-sufficient basis. He also reasoned that capital and labor would be immobile internationally, a proposition subsequently repudiated by advocates of globalization. (We return to the details of this argument in Chap. 8). Moreover Ricardo believed that such a redistribution of income would enhance the growth of the domestic economy as vibrant profit-seeking commercial farmers would reinvest their returns in improved techniques (what we would call today technology) that would reduce the overall cost of provisions and thereby improve society in general.

Thomas Malthus held the opposite position. He believed that frugal capitalists would over save, and that savings would not automatically find their way into investment. As a result the economy would lack the demand needed to realize profits and the economy would fall into a depression. Malthus' solution was the redistribution of wealth to the landed classes who would use it to build monuments and surround themselves with unproductive retainers, ensuring adequate overall demand. We save the details of the argument for the next section on the balancing of supply and demand, but it is important for the reader to see that many of today's most important economic arguments were developed by Malthus and especially Ricardo as they contemplated the effects of what we would call today free trade.

The Apogee of Classical Political Economy: Marx and Mill

Karl Marx is today mostly associated with communism, but he wrote and thought far more completely and insightfully, about capitalism.

His theory of distribution was closely linked to his theory of surplus value. Profits were the result of selling the output for a greater price than was paid for input. Capitalists would accumulate those profits as money capital to be reinvested in the process of producing goods more cheaply, hence garnering a larger market share. Money wages could fall with cheaper goods as the cost of wage goods fell, increasing the amount of surplus value to be capitalized, although workers would continue to be paid at the value of their labor power. As workers became more productive the benefits accrued primarily to capitalists and workers generally were no better off. Although Marx did envision the possibility of wages rising as capital accumulated he also subscribed to a theory of the "relative immizeration of the proletariat" meaning the lot of working people would become relatively worse off when compared to a tiny elite of capitalists.

Marx's analysis was cast in the quality of work life as well as wages. Economists who focused only on the quantitative aspects of lower prices and higher productivity overlooked the changes in the process of labor. Marx's critique of the existing political economy was grounded in terms of both qualitative and quantitative approaches to value. He believed that qualitative relations among people undergird the quantitative relations between people and things. The accumulation of capital depended upon the extraction of surplus value from immediate producers (i.e., workers), and the profit rate depended upon increasing the rate of surplus value, or labor productivity. To accomplish this, the character of work became stripped of its meaning. The mental work was separated from the manual work, first by organizational means such as the division of labor, and later by the application of fossil fuels to machinery. These changes had many social impacts. The worker became an appendage to the machine, no longer directing its application for the improved quality of the product, but rather the worker had to follow the dictates and pacing of the machine. The intellectual unity of head and hand was severed for all but a few very workers whose skills were sufficiently unique such that they could not easily be replaced by machines.

The resulting alienation that the worker felt from the products and processes of production would drive social change. Marx believed it was likely that wages could rise with economic growth, but that the changes in production and the degradation of the labor process could not be overcome with more money. This qualitative aspect formed a crucial part of Marx's theory of income distribution and inequality and the inevitability of social revolution.

John Stuart Mill's 1848 book, *Principles of Political Economy*, dominated the discipline until the 1870s, but offered little new in terms of value theory. Indeed he envisioned his own task as little more than updating Ricardo. Mill did offer a unique perspective, however, on income distribution. Production, according to Mill, was subject to natural law (i.e., the limitations of what we would call today resources), as envisioned by Smith, Ricardo, and the other classicals. But distribution was entirely a matter of the free will of human beings, and humans could change social institutions to accommodate a more equal distribution. Mill therefore showed concern about Irish peasants, industrial workers, and the position of women, and supported a series of reforms in order to increase their share of social wealth and elevate their status. Influenced by his wife, Harriet Taylor, Mill became a tireless advocate of the emancipation of women at work and in the home. Mill wrote that the time of Adam Smith, where the pursuit of self-interest would lead to social harmony, had come to an end as evidenced by the destitution of the working classes and significant social strife. Like Marx, Mill considered the qualitative aspects of social inequality and the future of society. The good life, for Mill, entailed a simpler and more equal society. "I confess that I am not charmed with the ideal of life held out by those who think that the normal state of human beings is struggling to get on; that the trampling, crushing, elbowing, and stepping on each other's heels, which form the existing type of social life are the most desirable lot of human kind, or any but the disagreeable symptoms of one of the phases of industrial progress" [18]. Thus for both Marx and Mill although it was true that industrialization brought greater material prosperity it also brought many undesirable and unpleasant aspects to the working class that they both were interested in overcoming.

Neoclassical Economics and the Marginal Productivity Theory of Distribution

The neoclassical vision of distribution could not have been more different from that of Mill. Rather than separating the mechanisms undergirding production and distribution, as Mill had done, the neoclassical theories of production and distribution are virtually identical. For 20 years, following the 1870s marginal revolution, neoclassical economics, based on scarcity and utility, was solely a theory of demand. Production was still based on classical principles of cost. But classical theory utilized an economic surplus approach, which entailed the possibility of exploitation, value created by one class is appropriated by another. Marginalism would become neoclassical economics only when production was placed on a marginal utility basis, and the possibility of exploitation was eliminated (at least in theory). The fundamental idea is that each factor of production [land (T), labor (L) and capital (K)] earns its marginal product (or incremental contribution to total output), no more and no less as rational individuals follow the price signals of the market. The result is equitable: one's reward depends solely upon one's contribution to society. The marginal product of labor therefore equals the wage rate (w), profits (π) are equated with the marginal product of capital, and rents (r) are determined by the marginal product of land. This can be added up to generate the total output (P), with MP_L being the marginal contribution of labor and so on.

Unfortunately this equation can be true only under a limited set of mathematical conditions. British economist John Hobson showed that if the marginal product of labor exceeded the average product (or the output elasticity is positive) the product of $MP_L \cdot L$ can exceed the total output to be distributed. But this is possible only if one

$$P = MP_L \cdot L + MP_K \cdot K + MP_T \cdot T.$$

to be distributed. But this is possible only if one or more of the factors (e.g., labor, capital) is not paid its marginal contribution. This problem was circumvented in the theoretical literature by Knut Wicksell and his student, Phillip Wicksteed, who developed "the law of variable proportions." Total output (P) equals the sum of the factors multiplied by their marginal products, and all rewards are distributed on the basis of their marginal products. The lack of a residual removes the possibility of the problem Hobson raised, but only when constant returns to scale are present.

Constant returns do exist when output expands proportionally with the increase in all inputs. In 1928 mathematician Charles Cobb and economist Paul Douglas published an article on long-term trends in income distribution that also contained their famous Cobb–Douglas production function, itself based on an equation initially specified by Wicksell:

$$Q = aK^{a}L^{1-a}$$

This equation says that the quantity of production (Q) equals the product of capital (K) and labor (L) raised to their relative factor shares. The mathematical outcome of the Cobb–Douglas function is that all inputs are substitutes. Land, which symbolizes all natural resources and which had been used in most previous assessments, was simply left out of the equation, as was energy. Both were subsumed, inappropriately, under the category of capital, as capital, as a productive asset, is essentially useless without energy. But if all input is substitutes the theory implies that society can maintain, and even increase, its level of output in the virtual absence of resources or energy, even were these included explicitly. This failure of neoclassical economics to include energy in their basic equations of production has bothered many biophysical scientists greatly, including Nobel prize winning chemist Frederick Soddy, anthropologist Leslie White, ecologist Howard Odum and his students Robert Costanza and Charles Hall, physicist Phillip Morowski, and some economists including Nicolas Georgescu Roegen. Nearly a century after the formulation of these neoclassical equations Cleveland et al. [19]

and Reiner Kummel [20] showed that 90% of productivity increases can be attributed to increases in net energy, that the productivity of labor is principally determined by the energy used to subsidize labors' muscles, and that capital is important because it is the means of using energy. More explicitly when energy is inserted into Cobb–Douglas type functions it is a far more important determinant of changes in production than is either capital or labor. Why this basic and empirically incontestable concept has escaped incorporation into general economic thinking is astonishing to us and to the distinguished scientists mentioned above.

The marginal productivity theory can be shown, mathematically at least, to produce equity, or fairness, but only under conditions known as perfect competition. This hypothetical market structure entails creating an abstract model in which equally powerless firms meet perfectly rational consumers in an impersonal market. In addition firms must be willing to accept zero economic profit in long-term equilibrium. In this model entrepreneurs earn only a "normal" profit, which is what they could earn in wages working for someone else. In 1934 economist Joan Robinson demonstrated that such outcomes are equitable only under conditions of perfect competition. In perfect competition what workers are worth (VMPP) is what workers are paid (MRP) because, according to neoclassical theory, marginal revenue equals price. But under real-world conditions of imperfect competition, where a firm has control over price, marginal revenue is less than price. In this case what workers are paid is less than what workers are worth. Robinson referred to this situation as exploitation. She, and we, believe this to be the normal, not exceptional, situation.

In summary, neoclassical economists built a mathematically elegant structure establishing a non exploitation theory of distribution. The functions that explain distribution are the same as those that describe production. The two theories are indistinguishable. However, the theory depended upon structures that do not occur in the real world: perfect competition, unlimited and reversible input substitution, and constant returns to scale. In addition they do not give energy any special role; it is just another commodity. Nonetheless students of economics are trained routinely and often exclusively on such models of perfect completion. It is the only market structure that has been conceptualized in which distribution is equitable and exploitation cannot exist, but it is contradictory to the reality in which humans operate.

John Maynard Keynes had little to say about income distribution, and what he did offer was contradictory. In Chap. 2 of *General Theory* he stated that the classical theory of employment rested upon two premises. First, that the wage equaled the marginal product of labor. This established the demand for labor as capitalists would hire labor only up to the point where the marginal product of labor equaled the prevailing equilibrium wage. At that point they would cease hiring additional workers. Second, he noted that the marginal utility of the wage equaled the marginal disutility of the work. In other words the prevailing wage is sufficient to bring forth the needed amount of labor. Although he rejected premise number two, Keynes accepted marginal productivity theory without reservation. But this implies that a reduction in wages can expand employment. Unfortunately this was inconsistent with much of Keynes' main point that the economy can balance at full employment only if the population has enough money to spend purchasing the products they have manufactured.

In Chap. 10 of "The General Theory" Keynes discusses the relation of savings versus spending in stimulating the economy. Specifically, he examined the role played by the propensity to consume (or the fraction of additional income that is spent). Keynes utilized R. F. Kahn's multiplier principle when he considered overall investment and employment, which states that income is expanded by an amount that equals propensity to consume; that is, the change in consumption changes with respect to the change in income. Mathematically: $k = \Delta C / \Delta Y$, where C symbolizes consumption and Y stands for aggregate income. But Keynes realized that savings came primarily from the wealthy, which he called "the saving classes." If the poor saved a smaller proportion of their incomes than do the rich, then a redistribution of wealth would result in greater total spending and a greater multiplier effect and a more rapid expansion of income and employment. But Keynes never came out for a policy of income redistribution. Rather he addressed the issue indirectly, calling for an expansion of public works [21].

Overall many economists, especially classical economists, thought deeply about the questions of distribution of wealth among the different classes of society. We can say that their discourse, and others like it, had a great deal of effect on the actual implementation of economic policy, at least until the last two or three decades. This was because tax and other government policies based on their thinking tended to result in a much greater equity in the distribution of the great wealth made possible by the industrial revolution, especially in the United States and Europe.

The Main Questions of Economics: #3. How Does the Economy Balance Supply and Demand?

Since the late 1700s most economists have focused on the possibility that the impersonal market forces of competition and flexible prices could balance the needs and desires of consumers with those of firms. Adam Smith wrote first of this possibility although he never drew a supply and demand diagram. His French popularizer, Jean Baptiste Say, codified Smith's vision of the "invisible hand" into "Say's law," which expressed the idea that the process of producing goods and services simultaneously creates the income to purchase them. This is better known as "supply creates its own demand." Neoclassical economics accepted Say's law as a fundamental part of their system. British neoclassical economist Alfred Marshall provided us with the modern supply and demand schema that we use currently.

Wicksell extended the analysis to the market for savings and investment, concluding that the overall economy would find its equilibrium at full employment. Keynes disagreed fundamentally with this proposition. Rather, he argued, the economy could reach equilibrium at a level of

output that was substantially less than full employment, and that it exhibited no internal tendency to change from that low-employment equilibrium. Keynes' arguments for governmental intervention in the economy remain hotly debated today, but there is no question that the cycles of boom and bust that followed the publication and at least partial implementation of his ideas have become much more subdued [22].

History of Supply and Demand

Adam Smith's genius lay in his ability to connect productivity increases made possible by the division of labor to events in the broader market. He believed that the natural price of any commodity could be found by the summation of wages, rents, and profits. Smith, however, also contended that commodities do not always sell at their natural prices. Rather, the short-term forces of supply and demand could result in a price that exceeded, or fell below, the natural price. The market price of any commodity was regulated by the quantity that was brought to the market and the willingness and ability of potential buyers to purchase the products. Smith termed this desire, backed by money, "effectual demand." If the quantity brought to market falls short of effectual demand those individuals seeking to acquire the goods will be willing to offer more money for them. Competition among these individuals will result in an increase in market price above the natural price. If effectual demand is less than the quantity brought forth then the market price may fall below the natural price. When the quantity brought to the market just equals the effectual demand the market price will equal natural price.

The Physiocrats had not worked out any theory of supply and demand, although the *Tableau Economique* can be thought of as an early circular flow model. What Smith took away from the Physiocrats was a confirmation in his belief in liberty. The market provided a mechanism by which the haggling of daily commerce would result in a tendency towards the balance found in natural law. This is most often known as the "invisible hand" and it is greatly admired by many economists today who resent government (or anyone) telling individuals what they should or should not purchase, for example, in response to concerns about climate change [23]. The other side of the coin is that in the absence of government regulation large powerful corporations have increasing power to regulate markets and affect individual freedoms.

Jean Baptiste Say argued in his *Treatise on Political Economy* that a market characterized by liberty would adjust automatically to produce an equilibrium in which all resources would be fully employed. Say held that every purchase was simultaneously a sale. No one would sell a commodity without the intent to buy another. Money would not be hoarded because it was simply a means of exchange and had no value unto itself. Because of this, supply creates a demand of equal magnitude. Furthermore, the means of purchase are created, in the form of factor payments (wages, rent, and profit) such that there is no shortage of effectual demand. Therefore, according to the principles of Say's law a general glut of unsold commodities, and a resulting depression due to lack of demand, are theoretically impossible. Say argued that an acute glut is certainly possible, but a glut in one sector would be matched by excess demand in another. Moreover, price fluctuations as described by Smith would assure that price changes born of competition would assure that the market price would equalize with the natural price. One could say that Say generated an idealized theoretical situation in which the free market would generate the best of all material worlds; many since have believed that to be true.

Malthus rejected Say's law, arguing that a general glut was a defining characteristic of a commercial economy. The years before the publication of his *Principles of Political Economy* were marked by severe depression. The subsequent riots alerted Malthus to the dangerous destabilizing effects of actually existing general gluts. In order for Say's system to work every class must spend its entire income. Although this was true of the working classes, Malthus realized that the components of price—wages, rent, and profit—were also the incomes of the various classes in England. He argued that capitalists limited their consumption in order to save. This meant that savings must equal investment. But he found that as capitalism progressed businesses could not find sufficient outlets in which they could receive profitable returns. As investment declined and savings were maintained a shortage of effectual demand would appear, heralding the onset of a depression due to lack of demand. The Malthusian solution was, as we have already seen, a redistribution of wealth and income to the landed classes. As gentlemen of leisure they would spend this income on unproductive personal retainers and monuments to themselves which would, according to Malthus, help maintain full demand. They would also patronize the arts, leading to an improvement in the character of society. Servants and artists would consume the material wealth produced by industry, but would not produce it. This would negate the cause of an overall lack of demand. Also, as we mentioned previously, the primary mechanism of income redistribution towards the aristocracy and gentry was the continuation of the Corn Laws.

Ricardo defended Say's law, and rejected the Malthusian solution of an expansion of unproductive laborers such as servants and retainers. He said that the support of unproductive personal servants would be as beneficial to future production as fires in the warehouses of the business classes. Ricardo believed that market forces would result in the balancing of savings and investment because of the behavior of investors. "No man produces but with a view to consume or sell, and he never sells but with an intention to purchase some other commodity, which may be immediately useful to him, or which may contribute to further production. By producing, then, he necessarily becomes either the consumer of his own goods, or the purchaser and consumer of the goods of some other person" [24]. Ricardo also criticized Malthus for focusing solely on consumption and failing to consider adequately investment itself as a component of effective demand. Ricardo's argument carried the day. His goal of enhancing accumulation by means of redistribution of income and wealth towards capitalists was finally realized in 1846, 23 years after his death, when Parliament repealed the Corn Laws.

Marx chided Ricardo for defending the automatic balance between supply and demand ("the childish babble of a Say, but hardly worthy of the Great Ricardo"). Marx argued that Say's law was applicable only to the stage of simple commodity circulation where an independent artisan enters the market with a commodity and sells it for money in order to purchase a different commodity. It was not applicable to an industrial capitalist society. The possibility of such an equilibrium occurring in a simple economy did not imply its inevitability in a modern one. Marx's writings on the balance of aggregate supply and demand in a modern economy can be found in the little-read Volume II of *Capital*, where Marx discussed the process of exchange. Here Marx begins with the abstract and highly unlikely possibility of a non-growing capitalist economy, where the entire surplus value is consumed and the economy goes on year after year at the same level and composition of output. He calls this "simple reproduction," as opposed to a growing economy that he terms "extended reproduction." To begin the analysis Marx divides the economy into two sectors or "departments." Department I produces means of production, known today as the capital goods industry. Department II produces means of consumption. In both sectors the total value (V) is composed of the sum of constant capital, variable capital, and surplus value. Equilibrium necessitates that the output of these two sectors is balanced.

In plain English, Marx believed that the combined demand of workers and capitalists in the department producing capital goods had to balance the demand for capital goods in the consumption goods sector. This is highly unlikely, however, because the driving force of capitalist competition is technological change in order to increase labor productivity. Capitalists simultaneously restrict their own consumption, while paying workers no more than the value of the subsistence wage, both in order to accumulate capital. Therefore there is no reason to believe that this abstract equilibrium condition will occur in

an actual economy. If the conditions of simple reproduction are not met in an actual economy crises can occur for a variety of reasons. The pace of technological change may result in a capital–labor ratio that increases faster than does labor productivity, precipitating a tendency for the rate of profit to fall. Slowly growing wages and technological unemployment may lead to insufficient effectual demand, and disproportionalities may develop as the capital goods and consumption goods sectors grow at different rates. For Marx, sectoral imbalances are the norm, and the possibility of a balance in aggregate supply and demand is but a highly unlikely theoretical possibility that contradicts the very essence of capitalism [26].

Most of what is taught as introductory microeconomic theory in English-speaking colleges and universities today is but an updated version of the neoclassical theory enunciated by Alfred Marshall in his 1890 *Principles of Economics*. Marshall was among the first to link marginal utility with demand, and he aggregated market demand curves from individual ones. By linking supply and demand, Marshall reasoned that equilibrium in the labor market occurred when individuals decided to supply hours to the market up to the point where the marginal utility of the wage equaled the marginal disutility of the work. This unrealistic point, although rejected by Keynes, still forms the theoretical core of modern labor economics.

Marshall also based his analysis on the substitutability of resources. Consumers will substitute one good for another based on the ratio of extra utility to the price that must be paid. Rational consumers substitute the relatively cheaper good for the more expensive one, as long as or utility remains nearly constant, and substitution ends when the ratio of marginal utility to price is the same for all commodities considered. This is known as the equimarginal principle. Marshall's mode of analysis applied equally to the firm as it did to the consumer. The theory of the firm began with the "representative firm" that exhibited no marketing, energy, or technological advantage over any other. He divided his analysis into periods. The short period was one in which

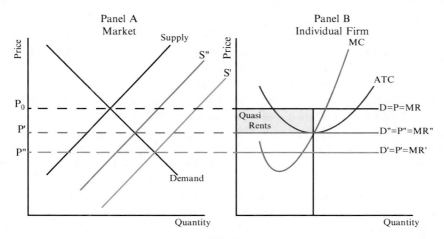

Fig. 4.1 Any profits in excess of the normal rate, which Marshall termed "quasi rents" would be eliminated by price competition among firms. (*Source*: tutor2u.net)

one factor (capital) was fixed but labor would be allowed to vary. This period was ruled by diminishing marginal productivity. If a firm applies increasing quantities of a variable input to a fixed input, eventually the rate of increase in output begins to decline. Ricardo first enunciated this idea in his debate with Malthus, but Marshall formalized it.

The onset of diminishing marginal returns implied an increase in marginal cost. If each additional laborer produces less output then, once diminishing returns have set in, a firm would need to hire additional workers, at additional costs, to produce the same incremental increase in output. The marginal cost curve, above the minimum point of average variable cost, becomes the supply curve. This became the basis of profit maximization for the individual firm. Profits would be optimized at the point where marginal cost equals marginal revenue, or the extra income derived from selling an additional product. Because the marginal cost curve is equivalent to supply and marginal revenue can be equated with demand, this point also represents the intersection of supply and demand. Any profits in excess of the normal rate, which Marshall termed "quasi rents" would be eliminated by price competition among firms (Fig. 4.1).

In Marshall's long period all factors of production are variable. Consequently diminishing

marginal returns, which require the application of variable input to fixed input, cannot operate. Long period costs were regulated by economies of scale. Traditionally classical political economists had posited that capitalists would add fixed and circulating capital up to the point of constant returns to scale, where output expanded proportionately with the application of inputs. But Marshall saw no a priori reason to assume constant returns. In the era where land played the primary role in production Marshall, following Ricardo, believed there was a tendency towards decreasing returns to scale. But in the era in which the restrictive role of nature was diminished, and the application of fossil energy could increase productivity dramatically, the tendency was towards increasing returns [27]. This rendered a parabolic long-run average cost curve where constant returns to scale represented the minimum achievable cost.

In Marshall's neoclassical synthesis the market will self-regulate to generate long-period equilibrium where marginal revenue = marginal cost = price = the minimum short-period average cost = long-period average cost. At this point profits are forced to the "normal" level and the outcome is *allocative efficiency*. Allocative efficiency occurs when the market price fully covers all underlying incremental costs and resources flow to their most lucrative use. Firms unable to achieve constant returns to scale can produce only at above average cost and will be forced into bankruptcy by price competition. Because the firm-level supply can be aggregated into market supply and market demand is simply the summation of individual demands, the market level balance of supply and demand is the most efficient allocation of resources. The idea that markets allocate efficiently is a deeply held belief of almost all economists, including most ecological economists.

By the 1920s supply and demand analysis had been extended to describe the workings of the primary sectors of the overall economy. According to Marshall the supply of labor, set by the disutility of the work, would come into balance with the demand for labor, which was determined by marginal productivity by means of subtle adjustments in the price of labor, or the wage rate. If wages were below the equilibrium rate shortages would occur, causing competing employers to offer a higher wage in order to attract workers. Unemployment was seen as a surplus of labor, caused by workers demanding wages in excess of equilibrium. Consequently the solution for unemployment was the reduction of wages. In many ways the neoclassical or market model provided logical rationales for management to pay labor as little as possible.

The aforementioned Knut Wicksell offered an analysis of the market for savings and investment, called loanable funds, based on the idea of the self-equilibrating market. Savings were specified as a positively sloped function of the interest rate (the price of money). Those with enough income to save would be induced to augment their savings by an increase in interest rates. Investment was negatively related to interest. At higher interest rates the costs of borrowing rose and less-profitable investment projects would be curtailed. The market would find its own equilibrium interest rate and savings would equal investment. These two conclusions about the functioning of aggregate markets served as the backdrop for John Maynard Keynes' critique of neoclassical economic policy. For John Maynard Keynes the question was not one of whether overall, or aggregate, supply would balance with aggregate demand, but whether the balance would occur at full employment. Keynes

began his 1936 opus, *The General Theory of Employment, Interest, and Money*, by accepting all the neoclassical postulates except two. He rejected Say's law and Marshall's idea that the supply of labor is determined by the interaction of the marginal utility of the wage and the marginal disutility of the work.

Whether this change in two initial propositions constituted a revolutionary change in the profession or was a matter of "moderate conservatism," as Keynes himself believed, has been, and probably will continue to be, a matter of considerable debate. But Keynes' conservatism was not about domestic spending. He saw the enterprise economy of the 1930s as being limited by internal and external factors. The internal factor was the persistence of severe unemployment and social dislocation that characterized the depression. The external factor was the presence of two alternate systems, Fascism and Bolshevism, which Keynes found highly distasteful. Keynes conservatism came from his desire to save and perpetuate the free enterprise system. His moderation came from a belief that leaving the economy to its own devices and awaiting the triumph of market forces would be insufficient to solve the problems created by the Great Depression.

The prevailing orthodoxy in the middle third of the last century was grounded in the notion that savings determined the level of investment. Furthermore, the balance of saving and investment was needed to achieve the overall balance of supply and demand. A simplified version modifies the circular flow model, (which is essentially a depiction of Say's law), to accommodate the reality that not all firms and household members spend all their money in current consumption. Money "leaked" out of the system flowed when individuals saved a portion of their income, when taxes were levied on income, and when purchases of foreign goods were made. On the other hand, income flowed into the system when businesses made investments, the government purchased goods and services, or when an economy sold goods in foreign markets and received the income from doing so. Consequently, the traditional circular flow model can be augmented with both leaks and injections (Fig. 4.2).

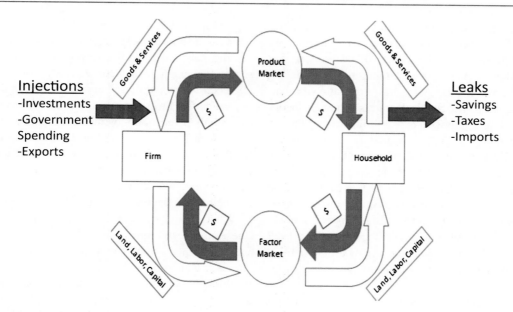

Fig. 4.2 Circular flow model with leaks

Given the conventions of the early twentieth century of a political commitment to a balanced budget that equated government spending and taxation, along with an international gold standard that balanced imports and exports, the main question facing Keynes was to what degree would savings balance with investment? Unless savings and investment balanced the aggregate supply of products (which were increased by investments) and the effectual demand for them (which were increased by consumptive expenditures) would not balance at full employment. He believed that finding adequate investment outlets for surplus savings, and not wage reductions, was the key to finding a macroeconomic equilibrium at full employment. The prevailing orthodoxy, on the other hand, was to treat the market for savings and investment as a market for loanable funds. Competitive market forces would lead savers and investors to vary the amount of funds with the price, leading the market to find an equilibrium rate that balanced savings and investment. Keynes disagreed vehemently that this was how it worked. Savings, in his analysis, depended upon income, and savings would increase only as income rose. Investment depended upon expected profit and the rate of interest. Savings and investment were

functions of different variables. Keynes believed there were no reasons for planned (or *ex ante*) savings to equal *ex ante* investment.

Keynes' greatest concern was not a shortage of savings, but savings that exceeded investment. The orthodox method of increasing savings was to increase the interest rate. This had the unfortunate effect of simultaneously depressing investment, thereby reducing the level of aggregate output and employment. As investment fell, so too did employment. Workers with less money buy fewer products, forcing business to reduce investment once more. The economy spiraled into depression and when it came to a balance the equilibrium was at a low level of output and a high level of unemployment. But if the interest rate is not determined in the loanable funds market, where is it determined? For Keynes interest was a monetary phenomenon, depending upon the interaction of the supply of money (determined politically by monetary authorities) and the preference investors have for holding their money as cash, or as balances to be invested in financial securities. Money plays an essential role in a modern economy, and the economy could not run without it. For Keynes, the fundamental problems of investment were those of uncertainty.

The present, when investments are made, lies between an unchangeable past and an unknowable future. Despite efforts of economists and mathematicians, the uncertainty posed by investment over the long term makes the rational calculations of neoclassical microeconomic theory essentially impossible. The future is sufficiently uncertain that the self-regulatory capacity of the laissez-faire economy is unlike that posed by neoclassical theory. Keynes believed that the object of the accumulation of wealth entailed investing now to receive rewards in the distant future. But our knowledge of the future is uncertain. In an oft-quoted passage from his 1937 *Quarterly Journal of Economics* article entitled "The General Theory of Employment," Keynes declared:

"The calculus of probability, tho mention of it was kept in the background, was supposed to be capable of reducing uncertainty to the same calculable status of certainty itself...By "uncertain" knowledge, let me explain, I do not mean merely to distinguish what is known for certain from what is only probable. The game of roulette is not subject, in this sense, to uncertainty; nor is the prospect of a Victory bond being drawn. Or, again, the expectation of life is only slightly uncertain. Even the weather is only moderately uncertain. The sense in which I am using the term is that in which the prospect of a European war is uncertain, or the price of copper and the rate of interest 20 years hence, or the obsolescence of a new invention, or the position of private wealth owners in the social system in 1970. About these matters there is no scientific basis on which to form any calculable probability whatever. We simply do not know" [27].

The use of money allows for a method to avoid all of one's assets being fixed in permanent and unchangeable assets. This ruled what Keynes called the speculative demand for money. But speculation is subject to waves of pessimism and optimism. Although the primary driving force of output, and therefore employment, was investment, the level of consumption was also important in determining the level of aggregate demand. The amount of consumption, like savings, was dependent primarily upon the level of income. The fraction consumed (the marginal propensity to consume) was subject to multiplier effects. Because the poor spend a greater fraction of their income than do the wealthy, Keynes believed that some augmentation of income growth could be affected by a redistribution of income. Given the uncertainty of investment, and the limitations of expanding the economy by means of money creation when interest rates are low, Keynes allowed for the state to spend in order to assure sufficient aggregate demand for the economy to balance at the full employment level of income. We return to his methods in the final question of this chapter.

What should we conclude about this main question of economics, about how economists view whether supply can possibly balance demand, and lead the economy away from the troubling boom and bust patterns that have characterized capitalism? The optimist might point out that most economists believe that firm-level supply is aggregated into market supply and likewise market demand is simply the summation of individual demands. Together these forces operating at the market level balance supply and demand well enough and in a way that is the most efficient allocation of resources. The idea that markets allocate efficiently is a deeply held belief of almost all economists. But cycles remain, although much less as a percentage of GDP following the publication of Keynes magnum opus and the partial implementation [22]. Even so today Keynes, as represented by arguments as to whether, or to what degree, governments should undertake deficit spending to restore ailing economies, is very much hotly contested. The cynic might say "economists throughout the history of economics often held strongly held beliefs that were in fact often contradictory to each other. Today we have little or no better idea than in the past as to which is correct." This is hardly a surprise to anyone who follows economics today.

What is missing from this and other economics questions is a consideration of what is likely to be the most important issue of economics today: issues of energy and other resources, of environmental degradation and (with the exception of Malthus and Marx) population. The issue was always how to take Nature's abundance and mobilize forces to turn that into wealth and employment. We can perhaps understand how this came about inasmuch as economics was mostly developed before the appropriate science, but the roots of economics have hardly budged with the new information we have now on resources and the environment, and probably the majority of economists today do not think there is any particular reason to worry too much about resource or environmental limitations.

The Main Questions of Economics: #4. What Are the Limits to Capital Accumulation?

The crucially important subject matter of economics from the time of the mercantilists was the accumulation of wealth, yet the methods of dealing with accumulation and growth changed substantially once the age of abundant and cheap fossil fuels began. All theorists who wrote in the age of solar flow developed theories of self-limiting accumulation. All classical political economists had growth theories that ended ultimately with society in a non-growing stationary state. But after the introduction of cheap oil the focus on the stationary state ended, replaced with the idea of indefinite growth as the result of efficiently functioning markets. However, the transition from classical political economy to neoclassical economics also saw a shift from the concept of long-term accumulation to that of static equilibrium. A neoclassical growth theory did not emerge until the 1950s in response to Keynesian views on the internal limits to growth and accumulation. As we enter the second half of the age of oil we are facing a new set of biophysical limits that interact with the internal limits found largely in the investment process. To address the role of biophysical limits adequately we first turn to the historical perspectives on the internal limits to accumulation.

The Classical Perspective and the Beginning of the Concept of Limits

For Adam Smith the process of economic growth began with the frugal saving capitalist and the workings of the "invisible hand." The desire to accumulate, which for Smith was innate in the human spirit, manifests itself as saving and investment. A frugal individual saves and invests the capital in expanding the division of labor and employment, along with improved equipment. The expansion of employment leads to rising incomes among all sectors of the population providing the means for the extension of the market. Smith wrote in preindustrial days thus he did not believe that machinery would replace labor. Rather it would expand its employment. But in the process lies the beginning of the stationary state. As employment and production expanded, so too would the demand for labor. This would serve to raise wages and diminish profits which would hinder further accumulation in the short term. The solution could be resolved only by the rather cruel operation of nature. Increased wages would lead to a greater number of surviving children. This would increase the supply of labor and result in the subsequent reduction of wages. But the reduction of wages would eventually decrease the labor supply as infant mortality would increase with less money to purchase food. But although nature would operate to regulate the labor market the long-term tendency was towards decline.

When a nation was fully complemented with people with respect to biophysical capacity to support them, wages would fall to the bare subsistence level. When the nation was fully stocked with all that the low level of wages could support, profits would fall as new investment opportunities vanished. Thus the ultimate trajectory of a vibrant system of perfect liberty was the stationary state. Smith saw this as unfortunate, as the quality of life in the progressive state was vibrant but life in the stationary state was melancholy. Life in the declining state was tragic. But for Smith, no nation was close to achieving its full complement of labor and capital, so the stationary state was a prospect for the distant future [26]. Smith's analysis of accumulation gave economists two methodological lessons that are still strong today. The lack of economic growth was stagnation which was to be avoided at all costs. Moreover the tragedy of the end of accumulation was found in the distant future. Today economists, politicians, and citizens alike tend to follow Smith's logic. Growth is the primary goal of most economic policies now, and many believe that the environmental consequences of growth will not occur for at least a 100 years.

Less than a decade after Smith's death his optimism, or that of his followers, was dashed. The arrival of the steady-state seemed imminent instead of not distant. British philosopher Thomas Carlyle surveyed the debate over the end of accumulation between Thomas Malthus and David Ricardo and dubbed political economy "the dismal science." The primary limit to accumulation for Ricardo was the existence of diminishing marginal returns. Given the existence of the Corn Laws the extension of cultivation onto poorer lands resulted in reduced harvests and increasing rents accruing to the landowners. The increases in rents and wages would diminish profits, resulting in the cessation of productivity increasing investments as soon as potential profits fell to the prevailing rate of interest. Only a suspension of the Corn Laws could remove the limit to growth.

Malthus saw the primary impediment to long-term accumulation in the increase in human population at a rate that would soon overwhelm the ability to provide sufficient food, resulting in mass starvation. Malthus advocated not only measures to limit population by "courting the return of the plague," but a transfer of wealth to the morally restrained landed classes. But Malthus too saw internal limits to accumulation. Capitalists tended to over save, thereby limiting effectual demand needed to extend the market and justify the increased level of production. As mentioned earlier, he advocated the redistribution of wealth to the aristocracy who would spend the income on retainers and monuments to themselves, eliminating the shortage of effective demand and perpetuating all that is good in modern society. For both Malthus and Ricardo questions of accumulation ultimately resolved the questions of distribution of wealth.

Marx did not have a theory of the stationary state. Rather he believed the internal contradictions of the capitalist system could result in its passage into socialism before the physical basis of the end of accumulation arrived. For Marx

only human labor created new value, although it was increased to an unprecedented extent by the application of coal to large-scale mechanization. Such mechanization reduced the per-unit labor content of commodities resulting in the reduction of their prices. Capitalists competed by means of mechanizing to reduce the price of their individual commodities below the social average. But as the expansion of constant capital increased faster than the increase in productivity profits would fall. This touched off an economic crisis, which could not, in the long run, be overcome by the mere addition of more fossil-fuel driven equipment. The resulting depression "solved" the tendency for the rate of profit to fall by decreasing the level of capital to labor, as bad debts were written off and factories shuttered, as well as by increasing the productivity of labor when desperate workers would work harder for less. Before the stationary state set in the increasing severity of periodic crises and a socialist political party would transform society by instituting rational planning in the investment process resulting in an end to economic crises and the true beginning of human history.

The Neoclassical Perspective of Indefinite Growth

Neoclassical economists held a very different opinion on the future of economies. The potential of continued dramatic increase in productivity (made possible by fossil fuels, although that was not mentioned) relegated questions of accumulation and growth to secondary status. Consumption was limited only by a budget constraint. However, rational consumers would maximize their well-being by substituting cheaper for more expensive goods, so that consumption could increase indefinitely. A similar process worked on the production side. Initially the optimal situation was the point at which supply balanced with demand. Although the possibility that profits could be reinvested resulting in growth was possible, the focus was clearly on static equilibrium. Only later, in the profound Depression of the 1930s, did a neoclassical theory of growth begin to emerge.

Sir John Hicks developed the idea of the *elasticity of substitution*, which meant in practice that expensive, unreliable labor could be substituted by cheaper, more reliable capital. He believed that a progressive society necessitated a positive elasticity. In other words, the price of progress was the redistribution of wealth from labor to capital. This would allow growth to continue indefinitely.

One conspicuous exception exists in the work of William Stanley Jevons. Before Jevons solidified his reputation as a marginalist he produced the previously mentioned empirical work, *The Coal Question*. Jevons argued that England's industrial domination of the world depended upon the development of heavy industry, and heavy industry depended upon an adequate supply of cheap coal. But Jevons believed there was no prospect for a reliable and cheap substitute for coal and that England's mines were slowly becoming exhausted. This would render much of England's population superfluous (and perhaps incapable of being fed) and essentially create the conditions for the return of the stationary state. Jevons offered no satisfactory solutions, however, his essay represents the initial exercise of the economic consequences of the absolute scarcity in the age of fossil fuels [27]. Jevons, however, never reconciled these two remarkable perspectives, although they would lead to rather contradictory positions on growth.

Keynesian economics gained prominence in the failed growth economy and Great Depression of the 1930s, but it was, perhaps surprisingly, not particularly oriented towards growth. Rather it focused on an explanation of the role of inadequate demand and uncertainty in producing a depression, as well as the futility of relying upon markets alone to produce sufficient demand to end the depression. A Keynesian growth theory was not to emerge until the very end of the Depression in 1939. Roy Harrod posited that a "natural growth rate" (G_n) consisted of the growth rate made possible by the existing level of population and the state of technology. At the same time the "warranted growth rate" (G_w) was the rate that would leave all parties satisfied that they had invested neither too much nor too little. The warranted growth rate depended upon the relation between the capital–output ratio and the marginal propensity to save. Harrod believed that the condition for steady state equilibrium, when $G_n = G_w$, would be highly unlikely. If G_w exceeded G_n the economy would slip into depression, and inflation would occur in the opposite case. The delicate balance between the natural and warranted growth rates is referred to as the "razor's edge." Any slight deviation produces economic instability. Harrod's work in England was supplemented by that of Evesy Domar in the United States. The essence of Domar's analysis was that the instability of investment was produced by the dual nature of investment. Although investment serves as a source of demand, it also increases productive capacity. If adequate spending outlets are not found for the augmented capacity investment will fall and economic growth will decline.

Neoclassical growth theory emerged as a critique of Harrod and Domar. In 1956 Robert Solow argued that the flaw in the Harrod–Domar approach was that economic instability resulted from the way they specified their equations. According to Solow the Harrod–Domar model used fixed proportions between labor and capital. When he replaced these fixed coefficients with a Cobb–Douglas function the instability disappeared and the functioning of markets would lead to stable growth trajectories. Solow managed to turn a social problem into a technical one, and maintained the neoclassical ideal of self-regulating markets over the long term.

In conclusion we find that there has been no general agreement among economists about either the proper rate of capital accumulation or about the existence of any limits to indefinite economic growth. In general we can say that at present neoclassic economics holds most of the situations accepted by the majority of the people who think about it, possibly because this is the type of economics that dominates our teaching in economics. Growth is the primary goal of most economic policy now, and many believe that the environmental consequences of growth will not occur for at least a 100 years. But if the arrival of serious environmental consequences is measured in decades not centuries then we will need to develop a theory in which the stationary state is not, in Smith's words, "melancholy". As economic growth is limited by resource availability and climate instability, how do we live well within Nature's limits?

The Main Questions of Economics: #5. What Is the Proper Role of Government?

Classical political economists stood for a limited role of government. These limited roles are embodied, in fact, into the U.S. Constitution. Governments should maintain property rights, enforce contracts, protect the nation from domestic and foreign enemies, and provide public goods. They should not intervene in market processes or regulate prices. Instead the invisible hand of the market would be sufficient to translate individual self-interest into social harmony. Say's law assured that the overall system would balance at full employment without the need for government direction. Thus our Constitution reflected the dominant economic thought of the time.

Neoclassical economists too accepted this proposition and translated it into mathematical propositions. The Walrasian core of neoclassical economics asserts that individual exchange on the basis of self-interest (in the form of equal marginal rates of substitution among trading partners) will satisfy not only the traders but result in the general equilibrium of the system as a whole at a point where no individual can be made better off without harming another. Prices serve as perfect carriers of information, and any intrusion of the government into market processes will distort the market's price signals and simply make the system not work.

Keynesian economics takes a very different position. The private operation of markets periodically produces insufficient demand and government action is needed to provide sources of demand that the private sector cannot do profitably. Although Keynes himself believed in the necessity for planning in the long term, there is little in Keynesian economics that justifies government intervention in the internal mechanisms of produc-

tion and profit-making itself. Nevertheless an increasing number of economists and politicians "bought into" government intervention more or less as Keynes had suggested. For many decades, from roughly 1930 through 1973, Keynesian demand management, or something like it, helped propel a long wave of economic growth that seemed to work extremely well as the U.S. government pumped more and more money into the economy in both war and peace and as the economy grew steadily year after year. Few paid attention to the fact that this was also an era of expanding supplies of cheap oil, which was, according to economics, "just another commodity."

But after the peak of U.S. oil production in 1970 long-term prosperity gave way to long-term stagnation in the midst of rising prices and a disenchantment with Keynesian economics and its attendant requirement for government intervention. This, and other factors, led to a return to political and economic conservatism in the nation as a whole accompanied by a conservative resurgence in the economics profession. Neoclassical economists were back in the saddle emptied of Keynesians and legislation reflected their free-market orientation. One can argue, however, the long-term result of these "reduce government intervention" policies resulted in the financial near-collapse in 2008. The election of 2010 seemed much like a contest between two sets of policies, neither of which worked in the recent past.

Summary

The reader should see that our deepest thinkers in economics, even those of the same general school, were frequently at odds with each other. Nevertheless our basic concepts of economics seem to have more or less "worked" over the past century or more, as economies expanded, jobs were made, fortunes were made and lost, and as many flocked to the countries run by economic thought seeking the good life. Why is this? Is it simply that each year for a century and a half the economies of the world tended to expand, more or less vindicating whatever economic theory was in ascendancy at that time? When, periodically, the economy stopped growing for a year or two scapegoats were needed. Often these scapegoats were the very same economists and economic theories (and their political advocates) that worked just a few years before. To your authors, trained in natural science, the scientific method, and the importance of the biophysical world, it seems extremely strange that these various economic theories have not been put forth as hypotheses that could be tested over time and accepted, or rejected, or accepted as "good under such and such conditions." But this seems barely to have happened, and whatever results might have been derived seem not to have one iota of impact on economics as it is taught and practiced.

Meanwhile it seems clear that many of the more basic economic theories (supply and demand, prices as determinants of some things, the basic concept of investments, price incentives to get people to work and to get things done, and probably many more) which were not in particular contention among the different economists or their different schools worked just fine, at least for some to many people. Who has generated a list of what economic principles we can accept as "proven," or at least as not yet disproven, as we have for various first principles in the physical sciences and even biology? We do not know.

Nevertheless there is a lot to be desired for our economics inasmuch as so often what seems to work in some places or times does not at others. We contend that fully understanding economics and the economic nature of policy entails a much deeper analysis of the patterns of energy use and its relation to economics, which of course is what most of this book is about. The role of energy, always ignored or underestimated, is likely to be much more important in the future as our supplies of oil and other premium fuels inevitably decline and as their prices, relative to our incomes and our national wealth, increase.

Speaking more generally, this perspective can integrate much more fully the discipline of economics with the natural sciences. As we noted energy was critical to the thinking of the earliest economists, although they could not use the language we would use today because the concept of energy was not clear to them or even physical

scientists at that time. For example, land was important in the eighteenth century because most of the energy available for economic production came from the sun, but the concept of photosynthesis as energy capture was barely understood. Likewise in the time of Adam Smith, factories were becoming increasingly important. These factories employed many workers whose muscles provided much of the energy to generate the transformations of raw materials into desired products. Then, as the industrial revolution came about, it was capital that allowed construction of the equipment that in turn allowed the use of the coal or oil to run machinery. In all cases a biophysical analysis shows that it is the energy that does the actual work in turning raw materials into useful goods and services. Therefore, although we agree that many factors contribute to the production of wealth, the critical element is and always has been energy. Without energy there would be no economies or economics because there would be no goods or services produced or moved from place to place or sold through markets. The more one controlled the most important energy source of the time, the more wealth production was possible, and because wealth often buys influence, the more political power the person or people who controlled that energy had. Economics and political economy have at their very base, energy.

Questions

1. Do you think that combining natural science and economics is a good idea? Why or why not?

2. How is a city like a natural ecosystem? How is it different?

3. What ideas did you get in this chapter from earlier economists that you think might be important for understanding our current situation?

4. Can you think of a "peak oil" situation that occurred 150 years ago? Does that have any relevance today?

5. Why do you think economists have tended to ignore energy in their basic equations? Were they justified in doing that?

6. From where did the early group of economists known as the Physiocrats believe that wealth came?

7. Define relative versus absolute scarcity.

8. What is economic surplus?

9. What are Heinberg's "five strategies for obtaining energy"?

10. Discuss one of the four main economic questions.

11. List four major schools of economics over time and one idea associated with each.

12. What is natural capital?

13. What is the source of wealth for a Physicocrat? A classical economist? A neoclassical economist? Yourself?

14. What was the *Wealth of Nations*? How does that relate to the title of this book?

15. Give one of the great economic ideas derived by David Ricardo.

16. What was the "diamonds versus water" paradox? How was it resolved?

17. How did Keynes think we could diminish the large swings in the capitalist economy?

18. What did classical political economy have to say about the distribution of wealth?

19. Discuss comparative advantage .

20. What is the "best first principle."

21. Was Karl Marx principally interested in communism?

22. Did Mill think about the distribution of wealth?

23. What important factor did the Cobb–Douglas production function leave out?

24. What are the main two views as to whether economies can balance supply and demand?

25. What earlier economist probably had the largest impact on what is taught today in basic economics courses?

Literature Cited

1. Campbell, C., and J. Laherrere. (1998) "The End of Cheap Oil". Scientific American. March: 78–83.

2. Heinberg, R. 2003. The party's over. New Society Publishers, Gabroiola Island, B. C. Canada.

3. Foster, J.B. and Magdoff, F. 2010. The great financial crisis. New York: Monthly Review Press. Von Liebig himself referred to this system of commercial agriculture as "robbery".

4. DeVroey, M. 1975. The transition from classical to neoclassical economics: A scientific revolution. Journal of economic issues. 9(3): 415–439.

5. Perelman, M. 2006. Railroading economics. New York: Monthly Review Press.

6. Shackle, G.L.S. 1967. The years of high theory. Cambridge: Cambridge University Press.

7. (DeVroey 430). ... without making another worse off. Government intervention could do no good, and much harm, as it would distort the signals of the market, which is seen as a perfect carrier of information.

8. Passinetti, L. 1977. Lectures on the theory of production. New York: Columbia University Press.

9. Smith, A. 1923. An inquiry into the nature and causes of the wealth of nations. New York: The Modern Library.

10. Smith 1776: 47.

11. Tucker, R. 1978. The Marx-Engels reader. New York: W.W. Norton and Company.

12. Marx, K. 1976. Capital. London: Pelican.

13. Marx, K. 1976.

14. Passinetti, L. 1979.

15. Klitgaard 2008.

16. (Smith 1923: 709–710, 96).

17. Ricardo, D. 1962. Principles of political economy and taxation. Cambridge: Cambridge University Press.

18. Mill, J.S. 1865. Principles of political economy. New York: D. Appleton and Company.

19. Cleveland, C. J., Costanza, R., Hall, C. A. S. and Kaufmann, R. 1984. Energy and the United States economy: a biophysical perspective. Science 225, 890–897.

20. Kummel, R. (1989) "Energy as a Factor of Production and Entropy as a Pollution Indicator in Macroeconomic Modeling". Ecological Economics. 1: 161–180.

21. Keynes, J.M. 1964. The general theory of employment, interest and money. New York: Harcourt Brace.

22. Hall, C., Cleveland, C. and Kaufmann, R., 1986. Energy and Resource Quality: The Ecology of the Economic Process. Wiley Interscience, New York.

23. Smith 1923: 56–57.

24. Ricardo 1962: 192–93.

25. Sweezy, P. 1942. theory of capitalist development. Monthly Review Press N.Y. 1942.

26. Sweezy, P.M. 1942: 75–79.

27. Keynes, J.M. 1964: 213–214.

28. Keynes, J. M. 1937. The general theory of employment. Quarterly Journal of Economics 209–223.

29. Spengler, J. 1972. The marginal revolution and concern with economic growth. History of political economy 4:481–482.

The Limits of Conventional Economics

5

Introduction

The last century has seen the ascendancy, indeed intellectual dominance, of neoclassical economics (NCE, also known as market or Walrasian or University of Chicago or, Washington consensus or, occasionally, neo welfare economics). The basic neoclassical model represents the economy as a self-maintaining circular flow among firms and households, driven by the psychological assumptions that humans act principally in a materialistic, self-regarding, and predictable way. Unfortunately the NCE model violates a number of physical laws and is inconsistent with actual human behavior. Consequently the NCE model is unrealistic and a poor predictor of people's actions, as an array of experimental and physical evidence and recent theoretical breakthroughs demonstrate. Despite the abundance and validity of these critiques, few economists seriously question the neoclassical paradigm that forms the foundation of their applied work. This is a problem because policy makers, scientists, and others turn to economists for answers to important policy questions. The supposed virtues of "privatization," "free markets," "consumer choice," and "cost–benefit analysis" are considered to be self-evident by most practicing economists, as well as many in business and government. In fact the evidence that these concepts are correct or do what most people believe

they do is rather slim and contradictory. Thus this chapter is a strong critique of economic theory, in this case NCE [1].

We offer a review and synthesis of NCE, paying particular attention to the lack of connection of NCE to biophysical reality and its inadequate characterization of human behavior. When all the criticisms are taken as a whole it is clear the NCE framework stands on an untenable foundation and that some other basis for interpreting economic reality must be found. It is clear that NCE is very limited in its usefulness and cannot guide us in our attempts to deal with the most important issues of our time, such as the depletion of oil and gas, climate change, financial crises, and the destruction of much of nature. We end by sketching alternative characterizations of human behavior and economic production.

Some Fundamental Myths of NCE

The edifice of NCE is built on myths and based on an outdated worldview. These myths are not merely harmless allegories because they provide the foundation upon which economic policy is made and cultural attitudes are distilled. Thus the worldview and policy prescriptions of most economists can only be described as "faith based" because many fundamental tenets of NCE are inconsistent with economic reality.

C.A.S. Hall and K. Klitgaard, *Energy and the Wealth of Nations: Understanding the Biophysical Economy*, DOI 10.1007/978-1-4419-9398-4_5, © Springer Science+Business Media, LLC 2012

Myth 1: A Theory of Production Can Ignore Physical and Environmental Realities

Real economies are subject to the forces and laws of nature, including thermodynamics, the conservation of matter and a suite of environmental requirements. NCE does not recognize or reflect the fact that economic activity requires the input and services of a finite biophysical world which is usually degraded by that activity.

Myth 1a: The Economy Can Be Described Independently of Its Biophysical Matrix

NCE is a model depicting abstract exchange relations considered only as goods and services and money within a world unrealistically limited to markets, firms, and households. Real economies also require material and energy from the natural world to allow that exchange, and are limited by the material and energy transformations necessary for economic activity. Students are introduced to the misleading circular flow model of the economy in the first days of Principles of Economics. This conceptual vision of the economy is one of a self-contained and self-regulating system independent of the biophysical system and its laws. There are but two sectors, households and firms, with goods and services going from firms to households, and productive input (land, capital, and labor) going from households to firms (Fig. 4.2). The model operates as a social model, where human wants and needs and responses operating in markets are all that are required for an economy to operate. Astonishingly the materials and energy required for the derivation of these goods and services are simply left out of the model.

Households serve as loci of consumption and possessors of property rights to the factors of production. Firms exist to produce and to hold property rights to the finished commodities. These property rights are willingly exchanged in markets for money. Neither monetary value nor physical materials are lost to heat or erosion as input is transformed into goods and services.

Thus the NCE "theory of production" is not a model of production at all, but rather a model of the distribution of productive input and the goods they had produced previously. No specific primary input from nature (such as energy or materials) is represented in this model, which is a representation of an economic system that cannot exist.

The NCE notion of scarcity is disconnected from biophysical reality for it is never absolute but only relative to unlimited wants. If we are confronted by the limits of one resource, the imaginative human mind, driven by the proper set of monetary incentives and protected property rights, will always create a substitute. No input is critical, therefore neither absolute scarcity nor the need of any particular resource is a problem in the long run. Thus in the neoclassical economics world the economy can simultaneously experience relative scarcity and infinite growth. Competitive prices, formed in markets, assure that resources flow to their best use.

Nicholas Georgescu-Roegen, and his student Herman Daly, were among the first to point out the absurdity of this depiction of production. Real economies cannot exist outside the global biophysical system, which is essential to provide energy, raw materials, and a milieu within which it can operate and assimilate wastes. [2, 3] Their first step to make an economic model consistent with reality is to put the economy inside the global biophysical system. Some natural scientists have gone several steps farther. Several writers [4–7] demonstrate clearly that the NCE model is unacceptable because: (1) its boundaries are drawn incorrectly and (2) the model is de facto a perpetual motion machine because it has neither energy input nor entropic loss. Many economists today, including many recent Nobel prize winners (e.g., Ackerman, Krugman, Sen, Stiglitz) have very serious reservations with the contemporary model. Most of the authors referenced in this paragraph, the authors of this book, and many other physical and social scientists are not interested in simply making corrections to the basic NCE models. Instead these scientists and others believe that the NCE model is incorrect at its core. For starters, although money may cycle seemingly indefinitely among goods and services, the real economic system cannot survive without continual input from, and output to, Nature.

Myth 1b: Economic Production Can Be Described Without Reference to Physical Work

The neoclassical economists' model of production does not require any specific physical input but is solely an exchange of existing input among firms. The economic process is driven not by the availability of physical resources, but rather by human ingenuity as depicted in the still widely used Cobb–Douglas function. The quantity of output produced (Q) is a function of only capital (K) and labor (L): $Q = A(K^\alpha L^{1-\alpha})$ where α represents capital's share of output, $1 - \alpha$ stands for labor's share, and $1 > \alpha > 0$, is multiplied by some constant A, considered "pure technological change."

Technology is independent of the input of land and capital and is calculated as a "residual" left when the contributions of the measured factors are subtracted from the growth rate of total economic output [8]. Not surprisingly the residual tends to increase over time. Thus most neoclassical economists believe that technology is an amorphous force that cannot be measured directly but can increase the productive power of the economy without limit. With the assumption that there are no diminishing returns to technology, there is no need to worry about physical work or the scarcity of any particular productive input.

The preoccupation with pure technological change as the driver of economic growth has caused earlier neoclassical economists to virtually ignore the critical importance of energy in powering the modern economy [8]. In contrast, many natural scientists and some economists have concluded that the explosion of economic activity during the twentieth century was due principally to the increase in the ability to do work through the expanding use of fossil fuel energy. In fact the neoclassical economist's technology residual disappeared when energy was included as an input. Energy as a factor of production was more important than either capital or labor for Germany, Japan and the United States in recent decades [6]. Ayers and Warr [9] further found that most improvements in "technology" have been simply an increase in the quantity of energy used or the efficiency of getting it to the point where the work is done. Although NCE models purport to show that technology alone has driven the industrial economy, historically, it has been a technology that mostly has found new sources of, and applications for, energy.

There are a number of additional criticisms that the natural scientist can level against the basic neoclassical model. These criticisms taken together mean that there is no possibility that we can assign any validity to the basic neoclassical model.

Criticism 1: Thermodynamics

Contemporary economics and its fundamental household-firm-market model (Fig. 5.1) pays only minimal attention to the first law of thermodynamics, and none at all to the second. In fact, the second law is completely incompatible with the conceptual model known as the circular flow. In the circular flow diagram there is never any value lost to waste or entropy. This is a serious conceptual flaw and an obstacle to designing economic policies that can meet the challenges of pollution, resource scarcity and depletion, and unemployment successfully. In effect the two laws say, "Nothing happens in the world without energy conversion and entropy production." The consequences are: (1) every process of industrial and biotic production requires the input of energy; and (2) because of the unavoidable entropy production the valuable part of energy (called exergy) is transformed into useless heat at the temperature of the environment (called anergy), and usually matter is dispersed, too. This results in pollution and, eventually, the exhaustion of the higher-grade resources of fossil fuels and raw materials. (3) Human labor, powered by food, can be, and was, replaced by energy-driven machines in the course of increasing automation.

Although the first and second laws of thermodynamics are among the most thoroughly

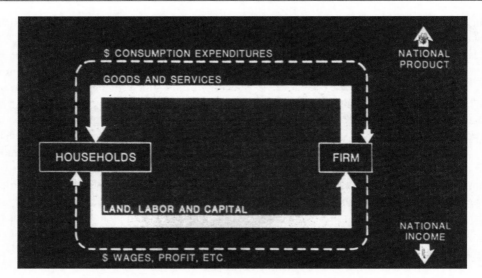

Fig. 5.1 The neoclassical view of how economies work. Households sell or rent land, natural resources, labor, and capital to firms in exchange for rent, wages, and profit (factor payments). Firms combine the factors of production and produce goods and services in return for consumption expenditures, investment, government expenditures, and net exports. This view represents, essentially, a perpetual motion machine

tested and validated laws of nature and state explicitly that it is impossible to have a perpetual motion machine (i.e., a machine that performs work without the input of energy), the basic NCE model is a perpetual motion machine, with no material requirements and no limits (Fig. 5.1). Most economists have accepted this incomplete model and have relegated energy and other resources to unimportance in their analyses. Rather than placing the economy within the confines of nature, this approach relegates all the limits of nature to a minor position within a system of self-regulating markets. This attitude was cemented in the minds of most economists by the analysis of Barnett and Morse, who found no indication of increasing scarcity of raw materials (as determined by their inflation-corrected price) for the first half of the twentieth century. However, their analysis, although cited by nearly all economists interested in the depletion issue, was seriously incomplete. Cutler Cleveland showed that the only reason that decreasing concentrations and qualities of resources were not translated into higher prices for constant quality was because of the decreasing price of energy [10]. Thus, it is only because of their historic abundant availability of many natural

resources that economics can assign them low monetary value despite their critical importance to economic production.

An apparent change in the perspective of the Nobel Laureate in Economics Robert M. Solow is interesting. In 1974 he considered the possibility that "The world can, in effect, get along without natural resources" because of the technological options for the substitution of other factors for nonrenewable resources [11]. More recently, Solow stated, "It is of the essence that production cannot take place without some use of natural resources." Clearly, there is need for more analytical and empirical work (some of which we provide in later chapters) on the relation between economic production and natural resources, especially energy, and how much of the resources are actually needed.

The conventional neoclassical view of the low importance of energy and materials goes back to the early days of neoclassical economics. Initially, the focus was not so much on the generation of wealth but rather on the "efficiency of markets" and the distribution of wealth. As a consequence, one started with a model of pure exchange of goods without considering their production. With a set of mathematical assumptions on rational consumer behavior, it was shown that through the exchange of goods in markets an equilibrium situation results in which all consumers maximize their utility. This benefit of (perfect) markets is generally considered the foundation of free market economics. It shows why markets, where greedy (or at least self-regarding) individuals meet, work at all. Later, when the model was extended to include production, the problem of the physical generation of wealth had to be inseparably coupled to the problem of the distribution of wealth as a consequence of the model structure. In the neoclassical equilibrium, with profit maximizing entrepreneurial behavior, factor productivities (e.g., the respective contribution of capital, labor, and energy) equal factor prices. This means that in conventional economic analysis the weights that the production factors contribute to the physical generation of wealth are determined by the factor cost shares. Thus energy's importance is assumed by most economists to be equal (only) to its cost.

Unlike their classical predecessors, neoclassical economists do not even bother to include the process of how things are actually made in their analyses. They just take the input prices, put them into a function, and the price and quantity of output is automatically generated. Here lies the historical source of the economists' underestimation of energy as a production factor, because in industrial market economies energy cost, on the average, is only 5–6% of the total factor cost (and of GDP). Therefore, economists either neglect energy as a factor of production altogether, or they argue that the contribution of a change of energy input to the change of output is equal only to energy's small cost share of 5–6%. This has led to a long-lasting debate on the impact of the two energy price explosions in the years 1973–1975 and 1979–1981 when the cost of energy increased to 14% of GDP even while supplying less physical energy. As we show below and more explicitly in [5 and especially 6], energy is more important in production than either labor or capital, although all three are needed. Curiously energy's low price is the reason for its importance, not its unimportance. For 200 years the economy has received huge benefits from energy without having to divert much of its output to get it. This is because basically we do not pay Nature for

energy, but only the cost of exploiting it. Likewise, the finite emission absorption capacity of the biosphere is more important to future economic growth than its present (nearly vanishing) price seems to indicate.

Neoclassical models built on the assumptions of Fig. 5.1 or 4.2 cannot explain the empirically observed growth of output by the growth of the factor inputs. There always remains a large residual (i.e., a statistical "leftover" that is not explained by the factors, in this case capital and labor, used in the analysis). This is formally attributed to what economists call either 'technological progress' or improvements in "human capital," which are long-term increases in skill and education of workers. Even Robert Solow stated, "This... has led to a criticism of the neoclassical model: it is a theory of growth that leaves the main factor in economic growth unexplained." [11] As we argue below, weighting a factor by its cost share is an incorrect approach in growth theory.

In fact the human economy uses fossil and other fuels to support and empower labor and to produce and utilize capital. Energy, capital, and labor are then combined to upgrade natural resources to useful goods and services. Therefore economic production can be viewed as the process of upgrading matter into highly ordered (thermodynamically improbable) structures, both physical structures and information. Where the economist speaks of "adding value" at successive stages of production, one may also speak of "adding order" to matter through the use of free energy (exergy). The perspective of examining economics in the "hard sphere" of physical production, where energy and material stocks and flows are important, is called biophysical economics. It must complement the social sphere perspective.

Criticism 2: Boundaries

The basic model used in neoclassical economics (Fig. 5.1) does not include boundaries that in any way indicate the physical requirements or effects of economic activities. We believe that at a bare minimum Fig. 5.1 should be reconstructed as Fig. 5.2 to include necessary resources and generation of wastes. Taking this assessment one step further, we believe that something like Fig. 5.3 is the diagram that should be used to represent in more detail the physical reality of an economy's working. It shows the flow of energy and matter across the boundary separating the reservoirs of these "gifts of Nature" from the realm of cultural transformation within which sub boundaries

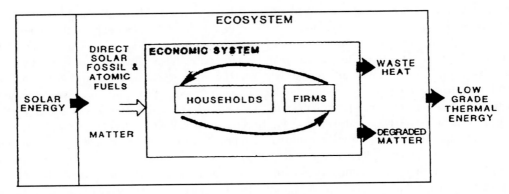

Fig. 5.2 Our perspective, based on a biophysical viewpoint, of the minimum changes required to make Fig. 5.1 conform to reality. We have added the basic energy and material input and output that is essential if the economic processes represented in Fig. 5.1 are to take place (*Source*: Daly 1977). See also Fig. 4.2

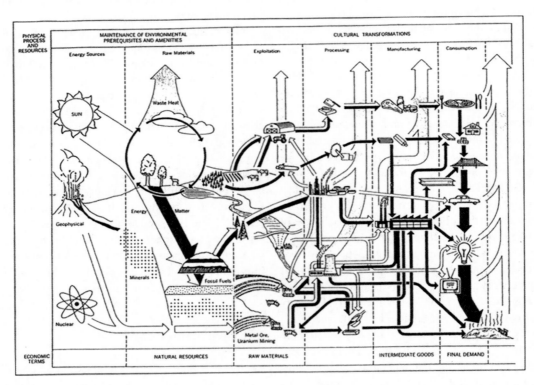

Fig. 5.3 A more comprehensive and accurate model of how real economic systems work. This is the minimal conceptual model that we would accept to represent how real economies actually work. Natural energies drive geological, biological, and chemical cycles that produce natural resources and public service functions. Extractive sectors use economic energies to exploit natural resources and convert them to raw materials. Raw materials are used by manufacturing and other intermediate sectors to produce final goods and services. These final goods and services are distributed by the commercial sector to final demand. Eventually, non recycled materials and waste heat return to the environment as waste [5]

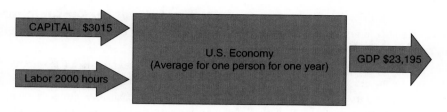

Fig. 5.4 A conventional economist's view (or perhaps a caricature of that) of one person's input and output to the process of economic production for 1990 (*Source*: Hall et al. 2008)

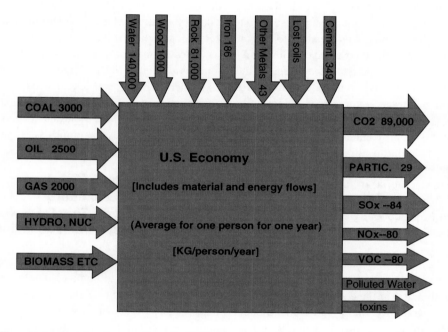

Fig. 5.5 The actual material and energy flows (in Kg/Year) associated with one person's involvement in the economy for the same year (*Source*: Hall et al. 2008 [12])

indicate the different stages of their further transformation into the goods and services of final demand. Such a diagram could be presented to every student in an introductory economics course so that the ways the economic process operates in the real world are properly understood. Another way of reflecting the necessary changes is that Fig. 5.4 shows the standard economist's view of one person's role in the economy, whereas Fig. 5.5 gives the biophysical perspective of what biophysical materials are actually needed to operate the economy for one person for 1 year.

Criticism 3: Validation

Natural scientists expect theoretical models to be tested before being applied or developed further. Unfortunately, economic policy with far-reaching consequences is often based on economic models that, although elegant and widely accepted, are not validated. Some economists do attempt to construct and verify hypotheses, they rarely, if ever, attempt this on the most basic assumptions of the neoclassical model. Validation also proves difficult or impossible because both classical and neoclassical theories were originally developed using concepts of production factors as they existed in agrarian societies [13]. These theories have been transferred more or less unchanged to applications in the modern industrial world. No provisions have been added to the basic theory for industrialization and its consequences. As the Nobel laureate in economics, Wassily Leontief, noted [14], many economic models are unable "to advance, in any perceptible way, a systematic understanding of the structure and the operations of a real economic system;" instead, they are based on "sets of more or less plausible but entirely arbitrary assumptions" leading to "precisely stated but irrelevant theoretical conclusions."

Most non economists do not appreciate the degree to which contemporary economics is laden with arbitrary assumptions. Nominally objective operations, such as determining the least cost for a project, evaluating costs and benefits, or calculating the total cost of a project, normally use explicit and supposedly objective economic criteria. In fact, such "objective" analyses, based on arbitrary and convenient assumptions, produce logically and mathematically tractable, but not necessarily correct, models.

The authority economists often assign to their "physics-based" models, starting with the basic neoclassical model of the economy, is somewhat curious. Unavoidably fuzzy economic models do not become precise just because they use, for example, Hamiltonians in analogy to the Hamiltonians in physics. In fact, in physics a Hamiltonian is an energy function representing the sum of kinetic and potential energy in a system from which one can derive the equations of motion of the particles of the system. In neoclassical production theory the price vector is given by the gradient of the output in the space of the production factors just as the vector of a conservative physical force is given by the gradient of potential energy in real space [15]. The quite imperfect economic analogy should not be confused with the thermodynamically rigorous model in physics.

Myth 2: A Theory of Consumption Can Ignore Actual Human Behavior

Just as neoclassical production assumptions violate principles of physics, its assumptions about human behavior are inconsistent with both a large body of psychological and neurological research, and even everyday human experience. It is well established that real human beings are other-regarding: that is, how one person values a certain economic outcome depends on how much it is valued by others. It is also well established that the consumption of market goods cannot be equated with an individual's happiness. Nevertheless, the fundamental behavioral assumptions of NCE require self-regarding consumers whose happiness depends upon their consumption of market goods. The cultural context of behavior is deemed irrelevant to neoclassical economic analysis as the emphasis is entirely on the behavior of the isolated individual.

Myth 2a: Homo Economics Is a Scientific Model That Does a Good Job of Predicting Human Behavior

At the heart of standard neoclassical economic theory is the model of human behavior embodied in *Homo economics* or "economic man." Economic texts usually begin with a very general statement about human nature that is soon codified into a set of rigid mathematical principles resting upon the idea that "people maximize their well-being by consuming market goods according to self-regarding, consistent, constant, well-ordered, and well-behaved preferences." The assumption that people are self-regarding has been falsified by considerable contemporary work in behavioral economics, neuroeconomics, and game theory [16–18]. For example, Henrich and colleagues, after examining the results of behavioral experiments in 15 societies ranging from hunter-gatherers in Tanzania and Paraguay to nomadic herders in Mongolia concluded: "[T]he canonical [NCE] model is not supported in any society studied."

In experimental settings and under real-world conditions, humans consistently make decisions that favor enforcing social norms over ones that lead to their own material gains [19]. Gintis describes several experiments showing that humans are both far more altruistic and far more vindictive than the NCE "rational" actor model allows. They will make decisions to punish persons they will never again encounter if those people "cheat" in experimental transactions, even if this means considerable monetary loss to themselves.

The centrality of the behavior of isolated individuals is reflected in the notion that consumers are sovereign in a market economy. Ackerman and Heinzerling [20] point out that the rise of economic orthodoxy put consumers at the center of analysis. The idea is that producers respond to consumer preferences rather than the reverse. Yet we all know that, in fact, consumer tastes are manipulated and that firms barrage us with advertising in order to increase their market share. Nonetheless, the centrality and pre-eminence of the individual in orthodox economic analysis precludes any analysis or emphasis on the context of individual behavior.

Myth 2b: Consumption of Market Goods Can Be Equated with Well-Being and Money Is a Universal Substitute for Anything

Most economic texts simply equate utility with happiness and assume that utility can be measured indirectly by income without any substantive or formal discussion of the matter [21]. The higher the income, the better off a particular individual (and hence society) is supposed to be. Yet there is considerable evidence that past a certain point income is a positional good; that is, if everyone's income goes up there is little or no long-term gain in social well-being. This implies that policies designed merely to increase per capita income may have little effect if the goal is to improve social welfare.

Psychologists have long argued and documented that well-being derives from a wide variety of individual, social, and genetic factors. These include genetic predisposition, health, close relationships, marriage, and education, as well as income [21]. It is generally true that people in wealthier countries are happier than people in poorer countries, but even this correlation is weak and the happiness data show many anomalies [22]. For example, some surveys show that people in Nigeria are happier than people in Austria, France, and Japan [23–25]. Past a certain stage of development, increasing incomes do not lead to greater happiness. For example, real per capita income in the United States has increased sharply in recent decades but reported happiness has declined [26].

When economists equate utility with income in the NCE model this effects the policy recommendations of economists which in turn effect the natural world. According to Arrow and colleagues [27], "sustainability" means simply maintaining the discounted flow of income over time. Leaving future generations the same or greater real income than the present leaves them at least as well off no matter what happens to specific features of the natural world. By this reasoning if the present discounted value of a rainforest is $1 billion in ecosystem services if left intact, but can generate a discounted investment flow of $2 billion if it is clear cut and sold, then it is the moral responsibility of the present generation to cut down the rainforest. With $2 billion the future generation could buy another rainforest or something of equal value and have $1 billion left over. This is the logic used by economists to justify the destruction of a substantial portion of the planet's ecosystems and species [28].

How the Neoclassical Model Fails to Deal with Distributional Issues

A different but extremely important and pungent critique of the neoliberal model comes from recent work by John Gowdy [28, 29]. Gowdy takes as his starting point the welfare model of John Rawls. (Here "welfare" is the same as "utility".) The basis of *welfare economics* is that each individual gains welfare proportional to her real disposable income increases. Thus a given individual will be "better off" by a factor of two if she has $2,000 rather than $1,000 to spend (or if prices are half as much). This concept also uses the idea of Pareto optimally. Both the Rawlsian and the Pareto approach assume that there is a linear relation between individual welfare and money. Thus if one individual becomes five times wealthier (say from $1,000 to $ 5,000) that is as great a social good as five people becoming twice as wealthy (say from $1,000 to $2,000 each). This is an important concept that lies behind welfare economics and has been used incessantly as a logic for developmental plans that tend to pay most attention to increasing GDP and relatively little attention to who gets the proceeds. This of course avoids the contentions within the developing world that development tends to enrich those who have while doing little, or even impoverishing, those that have not. By the Rawlsian–Pareto logic, or at least as employed by most contemporary neoclassical economists, if the total wealth is increased the distribution is not important, or at most is quite secondary. The entire economic perspective is often associated with social notions that people are well off or not in accordance with their own efforts rather than due to factors outside their control.

Gowdy's argument with the idea that distribution is not an important issue to economists, is that recent psychological investigation makes the case that human welfare and happiness do not increase linearly with income, but rather are curved downward. Curiously this is a conclusion also reached by thinking about the concept of marginal value—that the first units of something have much more value than additional units—a fact conveniently ignored by marginalist neoclassical economists! Hence supplying poor people with the basic necessities of life generates a greater deal of happiness and welfare with a given amount of money compared to much less happiness or well-being generated by the same amount of money in the hands of someone who is well off. Finally, according to Gowdy and Gintis the extensive social research done in recent years has completely undermined the "value neutral" assumptions that are the base of welfare and neoclassical economics, and calls into question all of the basic tenets of neoclassical economics.

Why Theory Matters

It is in the policy arena that the ideological nature of NCE reveals itself most completely. Most economists substitute the mythical neoclassical world of rational agents, certainty, and perfect information for the complex reality and uncertainty of real economies. Where reality and the neoclassical model disagree, reality is increasingly forced through policy to conform to the neoclassical model [30]. Neoclassical economists generally assume that people always respond rationally and consistently to price signals, therefore the goal of economic policy is to assign property rights and "get the prices right." The corollary assumption is that things of value to people have a price, and anything without a market-formed price must lack value. Prices are theoretically capable of reflecting all the relevant attributes of any good or service and all that people value. The rest of us are asked to take the validity of these assumptions and analyses on faith, and to turn our complex decision making increasingly over to barely regulated markets and cost–benefit analyses. This emphasis frequently leads to fundamental policy-related failures and problems that include the following:

1. The ultimate policy goal of NCE is not to correct any particular problem directly but rather to value correctly the problem in terms of everything else so that the "calculating machine" of the market can establish the pecking order of priorities. The focus on establishing "general market equilibrium" frequently means neglecting essential details of the policy problems under consideration, especially those for which it is difficult or impossible to determine a price (i.e., oil depletion, environmental degradation, and global climate change). Hence when we purchase a gallon of gasoline we pay only for getting that gallon to the pump, not for finding a new gallon to replace it, or something else if oil depletion makes replacement impossible.

2. The NCE model makes no qualitative difference between needs and wants, or among commodities produced, or among specific productive inputs, including energy. Everything we find useful is treated like an abstract commodity

substitutable for and by anything else. Absolute scarcity does not exist nor, within certain broad limits, are any specific conditions deemed necessary for human existence. Value is a relative matter expressed in relative prices. Because no single thing is essential, substitution among resources and commodities will occur until the marginal value of a commodity divided by its prices is the same for all commodities. At this point rational individuals have made optimal choices, and the sum of all optimal choices leads us to the "best of all possible worlds." Thus the tastes of affluent teenagers in malls for unnecessary but heavily advertised clothes or gadgets is given as much weight per dollar spent as health care or education for the less affluent.

3. The model assumes that aggregate income is a complete and sufficient measure of well-being. Operationally this means that total costs and benefits of policies can be determined by merely adding the monetary changes in the incomes of all isolated individuals affected. This implies that relative income effects don't matter to the individual; for example, a loss of $1,000 to a poor person can be more than compensated for by a gain in $1,100 to a billionaire. Similarly, preferences are considered to be exogenous to social context. Yet numerous studies have found that relative income effects matter and sometimes these effects can completely cancel out increases in total income which is always the primary goal of NCE. How much one person values a gain or loss depends on what others get, the income of each person relative to others, the "fairness" of the income change, and a variety of other social factors that are not included in the NCE model.

4. "Sustainability" in the NCE model means sustaining only the discounted flow of per capita income, not anything else such as biodiversity, oil stocks, human health, or social cohesiveness. This is known as weak sustainability. However, to live within nature's limits, we need to arrive at the conditions of strong sustainability, which requires that the profits from the depletion of a resource or degradation of an ecosystem are reinvested in developing alternatives or restoring degraded systems.

This entails looking at the bigger picture of how market systems function and interface with the biophysical world [12, 31]. Consequently one cannot arrive at a social decision to achieve an optimal macroeconomic scale by merely aggregating many separate efficient market outcomes.

NCE dominates policy making yet provides an inadequate toolbox for confronting the major problems of the present world: global climate change, biodiversity loss, oil depletion, loss of wilderness, and the recalcitrant problems of poverty and social conflict. It has been used as the basis for "the Washington consensus" which has been and continues to be exported to the developing world with essentially no assessment of its effectiveness or basis in reality and with enormous social and environmental problems [12, 31, 32]. We are led to believe that our most pressing environmental and social problems can be dealt with by simulating efficient market outcomes as if this alone provides the elixir for all that ails us. Yet we know that the concept of market efficiency rests on an untenable and faulty foundation and that the real market economy is not best described in this framework. The perpetuation of neoclassical economics, usually to the exclusion of other possible approaches, is essentially the substitution of faith for reason, science and empirical testing in many areas of economics. We must move beyond this faith-based economics and find a more illuminating way of understanding economic activity and informing decision making so that our policies will amount to something more than window dressing for the status quo.

Questions

1. What are some of the "myths" of neoclassical economics? Do you agree that these are myths? Why or why not?

2. Why is the circular flow model of the economy inconsistent with the laws of thermodynamics? Is that possible?

3. Nicholas Georgescu-Roegen, and his student Herman Daly are economists. Why are they such critics of conventional economics?

4. Economic productivity in neoclassical economics is usually represented as a function of capital and labor. Do you agree with that perspective? Why or why not?

5. What, in your opinion, should be the proper boundaries to be used in economic analysis? Can you draw a picture of how you would represent these boundaries?

6. What does validation mean? Why is this often difficult for economic models?

7. What are thought to be (within conventional economics) the main characteristics of *Homo economicus* (or "economic man").

8. Do you think that having greater amounts of money to spend will make you happier? Why or why not? Do you think wealthier people that you know are happier than poorer people?

9. Does an increase in income of, say, $1,000 have the same meaning for a wealthy person as for a poor person? How does that relate to the usual economist's position on Pareto optimality?

10. Why have neoclassical economists attempted to generate a "value neutral" approach to economics? To what degree have they succeeded, in your opinion?

11. Why does theory matter in economics?

12. What does sustainability usually mean within conventional economics? What might be some problems with that definition?

References

1. This chapter is derived in part from J. Gowdy, C. Hall, K. Klitgaard and L. Krall 2010. The End of Faith-Based Economics. The Corporate Examiner 37:5–11, and Hall, C., Lindenberger, D., Kummel, R., Kroeger, T. and Eichhorn, W., 2001. The need to reintegrate the natural sciences with economics. BioScience, 51: 663–673.

2. Georgescu-Roegen, N. 1975. Energy and economic myths. Southern Economic Journal, 41: 347–381.

3. Daly, H., 1977. Steady-State Economics. W. H. Freeman, San Francisco.

4. Cleveland, C., Costanza, R., Hall, C., and Kaufmann, R., 1984. Energy and the U.S. economy: a Biophysical perspective. Science, 225: 890–897.

5. Hall, C., Cleveland, C. and Kaufmann, R., 1986. Energy and Resource Quality: The Ecology of the Economic Process. Wiley Interscience, New York.

6. Hall, C., Lindenberger, D., Kummel, R., Kroeger, T. and Eichhorn, W., 2001. The need to reintegrate the natural sciences with economics. *BioScience*, 51: 663–673. Also: Kummel R., J. Henn, D. Lindenberger. 2002. Capital, labor, energy and creativity: modeling. Structural Change and Economic Dynamics. 3 (2002) 415–433

7. Wilson, E., 1998. Consilience: The Unity of Knowledge. Alfred Knopf, New York.

8. Denison, E.F., 1989. Estimates of Productivity Change by Industry, an Evaluation and an Alternative. The Brookings Institution, Washington, DC.

9. Ayres, R. and Warr, D., 2005. Accounting for growth: the role of physical work. Change and Economic Dynamics. 16: 211–220.

10. Cleveland, C.J., 1991. Natural Resource Scarcity and Economic Growth Revisited: Economic and Biophysical Perspectives. Pages 289–317. in Costanza R., ed. Ecological Economics: The Science and Management of Sustainability. New York: Columbia University Press.

11. Solow, RM. 1974. The economics of resources or the resources of economics. American Economic Review 66: 1–14. Also Solow, R.M., 1994. Perspectives on growth theory. Journal of Economic Perspectives 8, 45–54.

12. LeClerc, G. and Charles, H., (Eds.) 2008. Making development work: A new Role for science. University of New Mexico Press. Albuquerque

13. McCauley, J.L. and C. M. Kuffner. 2004. Economic System Dynamics. Discrete Dynamics in Nature and Society 1 (2004): pp. 213–220

14. Leontief W. 1982. Academic economics. Science 217: 104.

15. Mirowski, P. 1989. More Heat than Light. Cambridge: Cambridge University Press.

16. Gintis, H., 2000. Beyond Homo economicus: evidence from experimental economics. Ecological Economics. 35: 311–322.

17. Camerer, C., and Loewenstein, G., 2004. Behavioral economics: past present and future. In: Camerer, C, Loewenstein, G. and Rabin, M (Editors), Advances in Behavioral Economics. Princeton U. Press, Princeton, NJ and Oxford UK 3–52.

18. Henrich, J. et al., 2001. Cooperation, reciprocity and punishment in fifteen small-scale societies. American Economics Review, 91: 73–78.

19. http://www.youtube.com/watch?v=u6XAPnuFjJc

20. Ackerman, F. and Heinzerling, L., 2004. Priceless: On Knowing the Price of Everything and the Value of Nothing. The New Press, New York and London.

21. Frey, B. and Stutzer, A., 2002. Happiness and Economics: How the Economy and Institutions Affect Well-Being. Princeton University Press, Princeton, NJ.

22. Diener, ES., Diener, M. and Diener, C., 1995. Factors predicting the well-being of nations. Journal of Social Psychology, 69: 851–864.

23. Brickman, P., Coates, D. and Janoff-Bulman, R., 1978. Lottery winners and accident victims: Is happiness

relative? Journal of Personality and Social Psychology 36: 917–927.

24. Blanchflower, D. and Oswald, D. 2000. Well-Being over Time in Britain and the U.S.A. NBER Working Paper No.7481, National Bureau of Economic Analysis. Cambridge, MA.

25. Lane, R., 2000. The Loss of Happiness in Market Economies. Yale University Press, New Haven and London.

26. Meyers, D., 2000. The funds, friends, and faith of happy people. American Psychologist. 55: 56–6.

27. Arrow, K., Dasgupta, P., Goulder, L., Daily, G., Ehrlich, P., Heal, G., Levin, S., Goran-Maler, K., Schneider, S., Starrett, D., Walker, B., 2004. Are we consuming too much? Journal of Economic Perspectives. 18: 147–172.

28. Gowdy, J., 2004. The revolution in welfare economics and its implications for environmental valuation. Land Economics. 80: 239–257.

29. Gowdy, J. and J. Erickson. 2005. The approach of ecological economics. Cambridge Journal of Economics. 29 (2): 207–222.

30. Makgetla, N. and Sideman, R., 1989. The applicability of law and economics to policymaking in the third world. Jour. of Econ. Issues, 23: 35–78.

31. Hall, C., 2000. Quantifying Sustainable Development: the Future of Tropical Economies. Academic Press, San Diego.

32. Hall, C.A. S., Pontius, R.G. Coleman L. and Ko J-Y. 1994. The environmental consequences of having a baby in the United State. Population and the Environment. 15: 505–523.

The Petroleum Revolution II: Concentrated Power and Concentrated Industries

6

Introduction

In our first chapter we developed the link between the historical development of energy sources and the development of human society. More energy has allowed humans to do more work, including that of producing more humans. We use the joule, for those not steeped in physical science, as the standard measure of energy. One joule is the amount of energy needed to lift a mass of 1 kg a distance of 1 m on the surface of the earth. A joule is equal to about one quarter of a calorie. Our more familiar unit is the kilocalorie (often written as Calorie), and is found, for example, on the back of food packages. One kilocalorie is 1,000 cal, equal to about 4.18 kJ. Thus if you consume a drink that says it has 100 Calories it you will have consumed 418 KJ. Later, in Chap. 8, we explore the relation between energy and power from a scientific perspective. Power is the rate of doing work, and is commonly measured in watts. From the standpoint of physics, *power* is energy used or expended per unit of time, or the work that power causes or allows to be done. The most common unit of power is the watt, where 1 W = 1 J used/second.

But power means something else in a political and economic context, and here we want to extend the definition to ways in which power is used in the social sciences and day-to-day life more generally. English can be a difficult language for many people to learn because the same word, in this case power, can mean very different things. According to *The Oxford English Dictionary* power means, in addition to the sense used in physics, the "possession of control or authority over others, or a movement to enhance the status or influence of a specified group, lifestyle, etc." This definition seems equally appropriate to social realms, and this chapter reflects both perspectives—physical and social—on power. In most cases there is physical power behind any economic or social power. The latter cannot be measured as clearly and explicitly as can physical power but all are clearly related. When the physical power to run an economy was solar, the economic and political power tended to be more widely distributed. The increased use of fossil fuels, which are concentrated energy, tends to concentrate both economic and political power.

Petroleum and Economic Concentration

In Chap. 1 we developed the concept that control over energy and power, in the scientific sense, led to increased output and an increase in status, wealth, and power in the social sense. The development of petroleum fuels allowed a previously unimaginable increase in the ability to do physical work as well as unheard-of concentrations of economic power. This is true both for nations (the United States throughout the last century, Britain and Germany during the previous century) and for corporations or individuals. There has never been, and probably never will be, an energy source as concentrated, transportable and flexible as petroleum. At the same time, there have been few, if any, industries as concentrated in the economic sense as "the old house" of Standard Oil. Concentrated economic and physical power emerged together in the United States, and elsewhere. During the past century many hundreds of small oil companies coalesced into the "seven sisters" that essentially controlled global exploration and production. This revolution in industrial structure and the large monopolized firm occurred during the same historical time period, and not merely by coincidence.

Economic concentration is synonymous with the process of monopolization. We use the term *monopoly* not in the narrow context of an industry that consists of a single seller, but in the broader meaning of an industry being dominated by a few very large companies. (The technical term for this is *oligopoly*). In most of the developed world monopolized or concentrated industry is neither rare nor an anomaly. This is true despite the textbook model of businesses favored by mainstream economists: competitive industries of many powerless firms operating in impersonal markets that allocate resources with maximum efficiency. Rather, economic concentration is an explicit strategy on the part of firms themselves to control their economic environments and protect their opportunities to achieve profits in the long run [1]. The economic power controlled by a firm is regularly threatened by a host of internal and external forces: new products and markets, technological change, government regulation, and most importantly, the rise of excess capacity and ruinous price competition. If a firm expands its productive capacity and then fails to sell the products, or can sell the products only at lower prices, its profits evaporate. The history of big business is largely the story of coping with excess capacity and avoiding price competition, often by getting favorable consideration by government. Perhaps no person stated the desperation business felt for a strategy to protect profits from price cutting better than nineteenth-century steel magnate Andrew Carnegie.

"Political economy says goods will not be produced at less than cost. This was true when Adam Smith wrote, but it is not quite true today. ... As manufacturing is carried on today, in enormous establishments with five or ten millions of dollars invested and with thousands of workers, it costs the manufacturer much less to run at a loss ... than to check his production. Stoppage would be serious indeed. While continuing to produce may be costly, the manufacturer knows too well that stoppage would be ruin... . Manufacturers have balanced their books year after year only to find their capital reduced at each successive balance... . It is in soil thus prepared that anything promising relief is gladly welcomed. The manufacturers are in the position of patients that have tried in vain every doctor of the regular school for years, and are now liable to become the victim of any quack that appears. Combinations, syndicates, trusts—they are willing to try anything" [2].

Initially Carnegie was of the mind that combinations of firms to control prices by controlling markets (i.e., monopolies) were folly. Carnegie Steel was a technologically dynamic company that could benefit from price cutting because it could out-produce all its rivals at a lower cost. Initially the company sought competitive advantage by cutting its prices, and buying up its weakened competitors, not through monopolies. Yet Carnegie Steel would eventually become the core of "the steel trust" monopoly as U.S. Steel (itself absorbed by the interests of banker J.P. Morgan). As we show, the same phenomenon of concentration by means of price cutting would characterize the largest trust of the era and the champion of the petroleum revolution, Standard Oil.

Why Study Monopoly

Figure 1.13 indicates that we believe that a new set of abstractions and economic theory must be developed for the second half of the age of oil. All theories of how the economy works commonly used today were developed in the age of rising oil availability and high energy returns on investment. To build a new theory we need not abandon everything from the past. Rather we need to refine prior approaches and adapt them to a new era of biophysical constraints and limits to growth. But, more than anything, we need to begin this theoretical development from the point of understanding the economy as it actually exists, which is not simply a collection of small powerless companies who accept passively the impersonal forces of the market and forego large economic profit in the interests of low consumer prices and a stable general equilibrium. Rather the economy as it actually exists is dominated by giant corporations, operating on a national and international basis. These companies want to control market forces that threaten not only short-term profits but also their long-term growth in profits. These forces include ruinous price competition, rising costs of production, periodic recessions, excess capacity, unwelcome taxation and regulation, and the destabilizing effects of rapid technological change.

The study of the concentrated economy is important beyond the microeconomic level of the individual producer or consumer; the effects of monopolization are equally, if not more, important for the overall, or aggregate, macroeconomy. Scholars such as Paul Baran and Paul Sweezy argue that a monopolized economy tends to stagnate rather than grow because of the internal dynamics of capital formation, as well as pricing and output decisions on the part of the large-scale firm in a concentrated industry. In simpler terms, a concentrated economy cannot always create the growth needed to provide other laudable social goals such as full employment and poverty reduction. The solutions to Depression-borne problems of the nineteenth century, which often favored corporate growth, have become the cause of different economic and social problems in the twentieth and twenty-first centuries. Market economies suffer from essentially two sets of limits. The first is the familiar set of internal limits revolving around the process of capital formation and investment, business cycles, and the uncertainties of competition with other firms. Strategies of industrial concentration were first developed to transcend this set of limits. The second half of the age of oil will see an increased impact of the second limits: a suite of external, or biophysical, limits to growth as the raw materials necessary for the earlier strategy of growth become increasingly limited. Economics in the second half of the age of oil will require an understanding of the interaction of both the internal and the biophysical limits to economic activity. In this new era, continued growth is highly unlikely because of biophysical constraints such as peak oil, declining EROI, climate change, and degradation of our oceans and soil fertility. Let us begin with the study of the petroleum revolution in the context of the concentration of economic power and the development of large-scale industry.

Petroleum and the Social Revolution

In 1850 the "civilized" world was illuminated at night principally by whale oil, which was undergoing its own peak and decline as species after species of whales were hunted to near extinction (Fig. 3.4e). By the late 1850s kerosene was being refined regularly in Europe from crude oil obtained from hand-dug pits. The invention of a lamp with a glass chimney that would reduce the smoke and brighten the flame contributed greatly to the demand for kerosene. But kerosene could become "the new light" only if adequate and cheap supplies could be located. The limiting factor was the cost of hand-digging pits; the solution was to be found in well-boring, soon to be known as drilling. The first commercially viable oil well in the United States was drilled by a promoter named Edwin Drake, given the appellation of "Colonel" by his supporting bankers to impress the rural population. Drake and his drillers struck oil in August 1859. Within a year and a half of Drake's successful well another 75 wells were producing oil. Early successes created new boom towns such as Pithole and Oil City. Production in the oil regions of Pennsylvania soared from 300 barrels per day in 1859 to 3 million in 1861. As a result of the surge in supply prices fell from $10 per barrel in January of 1861 to ten cents per barrel in June of 1861. Within a year demand expanded and oil prices rose again to over $7 per barrel.

As we enter the second half of the age of oil, an age characterized by increasing shortages, declining energy returns on investment, and rising prices, one should not forget that the history of the first half of the age of oil was quite the opposite: high EROIs, periodically plummeting prices, and overproduction. During the 1860s and 1870s many small producers began to merge. This increasing monopoly concentration appears to have been a strategy to cope with the falling profits, prices, and bankruptcy caused by overproduction of an easily obtainable resource. Moreover, the legal basis of the new industry stemmed from the old English common law principle of the "Rule of Capture." The petroleum beneath the ground belonged to the owner of the land above. But because the oil beneath was part of a common pool that could be depleted by a few, the incentive was to extract as much as possible as soon as possible in a process known as "flush production." No place in the oil regions serves as a better example of the excesses of flush production and speculation than the town of Pithole. With the discovery of oil, property values soared, especially as oil production increased to over 6,000 barrels per day. Derricks were erected on myriad tiny lots. Rapid extraction damaged the underlying strata leaving a large share of the petroleum unextractable, due largely to the collapse of underground pressure. Property values and the town too collapsed.

Despite the demise of Pithole, production in the Pennsylvania oil region as a whole continued to increase reaching 3.6 million barrels a year by the end of the Civil War. Given this much production producers struggled to find adequate markets for the output, another problem that characterized the industry in its prepeak years. Crude pipelines were constructed to avoid the bottlenecks imposed by poor roads and recalcitrant teamsters, and the Titusville Oil Exchange opened in 1871 in an attempt to shorten the link between supply and demand. It was on this exchange that the present structure of long-term contract prices, short-term "spot market" prices, and very long-term futures markets were established [3].

Once the chimney lantern became common in the United States the expansion of demand for kerosene was largely a function of the general economic expansion and political stability that emerged at the end of the Civil War. This rise in economic activity affected many of the country's primary industries and would be accompanied by an increase in the scale and scope of both manufacturing and transportation. The postbellum period saw the creation of the national corporation, the expansion of long-lived fixed capital, and the replacement of the craftsperson operating on a local scale with semiskilled operative labor and centralized management. It was also the beginning of the nation's dependency upon fossil fuels. The energy density of concentrated fuels combined with the new organization of labor produced dramatic increases in productivity and output. The new large-scale industry opened opportunities for large-scale businesses to control factors often left to chance in the older competitive economy.

The Rise of Standard Oil

No company is as closely associated with the concentration of economic power as Standard Oil. Standard Oil began modestly as a trading partnership in post-Civil War Cleveland, Ohio, and rose to become the largest and most powerful company in the nation and the world's first multinational corporation by the end of the nineteenth century. By the middle of the twentieth century it was the largest corporation in the world. Standard Oil originally rose to power in the first stage of the petroleum revolution: the provision of kerosene for illumination.

The construction of a new rail line into Cleveland, Ohio, which had access to the Great Lakes and proximity to the Pennsylvania oil regions made Cleveland an ideal center for petroleum refining. By 1865 a young general merchant by the name of John David Rockefeller had become the largest refiner in the city. The refining industry was still competitive and the techniques of refining simple enough to preclude advanced technology as a barrier to entry. The result was a large number of small producers and intense price competition. As Rockefeller's refining capacity grew he realized he needed to find markets to absorb the output. To assure profitability Rockefeller developed a multipronged strategy, centered upon the production of a high-quality product at a lower cost than his competitors. The very name Standard Oil stems from the quality of the company's kerosene. Standard Oil was able to control quality so that Standard's kerosene contained a negligible quantity of the dangerous by-product, gasoline. Cost control was accomplished by a combination of large-scale production, reduction of transportation costs, and *vertical integration*, that is, amassing all stages of an industry from refining, to marketing, to transportation on an in-house basis. It was only later, when new oil fields were discovered, that Standard integrated backwards into oil extraction.

The primary method Standard used to reduce transportation cost was the use of the railroad rebate, which was enabled by Standard's scale of production. Business historian Alfred Chandler reports that in 1872 the first railroad approached the Lake shore running from Cleveland to New York City, and willingly reduced transportation costs per barrel from $2.00 to $1.35 in return for a guarantee that Standard would supply 60 carloads of oil per day to be transported. The increased output benefited the railroad as well as Standard by allowing a more consistent use of the railroad's capacity [4]. Standard then extended the policy of extracting rebates on the shipment of their oil to receiving a rebate, or drawback, of 25 cents per barrel, on the shipment of their *competitor's* oil. According to energy analyst Daniel Yergin: "For what this practice really meant was that its competitors were, unknowingly, subsidizing Standard Oil. Few of its other business practices did as much to rouse public antipathy toward Standard Oil as these drawbacks—when eventually they became known" [3].

The problems of price instability, cost control, and capacity utilization, a regular feature of the industry since its inception, were exacerbated by the decline in overall economic activity following a financial panic in 1871 and the subsequent depression that lasted from 1873 to 1879. Chandler reports that the index of wholesale commodity prices, which stood at 151 in 1869 fell to 82 in 1886 [4]. Standard's production of

refined kerosene continued to rise over the course of the 1870s but its ability to market its product at a profitable price did not. Standard's strategy to address the threat of ruinous price competition was consolidation. In today's technical terms this is called *horizontal integration* or the absorption of potential competitors for the purpose of controlling market price. Thus Standard undertook both vertical and horizontal integration, and became increasingly the only game in town.

Merger was Standard's favored means of consolidation, and price cutting was its tactical method. Lower costs of production, made possible by economies of scale and cheaper transportation costs, allowed the company to undersell potential rivals. When faced with an independent producer that would not sell willingly, Standard subjected them to "a good sweating." They would increase output until the market price dropped below the rival's cost of production. Standard would then purchase the nearly bankrupt company at a favorable price, and then restrict output so the price would once again climb. In the process they brought the most able executives into Standard management. By 1881 Standard controlled 90% of the kerosene market and sold 70% of its output in Europe. By the mid-1880s Standard controlled 80% of marketing as well [4]. Despite the greatest degree of monopoly control that the nation has ever seen, the Standard alliance remained vulnerable to outside forces and reacted in a number of different ways to dissipate those threats and bring stability and control to the market for petroleum.

Price competition was not the only threat that faced Standard. Others included new sources of supply and new modes of transportation, as well as legal challenges. One threat was the attempt of independent producers to break Standard's hold of railroad transportation by building their own pipeline from the oil regions to the markets in the eastern United States. Standard then quietly acquired an interest in the Tidewater pipeline in 1879 and gained effective control of pipeline transportation within 2 years. Another problem was the discovery of fields outside of the Pennsylvania oil regions, first in Lima, Ohio. The additional production flooded the market and resulted in a price decline. After much debate, Standard's interests became directly involved in production, circumventing the oil exchanges. By 1891 Standard controlled approximately 25% of oil production. Standard had succeeded in building a truly integrated company, from extraction, to refining and transportation to marketing [3].

By the mid-1890s Standard had become a fully consolidated and vertically integrated company. This form of business organization allowed Standard to withstand legal challenges to its strategy of price control by means of merging with competitors and fixing prices. Control over prices and ruinous competition was codified beyond a mere association of producers in 1882 when Standard formed the perfectly legal Standard Oil Trust. Stock shares of the various operating companies were ceded to Standard Oil of Ohio in return for trust certificates. Decisions about the direction of the company were made by a set of directors acting on behalf of the shareholders of the Standard Oil Trust rather than in the interests of the separate operating companies. Although popular lore focuses upon price fixing, the first actions of the trust were to control costs. They reduced the number of refineries and concentrated production. Forty percent of output was produced by three refineries and the average cost per gallon of refined oil fell from 1.5 cents to 0.5 cents. Standard expanded their marketing apparatus to assure adequate outlets for the newly expanded production, establishing wholly owned subsidiaries Continental Oil and Standard Oil of Kentucky as marketing companies [4]. Popular opinion and outrage led to the passage of the Sherman Anti-Trust act in 1890, which banned conspiracies in restraint of trade. However, the Sherman Act was not intended to address the benefits of cost reduction by means of vertical integration, only price fixing due to horizontal integration. The cost cutting by expanding scale and controlling market allowed Standard to survive three significant challenges to become, by the mid-twentieth century, the largest and most profitable corporation in the world.

Beginning in the 1890s several states filed suit against the Rockefeller interests, as well as against John D. Rockefeller himself. In 1907 the Federal government filed suit in the circuit court alleging that Standard Oil was in violation of the Sherman Anti-Trust Act. The circuit court found in favor of the government and Standard Oil appealed the case to the Supreme Court. In 1911 the Supreme Court validated the decision of the Federal Circuit Court: Standard Oil had conspired to restrain trade. The Standard Oil trust was dissolved into 34 separate operating companies, the most prominent being Standard Oil of New Jersey, Standard Oil of New York (Socony-Vacuum), and Standard Oil of California. Despite the break-up, Standard of New Jersey (later Exxon) remained the second largest industrial corporation in the country [4]. Jersey Standard is of particular interest. In an attempt to circumvent state-level legal challenges, popular opposition to the trust, and lackluster acceptance of the certificates of trust by financial markets, the company took advantage of holding company legislation, recently passed in New Jersey in 1889. The holding company legislation allowed manufacturing companies to purchase the stock of other corporations and issue its own securities for the acquisitions. The holding company replaced the trust as the legal vehicle for consolidation and merger, and provided for even tighter control over the pricing and output decisions than did the trust. More effective and consolidated management was able to exert control over all phases of an operating company [5]. The Standard Oil Trust reincorporated in 1899 as a holding company: Standard Oil of New Jersey. Its capitalization increased from $10 million to $110 million, and it controlled the stock of 41 other companies [3].

Further Challenges to the Standard Empire

A new legal form, vertical integration, and virtual control of the world market for kerosene did not insulate Standard Oil entirely from external threats to their control and profitability. They were to face new challenges at the twilight of the nineteenth century. These challenges came from new and substantial sources of supply, both foreign and domestic, new rivals to production and marketing. The new domestic sources of supply were discovered in Texas, Oklahoma, and California. Along with these discoveries came large and powerful new companies which are today as recognizable as is the name Standard: Texaco, Gulf, and Unocal. Other abundant sources of supply came into production in Russia, Romania, Indonesia, and by the early 1900s, Persia. New international companies such as Royal Dutch/Shell and BP were born of these discoveries. Another fundamental transformation of the petroleum industry occurred in this same period: the eclipse of kerosene by the electric light. Next another new innovation, the gasoline-powered automobile, would give vast new sources of growth and profit to the petroleum industry.

New Sources and New Rivals

Standard Oil initially satisfied domestic and world demand from its Pennsylvania oil fields. That was to change in the latter decades of the nineteenth century. The existence of oil on the shores of the Caspian Sea had been chronicled by Marco Polo. The first wells replaced hand-dug pits by 1872 and by 1873 some 20 small refineries were located in the Russian city of Baku. The industry expanded rapidly, from less than 600,000 barrels in 1874 to 23,000,000 barrels in 1888, aided by the financing of the Nobel family. The Nobel Brothers Petroleum Producing Company was fully integrated, both backwards into wells, tankers, and storage facilities and forward into refining and marketing. The demand for kerosene in Russia alone was insufficient to absorb the output of the Baku refineries. The short winter days and need for illumination could not overcome the poverty of the Russian peasantry. Nobel's success brought new competitors in the form of the Rothschilds, who purchased the railroad from Baku to the port of Batum on the Black Sea. Russian kerosene was now able to compete with that of Standard, which had previously controlled European markets. The American company then launched the type of price war that allowed them to consolidate their domestic empire. But the Russian-based compa-

nies fought back. The Nobels established a marketing company in the United Kingdom and the Rothschilds improved the Baku–Batum railroad technically and eventually constructed a pipeline. By 1891 the Russian share of the world's kerosene exports rose to 29%, with a commensurate decline in U.S. exports [3].

The Rothschilds, especially, were plagued with the age-old problem that characterized the industry in the first half of the age of oil: how to market the surplus resulting from the expanded production and refining of the new sources of supply. They turned their sights to East Asia and found an agent by the name of Marcus Samuel to sell their product to a wide network of merchants and traders. In the early 1890s Samuel had developed the bulk tanker to reduce shipping costs, and by 1893 achieved access to the newly opened Suez Canal, cutting 4,000 miles from the traditional route to Asia around the Cape of Good Hope. In the same year Samuel founded a tank syndicate to reduce ruinous price competition in oil storage. By 1902 more than 90% of the oil transported through the Suez Canal was under the control of Samuel's company, Shell Oil. Another threat to Standard's control came after the discovery of oil on the Indonesian (then Dutch East Indies) island of Sumatra. In 1885 the first successful wells were completed and production was concentrated under the auspices of the Royal Dutch Company in 1890. By 1892 Royal Dutch constructed a pipeline from the oil fields to coastal refineries and by 1897 output had increased by five times from a mere 2 years earlier. Standard had previously marketed kerosene in Indonesia and considered Royal Dutch a threat which they desired to incorporate into the Standard operation. Instead, Standard was spurned and negotiations commenced to amalgamate the company soon to be known as Royal Dutch Shell. The Asian producers and marketers wanted a greater degree of concentrated power to withstand what they perceived to be the imminent Standard tactic of price cutting [3]. The new company would survive to become one of the world's majors.

In addition to international challengers to its foreign markets, Standard was subject to declines in its domestic reach two decades before the Supreme Court ordered its dissolution in 1911. First, Pennsylvania independent oil companies, united under the name of Pure Oil, constructed a pipeline to market their output on the east coast of the United States. Second, as early as 1885 it was clear that the output of Pennsylvania fields had peaked and begun serious decline. The state geologist of Pennsylvania stated that "the amazing exhibition of oil is only a temporary and vanishing phenomenon—one which young men will see come to its natural end" [6]. The oil boom of the entire Appalachian basin was already over by 1900. Third, in the early 1890s large fields were discovered in Southern California. By 1910 California's 73,000,000 barrels represented 22% of the world's output, mostly controlled by the independent company Union Oil (now Unocal). Standard finally commenced operation in the California fields, establishing Standard Oil of California (now Chevron) in 1907. However, the monumental change in the oil business occurred in January 1901 with the discovery of the Spindletop field previously mentioned in Chap. 1. The original gusher produced 75,000 barrels per day and a new oil boom had begun. Land values skyrocketed and population soared from 10,000 to 50,000. In an experience similar to the one that occurred in the Pennsylvania oil regions, numerous tiny leases led to more than 400 wells on Spindletop itself. Prices collapsed to 3 cents per barrel. The original promoters needed markets for their oil and found a likely buyer in Marcus Samuel's Shell Oil at a long-term price of 25 cents per barrel. The glut caused by the Spindletop find was augmented by another discovery in Oklahoma. The common problem of overproduction led not only to falling prices but in this case, as in Pithole, flush production depleted the well. Underground pressure gave way in 1902, the year after discovery.

The stabilization of the Texas industry would fall to the Pittsburgh bankers (the Mellons) who had financed the initial operation. The original promoters were dismissed, the contract with Shell renegotiated, and the Mellons began the development of a vertically integrated company based on the extraction and refining of petroleum. Their first task was to come to terms with the overcapacity that the construction of the new refinery and pipeline network created. The corporation that restructured and further integrated into nationwide marketing became known as Gulf Oil. In addition another significant corporation, Texaco, was built upon the expansion of transportation, storage, refinery capacity, and the currying of important political connections. Every discovery would bring a glut of new oil and price declines into the market. This, in turn, created the need for constant expansion into new markets. Standard's control of the industry was clearly in decline. In 1880 Standard controlled 90% of kerosene refining in the country. By 1911, the year of its dissolution, the former monopoly controlled but 65% of domestic kerosene output and its international markets likewise declined in the face of new discoveries and new competitors [3]. Yet while Standard's control was declining its profits and output increased. The new century was to bring the end of the kerosene era but the dramatic expansion of oil demand as we entered the age of the internal combustion engine and the automobile.

Markets Lost and Markets Found

As we have said the primary use of oil in the first stage of the petroleum revolution was for illumination purposes. The market for kerosene, however, was to all but disappear at the end of the nineteenth century. In 1879 Thomas Edison perfected the incandescent light bulb and began operations of a generating plant in 1882. Edison made sure to price electricity competitively. Electricity overcame many of the drawbacks of kerosene such as smoke, soot, and oxygen use. But the adoption of electricity was not immediate. The original generating plants, located near load centers until the adaption of alternating current, were powered by coal-fired piston engines which were very noisy and dirty. Moreover, electricity was considered dangerous and the cause of myriad great fires that swept the urban centers of the Northeastern United States at the dawn of the twentieth century. While a young man, Klitgaard spent many years as a restoration carpenter and saw the reason. He observed and corrected many situations where electricity entered urban dwellings at 240 V over bare wires, with only ceramic insulators separating the wires from the dry roof beams upon which they were placed. But once these safety constraints were overcome technically the use of electricity for light and power caught on quickly. In the time period from 1885 to 1902 demand for lightbulbs soared from 250,000 to 18,000,000 per year. In 1890 only 15% of urban railways and streetcars were powered by electricity. By 1902 94% used electricity as a motive force [4]. Problems with carbon emissions as greenhouse gases had barely been recognized theoretically. The switch to electrical power virtually eliminated the very serious public health problems associated with the use of horses as beasts of burden.

Electricity fundamentally changed the process of production. When factories were powered by a central source, steam or water, the layout of the factory was dictated by distance from the central source, and power was delivered to places of use by a dangerous and inefficient system of pulleys and belts. Factories had to be multistory affairs on a small footprint. Much time was lost to the movement of semifinished goods between floors. The advent of the electric motor allowed sprawling single-story sheds with the power source decentralized to the individual machine. Here again, we see the role of energy in the improvement of productivity. The same process of industrial concentration occurred in the electrical industry itself. In 1892 the New York banker J.P. Morgan consolidated Edison Electric with Thompson–Houston to form General Electric which shared the market only with Westinghouse. In the type of corespective behavior common to oligopolies Westinghouse and GE regularly shared patents [5].

The Age of Gasoline

In the first phase of the petroleum revolution gasoline was a dangerous by-product. But gasoline become the primary petroleum product with the invention of the automobile powered by the internal combustion gasoline engine. Automobiles gained acceptance in Europe by 1895, and soon after began to sweep personal transportation in the United States. Eight thousand cars were registered in 1900. By 1912 nearly a million vehicles were on the road [3]. One year later Ford took advantage of the possibilities afforded by the electric motor and single-story shed production when he built his first assembly line in Highland Park, Michigan. In the early days of the industry automobiles had been assembled by teams of skilled workers, often bicycle mechanics, who built each car from the wheels up. Automobiles were little more than luxury items for the affluent. Ford's Model T, introduced in 1908, sold for $850, then an enormous sum. After the construction of the Highland Park plant cars were assembled by semiskilled operatives on a continuous line. The price of a Model T fell as the cost of production fell with the expansion of scale and an increase in the throughput of materials and labor. By 1925, the peak of the first automobile boom, a Model T sold for $240. Mass production changed the automobile from a luxury item to one that workers could afford. Ford workers were paid above the industry average. Ford nearly doubled industry standard wages when he commenced his famous "$5 day" in 1915, essentially as a cost-saving measure. Previously assembly line work was seen as so degrading that the Ford plants had a difficult time retaining an adequate workforce. Absenteeism was 10.5% and turnover reached 470% in 1913. Turnover costs in 1913 alone were nearly $2 million. So Ford raised wages to keep his workers. "There was… no charity involved…. We wanted to pay these wages so that business would be on a lasting foundation. We were building for the future." A low-wage business is always insecure. The payment of $5 for an 8-hour day was one of the finest cost-cutting moves we ever made (Ford, quoted in Perelman, 2006: 135) [7].

As the price declined, and credit was offered, sales and registrations of automobiles increased steadily, reaching 23 million in 1925. Registrations fell during the Depression, and new cars were not produced during the Second World War, as auto plants were converted to produce tanks and airplanes. Moreover, gasoline and tires were rationed during the war. The second automobile boom commenced following the war and produced lasting effects upon the nation. In 1950 40 million cars were registered in the United States, a figure that climbed to over 65 million in 1962 to more than 250 million by 2007.

The automobile qualifies as what economists call an *epoch-making innovation*. Few other such technological changes qualify. An epoch-making innovation must not only absorb large amounts of capital investment, but must create more opportunities for investment in other industries. Baran and Sweezy contend that only three innovations transformed society, absorbed sufficient capital, and created new industries and processes: the steam engine, the railroads, and the automobile. To this DuBoff adds electrification and Perelman contends that computerization must be considered [5, 7, 8]. The automobile not only absorbed tremendous amounts of fixed capital, accounting for 6.3% of all value added in manufacturing in 1929, but also created myriad peripheral industries. Repair shops, drive-in movies, motels, gas stations, and the fast-food industry owe their existence to the automobile. The automobile itself is dependent upon petroleum for energy. Indeed all epoch-making innovations have been energy-intensive, indeed among the most energy-intensive products of their day. Moreover, these innovations have been subject to a similar degree of industrial concentration as was the petroleum industry, largely for the same reasons: the need to rationalize production, reduce costs, expand market share, and avoid ruinous price competition.

Industrial Concentration as a Consequence of Concentrated Energy

Before the massive use of fossil fuels, production was essentially organized on the basis of small shops using skilled labor. Skilled master craftspeople were generally responsible for all or many stages of production, and agreed to be responsible for the training of apprentices. Upon completion of their apprenticeships new craft workers were deemed fit by the society of masters to travel to obtain independent unsupervised work. In fact they were called journeymen. After a long period of learning not only the myriad skills needed of an all-round craft worker, but also the business aspects of the trade, journeymen could rise to the rank of master. Societies of masters, which were called guilds, decided collectively upon prices and standards of quality. This world of small business did not display the type of price competition found in microeconomics texts. As an institutional structure, guilds limited the type of competition that could ruin a master's fortune. Instead the guilds brought stability to the preindustrial economy. Thus the modern concept that competition is necessary for efficient operation of businesses was not the historical norm.

Few examples existed of alternative organizations. Large-scale textile mills appeared along the swiftly flowing rivers of New England at locations such as Lowell and Lawrence, Massachusetts and Manchester, New Hampshire by the 1820s. These mills not only employed larger numbers of workers than the typical small shop, but they were not organized around the principle that every entry-level worker would become eventually a master. The labor force of the early textile mills consisted mostly of young women recruited from the hardscrabble New England farms, whose employment, frequently boring and even brutal, was expected to be temporary.

In the decades after the Civil War the U.S. economy went through a process that economic historian Richard DuBoff termed "the Grand Traverse" and what we call industrialization or the development of the hydrocarbon economy. This transition entailed the transformation of a primarily local and regional economy utilizing local natural sources of energy into an economy based on large-scale industry, mass production, and the use of fossil energy, generally derived from far away. The railroads were the nation's first big business. Railroad building commenced in earnest in the late 1840s, following the nation's first Depression. There were only about 2,300 miles of track when the decade of the 1840s began. Another 5,100 miles of track were added in the 1840s and 21,400 in the 1850s. After the

Civil War, track building increased significantly. In the 1880s additions to track construction peaked, when another 74,700 miles were built. By the time railroad travel was supplanted by the automobile and freight was hauled primarily by truck, the railroads had established themselves as the nation's first large-scale enterprise. Railroads accounted for 15% of all gross private domestic investment in the 1850s and 18% in the 1870s and 1880s [5]. Moreover railroads helped develop the communications networks, as telegraph wires were built along railroad rights-of way. The construction of a viable transportation and communications infrastructure was vital for the transformation of the economy as a whole. Recall how Standard consolidated its hold on refining by achieving lower cost transportation by means of an existing railroad network. The ability to manage a nationwide market was greatly enhanced by a functioning transportation and communications infrastructure.

The economy was transformed fundamentally in the years following the Great Depression of the 1870s as industrialization increased more and more. Not only did the scale of production increase, but so did the organization of labor. As in the case of Standard Oil the control of costs became a fundamental element in the competition between large enterprises. Jobs were subdivided in a way that Adam Smith himself could barely imagine. The essence of competition became based on increasing productivity. Craft workers were supplanted in manufacturing by an immigrant force of unskilled and semiskilled labor who were willing to do boring repetitive piecework for secure wages. Behind the ability to mechanize, transport, and detail, labor was the access to cheap energy. Business historian Alfred Chandler states the matter succinctly: "Cheap coal permitted the building of large steam-driven factories close to commercial centers and existing pools of labor. In the heat-using industries the factory quickly replaced the artisan and the craftsman.... Coal, then, provided the source of energy that made it possible for the factory to replace the artisan, the small mill owners, and the putting-out system as the basic unit of production in many American industries" [4].

Threats and Opportunities

Chandler also makes the important point that the revolution in transportation, itself based upon cheap energy, further transformed the distribution of products. The modern corporation was not born with the advent of mass production but rather necessitated the unification of mass production with mass distribution. If a company produces more than it can sell, the incentive to produce even more output or invest in capital equipment, declines. This will be a theme that recurs through subsequent years of economic development. Capital accumulation, brought about by investment in capital goods, is the engine of growth in a private enterprise economy. Periods of lagging investment bring about economic downturns, and the low profit potentials of a sluggish economy further reduce the ability to find profitable outlets for one's investment capital. The percentage of Net National Product (or Gross National Product minus Depreciation) that went into investments climbed steadily over the course of the nineteenth century from 10% in the 1840s to 18% in the 1870s to 20% in the 1890s [5]. This growing level of investment aggravated the problems that can occur from producing more than can be consumed. As Andrew Carnegie had realized, large-scale companies would attempt any alternative to shutting down; the consequence of walking away from the considerable costs embodied in the capital equipment was unthinkable.

Various forms of economic concentration, such as vertical integration, horizontal integration, trusts, and holding companies were responses to a number of chronic problems that plagued American enterprises operating in the new world of expanding markets, rapid technological change, financial uncertainty, and the availability of cheap energy. Concentrated fuels certainly opened up vistas of low-cost production and transportation unheard of before the harnessing of fossil fuels, but cheap energy alone was insufficient to protect producers from a set of internal limits to capital accumulation. Viewed in this light, monopoly is not a minor aberration to an otherwise competitive economy. Rather it is the eventual outcome of a competitive process as companies attempt to control their economic environment and protect profits and potential growth by avoiding the type of competitive behavior that could perhaps ruin them. In essence the history of the American industrial revolution is the history of both cheap energy and monopoly concentration, and is understood best as a combination of these factors.

Thus economic concentration emerged not as a mistake in the competitive process, as today's mainstream microeconomic theory would have us believe, but as an explicit strategy.

Even as neoclassical economists were perfecting the elegant theory of the "perfect competition" industrialist, Carnegie, Rockefeller, and other captains of the oil industry were decrying the ruinous effects of "cutthroat competition." For the theorist, price competition was necessary for their view of economic perfection. Resources flowed to their most lucrative use while the market system forced competing firms to produce at the lowest possible cost and pass the savings on to consumers in the form of low prices. In the end the system balanced in a stable equilibrium. The only way to ensure a perfectly competitive equilibrium, however, is to ignore the problem of fixed cost. In fact the initial assumption of the economists of no barriers to entry precludes the analysis of the cost of long-lived fixed productive assets. But industrialists operated in the real world where large-scale industry required substantial investment in fixed capital. If, at the same time, the cost of producing one more unit of output (what economists call marginal cost) is low, real-world producers face a dilemma.

The theory of perfect competition asserts that competition will bring prices down to the level of marginal cost. Theoretical entrepreneurs are willing to accept the going rate of normal profit as all else is competed away. Moreover, the system is stable and there is no tendency to change. But in the real world of business, managers who earned no profit and had no prospects for profit growth would quickly be out of a job. If a real-world industrialist borrowed money to purchase large-scale equipment and then found prices competed

down to the level of producing one more unit of output, the company would never be able to generate revenue sufficient to repay their bondholders and bankers. One may think of railroads, where most of the cost is in tracks and locomotives and little of the cost is in cheap fuel or labor, or in the modern world airlines, for such real-world examples. Chandler summarized the position of the railroads when he said: "Competition between railroads bore little resemblance to competition between traditional small, independent unit commercial or industrial enterprises. Railroad competition presented an entirely new business phenomenon. Never before had a very small number of large enterprises competed for the same business. And never before had competitors been saddled with such high fixed costs. In the 1880s fixed costs, those costs that did not vary with the amount of traffic carried, average two-thirds of total cost. The relentless pressure of such costs quickly convinced the railroad managers that uncontrolled competition for through traffic would be ruinous.... . To railroad managers and investors, the logic of such competition appeared to be bankruptcy for all" [4]. Additional information on this period is given in References [5–10].

The Loss of Worker Power, and the Gain in Financial Power

Productivity continued to rise as the result of the prolonged investment boom and the increase in the energy subsidy to each worker [11–12]. Productivity growth averaged only 1.6% per year from 1889 to 1919. After the 1920–1921 recession until the late 1950s it averaged 2.3% annually. New processes such as electrification increased industrial efficiency and the new technologies of the automobile further reduced the costs of transportation. These innovations, of course, depended upon an ample supply of cheap fossil energy, much of it from the newly discovered sources in California, Texas, and Oklahoma. Unfortunately consumer demand did not increase as rapidly as productivity or organizational innovations such as scientific management, resulting in wage growth that did not keep up with production. The lack of purchasing power combined with the ebbing of the investment boom created the conditions underlying the Great Depression. Automobile sales peaked in 1925, the year before the peak in investment as a whole. Construction of skyscrapers in major Eastern cities ground to a halt. The decline in demand for autos and skyscrapers reduced the demand for steel, and declining demand for steel further reduced the demand for coal. In another blow to investment, a hurricane devastated South Florida, destroyed the railway through South Florida and the Keys promoted by John D. Rockefeller's early partner, Henry Flagler and brought a speculative boom in suburban housing to a close.

Yet even while the real economy was "softening," the demand for financial securities continued to rise, fueled by margin buying. Investors could purchase a stock by putting up only a fraction of the value of the stock (the margin), and borrowing the remainder from their brokers. (This is called leverage today.) The volume of such loans (the broker's call market), according to John Kenneth Galbraith, was the most accurate index of speculation. In the early 1920s the volume of these loans was approximately one to one and a half billion dollars. By 1927 the market increased to a volume of three and one half billion. 1928 saw broker's call loans increase to four billion, and by 1929 six billion dollars. With all this debt-fueled buying, stock prices registered impressive increases throughout the summer of 1929, enhancing the optimism of the market and increasing further the demand for call loans. But reports of the underlying weakness in the real economy began to sap the confidence of some knowledgeable investors throughout the fall of 1929. By October the markets were wavering, although the confidence of investment bankers remained high. Charles Mitchell of National City Bank believed the underlying fundamentals of the economy were sound, and that too much attention was being paid to broker's call loans. Nothing, according to Mitchell, could arrest the upward trend [12].

The Great Crash

On October 29, 1929, however, the stock market collapsed. Stock values plummeted by $26 billion. In relative terms the stock market lost approximately one-third of its September value. The economy was soon plunged into depression. GNP declined by 12.6% from 1929 to 1930, and unemployment increased from 3.2% in 1929 to 8.7% in 1930, peaking at 24.9% in 1933. But how did this happen given that less than 2.5% of Americans owned stock [12]?

The answers lie partly in the weakness of America's banking system. Rural banks, in particular, were chronically undercapitalized and more than 500 per year failed even in good economic times. However, the crisis of bank failures climbed after the stock market crash to include urban money center banks. After the collapse of the stock market, heavily leveraged investors could not repay their brokers who, in turn, could not repay the banks. An additional 1,352 banks (above the normal 500) failed by the end of 1930. Policy decisions exacerbated the failure of the banking system as the Hoover administration tightened credit and raised interest rates, partly to punish speculators and partly to shore up the British pound. Moreover, the international gold standard was rendered unworkable after the stock market crash and wave of bank failures. According to the dictates of the gold standard at the time, all trade deficits had to be paid in gold at the end of the year. But gold also functioned as the domestic currency. Squaring international accounts under the prevailing institutional arrangements meant reducing a nation's domestic money supply. This exacerbated the deflationary tendencies already touched off by the collapse of banks and financial markets. In addition, the Versailles Treaty ending World War I had imposed $2 billion worth of reparations on Germany. Germans borrowed heavily from U.S. banks to pay their reparations to France. England and France used the reparation payments to repay their loans to U.S. banks. The collapse of the U.S. banking system precluded more loans to Germany. Germany thereby defaulted on their reparation payments, and England and France suspended payments upon their war debts. The international trade system simply collapsed hastening the re-emergence of hostilities in a world shaken by long-term depression.

The world that emerged from the Great Depression and subsequent world war was a world fundamentally transformed. The ideology that markets would find their own efficient equilibria was dealt a near-fatal blow by the depth of the Depression. The New Deal and Keynes' *General Theory of Employment, Interest, and Money* were to establish the role of government intervention in the economy. Commodity money in the form of the gold standard would give way to government-generated fiat money. International oil supplies would remain in the hands of the Allied powers, and oil would soon become officially denominated in U.S. dollars. In short the postwar social and economic order would soon become dominated by the United States as a political power, the large-scale corporation as an economic power, and by petroleum as a source of energy and power.

Conclusion

In the years following the Civil War the American economy was transformed from a small-scale, regional endeavor based on skilled labor, hand tools, and natural sources of energy such as wood and grass into a large-scale, national economy powered by cheap fossil energy, long-lived fixed capital in the form of machines, and factories utilizing deskilled operative labor. Long before the peak of U.S. oil production the economy experienced myriad periodic downturns, including three Great Depressions in the 1870s, the 1890s, and the 1930s. During these times the pressure on the large-scale industries became intense, and many were driven towards bankruptcy by competitive price devaluations. Facing bankruptcies the favored strategy was the concentration of industry by means of consolidation and merger. By the 1890s two merger movements had produced most of the characteristics of big business

we recognize today, from a few firms controlling the majority of an industry's output to the rise of nonprice competition. Horizontal mergers were designed to eliminate ruinous price competition and vertical integration reduced costs by bringing all aspects of production, distribution, and marketing within the control of a central management and creating the economies of scale. By the end of the century these concentrated industries had devised mechanisms to cope with the chronic problems of overproduction and excess capacity that accompanied price competition.

The evolution of the large corporation and the concentrated industry was a fundamental part of the industrial revolution itself. Many economic historians have chronicled the role that the rise of monopoly concentration played in the American economic experience. Few, however, have focused on the role played by cheap energy. Because we believe that economics should be both a social and a biophysical science it is important to link the development of energy and power as physical entities with the social and economic factors that they allowed and generated. We can achieve a better understanding of how the economy works, historically as well as contemporaneously, by viewing the development of economic power in the context of power in the physical sense

The economy, however, still experiences a roller coaster of expansion followed by depression or recession despite the existence of dramatic technological change, the availability of cheap energy in the form of coal and then petroleum, the economic concentration, and the organizational innovation. Even in times of abundant cheap energy, such as the 1930s, the economy experienced a downturn due to the internal dynamics of technology, investment, productivity, demand, and excess capacity. Historically this internal tendency is periodically reversed by the introduction of epoch-making innovations such as the steam engine, the railroads, electrification, and the automobile, allowing for the long-term expansion of productivity, investment, and economic growth. All of these innovations were energy-intensive and depended upon the availability of cheap energy. The digital revolution, energy intensive in its collective impact, may or may not qualify as a major epoch-making innovation, but it seems not to have resolved the problems inherent with the others, as the major economic downturns of 2000 and 2008 seem to indicate [13].

What is the fate of the concentrated economy if and as the age of cheap energy comes to an end? In other words, will the biophysical constraints combine with the already existing internal limits to bring about the end of the growth economy? What are the chances that another epoch-making innovation will usher in another buoyant era of economic growth? Can some kind of "green" energy do this? Could this take place while nearly every scientific measurement of the human impact upon the planet indicates we are already in overshoot. If we are already exceeding the biophysical limits of the planet, we doubt severely that humans can grow our way into sustainability. But economic growth is at the heart of a monopolized economy. How do we reconcile the need for living within our biophysical limits with the need to produce jobs, opportunity for the next generation, and reduced poverty? Much of the rest of the book focuses on that question.

Questions

1. How did the emergence of the fossil fuel age result in a concentration of political and economic power?
2. What is an oligopoly?
3. What was the first large-scale use of petroleum? What resource was it replacing? Why?
4. What is vertical integration?
5. What is horizontal integration? How was it accomplished by Standard Oil?
6. We see kerosene replacing whale oil, and electricity replacing petroleum, both fairly rapidly. What do you think will replace electricity, if anything?
7. Why didn't the end of the kerosene age mean the end of Standard Oil?
8. What was Henry Ford's idea about guaranteeing sales for his Ford automobiles?
9. What is an epoch-making innovation? Can you give three examples and tell how each is related to energy, and do you believe there are any happening now?
10. What was the relation of the rise of coal to skilled labor?
11. Can you give several perspectives on the role of competition in the economy?
12. What was the objective of the Sherman Anti-Trust Act in 1890?
13. Do you think the basic business conditions of the early 1900s were very different from those of today? Why or why not?
14. "The ideology that markets would find their own efficient equilibria was dealt a near fatal blow by the depth of the 1930s depression. The New Deal and the *General Theory of Employment, Interest, and Money* established the role of government intervention in the economy, as well as a focus on the inability of the private sector alone to create sufficient overall demand to maintain full employment." Discuss these two sentences in light of today's economy.
15. A general problem of industrial capitalism is that the economy is usually unable to absorb all that is produced by the very productive fossil-fueled economy. What were some of the approaches used in the 1950s to deal with this problem?
16. How might the end of cheap oil change the way that our industrial economy operates?

References

1. J. K. Galbraith. 1967. The New Industrial State. Princeton University Press, Princeton NJ.
2. Carnegie, A. 1889. The bugaboo of the trusts. North American Review.148: 387.
3. Yergin, D. 2008. The prize: the epic quest for oil, money, and power. New York: Free Press. p. 6.
4. Chandler, A. 1977. The visible hand: The managerial revolution in American business. Cambridge, Ma: the Belknap Press. P. 321.
5. Duboff, R. 1989. Accumulation and power. Armonk, New York. M.E. Sharpe.
6. Hall, C. and C. Pascualli. in press. Colin Campbell, Jean Laherrere and the science and implications of peak oil. Springer

7. Perelman, M. 2006. Railroading economics. New York: Monthly Review Press

8. Baran, P. and P. M. Sweezy. 1966. Monopoly capital. New York: Monthly Review Press.

9. Piore, M. & C. Sabel. 1984. The second industrial divide. New York: Basic Books. Pp. 49–72.

10. Hacker, L. 1940. The triumph of American capitalism. New York: Simon& Schuster.

11. Hall, C. A. S., C. Cleveland and R. Kaufmann. 1986. Energy and resource quality: The ecology of the economic process. Wiley Interscience, N. Y.

12. Galbraith, J. 1988. The great crash of 1929. Boston: Houghton-Mifflin. Pp. 68–107.

13. Kennedy, David. Freedom from fear: The American people in depression and war. Oxford University Press.

The Postwar Economic Order, Growth, and the Hydrocarbon Economy

7

Introduction

A recurring theme of this book is that economics should be approached both from a biophysical and a social perspective. This is especially important when viewing economics through the contours of history. For the vast majority of time humans lived off solar flow. For a very brief moment in time we have been able to appropriate fossil hydrocarbons to power our economy, and the result was a tremendous increase in productivity and the amount of material goods available to humans. Fossil fuels enabled the industrial revolution and beyond. At the same time, the increase in energy does not automatically determine the course of economic history. The industrial revolution consisted of more than simply more energy and more machines. It also entailed a fundamental reorganization of work and the general institutional arrangement of society. The economy of the early twenty-first century is not just a larger version of the economy of the early nineteenth century. It is fundamentally different. This chapter views the development of the American economy from the middle of the twentieth century through the financial crisis and recession of 2008. In 2008 Barack Obama was elected president of the United States with a great deal of optimism. But 2010 saw a conservative resurgence based on poor economic growth. We pose a question. Can the progrowth agenda that dominated the twentieth century withstand the biophysical limits that will be imposed by peak oil and climate change?

To answer this crucial question we need to look carefully at the patterns of history as well as viewing carefully the scientific data, which we do with the remainder of this chapter.

The years following the end of the Second World War were a time when the wealth and power of the United States were on the rise. After the stagnation of the Depression and the sacrifice of the war years there was, once again among a large proportion of the American population, a belief in abundance. From the depths of the Depression was born the "golden age" of the American economy. The era was characterized by the growing international power of the United States, both economically and militarily. The wealth that flowed in from the rest of the world was shared more broadly, and with a greater segment of the working population, than at any time since the Industrial Revolution. Home ownership became a reality for a greater share of the population, and it could be achieved upon a single income. The days of conservation and sacrifice were gone. Spacious automobiles traversed newly constructed freeways to arrive at Disneyland in Anaheim, California from far-flung suburbs. And they brought kids, lots of them, as the "baby boom" was just gaining headway. Disney's "Tomorrowland" showcased "the house of the future" replete with all-electric appliances, futuristic design, and virtually no attention to insulation or energy conservation. The future looked promising. It was a future based on cheap oil and economic growth.

But the year following the opening of Disneyland in 1955 was a year of warning.

C.A.S. Hall and K. Klitgaard, *Energy and the Wealth of Nations: Understanding the Biophysical Economy*, DOI 10.1007/978-1-4419-9398-4_7, © Springer Science+Business Media, LLC 2012

In 1956 the nationalization of the Suez Canal by Egypt briefly halted the shipments of oil to Europe and threatened the existing international order. In the same year Roger Revelle and Charles Keeling first began to measure carbon dioxide concentrations in the atmosphere, and M. King Hubbert wrote his famous paper predicting the peak of domestic oil production a mere 15 years in the future. But the academics were ignored and the crisis in North Africa was quickly brought under control. It was a time when Americans could seemingly do anything, including building the dream of happiness through material abundance and perpetual growth. Economically and politically the prosperity was built on the five basic pillars of the postwar social structure of accumulation we introduced in Chap. 1. To recap:

It was a time of peace, and peace on American terms, Pax Americana. The capabilities of other industrialized nations were decimated. But the war rekindled U.S. industry from the depths of the Depression. No other nation could match U.S. industrial output. Rather than seeing the European nations as serious competitors, national and international policy sought to shore up their devastated infrastructures and restore their demand for goods, particularly U.S. goods. The U.S. dollar replaced the repudiated gold standard. Essentially, the rest of the world was willing to give the United States interest-free loans in their own currencies to hold the dollar. Inasmuch as the world's resources, including oil, were denominated in dollars, the country could buy in a buyer's market and sell in a seller's market as the terms of trade (or ratio of export price to import price) consistently favored the United States.

The labor strife that characterized the later years of the Depression and year following victory in Europe and Japan began to dissipate after the conservative General Motors and the militant United Auto Workers signed a contract in 1948, known in the annals of labor history as the "Treaty of Detroit." The union gave up claims to determining output and technology and received cost of living adjustments, medical care, retirement, and most important, wages that increased as productivity rose. Following the 1948 UAW-GM contract this "productivity bargaining" became

standard practice in the nation's large-scale, mass-production industries. Although this capital–labor accord was limited to mostly white men who worked for large corporations, the sharing of productivity gains was a new phenomenon, and when added to the savings that accrued during the war, became the economic basis for the mass consumption that characterized the "golden age." Business may have paid billions in wages and benefits but they got a good deal, as the capital–labor accord provided both stability on the job and a source of demand for their products.

Never had the citizens of the United States exhibited such a faith in the idea that government could enhance the welfare of its people. The "New Deal" helped combat the devastation of the Great Depression, and the government had just mobilized a formerly isolationist nation to win the war. A broad capital–citizen accord united under the banners of economic growth and anti-communism. The civil rights and labor movements became as much a part of the postwar growth coalitions as did business. Congress passed the Employment Act of 1946, mandating reasonably full employment, stable prices, and economic growth. Growth would become the vehicle to achieve the other, laudable, social goals including a war on poverty, expanded civil rights, and the funds for military and economic expansion.

Little foreign competition existed to threaten the nation's large oligopolies, the dollar was the international currency, and U.S. demand was stable and rising. Antitrust policy seemed to be more directed towards keeping new firms from upsetting the industrial balance than to breaking up the older concentrated industries that had just helped win the war. Industry after industry such as automobiles, breakfast cereals, and petroleum refining settled comfortably into "Big Threes" or "Big Fours." In fact, a new merger movement was about to begin. Finally, it was the age of cheap oil, and the United States was still the dominant oil producer in the world. The great finds of the 1930s had found little use during the Depression, but later allowed the United States to supply 70% of the oil for the Allied war effort. Cheap oil, in conjunction with the aforementioned structural changes, helped fuel the mass consumption, economic growth, and military muscle for years to come.

All that was soon to change. By 1970 Hubbert's ominous prediction turned out to be accurate as U.S. oil production for the "lower 48" peaked. Oil price shocks buffeted the economy in 1973 and 1979 threatening both the mobile lifestyle and economic growth. American producers no longer had the spare capacity to keep foreign producers from using "the oil weapon." This was the era that saw the rise to power of OPEC, the Organization of Petroleum Exporting Countries. The 1970s and early 1980s were the time of stagflation, or simultaneous inflation and recession. Under mainstream Keynesian theory inflation was only supposed to appear if demand continued to expand past the level that would support full employment. But in the 1970s prices rose even in the presence of substantial unemployment. Keynesian policies no longer seemed to work. If the government pursued an expansionary policy inflation worsened. If it cut its spending or raised taxes to reduce budget deficits, or made money harder to come by, unemployment soared to politically unacceptable levels. Moreover, the international monetary accords conceived and born in Bretton Woods, New Hampshire, in 1944 collapsed under their own weight and U.S. policy. The accords had been built upon the willingness of the United States to convert holdings of dollars to gold at $35 per ounce. By the early 1970s foreign claims exceeded the magnitude of the gold supply. President Richard Nixon closed the gold window in 1973, ushering in a new era in international monetary politics: one that was far less favorable to the growth of the United States.

Part of the expansion of foreign dollar holdings was based on the expansion of American business abroad and part was attributed to increased military expenditures. The war in Southeast Asia was not going well. Rand Corporation systems analyst and respected neoclassical economist Daniel Ellsberg expressed dismay after briefing high-level government officials as to the conditions on the ground only to have them turn around and tell a far more optimistic story to the nation. In 1971 Ellsberg released the "Pentagon Papers" which

showed the disconnect between the assessment of war planners and public officials to the *New York Times*. This earned Ellsberg a spot on Richard Nixon's "enemies list" and the honor of being called "the most dangerous man in America [1]." Ellsberg was correct in his assessment and by May Day, 1975, North Vietnamese tanks broke down the gates of the American Embassy heralding the end of America's longest war to date. By 1979 the "friendly" government of the Shah fell in Iran, replaced by an anti-American government of Shia clerics. Oil prices soared and *Business Week* lamented "The Decline of US Power" in their Special Issue of March 12, 1979 [2]. Terms of trade, along with corporate profits, fell [3].

By the late 1970s America, along with much of Europe, elected conservative governments. Ronald Reagan in the United States and Margaret Thatcher in the United Kingdom gained power and began to develop new economic policies based on low taxes, remilitarization, antiunion campaigns, the reduction of domestic spending growth, deregulation of business and finance, and restrictive monetary policies to reduce inflation. Social democratic governments in Germany, Sweden, France, and Italy were replaced by conservatives as well. The Soviet Union, crippled by falling oil prices and cold-war military spending did not achieve the state of advanced socialism called for by the politbureau in the post-Breshnev days, and the openness (Glasnost) and restructuring (Perestroika) called for by Michail Gorbachev led to the break-up of the USSR. The Chinese Communist Party began openly to court entrepreneurs. The cold war was won, and there were no viable alternatives to multinational capitalism.

Yet economic growth did not respond over the long term, despite great new finds of oil in the North Sea and the North Slope of Alaska. Without the revenues from the North Sea oil Thatcher's austerity program would have never worked. Debt swelled as well, with the United States changing from the world's greatest creditor to the world's greatest debtor in less than a decade. The Clinton administration completed the work of the "Reagan Revolution," deregulating fully the U.S.

financial services industry, trading carbon limitations for a North American Free Trade Agreement, and ending "welfare as we know it." Eight years of the administration of George W. Bush saw 9/11 leading to two inconclusive wars and the explosion of a debt economy that ended with the financial meltdown and housing crisis of 2008. Oil prices rose to historic highs in the same summer. As this book goes to press, the financial crisis has turned into the worst economic downturn since the Great Depression. But neither oil price spikes nor recessions are new phenomena in the American economy. What were the transformations that occurred prior to the current crisis, and what lessons can be learned from them?

Historical Antecedents: Depression and War

Many, if not most, analyses of the trajectories of the twentieth century economy focus on social and economic forces alone. It is our contention that including the role of energy will provide a better analysis because the role of energy is generally missing from economic analysis. Changes in the social structure of accumulation and changes in energy should be analyzed in conjunction. Again, historically the U.S. economy has experienced three major depressions in the hydrocarbon era: 1870s, 1890s, and 1930s. All came after the discovery and exploitations of fossil hydrocarbons. Despite the ability to increase productivity by applying energy-dense fuels, business still needs to sell the products at a profit, expand markets, and realize the gains of productivity. When this does not occur the economy slips into Depression. The end of the twentieth century, from the 1950s until the present was characterized as an age of economic growth. The 1950s and 1960s were golden years, and the 1970s were an age of stagnation. Economic growth revived somewhat in the 1980s, but the burden of debt soared. The long-term consequences of a debt-driven and speculative "casino economy" came due in 2008. But what does the future portend? Will we, through social reorganization, transcend our current problems, or will a set of external and internal biophysical limits augment the pre-existing social ones to produce an age of austerity?

As we saw in Chap. 6 the world economy collapsed into depression for the entire decade of the 1930s. In the United States the presidential election of 1932 pitted two candidates with opposite opinions as to the Depression's origins. Incumbent president Herbert Hoover believed the cause stemmed from the Great War and subsequent Treaty of Versailles that ended the war. The victorious Allied powers redrew the map of the Middle East as they dismembered the Ottoman Empire, which had sided with Germany and the Austro-Hungarian Empire in the war. The new map showed a curious phenomenon. Places with large populations had little oil and places with abundant oil reserves had very few people. The Austro-Hungarian Empire was also broken up, and Germany was stripped of its African colonies, forced to accept sole responsibility for the war, and pay some $33 billion in reparations to Britain and France. Germany was also deindustrialized and the area to the west of the Rhine River was demilitarized. Without the industrial wherewithal to pay the reparations the German economy was essentially crippled, and in order to pay the reparations the Germans borrowed money from banks in the United States. The British and French then used the reparation payments to repay their wartime loans from the United States, who had emerged from the war as an international creditor. In turn, the U.S. banks then loaned the money back to Germany. The stock market collapse of 1929 and subsequent banking collapses of the early 1930s disrupted this precarious and unstable system. Unable or unwilling to continue, U.S. banks stopped the loans to Germany, who then defaulted on their reparation payments to England and France. The British and French no longer had the funds to repay their loans to U.S. banks. Without the infusion of funds from the United States the system collapsed and world trade evaporated. The United States Congress passed high protective tariffs of up to 67% on selected agricultural commodities to protect their own markets. President Hoover reluctantly signed the Hawley–Smoot Tariff despite the opposition of the nation's most prominent economists. The British created an Imperial Preference System to limit trade within its empire, and Germany contemplated a policy of economic self-sufficiency. World trade, which stood at $36 billion in 1929 dropped to $12 billion in 1932 [4].

The tariff and trade situation was exacerbated by the international gold standard. Under its provisions a nation was obligated to pay off any trade deficit in gold on an annual basis. However, because gold also functioned as a domestic currency, nations often had to drain their domestic currencies in order to square their international balances. Theoretically this was supposed to reduce prices and make a nation's exports more attractive to potential importers. In practice the reduction of money touched off not only falling prices (deflation) but also unemployment, recession, and international speculation of debtor nation's currencies. Panicked investors in the United States withdrew their deposits, precipitating a banking panic in 1930. Faced with just such a gold drain the British suspended the gold standard in 1931, adding to the predicament of banks with the withdrawal of international deposits. In addition Hoover advanced legislation to increase U.S. tax rates in order to enhance revenue and balance the domestic budget. He believed that balancing the nation's budget would provide the banking system with desperately needed liquidity. However, the economy slipped deeper into depression as wealth creation declined, along with tax revenues. The Federal budget slipped into a deficit of $2.7 billion, which was the largest peacetime deficit in American history. Much of this deficit resulted from Hoover-era policies to stimulate the economy by means of injecting funds into the struggling sectors of the economy.

Congress passed the Glass–Steagall Banking Act in 1931 which not only made the banking system safer by separating speculative securities trading (investment banking) from taking deposits and making loans (commercial banking) but also made it possible for the Federal Reserve to release large amounts of gold from its holdings thereby expanding the monetary base. In 1932 Congress passed the Federal Home Loan Bank Act which allowed banks to present mortgage paper for rediscounting at the Federal Reserve and allowed banks to use mortgages for collateral in obtaining loans of badly needed capital. Finally Hoover proposed the creation of the Reconstruction Finance Corporation (RFC) which was designed to allow the government to loan taxpayer dollars directly to struggling financial institutions. Congress initially capitalized the RFC at $500 million and authorized it to borrow up to $1.5 billion. The RFC was the progenitor of the Troubled Assets Relief Program (TARP) created in the waning days of the administration of George W. Bush to deal with the financial collapse of 2008. The reaction in 1932 was as mixed and varied as was the reaction in 2008–2009. Progressives called it "socialism for the rich." *Business Week* hailed the RFC as "the most powerful offensive force that governmental and business imagination has, so far, been able to command [5]."

However, given Hoover's position that the Depression was of foreign origin, his domestic policies were both tepid and hamstrung by his view of how the international economy functioned. Hoover remained committed to the principle of voluntarism and only begrudgingly accepted institutions such as the RFC. But more important, he was more strongly committed to two of the most sacred principles of classical economics: the belief in balanced budgets and an unwavering fealty to the gold standard as the linchpin of the international economy. He raised interest rates and taxes when the system cried out for increased credit and increased spending, largely because he believed that not doing so would increase the gold drain and jeopardize the position of allies and trading partners such as Great Britain.

Hoover's Democratic rival in 1932, New York Governor Franklin Delano Roosevelt, had an entirely different conception of the causes of the Depression. He believed its cause was primarily domestic. While a candidate FDR surrounded himself with a number of Columbia academics that was branded "the brains trust" by a *New York Times* reporter. Chief among his economic advisors was Rexford Tugwell, who was an adherent of the "stagnation thesis" advocated by economists such as Alvin Hansen and Paul Sweezy (see Chap. 6).

Roosevelt came to accept Tugwell's arguments that the mature economy had reached its frontiers, and that no great epoch-making innovations would be forthcoming. (FDR was well aware of the "frontier thesis" for he had taken classes, while a student at Harvard, from Frederick Jackson Turner who initially advanced the idea.) The problem was one of overproduction of capital and not a shortage of it, along with the flip side of underconsumption. Roosevelt enunciated his belief in underconsumption in two 1932 speeches while a candidate: one at Oglethorpe University in Atlanta, Georgia on May 22, and another the Commonwealth Club of San Francisco in September. The Commonwealth Club speech is worth quoting at length, as it foreshadowed the tenor of New Deal programs to come. The new Deal was to be about consumption instead of production, and equity instead of growth.

> Our industrial plant is built; the problem just now is whether under existing conditions it is not over-built. Our last frontier has long since been reached, and there is practically no more free land... . We are not able to invite the immigration from Europe to share our endless plenty. We are now providing a drab living for our own people. Clearly this calls for a reappraisal of values. A mere builder of more industrial plants, a creator of more railroad systems, an organizer of more corporations is as likely to be a danger as a help. The day of the great promoter or financial Titan, to whom we granted everything if only he would build, or develop, is over. Now our task is not discovery, or exploitation of natural resources, or necessarily producing more goods. It is the sober, less dramatic business of administering resources and plants already in hand, of seeking to reestablish foreign markets for our surplus production, of meeting the problem of underconsumption, of distributing wealth and products more equitably [6].

The New Deal was neither a well-enunciated program nor a manifesto for economic growth. Rather it was a set of sometimes contradictory experiments to pursue the goals of rescue, recovery, reform, and restructuring. Rescue came first. According to economic historian Ranjit Dighe, FDR's lieutenants, acting on incomplete information and in collaboration with Hoover's financial advisors, declared a national bank holiday, closed insolvent banks, recapitalized them through the RFC and reopened them for a trusting and

Table 7.1

Year	Unemployment rate
1929	3.2
1930	8.7
1931	15.9
1932	23.6
1933	24.9
1934	21.7
1935	20.1
1936	16.9
1937	14.3
1938	19.0
1939	17.2
1940	14.6

Source: *Historical Statistics of the United States*, p. 73

newly confident public. FDR's "Fireside Chat" helped to restore confidence among a battered and beleaguered public. Chief advisor and organizer of the Brains Trust, Raymond Moley, held the belief that the efforts essentially saved capitalism in eight days [7].

Since that time the administration of Franklin Roosevelt has set the standard for presidential performance. He passed 16 major bills in his first 100 days in office, most reflecting his concerns about overproduction and his fiscal orthodoxy which entailed a belief in balanced budgets. In retrospect this fiscal orthodoxy accounts partially for the fact that unemployment remained stubbornly high throughout the course of the Depression (Table 7.1).

In addition to the banking bill the first 100 days saw the Beer and Wine Act, which was designed to raise revenue in anticipation of the repeal of Prohibition, and the Economy Act designed to cut $500 million from the federal budget. FDR also advanced two bills to deal with the stubbornly persistent problem of unemployment. The Civilian Conservation Corps (CCC) put a quarter million young men to work beautifying the nation's countryside and working on flood control and forestry projects. The Federal Emergency Relief Act injected federal money directly into depleted state coffers for the purpose of unemployment assistance. Concerns over energy were also a crucial component of the legislation of the first 100 days when Congress created the Tennessee Valley Authority (TVA). The federal government had built a dam at Muscle Shoals, Alabama in 1918 to provide power for the production of nitrates, which are the basis of not only explosives but fertilizer. After the dam was completed too late to be of use for the war effort, a cohort of private utilities successfully blocked the efforts of progressive Republican George Norris to have the federal government operate the dam. The Tennessee Valley Authority not only created and federalized the dam but also charged the TVA with flood control, the combating of soil erosion and deforestation, and the construction of additional dams to bring electricity to southern Appalachia. Faced with a 95% decline in home construction since 1929 Congress created the Home Owners Loan Corporation, rather than committing to the large-scale expansion of public housing, as recommended by New York Senator Robert Wagner. The HOLC stopped the surge of foreclosures (up to 1,000 per day) and introduced standard accounting practices into mortgage lending. This followed the creation of the Federal Housing Administration in 1934. Traditionally mortgages required a 50% down payment and a short-term, interest-only loan. If the homeowner was diligent with his or her payments the note would be refinanced for another 5 years. But when the banking system collapsed between 1929–1933 banks were simply not in a position to refinance the loans even if the homeowners were able to make the interest payments. The FHA replaced these traditional mortgages with low down payment, long-term (up to 30 years), low interest, amortized loans where both principal and interest were repaid in equal monthly payments. Moreover, the FHA insured these mortgages from default. Despite the insurance, bankers were reluctant to write FHA loans. Some were worried about government intrusion whereas others were concerned about holding on to a low-yield asset for some 30 years. To allay the fears of the bankers, Congress subsequently created the Federal National Mortgage Association (FNMA, better known as "Fannie Mae") to bundle the mortgages into securities that could be sold on short-term markets. FNMA functioned successfully as a government corpo-

ration until it was largely privatized in 1968 [8]. As a semiprivate corporation, it collapsed again in the financial crisis of 2008.

The hallmark of the first 100 days was the passage of the National Industrial Recovery Act. The NIRA, along with the Agricultural Adjustment Act, were aimed not just at recovery but also restructuring of the economy on the basis of rational economic planning to replace the newly failed market system as the basis for regulation of prices and output.

The National Industrial Recovery Act established the National Recovery Administration. It provided a series of complex codes by which business would comply with the need to combat overproduction in order to receive funds. The act also allowed for labor unions to bargain collectively and it established minimum wages and maximum hours. The law virtually suspended antitrust laws. Economic theory holds that monopolies restrict output and raise prices, a strategy tailor-made for remediating falling prices and overproduction. This allowed the federal government to plan rationally the output and prices for whole industries. Congress also passed the Agricultural Adjustment Act on underconsumptionist grounds. The bill was designed to restore the balance between industry and agriculture and raise farm incomes by restricting crop output in order to raise agricultural prices. Increased rural incomes would provide the wherewithal for the purchase of the output of industry. The bill was paid for by increased taxes on agricultural processors. The NIRA also established the Public Works Administration (PWA) designed to administer a large-scale and ambitious infrastructure construction agenda. The PWA was charged not only with the construction of energy-related projects, but it also assumed the duties of stabilizing the near-anarchy of the oil fields of the Southern Plains [9].

After the First World War fears of oil shortages surged. These fears were allayed by two large oil discoveries. In 1926, interestingly enough the peak of the 1920s automobile boom, oil was discovered in the Permian Basin in West Texas and Oklahoma. As usual, large new additions to the supply of oil depressed prices.

Oil that was selling at $1.85 per barrel in 1926 averaged only about $1 per barrel in 1930. Then, in 1930, another huge discovery was made in East Texas, one that dwarfed the combined output of Pennsylvania, Spindletop, and Signal Hill in California. The East Texas wells added another half million barrels per day to the oil supply. Consequently prices dropped again to as low as 10 cents per barrel further aggravating the already low price levels precipitated by the Depression. The Texas Railroad Commission, established in the Populist era to exert control over railroads assumed the responsibility (despite dubious legality) of regulating oil production by regulating its transport. The strategy of the Railroad Commission was one of "prorationing" or limiting oil shipments to a fraction of oil reserves. Problems arose in Texas and Oklahoma (where the Commerce Commission employed a similar strategy), when producers exceeded their allotted shares, shipping illegally what came to be known as "Hot Oil." The problem became so pronounced that Texas Governor, Ross Sterling, declared that East Texas was in a state of insurrection, and called upon the Texas Rangers and the National Guard to quell the problem.

The NRA was first called upon to impose its codes to reduce competition and stimulate economic recovery. The oil supply problem was severe enough, however, that newly appointed Secretary of the Interior Harold Ickes, brought the regulation of the East Texas fields under the aegis of the interior department when he was informed, in August 1933, that oil prices had fallen to three cents per barrel. The Oil Code, established under the NRA gave Ickes the power to set monthly quotas for each state. The anarchy in the oil fields abated under the auspices of the NRA and Interior Department. However, the Supreme Court found the NIRA and AAA unconstitutional in 1935. The conservative bloc was joined by liberal antimonopoly crusader Louis Brandeis, who objected to the suspension of the anti-trust provisions.

When the NIRA was declared unconstitutional in 1935 a separate law, the Connally Hot Oil Act was established to maintain price stability [10]. The Texas Railroad Commission remained effective at reducing cutthroat competition and falling prices until the 1970s [11]! The Roosevelt Administration responded to the Supreme Court's decision that the NIRA and AAA were unconstitutional by launching a broad and progressive agenda of reform, restructuring, and redistribution in 1935, often called "The Second New Deal." That year saw the passage of the Social Security Act, providing pensions for the elderly. It was ostensibly devised to reduce unemployment by removing the aged from the labor force to reduce unemployment, and was constructed on the principle of private insurance, rather than as a dole. Once again, FDR's fiscal orthodoxy necessitated that the program be funded by regressive payroll taxation rather than from the treasury. The Social Security Act also provided for Aid to Dependent Children, later modified to become Aid to Families with Dependent Children (AFDC) soon to become the backbone of the Great Society welfare programs of the 1960s. The government also became an employer with the creation of the Works Progress Administration (WPA). The WPA created jobs for construction workers who built miles of highways, public buildings, and university campuses. The WPA also employed writers and artists. In the first year of the program the WPA employed more than three million people, and 8.5 million over the life of the agency [12].

Further legislation was passed to structurally reform the nation's financial system. The Federal Reserve was given increased powers to conduct open-market operations which entail the buying and selling of pre-existing Treasury securities, needed now that the gold standard was abandoned. Moreover a tax bill created a strongly progressive income tax in order to achieve the goal of fairness embodied in the New Deal philosophy. These rates, up to 79% for the top incomes, were accompanied by high inheritance taxes which were designed to reduce the intergenerational transmission of wealth. Perhaps the most important law of the New Deal era for working people was the creation of the National Labor Relations Board. The National Labor Relations Act placed collective bargaining provision (Section 7a) of the now unconstitutional NIRA as a separate law in and of itself. Not only did the administration believe that collective bargaining would increase wages and serve the goals of redistribution, but it would also bring about labor peace. The new board would replace the organizational strike with a monitored election. It was also the vehicle that enabled the development of the capital–labor accord that would become a crucial pillar of postwar prosperity. The New Deal ostensibly came to an end in 1938 with the passage of the Fair Labor Standards Act. This act established 40 h and the standard work week and further solidified minimum wages [13]. Although the New Deal was successful in establishing significant structural reforms and developing a faith in government that has not been seen since, it was never successful in eliminating the stubborn specter of unemployment. Moreover, New Deal policies were not directed towards economic growth. However, the focus of government policy would change significantly with the advent of the Second World War.

The United States officially entered World War II on December 8, 1941. However, the country had been providing food, armaments, and much-needed oil to embattled Britain for more than a year. President Roosevelt officially declared the United States to be the "Arsenal of Democracy" in December of 1940, but the country had been supplying war materiel to the Allies since 1939. Historian David Kennedy states the matter concisely: the war was won with Russian lives and American machines, "… the greatest single tangible asset the United States brought to the coalition in World War II was the productive capacity of its industry [14]." The war ended the Depression, however, the conditions of the Depression were also instrumental in mobilizing for the war. At the war's onset nearly nine million workers were unemployed and half of the nation's productive capacity was idle. By war's end the American economic machine produced nearly 300,000 aircraft, 5,777 merchant ships, 556 naval vessels, nearly 90,000 tanks and over 600,000 jeeps. Of the 7.6 billion barrels of oil used during the war six billion came from the United States. Given the tremendous finds of the late 1920s and early 1930s the United States possessed an enormous surplus of one million barrels per day out of a total production of 3.7 million barrels. By war's end oil production had risen to 4.7 million barrels per day. Moreover the technological change of replacing thermal cracking with catalytic cracking, along with a guaranteed market for the expensive process, allowed petroleum engineers to refine 100 octane aviation gasoline. This allowed American planes to fly farther, maneuver more agilely with up to 30% more speed and power than their German and Japanese rivals. The United States supplied more than 90% of the 100 octane aviation fuel used by the Allies. The development of long-distance warplanes allowed for escort cover in the all-important trans-Atlantic tanker routes, which had previously been decimated by German U-boat activity. In addition, the new long-distance bombers destroyed the German coal gasification (Fischer–Tropsch) plants.

The United States was to be much changed by the war. It was the only belligerent nation in the history of the world to see its standard of living rise during wartime. Economic concentration would increase, labor union militancy would be tamed in support of the war effort, and women and African Americans would enter the ranks of industrial production and clerical work in unprecedented numbers. In 1939 the unemployment rate stood at more than 17%. By 1944 it had fallen to 1.2%. Not only did the rate fall, but the war-driven economy absorbed an additional three million new labor force entrants along with more than seven million workers, mainly women, who were previously excluded from active labor force participation. Perhaps most important, from a perspective of economic policy, the agenda of the Roosevelt administration turned from one of stability and social equity to one of more and more production. World War II saw the sudden rise of growth economics.

The Postwar Economic Order

The United States emerged from the war in an unprecedented position of economic, political, and military power. The nation was the only intact industrial power in the world, and it supplied the majority of the world's oil. European cities were in ruins. The Allies were deeply in debt and the United States was the world's greatest creditor. In June, 1944 the Allies met in Bretton Woods, New Hampshire, to reconfigure the international monetary system. Unlike the aftermath of the last Great War, no pretense was made of returning to the gold standard which had worked so poorly and helped create the conditions of poverty and political chaos that led to the war. The dollar was considered "as good as gold" and tremendous advantages flowed towards the United States, consolidating its dominant position.

The United States agreed, in return, to redeem foreign currencies in gold at $35 per ounce. To rebuild war-torn Europe the International Monetary Conference created the International Bank for Reconstruction and Development, better known as the World Bank. They were to make large-scale loans for the rebuilding of infrastructure: roads, bridges, power plants, refineries, office buildings and factories. To provide adequate liquidity, or readily available money, the International Monetary Fund was created. In addition the Fund was charged with buying and selling currencies in order to keep them in balance with the dollar at the agreed-upon rate. Because the use of protective tariffs and beggar-thy-neighbor policies had dried up world trade and helped transmit the Depression internationally, the conference also created a General Agreement on Tariffs and Trade (GATT) to encourage free and open trade. The belief was that nations that trade with one another do not go to war. Although the conference proceeded on Keynesian lines, the plan of British delegate John Maynard Keynes for an international clearing union was not accepted. Keynes' plan provided a framework whereby nations with large trade surplus would redistribute money to nations with large trade deficits in order to keep trade balances within reasonable bounds. The United States was not only the world's most powerful nation; it was also the world's largest creditor. American representatives, who were in no mood to adopt Keynes' plan, had the power to prevent its implementation. The GATT would have to suffice, although those present hoped for a more fully functional World Trade Organization. The

WTO was finally created in 1995. However the United States did supplement the World Bank funds with its own initiative known as the Marshall Plan.

The theoretical ideas behind the Marshall Plan, conceived by General George C. Marshall and President Harry S. Truman, were economic and political. Many political parties in Western nations such as Italy, West Germany, France, the Netherlands, and even Britain found socialism and social democracy appealing in the chaos that followed the war. Conservative Prime Minister Winston Churchill returned from the final meeting of the "Big 3" with Truman and Stalin to find that he had been deposed. The Labor Party triumphed over the Conservatives advancing ideas such as redistribution and a national health service. The framers of the Marshall Plan realized that no single market economy could thrive in a sea of economic stagnation. Under the Marshall Plan the United States provided almost $9 billion to the European economies to ward off the growth of indigenous socialist movements by strengthening the financial markets and production capacity of European democracies. Most of that money (up to 80%) was used to purchase U.S. exports. It also ensured that American corporations would gain entry into formerly protected colonial markets. The United States also agreed to sacrifice some of its declining domestic industries to the greater good of free trade. At the time this was highly favorable to the expansion of American business. United States foreign direct investment increased from $11.8 billion in 1950 to $76 billion in 1970. The share of total profits from foreign operations also rose from 7% in the early 1950s to 21% by the early 1970s. At the same time up to 46% of all deposits in major New York banks were derived from foreign sources [15].

Structural Conditions Following the War

Back at home the economic scene changed on the domestic front. The American public exited the war with the greatest accumulation of savings relative to income at any time in the country's history. Wages rose and unemployment fell. The prominent economist John Maynard Keynes reasoned, in *The General Theory of Employment, Interest, and Money,* that the build-up of excess savings was a primary cause of the Great Depression. But such was not the case in the postwar United States. Deprived of consumption by ten years of depression and five years of war, Americans were again on the verge of being major consumers once again. Economists called this "pent-up demand."

The strong position of international power allowed corporations to address the labor situation at home. They now had the wherewithal to share the fruits of productivity growth with workers in order to achieve labor peace and create a domestic source of demand for their products. They could have both rising profits and rising wages. After the "Treaty of Detroit" productivity bargaining became the pattern in large industry. Because wages increased with productivity, labor had a strong incentive to increase productivity. Moreover, wages were supplemented with retirement pensions and health care benefits, once-militant workers now had a strong stake in maintaining the system they once struggled against. Productivity (or output per worker) grew at 2.9% per year in the 1950s and 2.1% per year in the 1960s. Wages rose by an average of 2.9% per year in the 1950s and 2.1% per year in the 1960s, and the gross national product grew at an annual rate of 3.8% and 4.0% in the same time period. Corporate profits remained strong. From the late 1940s when the Marshall Plan was implemented until the oil boycott of 1973, after-tax profits grew at 7% annually [16].

Accrued savings plus the additional worker and business income translated into growing consumption expenditures, especially as regards gasoline, automobiles, and housing. Total consumption expenditures increased dramatically from $70.8 billion in 1940 to $191 billion in 1950 to $617.6 billion in 1970. In 1943, the year the last automobile was constructed for the duration of the war, only 100 cars were sold in the United States, but by 1950 more than 6.6 million cars received new tags. The prestagflation-era figure peaked in 1965 when more than nine million cars

left the showroom floor. But by 1970 one could tell something ominous was happening for the automobile-crazed population as passenger car sales declined to less than the 1950 level. A similar pattern existed in housing. In the depths of the Depression only 221,000 new dwellings (public and private) were started. In 1950 the nation's building contractors and trade workers constructed close to two million homes. After that a high level, exceeding one million new homes per year existed in every year whether prosperity or recession. However, by 1970 only 1.5 million new homes were started. Gasoline prices remained cheap, as the United States, which at the time still produced 52% of the world's oil was relatively unaffected by world events and price spikes such as the one caused by the Suez Crisis of 1956. In 1950 the price per barrel of oil was $2.77, or an inflation-adjusted price of $25.10. The real price of oil did not exceed this level until 1974, during the first oil crisis of the 1970s [17].

U.S. oil companies strengthened their position in the years following World War II in the all-important Arabian peninsula, soon to become the world's largest source of crude oil. The original concession was given to Standard of California in 1933 for an up-front payment of $175,000 and an additional $500,000 to be given to King Ibn Saud if oil was found. Standard of California brought Texaco into the consortium to form Aramco (the Arabian-American Oil Company). In 1933 Gulf Oil received a 50% share of the oil newly found in Kuwait, a concession they would share with Anglo-Iranian Oil Company (soon to become British Petroleum). After the war Aramco found that they had insufficient marketing operations to dispose of all the oil being pumped from the Saudi fields. They entered into a broader consortium with Standard Oil of New Jersey (soon to become Exxon) and the Standard Oil Company of New York (soon to become Mobil). Aramco was able to overcome the stranglehold of Shell and Anglo-Iranian for marketing in Europe, and fears of overproduction were allayed. Gulf Oil, which was long on crude and short on markets, entered into a consortium with Shell, which was long on markets and short on crude. The basic conditions for expansion, increased production, and increased marketing capabilities were in place. The era of economic growth, based on a social structure of accumulation amenable to business ascendency and lots of cheap oil were in place.

The immediate postwar period was also the era of decolonization, and the new spirit of independence changed the world of oil production throughout Africa and Asia, and the Middle East. Oil-producing nations moved to increase the share of Ricardian rents for their precious resource. The original concessions of the late nineteenth and early twentieth century's gave the international oil companies ownership rights of the oil for initial payments and an agreed-upon royalty per barrel. Countries that granted concessions were interested in having the oil companies lift as much oil as possible as it enhanced their revenues. The oil companies, however, were ever-mindful of the industry's history of gluts and falling prices. The companies, therefore, had an incentive to limit production to what they could market, and the companies were in charge of production. The aforementioned oil deals resulted in a tight oligopoly, And oligopolies, as you may recall, pursue a strategy of maximizing profits in the long term by means of limiting output, maintaining stable prices, and enhancing control over production, marketing, and distribution. Fearing nationalization of their Venezuelan concession Standard Oil of New Jersey agreed to split the rents on a "fifty–fifty" basis. The deal soon became the model for Middle Eastern producers, and the potential instability abated, albeit at higher costs to the oil companies. Royalties were to be paid at an official "posted price" that could differ from the market price. At the time of the deal the posted price generally exceeded the market price, which was kept low by the tremendous surplus capacity of oil. This transmitted an even greater share of the rents to the producing counties. However U.S. oil companies were aided by their government, as cost increases were softened by a provision in the U.S. tax code that allowed them to count the new rent payments as taxes and deduct them from their U.S. obligations. Essentially the stability of the oil industry was paid for by U.S. citizens. But oil was cheap and plentiful and incomes were rising. There was no tax rebellion in the United States. However, as we saw in Chap. 6, new forms of competition can destabilize an oligopoly structure. Independent oil companies such as Getty Oil in the United States and Italy's AGIP, wishing to break into Middle Eastern production, simply offered a greater share of the rents as the price of entry. The era of colonial subservience on the part of producing nations was beginning to end. Yet the acquiescence of oil companies and governments to the new rent sharing plan provided stability for years to come [18].

The Age of Economic Growth

At the end of the war all the pieces for a renewed era of prosperity were in place. American companies gained vast and profitable international markets. Few, if any foreign corporations were in a position to compete effectively. The United States was the most powerful nation in the world, economically and militarily. The world monetary system was based on the dollar. Productivity growth, much of it derived from the application of cheap oil [19], fueled increased profitability, and the increased wages, along with historically unprecedented savings and the expansion of consumer credit, served as the basis for an explosion in consumption. The war showed more than anything that Keynesian economics, based on deficit spending and public funding of infrastructure, worked. In this era American Keynesian, now calling their approach "The New Economics," began to transform and sanitize the works of Keynes from a theory based on the problems of uncertainty and speculation into a herald call for, and mechanism of, economic growth.

The years immediately following the war brought both labor unrest and fears that the economy would slip again into Depression. Congress moved, on the advice of the New Economists, to deal with the fears that large-scale unemployment would emerge once the stimulus of the war ended by passing the Employment Act of 1946. The measure started originally as Senator Robert Wagner's "Full Employment Bill." Wagner's proposal gave every American the statutory right to a job. If they could not find one in the private sector the government would create one for them, as they had during the Depression under the auspices of the Works Progress Administration (WPA). The bill was to be paid for by a tax on employers. Not surprisingly American business opposed the bill. Not only did they dislike the taxes to be levied on them, but the general belief was that the absence of the power to dismiss workers would make labor discipline and productivity increases impossible. The eventual legislation was the result of political compromise. The Act directed the government to pursue policies that would result in "reasonably" full employment, stable prices, and economic growth. Growth would be the mechanism that enabled the other two goals. Economists Samuel Bowles, David Gordon, and Thomas Weisskopf argued that this stalemated the traditional goals of the labor movement, full employment and income redistribution, and replaced them with economic growth [20]. The Act also obligated the president to give an annual economic report to the Congress, as well as mandating the creation of a Council of Economic Advisors.

The movement towards a strategy of economic growth began in earnest with the work of the Council of Economic Advisors (CEA), especially after Leon Keyserling advanced to the chairship in 1949. Population grew with the baby boom, military technologies began to affect the civilian world, and new frontiers emerged as former farmland was converted to suburban homes, all fueled by a tsunami of cheap oil. In Keyserling's imagination, growth could achieve two goals beyond the attainment of reasonably full employment. If the economy grew, more could be given to those at the bottom of society's income distribution without raising taxes, which might adversely affect production and profits. The Council firmly believed that only growth could "reduce to manageable proportions the ancient conflict between social equity and economic incentives which hung over the progress of enterprise in a dynamic economy [21]." The National Security Council prepared a document, NSC-68, in which economic growth was at the heart of the strategy. Only through economic growth could

the United States meet its domestic priorities of achieving reasonably full employment and stable prices, yet at the same time fund its new military objectives of the "containment of communism" by arming and locating military bases in "friendly" client states. Despite these arguments, Truman was somewhat tepid in his acceptance of economic growth and the next president, Dwight Eisenhower, was rather indifferent, preferring a strategy of price stability. The true era of the liberal growth agenda would come during the presidencies of John Kennedy and Lyndon Johnson.

The growth-oriented economists of the era believed they had conquered the business cycle such that recessions and depressions would be a thing of the past. By means of fiscal policy (taxing and spending) and monetary policy (money supply and interest rates) the New Economists could fine tune the economy as if it were a well-oiled machine. If the economy performed sluggishly the government could stimulate the economy and the increased spending would translate into an expansion of output and jobs. If prices rose to uncomfortable levels, inflation would be controlled by subtle downward adjustments in spending or the amount of money available to the economy. Theoretically, they believed inflation only occurred once full employment had been achieved, and resulted from demand that was in excess of what the economy could produce at full employment. So any reduction in demand was supposed to decrease prices but not employment. In terms of policy the liberal growth agenda rested upon four main pillars. Current production had to be balanced with existing productive capacity.

This was accomplished by expanding demand via the Kennedy–Johnson tax cuts. Costs were kept in line with wage–price guideposts and the use of presidential authority to convince union leaders to mediate their wage demands. This was known as "jawboning." Finally, growth was stimulated by encouraging investment. Policy instruments included accelerated depreciation and Kennedy's famous investment tax credit. In addition the expansion of cold-war military spending added to the economy's overall (or aggregate) demand. In the 1960s their policies resulted in impressive outcomes. Unemployment rates were less than 4% by 1966, real (or inflation-adjusted) gross national product grew at 5% per year, and the inflation rate remained low. The number of Americans in poverty fell from 22.4% of the population in 1960 to 14.7% in 1966. The boom was driven by an increase in investment, with inflation-adjusted gross private fixed investment rising from $270 billion in 1959 to $391 in 1966. The only stubborn inconsistency was the degree of inequality, with the U.S. distribution of income being more than four times as unequal as that of Sweden and twice as unequal as that of the Soviet Union. But policies of growth were to take precedence over those of distribution. President Lyndon Johnson believed that redistribution policies were doomed to failure because they were counter to the Puritan work ethic, they would be a political disaster, and they were counter to the growth agenda. Consequently the direction of the War on Poverty was towards productivity enhancement of the poor rather than towards income maintenance programs. The postwar prosperity was built on a series of growth coalitions with organized labor, the civil rights movement, and the women's movement basing their strategies of reaching the top on economic expansion sufficient to include them. It was a time when a far greater proportion of the population believed that the wise actions of the government could benefit them than is commonplace in the early twenty-first century [22].

As long as the material conditions of prosperity (international hegemony, labor peace, rising productivity, cheap oil, and the domestic limitation of cutthroat competition) remained in place, expansionary monetary and fiscal policy could produce growth with stable prices. However, by the 1970s the very success of earlier action led to the demise of the postwar social structure of accumulation. By the 1970s domestic oil production peaked, Europe and Japan caught up in terms of productivity, inflation gripped the nation in conjunction with rising unemployment, wages fell, and jobs began to leave as the economy became both globalized and more competitive.

Peak Oil and Stagflation

A great deal has already been written about the era of stagflation, some of which is reviewed in this chapter. However, what tends to be missing in the economics literature is the advent of external biophysical limits. It was in the 1970s that the biophysical limits, in the form of peak oil, began to affect world economics and politics. As per M. King Hubbert's prediction, domestic oil production peaked in 1970. Yet demand for oil to fuel transportation, heating, and industrial growth still continued to grow at about 3% per year. The era of rapid and sustained economic growth based upon cheap oil came to a temporary end, giving rise to a decade of malaise in the United States and elsewhere, characterized by not only economic stagnation and high unemployment, but rising prices as well. The shock of rising oil prices did not come all at once, but in 1973 a series of events that had been building throughout the postwar period culminated in the first energy crisis seriously to affect the United States.

In the 1950s the world oil industry was destabilized by the same forces that historically destabilized the oil industry in the United States: large new discoveries, glutted markets, and falling prices. Crude oil production in the non-socialist world rose from 8.7 million barrels per day in 1948 to 42 million barrels per day in 1972, mostly as a result of discoveries in the Persian Gulf area. Consequently, although U.S. production increased, the U.S. share of world production fell from 64% to 22% in the same time period. Proven reserves increased from 62 billion barrels to 534 barrels, excluding the socialist nations. By 1960 Soviet production was nearly 60% that of the Middle East. This exceeded domestic demand and the oil entered the world market, putting additional downward pressure on market prices. In April 1959 huge new discoveries of high quality, low sulfur oil (light sweet crude) were made in Libya, and by 1965 Libya was the world's sixth largest oil producer. The result was more cutthroat competition and falling prices. But, as per their agreement, oil companies had to pay royalties to producing nations on the basis of the official posted price, which was not falling with the increased supply. Consequently their profit margins fell. In August of 1960 Standard of New Jersey unilaterally cut the posted price by 7%, enraging the oil-producing nations. Spurred on by the oil ministers of Saudi Arabia and Venezuela the producers met with the intention of forming a body similar to the Texas Railroad Commission which would prorate shipments and help control the decrease in prices. In September the Organization of Petroleum Exporting Countries (OPEC) was born [23].

Political turmoil hit the Mideast in the late 1960s. In 1967 Israel pre-emptively attacked Syria, Jordan, and Egypt in what is known as the Six-Day War. Saudi Arabia withdrew their oil from the world market in an attempt to create shortages and economic discord among Israel's supporters in Europe and the United States. However, the strategy was ineffective and led primarily to declining revenues for the Saudis. There was still sufficient spare capacity in world oil production and in the United States to make up for the difference. That was soon to change. In 1969 a coup led by Colonel Muammar al-Quaddafi, overthrew King Idris. The new government demanded a large increase in the posted price and ordered oil companies to cut production. With the Suez Canal still out of service, the short delivery trip to Europe across the Mediterranean enhanced Libyan power. Furthermore the Trans-Arabian Pipeline (or Tapline) was ruptured by a bulldozer making oil transportation even more difficult. This set the stage for competitive price increases among producing nations. Iran increased its price in 1970 followed by Venezuela and Libya again. By the time negotiations came to an end the posted price had increased by 90 cents per barrel. By 1970 the United States was essentially powerless to control the situation, as it no longer possessed sufficient spare capacity to overcome events in the Middle East. United States oil production peaked in 1970 at slightly more than 11 million barrels per day, and has fallen ever since, despite increased drilling effort, new discoveries, and tremendous political pressure.

The Fateful Year of 1973

In September Colonel Quaddafi nationalized 51% of the remaining oil companies not expropriated in the original coup. He worried little about retaliation as the spare capacity to overcome his moves no longer existed. Europe was simply too thirsty for Libya's light sweet crude. But this effort was dwarfed by the events of the following month. Still hurting from the humiliation of the 1967 defeat, new Egyptian President Anwar Sadat, in conjunction with Syria, launched a surprise attack on Israel during the holy month of Ramadan in the Islamic world but also on the highest Jewish religious holiday of Yom Kippur. Sadat's forces were on the verge of defeating the Israelis, who were running short of munitions and materiel. Israel regrouped and staved off defeat but events were soon to grow in scope. Outraged by the American resupply efforts Saudi Arabia called for a boycott of oil to the supporters of Israel, particularly the United States and the Netherlands. The Saudis called for production cuts of 5% per month for the entire world, and a complete cutoff to the Americans and the Dutch. They threatened the partners in Aramco with loss of the concession if they sent as much as one drop of oil home. Ironically, it was the U.S. oil companies themselves who carried out the boycott, not the Saudi state. As recently as 1967 the removal of oil from the world market had not worked as a political weapon for the Arab states, as sufficient spare capacity existed in the world market to overcome their efforts. This was no longer the case once the production of the world's major swing producer, the United States, had peaked. The Saudis withdrew about 16 million barrels per day from the world oil supply and other producers had insufficient spare capacity to make up the difference. Iran increased oil exports by some 600,000 barrels but they, and some others could not compensate for the Saudi withdrawal. All in all, the world's oil supply fell by about 14%.

In the United States gasoline prices quadrupled as the world price of oil increased with success of the Saudi boycott. The nation's oil imports had nearly doubled from 3.2 million barrels per day when domestic production peaked in 1970 to 6.2 million barrels per day in 1973. Before the October war the posted price was $5.40 per barrel. By December oil was selling for as much as $22 per barrel. Gas lines became a feature of American life. Calls for action to increase the supply echoed from all corners of the nation. However, the oil companies were no longer just American enterprisers, but multinational corporations who tried to apportion the hardship equally among their various markets. There would be no special treatment for any particular nation, especially the United States. Patriotism did not include the potential loss of the Saudi concession for the American partners in Aramco. President Richard Nixon was essentially powerless to do much of anything, embroiled as he was in the loss of his own job owing the revelations of the Watergate scandal. However, the effects of the oil price run-up wreaked havoc with his new economic policy, intended to stop the gathering stagflation that had been emerging for years and was seriously aggravated by the increase in oil prices [24].

The End of the Liberal Growth Agenda

The 1973 energy crisis was not the only force that crippled the U.S. economy. Internal economic factors were at work too. The pillars of postwar prosperity were all crumbling by the late 1960s. The oil price shocks exacerbated an already-existing decline, and a series of events unfolded from the mid-1960s until the late 1970s that illustrated the decline. The rising power of the oil-producing nations was only one sign of the end of Pax Americana. There were many others. Europe and Japan, once war-torn nations is a state of shock, caught up to, and even surpassed, the United States in terms of productivity growth. Despite the rising cost, imports increased from 4% of GNP in 1948 to 10% in 1972. The U.S. share of total world exports in 1955 was 32%. It stood at only 18% in 1972. The postwar monetary

system was based on fiat money, where the value of a nation's currency depends upon productive power and political stability. American productivity growth, which averaged 2.7% per annum in the 1950s, fell to 0.3% per year in the 1970s. The United States no longer bought in a buyer's market and sold in a seller's market. Moreover the expansion of cold war military spending plus the outflow of funds directed towards foreign investment worsened the U.S. balance of payments situation. The Bretton Woods Accords mandated the United States convert holdings of foreign currencies to gold at the price of $35 per ounce. By 1973, with outstanding claims exceeding the American gold stocks, Richard Nixon "closed the gold window" and the Bretton Woods Accords collapsed, thereby ending the dominant position of the dollar and all its economic benefits. Soon, the world was to open up to an unprecedented increase in global competition. The days of the insulated oligopoly position of U.S. business were nearing their end, and the demise was reflected in the decline of corporate profits. After-tax corporate profits for the nonfinancial sector, which averaged 10% in 1965 dropped to less than 3% in 1973.

With productivity growth on the decline and international dominance eroding, American corporations could no longer "afford" the expensive mechanisms of labor peace to support the capital–labor accord. An open shop movement began in housing construction by the 1980s and myriad consulting firms specializing in "managing without unions" emerged. Union membership, which stood at 35% in 1955, began its secular decline. By 1973 unions represented slightly more than 25% of the nation's private sector workers. The economic benefits that corporations shared with workers began to erode as the economy entered the long slump of the 1970s and early 1980s. Hourly income, which grew at 2.2% per year in the long expansion of 1948–1966, grew only at 1.5% year in the time period between 1966 and the 1973 oil boycott. Unemployment rates, which had been as low as 3.6% in 1968, began to rise as well, reaching 5.6% by 1972. Federal budget deficits increased from $2.8 billion in 1970 to $23.4 billion in 1973. The Federal Reserve

System accommodated the booming economy by keeping interest rates low and credit readily available. The government also reduced business taxes to keep the economy expanding and spur further investment. The result of this economic stimulus was classic Keynesian "demand–pull inflation." There was simply "too much money chasing too few goods," and prices began to rise from 1.3% per year in 1964 to 3% in 1966. By 1965 President Lyndon Johnson's advisors were recommending either a tax increase or a decrease in spending. Neither strategy fit with Johnson's political or economic objectives, so inflationary pressures continued to build.

In 1970 Richard Nixon began to engineer a mild recession and unemployment began to rise. However the recession was short lived, and Nixon soon returned an expansionary fiscal policy, annoying his conservative backers by announcing that he was now a Keynesian. Government deficits rose from $11.3 billion in 1971 to $23.6 billion in the quarter preceding the 1972 election. Unemployment declined and Nixon was re-elected. However, the brief and mild recession did not wring the inflationary pressures from the economy. Prices continued to rise, but a new phenomenon was about to occur: rising prices in the context of high levels of unemployment. Upon succeeding Richard Nixon as president in 1974, Gerald Ford and his advisors pursued a contractionary policy under the guise of "Whip Inflation Now." Spending was reduced and taxes were increased to produce a budget surplus that exerted a downward force on aggregate demand. In addition the oil price increases (commonly referred to as the OPEC tax) removed another $2.6 billion of purchasing power from the economy. Despite the reduction in spending, prices continued to rise, with inflation averaging 11% by 1974. The Federal Reserve tightened credit as well. The inflation rate abated slightly, to 9.2% in 1975, and then further to 7.8% by 1978. But unemployment increased to 7.7% in 1976 in response to the contractionary policies [25].

Traditional demand management practices were no longer working because they were treating the wrong economic "disease." If the government expanded the economy inflation worsened without achieving full employment. If the government conducted contractionary policies unemployment soared without eliminating inflation. Political economists concluded that the economy was suffering from an entirely different form of inflation known as cost–push, where rising prices led to rising business costs, which were passed on to consumers in the form of higher prices. Oligopoly power remained strong and business was able to pass on rising energy costs as higher prices. The last vestiges of the capital–labor accord took the form of cost of living adjustments (COLA) provisions in union contracts. When business passed on costs as higher prices workers received an automatic increase in wages. In addition oligopolies had long ago stopped relying upon the market to determine prices. Rather, they set a target profit rate and marked up costs in an attempt to achieve their targets. When the nation's monetary authorities raised interest rates, the businesses that were able to simply raised prices. Consequently restrictive monetary policy and high interest rates exacerbated the inflationary spiral rather than reducing it [26].

The decade of the 1970s remained a stagnant one. The ineffectiveness of the policies of the pro-growth New Economists did not change with the election of a Democratic president, Jimmy Carter, in 1976. The economy simply could not produce growth. Carter attempted to deal with the problem of structurally embedded inflation by deregulating the airline industry And appointing a conservative chair of the Federal Reserve Board, Paul Volker, who was to pursue contractionary policies that would drive interest rates up to 20% in the coming years. Yet inflation varied between 5.75% and 7.6% until 1978, which were, themselves, historically high levels in the postwar era. But things were to change rapidly, and for the worse, once again driven by oil prices, in 1979.

The Fateful Year of 1979

In 1953 the Central Intelligence Agency engineered the overthrow of the elected Prime Minister of Iran, Dr. Mohammed Mossadeq. They installed Reza Pahlevi as the Shah of Iran, who engaged in a rapid modernization program. This modernization led to many of the economic problems associated with rapid growth: traffic-clogged streets, rising prices, urban pollution, and income inequality. Moreover the Shah's secret police (SAVAK) forcefully and brutally repressed any and all opposition. They moved especially against the traditional religious leaders who had opposed the Shah's father when he centralized the modern nation of Iran. By 1979 the Shah's empire crumbled, and he fled to the United States just before the arrival of Ayatollah Ruhollah Khomeini who subsequently proclaimed the Islamic Republic of Iran. In the waning days of the Shah's regime Iranian oil workers struck, disabling production. Exports fell from 4.5 million barrels per day to less than one million. By Christmas 1978 Iranian oil exports stopped entirely. Oil prices increased by 150%, stimulating a panic. This led to further speculative increases. Saudi Arabia and other OPEC nations increased their own production but the shortage was real [27]. When the Saudis, worried that the increased production would damage their wells, reduced production, prices spiked again. Then Iranian students seized the American Embassy. The responsibility for a failed rescue attempt fell upon President Carter, who had tried to govern in the center while imposing an austerity plan. He spoke to the American people that "life was not fair," placed solar panels on the White House roof, turned down the thermostat, and urged his fellow citizens to do the same. Many were in no mood to listen. Earlier in the year Carter had to deal with the partial core meltdown of a nuclear power plant in the Susquehanna River on the outskirts of Harrisburg, Pennsylvania at Three-Mile Island. America's energy future was highly uncertain, and the economy was on the verge of plunging into another oil-price driven recession. Carter was to be a one-term president, learning the hard lesson that, in times of austerity, the center moves to the right. The 1980 election pitted the incumbent Carter against former actor and governor of California, Ronald Reagan. Reagan won in a landslide, promising the return of "Morning in America." His economic plan was sold as one designed to restore the lost American hegemony, control labor and energy costs, and boost corporate profits.

The Emergence of Supply-Side Economics

The Reagan-era economic program was designed to raise corporate profits, reduce inflation, and restore American power in the world. Their record was mixed. Profits never increased and the price was an explosion of public and private debt, and an increase in inequality along with a major recession. The focus was to be on the supply side of the economic balance. Stimulating aggregate demand alone had led to inflation without reducing unemployment. The idea was that if business costs were reduced and access to capital increased, the increase in aggregate supply would expand output while reducing prices. Basically this was Say's law in a new package. To accomplish this goal the Reagan administration launched an interrelated five-point program. The conservative social program, in conjunction with the latest oil price run-up touched off the worst recession since the Great Depression in 1981–1982. The elements of the supply-side program included:

- The use of a restrictive monetary policy to generate high interest rates and engineer another recession, largely in order to raise unemployment to discipline labor.
- Further intimidate or eliminate labor unions in order to reduce wage-based inflation and enhance the ability of business to appropriate the gains of productivity.
- Deregulate business, especially finance, in order to restore competition. This also entailed the elimination of environmental laws and worker safety laws to further reduce costs to business.
- Increasing the degree of inequality in order to redistribute income and wealth towards the wealthy and corporations. This was accomplished by means of changing the tax code.
- Remilitarization and the return to an aggressive, unilateral, anticommunist military policy.

Each of these points deserves some explanation in greater detail. To begin with, Jimmy Carter had appointed a conservative central banker, Paul Volker to the Federal Reserve Chair in an attempt to restrain inflation and he moved to increase interest rates. In 1978 the rate that banks charge one another for overnight loans (called the federal funds rate) stood at 7.9%. Volker remained as chair of the Federal Reserve Board during the Reagan administration and did not ease his restrictive monetary policies until the mid-1980s. By 1981 the federal funds rate rose to 16.4% and rates for home mortgages rose to nearly 20%. Unemployment increased from 5.8% in 1979 to 9.5% in 1982. Failures per 10,000 businesses rose from 27.8 to 89.0 in the same time period. Economists Samuel Bowles, David Gordon, and Thomas Weisskopf termed this policy "the Monetarist Cold Bath."

One of Richard Nixon's goals was to "zap labor." The Reagan administration continued this policy with elevated enthusiasm. As a candidate Reagan, who was once a union president himself (the Screen Actors Guild) courted the more conservative labor unions, including the Professional Air Traffic Controllers Organization (PATCO). Labor relations had been strained during the Carter administration. Air traffic control was the most stressful job in America, managers paid little attention to the problems of the rank and file, and equipment was aging and inadequate. Carter ordered a comprehensive study of the industry but ignored implementing upgrades on the grounds of the expense in times of austerity. PATCO struck and its members voted overwhelmingly for Reagan, helping make him the only labor union president ever to become president of the United States. As President, Reagan ordered the air traffic controllers back to work, and when they did not do so he jailed the union president and decertified the union. A month later when federal unions were set to renegotiate their contracts, they buckled and significantly reduced their wage demands. Moreover the National Labor Relations Board was staffed with leaders who were hostile to the very notion of collective bargaining. Labor union membership continued its decline and wages began to fall.

The Reagan administration also continued the Carter era experiments with deregulation, launching a public campaign to convince the nation's citizens that regulations were outmoded and cumbersome. The older regulatory agencies, such as the Interstate Commerce Commission, were created at the behest of business to control cutthroat competition. During the Great Depression the nation's banks were regulated in an attempt to stem the financial crisis. The Reagan administration turned to the dismantling of the newer regulatory agencies such as the Environmental Protection Agency (EPA) and the Occupational Safety and Health Administration (OSHA) which they believed to be a primary cause of the increases in business costs. Their staffs were cut and their budgets slashed. Spending on regulation declined by 7% from 1981 to 1983 and staffing was reduced by 14%.

The high interest rates that resulted from the monetarist cold bath led to the problem of financial disintermediation. During the Great Depression thrift institutions such as savings and loans were allowed to pay higher interest rates on deposits than were commercial banks (Regulation Q). In return they were to loan money only for purchases of homes and apartment buildings. But the increase in interest rates made Regulation Q irrelevant, as deposits left the savings banks to find more lucrative returns in other financial markets. The Garn-St. Germain Depository Institutions Act of 1982 allowed savings banks to pay market interest rates and to invest their funds in more speculative housing projects with little or no oversight. The system was perilously close to collapse as the election neared and came to a crashing halt during the presidency of Reagan's successor, George H.W. Bush, necessitating the need for a multibillion dollar bail out.

As a candidate Ronald Reagan stood on the steps of the state capitol in Concord, New Hampshire and proclaimed that for America to get richer the rich need to get richer. This was accomplished by reducing the progressivity of the tax codes, whereby the wealthy pay a proportionately larger share of their income in taxes. The effective corporate tax rate dropped from 54% in 1980 to 33% in 1986. The Economic Recovery 1981, better known as the Kemp–Roth tax cut reduced the top marginal tax rates of top income earners from 70% to 50% and cut overall taxes by 23% over the course of 3 years. It also reduced estate taxes, allowed for accelerated depreciation, and reduced corporate taxes by some $150 billion. Government revenues fell by $200 billion. As a result the income distribution of the United States changed, becoming more skewed towards the top. The Gini coefficient, which measures overall income inequality, rose from 0.406 to 0.426 over the course of the Reagan administration. The higher the coefficient, the greater is the degree of inequality. The share of income accruing to the top 1% of the population rose from 8.03 in 1981 to 13.17 in 1988. The share that went to the top 0.01% tripled. This was supposed to free up funds for investment in the newly deregulated economy. But investment simply remained stagnant despite the redistribution towards the top.

Finally, the last component of the supply-side agenda was an increase in military spending. Military spending as a percentage of gross national product peaked at 9.2% during the height of the Vietnam War but had declined since then. But between 1979 and 1987 inflation-adjusted military spending increased by 57%. The Reagan administration had clear cold-war objectives. They believed that the Soviet Union would bankrupt itself trying to keep up with American spending. Soviet planners were themselves projecting rapid economic growth, and believed Soviet growth would outstrip that of the stagnant United States within two decades. The United States got lucky. Increased military spending plus declining oil revenues were the primary economic cause of the collapse of the Soviet Union at the end of Reagan's presidency. But the increases in military spending were also designed to increase U.S. power in an increasingly militant world. The United States conducted military operations, in a number of places. It was hoped that the increased military power would restore the days of Pax Americana, and bring the benefits of a strong dollar and low raw materials prices back to the country [28].

Supply-side economics was sold to the American public through a series of editorials in the *Wall Street Journal*. It was based roughly on a philosophy that has come to be known as neoliberalism (liberal in the sense of classical liberalism, not in the sense of New Deal liberalism). The Reagan administration had a firm belief that in the absence of government regulations and participation in capital markets (i.e., borrowing

to cover deficits) the competitive market would burst forth with a veritable gale of investment activity. Freed from the burden of onerous taxes corporations would increase their level of job creation. Furthermore, the market was such a perfect carrier of information and incentives that the market forces of competition and flexible prices would be sufficient to regulate the economy by themselves. They said American workers had become soft and protected by unions that both limited productivity growth and insisted that the benefits of rising productivity be shared with workers in the form of rising wages. The power of unions needed to be broken according to the neoliberal philosophers in order that productivity growth could begin anew. Moreover, social spending and income maintenance programs should be cut to provide incentives for workers to fill the ranks of the newly created low-wage jobs. Finally, the United States must attain, once again, its position of global power. This is the one area where conservatives willingly bypassed their reliance on the market. Although discretionary spending should be cut, military spending was beyond reproach.

Given these objectives, the macroeconomic performance of the Reagan years produced mixed results. Inflation rates fell, dropping into the 3–4% per year range by the mid-1980s from a high of 13.6% in 1980. Much is made of the effectiveness of the assault on labor unions in lowering the rate of wage growth and the decline in interest rates since the zenith of the "cold bath" policy. What is rarely mentioned is the role falling oil prices played in both controlling cost–push inflation and bringing about the demise of the Soviet Union. In the mid-1970s additional sources of oil were discovered in Mexico and in the North Sea between the United Kingdom and Norway, all beyond the control of OPEC. The first oil from the North Sea flowed into England in 1975. In the period from 1972 to 1974 oil was discovered in the Bay of Campeche in Eastern Coastal Mexico. The wells were prolific enough such that Mexico met its own needs and began to export to the world market. The Trans-Alaskan pipeline, on hold since the late 1960s, was completed in 1977. With the completion of the pipeline Alaskan oil production soared from a mere 200,000 barrels per day in 1976 to slightly more than two million barrels per day in 1988. (Since the 1988 peak Alaskan oil production has subsequently fallen to only 700,000 barrels per day as of 2008.)

Further downward pressures on price came from the development of alternative energy sources: nuclear power in Europe, natural gas and coal, and the conservation that resulted from increased energy prices. By the mid-1980s a spare capacity of 10 million barrels per day emerged. These forces caused OPEC to reduce its prices. By 1985 the price of oil had fallen to $10 per barrel, almost eliminating cost–push inflation [29]. The Soviet Union, deprived of oil revenue, which accounted for a third of its income, could no longer maintain its military spending, especially after its defeat in Afghanistan. The collapse of the Soviet system was soon to follow.

The Federal budget deficit increased dramatically over the course of the 1980s, driven by the reduction in tax revenues, the high interest rates associated with the monetarist cold bath, and the expansion of federal spending. Between 1981, when the Kemp–Roth tax cut became law and 1988 (the last year of the Reagan administration) tax revenues as a percentage of gross national product fell from 15.7% to 14%. In the same time period, military spending increased from 5.3% to 6.1% of GNP while interest obligations rose from 2.3% to 3.2%. Federal spending on education and infrastructure declined. Given the increase in military and interest spending the size of the federal government did not decline, nor did the "supply side" tax cuts deliver the promised explosion of economic growth. Rather, the deficit increased from 20% of GNP in 1979 to 22% in 1981, where it stayed until 1987 [30]. Furthermore the push towards financial deregulation allowed banks and other financial institutions to increase their own indebtedness, although the structural changes of the Reagan administration would give way to a much greater financial explosion by the early twenty-first century.

Reagan's successor, George H. W. Bush, attempted to carry on the same policies, especially in the area of keeping taxes low. However, deficits kept mounting and the new president was constrained further by the passage of the Gramm–Rudman–Hollings Balanced Budget and Emergency Deficit Control Act of 1985. The Act imposed binding constraints upon federal spending and limited the creation of further deficits. Bush campaigned in 1988 on the promise of no new taxes, but the military spending needed to pursue a war in oil-rich Iraq and the bailout of the failed savings and loan industry threatened to expand the deficit beyond the Gramm–Rudman–Hollings limits. Reluctantly Bush agreed to raise taxes, and consequently the conservative wing of the Republican Party abandoned him. This set the stage not only for the election of Democrat Bill Clinton, but also for the resurgence of the conservative influence upon the Republican Party. Clinton was destined to carry out the legacy of the Reagan Revolution. Running as a centrist Clinton campaigned on the basis of renewing economic growth by means of supply-side measures to increase labor productivity. Primary among them were public investments in education and infrastructure. However, there was a competing agenda among the Clinton advisors to reduce the size of the budget deficit in order to protect the integrity of the nation's financial markets, increasingly susceptible to international demands and pressures. The deficit hawks argued that large deficits limit long-term growth and scarce appropriate international capital, resulting in rising interest rates and a greater portion of the federal budget being devoted to interest payments. The deficit hawks, led by new Federal Reserve chair Alan Greenspan, won the day. No large-scale fiscal stimulus by means of public investment would be forthcoming. Instead there was a modest tax increase that received no Republican votes in either the House of Representatives or the Senate. Although the title of Clinton's campaign pamphlet was entitled *Putting People First*, his policies put the needs of the bond markets first. The growth path was to be fine-tuned by monetary policy alone, and the Federal Reserve pursued an essentially "accommodative" expansionary "easy money" policy.

In Clinton's second term the deficits turned to budget surpluses, rising from $69.3 billion in 1998 to $236.2 billion in 2000. In 1999 Clinton also signed the Financial Services Modernization Act, which repealed the Glass–Steagall Act of 1933. Commercial banking was no longer separated from investment banking. The act provided the impetus for yet another merger movement, this time involving the consolidation of financial services. Citibank merged with Travelers Insurance to form Citigroup. Wells Fargo merged with Norwest to provide myriad financial services, and American Express expanded their product line into nearly every aspect of money management. The bill also ensured that hedge funds would remain unregulated forever! As a result of the deregulation of banks and financial services, debt began to expand. Wage growth remained low, averaging only 0.5% per year throughout the 1990s. Moreover the economy was expanding on the technological changes brought about by computerization and the early days of the Internet. Most technology stocks were traded on the National Association of Stock Dealers Automated Quotation Index (or NASDAQ). In 1994 the NASDAQ index stood below 1,000. By 2000 it had climbed to over 5,000 in what is often called the dot.com bubble.

Despite the federal budget surplus, the expansion of debt begun in the Reagan years continued to climb. When wages and incomes of the vast majority of the population are growing slowly the only way to increase spending is to increase access to credit. From 1990 to 2000 gross domestic product increased from $5.8 trillion to $9.8 trillion. However, outstanding debt increased from 13.5 trillion to $26.3 trillion. Household debt nearly doubled during the period, from $3.6 trillion to $7 trillion, but financial firm debt more than tripled from $2.6 trillion to $8.1 trillion. The economy seemed to be running on financial speculation fueled by easy access to credit, as well as by relatively cheap oil. Oil prices generally remained stable throughout Clinton's years as well as relatively cheap allowing for revenues to be directed towards deficit reduction rather than increasing oil costs. Oil prices were less than $20 per barrel

when Clinton took office and remained at the $30 per barrel level when he left. Oil production also remained high, ranging between 25 and 30 million barrels per day. Clinton's years saw neither spikes in gasoline prices nor energy crises.

Clinton, trying to burnish his centrist image, also pledged to end "welfare as we know it," and did so by signing The Personal Responsibility and Work Opportunity Reconciliation Act of 1996, an election year. The act essentially ended welfare (or Aid to Families with Dependent Children) as an entitlement program. AFDC was replaced by Temporary Assistance for Needy Families (TANF), and recipients needed to work in order to earn their checks. The new law was supposed to restore America's work ethic and also be helpful in deficit reduction. Average monthly welfare payments (AFDC or TANF) adjusted for inflation in 2006 dollars fell from $238 per month in 1977 to $154 in 2000. Not surprisingly with increased financial mergers, a technology bubble in the stock market, rising access to debt, reduced welfare benefits, and slowly growing wages, the degree of inequality increased as well. The Gini index increased from 4.54 in 1993 to 0.466 in 2001. This meant that every year of the Clinton administration exhibited greater income inequality than any year of the Reagan administration. The share of aggregate income accruing to the top 1% increased from 12.82% to 15.37% over the same time period, and the share going to the top 1/100 of a percent rose 55%, from 1.74% when Clinton began his term to 2.4% when he left office.

A recession began shortly after Clinton left office, driven by the build-up of excess capacity in the computer industry and the subsequent fall in NASDAQ values known as the dot.com bust. During the first term of George W. Bush the unemployment rate increased from 4% in 2000 to 6% in 2003. Following attacks on the World Trade Centers and the Pentagon in September of 2001 the Bush administration pursued wars in Afghanistan and Iraq. Oil prices rose from approximately $30 per barrel in 2003 to nearly $150 per barrel in 2008, driven largely by the dislocation of war.

Warning Signs in the Early Twenty-First Century

The economy of the twenty-first century grew, albeit at a slower rate than in the non-Depression years of the twentieth century. Without cheap oil as a basis of economic growth other factors must be called upon to explain economic performance. The primary drivers of economic growth were the creation of ever-increasing levels of debt in all sectors of the economy, an accommodating central bank policy of low interest rates, and financial deregulation. These factors led to a series of speculative bubbles, particularly in housing, debt-fueled growth in military spending, and rising inequality. The limits of this strategy became clear in 2008 when the financial system virtually collapsed, to be saved only by several trillion dollars infused into the reeling financial sector, known as the Troubled Assets Relief Program (TARP), and the Federal Reserve. The financial panic translated into the real economy and unemployment rates rose into the 10% range and capacity utilization fell from 81.3 in 2007 to 70.0% in 2009, before "recovering" to 74.2% in 2010 [31] (Fig. 7.1).

The housing sector was particularly hard-hit just as it was in the Depression. Vastly overinflated housing values collapsed as much as 40% in particularly speculative markets such as Las Vegas, Miami, and the major cities of Southern California. Unemployment in the building trades rose to 20%. The third phase of the crisis is just beginning as this book goes to press, and can be found in a fiscal crisis among states. Most states have a balanced budget provision in their constitutions, and the fall in revenue from lost housing values and taxes upon financial assets forced drastic cuts in many programs and services. Layoffs of public employees are soaring and many states are removing the rights to collective bargaining for public employees. More such attempts to reduce costs to state and local governments and instill the "flexibility" of having employees pay for the effects of the economic downturn are likely to occur in the future.

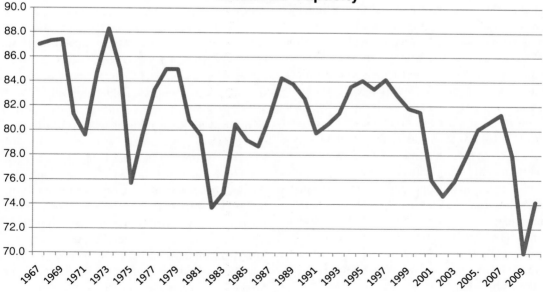

Fig. 7.1 *Source*: Economic Report of the President, 2011, Table B-54, http://www.gpoaccess.gov/eop/tables11.html

The Housing Bubble, Speculative Finance, and the Explosion of Debt

The economic downturn of 2008 began much as did the Great Depression of the 1930s: with a major hurricane and a collapse of speculative housing. Although the events of the late 1920s were centered in Florida, the antecedents of the 2008 crisis were truly global. Throughout the latter years of the twentieth century a global pool of money, or a glut of savings, were building from sources as diverse as sovereign wealth funds based on petroleum profits, to Chinese trade surpluses, to individual accumulations in high-saving nations. By the middle of the first decade of the twentieth century these funds had grown to over $70 trillion. Traditionally such funds had been invested in safe assets such as U.S. Treasury securities. However, by 2004 the Federal Reserve Board of the United States had driven interest rates down to the 1% range by purchasing Treasury securities from banks, thereby releasing more money into the system following the collapse of the high-tech bubble. Investors were forced to look elsewhere for better rates of return. One location they found was the housing market in the United States, as well as other housing markets. Prices were rising and the instruments created in the Great Depression such as insured long-term, amortized, mortgages and the creation of a secondary market where mortgages could be bundled and sold as short-term securities made the market appear safe from risk. Rates on mortgages of 5–7% were far more appealing than were 1% returns on Treasury bonds. The demand from global investors was sufficient, and the system of commissions and bonuses so lucrative, that standards for qualification based on income, assets, and employment stability were systematically lowered, then ultimately ignored. By 2006 mortgage brokers were no longer asking for documentation of income, employment, or other assets. The famous NINJA loan, or liar's loan, was born: No income, no job or assets [32]. By 2006 fully 44% of mortgage loans required no documentation. In addition the average loan-to-value ratio increased to 89% by 2006, as the number of no down payment (100% financing) mortgages climbed from 2% in 2001 to 32% in 2006 [33].

The process was abetted by the general climate of financial deregulation that had characterized the U.S. economy since the 1980s, and specifically by the policies of the second Bush administration that allowed a 5% (or less) equity position for lenders. The secondary market, created during the Depression at the insistence of banks, allowed for the pooling of mortgages into mortgage-backed securities. As long as the potential for default was low, because the standards for qualification were high, these securities were fairly risk free, as they had been historically. However, the emerging, and unregulated, sectors of the financial securities industry created even more exotic instruments by which to finance housing. Groups of mortgage-backed securities were themselves bundled into collateralized debt obligations (CDOs) and they were further divided into slices (or, to use the French word, *tranches*). Ratings agencies, usually hired by the issuers, declared these CDOs to be investment grade (Aaa).

On the basis of the investment grade rating mortgage security investors were able to purchase insurance policies against possible default known as credit default swaps. In the deregulated climate of 2007–2008 one did not even need to own an asset in order to purchase an insurance policy. Thus some "assets" were often insured by multiple parties at 100% each. This proved fatal to insurers such as AIG. The existence of global surplus savings and a lightly regulated climate served as an incentive for mortgage brokers, who would sell the loan immediately, to offer more mortgages to more people who simply did not have the income to pay the loans. But the risk would be managed further up the chain, by regional banks and in the money center banks in the world's financial districts. The nation's central bankers (e.g. Alan Greenspan, Ben Bernanke, Timothy Geithner) assured the public that the new financial innovations would reduce systemic risk. Yet beyond the housing bubble and the Wall Street casino another problem was brewing beneath the surface, the problem of unsustainable levels of debt.

Table 7.2 Domestic debt and GDP (trillions of dollars)

Year	Gross domestic product	Total debt	Household	Debt Financial firm	By sector Non-fin'l business	Gov't (local, state and federal)
1976	1.8	2.8	0.8	0.3	0.9	0.8
1980	2.8	4.5	1.4	0.6	1.5	1.0
1985	4.2	8.4	2.3	1.2	2.6	2.3
1990	5.8	13.4	3.6	2.6	3.7	3.5
1995	7.4	17.9	4.8	4.2	4.2	4.7
2000	9.95	26.3	7.0	8.1	6.6	4.6
2005	12.6	39.8	11.7	13.0	8.5	6.6
2010	14.7	50.5	13.4	14.2	11.1	11.8

Source: Economic Report of the President 2011, Table B-1
http://www.gpoaccess.gov/eop/tables11.html
http://www.federalreserve.gov/releases/z1/Current/Coded/coded-2.pdf
http://www.federalreserve.gov/apps/fof/Guide/P_9_coded.pdf

The purchase of everything from innovative financial instruments to bundles of loans was highly leveraged, that is, purchased with borrowed money, often at ratios of 50:1 or more. The system remained solvent as long as housing prices kept rising. Consumers could treat their houses as ATMs. In 2004 and 2005 Americans withdrew $800 billion in equity each year. This allowed for the purchases of more home improvement products, automobiles, and exotic vacations, as well as mundane purchases of daily life. More than 7,000 Walmarts and 30,000 MacDonalds were constructed to meet the growing demand. New television shows such as "Flip This House" advised potential real estate speculators as to which improvements would result in easy financial profit. Wharton School senior strategic planner James Quinn estimates that without these withdrawals economic growth would have been no more than 1% annually between 2001 and 2007. Homebuilders followed suit, constructing 8.5 million homes in 2005, about 3.5 million more than could be justified by historical trends [34]. However, by 2006 home prices began to fall. This touched off the downward cascade typical of a positive feedback loop. As homeowners found themselves "underwater," or owing more on their mortgage than the house was worth, mortgage defaults began to increase. CNN estimated that by the last quarter of 2010, 27% of all homeowners were in this situation. As defaults escalated, increasing 23% from 2008 to 2009, the bundled securities that were constructed from these pools of allegedly safe, investment grade securities began to lose value. Worse, so many of them were highly leveraged, the falling prices of homes and bundles of mortgages created a panic. Because the financial instruments were so complex, even banks could not figure out what their portfolios were worth. Consequently, the mortgage crisis could not be isolated in the riskier "subprime" market but spread to the entire economy. Major investment banks were crippled as well. Two Bear Stearns hedge funds collapsed, precipitating the general financial panic, 157-year-old Lehman Brothers went bankrupt, and Merrill-Lynch was absorbed by Bank of America, under considerable pressure from the Treasury.

The debt problem, however large, was not limited to housing. As the data in Table 7.2 indicate, debt was expanding in all sectors of the economy. By 2008 household debt, including mortgage debt and consumer credit, amounted to $13.8 trillion, equivalent to the nation's gross domestic product. By 2005 consumer debt exceeded income after taxes, standing at 127% of disposable income. Debt service ratios, or the percentage of disposable income used to pay principal and interest on contracted loans rose from about 11% in 1980 to nearly 14% in 2005 as the crisis loomed. The burden was felt highly unevenly. Those in the top fifth of the income distribution paid only 9.3% of their incomes in debt service by 2004, whereas those in the middle two

Fig. 7.2 *Source*:
Economic Report of the
President 2011, Table B-1,
http://www.gpoaccess.gov/
eop/tables11.html, http://
www.federalreserve.gov/
releases/z1/Current/Coded/
coded-2.pdf, http://www.
federalreserve.gov/apps/
fof/Guide/P_9_coded.pdf

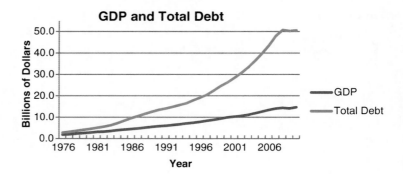

fifths paid between 18.5% and 19.4% [35]. The debt of nonfinancial corporations increased twentyfold between 1970 and 2007, and the debt of financial firms (banks, insurance companies, mortgage brokers, etc.) expanded by a factor of 160! (Fig. 7.2).

Banks had long been seen as recipients of deposits and lenders of money, safe and conservative in their outlook. But in the new world of deregulated finance the financial service industry became the largest borrower in the economy. It was this leverage that transformed the financial structure and made it vulnerable to disruptions. In the words of John Maynard Keynes, "Speculations may do no harm as bubbles on a steady stream of enterprise. But the position is serious when enterprise becomes the bubble on a whirlpool of speculation [36]."

The Deficit and the National Debt

The federal government was another participant in the increase in debt. Budget deficits climbed from $3 billion annually in 1970 to $1.414 trillion in 2009, while the national debt, or sum of yearly deficits, which was less than $1 trillion during the Carter administration, ballooned to over $14 trillion. The primary drivers of this increase were a reduction in taxes, especially at the top of the income distribution, and the expansion of government spending, primarily for the military and for entitlement programs such as Social Security, Medicare, and Medicaid. Military spending in 1970, the year of peak domestic oil production and the beginning of the era of

stagflation, the government brought in $192.8 billion in receipts and spent $195.6 billion, for a deficit of $2.8 billion. In the last year of the Reagan administration, whose economic policy was built upon increased military spending and tax cuts, the annual deficit soared to a historically unprecedented $155.2 billion. The last 2 years of the Clinton administration actually saw modest budget surpluses, as the growth rate of military spending declined and tax receipts increased with the high-tech boom. Deficits began to climb with the second Bush administration, beginning with a $3 trillion tax cut (over time), raising to $458.6 in 2008. By 2009 the annual difference between receipts and outlays was $1.4 trillion. Income tax revenue dropped from $1.635 trillion in 2007 to $898 billion in 2010. Military spending, which stood at $294 billion per year when the Bush administration took office rose to $616.8 billion in 2008. It has continued to climb during the Obama years, reaching the level of $693.6 billion in 2010. The Office of Management and Budget estimates that 2011 military spending will exceed $768 billion. In 1970, at the height of the Vietnam War, military expenditures were 8.1% of gross domestic product, and total government spending was 19.3%. By the end of the Clinton administration military expenditures had fallen to 3% and total spending remained about the same, at 18.5%. By 2010 military spending stood at 4.8% of GDP, and total spending rose to nearly 24%. Mandatory expenditures, such as those on health care (Medicare for the aged and Medicaid for the poor), along with Social Security and other income support programs (unemployment insurance, Supplemental Security Income for the disabled, food stamps, etc.) increased from $60.9 billion, or 6% of GDP, in 1970 to more than $2 trillion, or 14.7%, in 2009 [37].

Despite the increase in mandatory expenditures for entitlement programs the income distribution grew more skewed, largely as a result of a stock market boom and subsequent bailout, along with tax cuts at the top of the income distribution. In 1980, the beginning of the neoliberal economic strategy, the Gini coefficient was 0.403 and the top 20% of the income distribution claimed 16.5% of aggregate income. The top 1% received 8% of income and the top 0.01%, 0.065%. By the end of the second Bush administration the Gini coefficient increased to 0.466, indicating a greater degree of overall inequality, the top 1% claimed 17.67% of aggregate income, and the share of the top 0.01%, which amounts to about 14,000 families out of a population of 300 billion, rose to 3.34% [38].

The 2010 Congressional elections saw a large enough segment of the population expressing concern that the Democratic majority in the House of Representatives was unseated. Despite all the claims and counterclaims one question hangs over everything: is the economy nearing peak debt as well as peak oil? The political will to expand more debt within the United States is clearly shrinking, and the willingness of other economies and investors to purchase Treasury securities is also in decline. But what are the potential effects of declining government participation in the economy? And what are the debt resolution options: more debt expansion or economic collapse. Since the 1980s the economy has been driven primarily by cheap oil and cheap debt. What happens when they are both removed? If one believes that the market economy is resilient and self-regulating then a decrease in government spending will simply free money for spending in the private sector and the economy will prosper. If, on the other hand, one believes the explosion of financial speculation and debt was due to investors seeking financial profits in an otherwise stagnant real economy, as indicated by declining rates of capacity utilization, then the reduction of government spending, coupled with the rise of inequality, might cripple the economy

by reducing its overall level of demand. The age of peak oil may well be the age of degrowth and of peak debt as well. It is likely to be an age of austerity.

Conclusion

The world economy collapsed into depression in the 1930s. Governments faced few ways out: Fascism, Communism, or Social Democracy. John Maynard Keynes wrote his classic text, *The General Theory of Employment, Interest, and Money* as a guidebook for saving capitalism from itself in order to avoid the other outcomes which he detested. In the United States the program took the form of the New Deal. Although Franklin Roosevelt was able to restore confidence among a shattered population and put millions back to work, the New Deal did not engineer an economic recovery. It took the Second World War to do that. The United States exited the war in a clear position of economic and military power. Its corporations expanded into former colonies, the terms of trade were positive, and the prospects bright enough that corporations could share the gains from rising productivity with workers, ensuring adequate income to buy their products while increasing profits at the same time. Oil was cheap and plentiful, and the American consumer could utilize the rising quantities of cheap and available oil to live the American dream of a house in the suburbs, good schools, a steady job, and a cornucopia of consumer goods.

All that was to change in the 1970s. Domestic oil production peaked and the United States no longer possessed the spare capacity to cushion events in the world oil market. Beginning with two oil supply disruptions, in 1973 and 1979, the citizens of the developed world saw rising prices and constricted supplies. At the same time the era of stagflation commenced, and mainstream Keynesian policies no longer worked. Efforts to expand the economy resulted in rising inflation, and efforts to control inflation made unemployment rise to politically unacceptable levels. After a period of impasse a neoliberal agenda was consolidated during the Reagan years, consisting of a

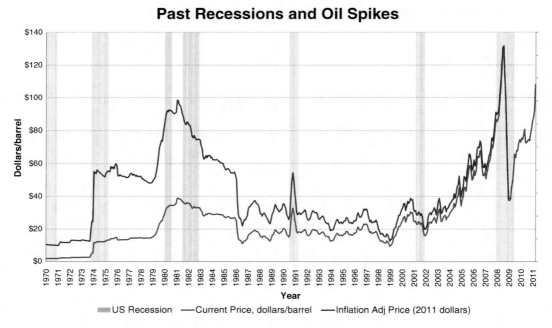

Fig. 7.3 *Source*: After Steven Kopits, personal communication. Original data from the Energy Information Administration, short-term energy outlook, monthly data, through March 2011

belief in small government, deregulation, low taxes, and a strong military. The economy did recover by the end of the 1980s but the price was ever-increasing levels of inequality and a rising debt burden. The neoliberal approach continued through the Democratic administration of Bill Clinton, where it was consolidated further. The bill came due at the end of the second Bush administration as the soaring debt burden and lax regulatory climate led to a near collapse of the world financial structure. But throughout this period, the time after the peak of U.S. oil production, new phenomena began to occur with monotonous regularity. Oil prices would rise in a prosperous economy. The rise in oil prices would contribute to the next recession. The recession would reduce oil prices, then the process would repeat itself (Fig. 7.3).

Most explanations of the postwar social order focus on the internal dynamics of the world economic system: its overall demand, technology, and the distribution of income. How do these factors affect the aspirations of the world's population for a decent income and a meaningful life?

But we contend that the world economic system is limited not only by its internal dynamics but also by the external biophysical conditions posed by the availability of energy and the consequences of using it.

As this book goes to press the Middle East is afire with democracy movements. Oil prices rose and fell once more (from $110 per barrel as of mid-March 2011 to $91 per barrel at the end of June to $80 in September) and economists, pundits, and politicians alike express concern as to the effect on the economic recovery. On March 11, 2011 an earthquake of magnitude 8.9 on the Richter scale, the most powerful one in recorded history, struck the northern coast of Japan. The nation was devastated by the quake and subsequent tsunami. The lack of electricity shut down the cooling systems of the Fukushima nuclear reactor complex. The latent heat from the fuel rods boiled away the water, resulting in a partial core meltdown. The heat also liberated the hydrogen from the oxygen leading to the build-up of flammable hydrogen gas. On March 14, 2011 the second of the reactors exploded. The viability of the third is in question. If the future of nuclear power is again in question, what will the world's energy options consist of?

At some point the production of oil on a world basis will peak. Problems of instability and rising prices will cease to be just cyclical and political, but will become secular, geological, and permanent. What does that portend for the economic system? Will peak oil exacerbate the inherently stagnationist tendencies of the monopolized economy as Baran and Sweezy argue? How can we generate employment and reduce poverty, advocate democracy, and rebuild after natural disasters when the energy base to do so is in decline. If every scientific measurement, from ecological footprinting, to biodiversity loss, to peak oil, to carbon dioxide concentrations in the atmosphere, show that humans have overshot the planet's carrying capacity, then how can we grow our way into sustainability? We can't, but how do we deal with the consequences of a nongrowing economy which historically manifest themselves as periodic depressions. We return to these questions in the final section of our book.

Questions

1. What was the "Treaty of Detroit?" How did it affect postwar labor relations in the United States?

2. What were the four "pillars of postwar prosperity?" Explain how each helped set the stage for the long economic expansion of the 1950s and 1960s.

3. What was the New Deal? What problems did it try to address, and what was its major legislative accomplishments? How successful was the New Deal in restoring American prosperity?

4. What was the role of the Second World War in transforming the U.S. economy?

5. How did the world oil industry, and the U.S role in it, change in the years after the Second World War?

6. Why could the period from the end of the Second World War be characterized as the era of economic growth?

7. Why was the "New Economics" of the 1960s successful in stimulating economic growth?

8. What is stagflation? What was the role of peak oil in bringing about stagflation in the United States?

9. What other factors led to the erosion of the pillars of postwar prosperity?

10. Why was the "New Economics" unsuccessful in eliminating stagflation?

11. What are the major tenets of the conservative growth agenda, also known as neoliberalism?

12. To what degree was the neoliberal program of the Reagan era successful? What were the economics and social costs of this success?

13. How did the Clinton administration carry on the neoliberal agenda? How did low oil prices during the 1990s affect U.S. economic performance?

14. How much did debt expand in the first decade of the twenty-first century? What were the economic outcomes?

15. How might biophysical limits affect economic performance as we enter the second half of the age of oil?

16. In the 1950s and 1960s economic growth was driven by cheap oil; as the oil ceased to become less cheap something else had to drive economic growth: cheap money and the expansion of debt. Is this sustainable?

References

1. Chandler, Michael, Ehrlich, Judith, Goldsmith, Rick, and Lerew. Lawrence. 2009. The most dangerous man in America: Daniel Ellsberg and the pentagon papers. Kovno Communications.

2. Editors of Business Week. "The Decline of U.S. Power." Business Week. March 19, 1979. Pp. 37–96.

3. Bowles, Samuel, David Gordon and Thomas Weisskopf. 1990. After the wasteland. Armonk, NY: M.E. Sharpe, Inc.

4. Kennedy, David. 1999. Freedom from Fear. London: Oxford University Press. p. 77

5. Business Week quoted in Kennedy 1999. P. 84.

6. Franklin D. Roosevelt quoted in Kennedy 1999, p.123and Collins, Robert M. 2000. More: The politics of economic growth in postwar America. New York: Oxford University Press. p. 5.

7. Dighe, Ranjit. 2011. "Saving Private Capitalism: The U.S. Bank Holiday of 1933. Essays in Economic and

Business History. Vol. 29. P. 41. Kennedy. 1999. P. 136.

8. Klitgaard, Kent A. 1987. "The Organization of Work in Residential Construction. (Unpublished Ph.D. Dissertation) University of New Hampshire.

9. Kennedy 1999. Chapter 5: pp. 131–159.

10. Yergin, Daniel. 1991. The Prize: The epic quest for oil, money, and power. New York: Simon and Schuster. Chapter 13. Pp. 244–259.

11. Deffeyes, Kenneth S. 2001. Hubbert's Peak: The impending world oil shortage. Princeton, N.J.: Princeton University Press. pp. 4–5. Petroleum Geologist Kenneth Deffeyes, a colleague of Hubbert, realized that the U.S. oil supply had indeed peaked when in 1971 he read in San Francisco Chronicle the that the Commission instructed oil companies that they could produce at 100% of capacity

12. Kennedy 1999. Pp. 252–253

13. Kennedy 1999 Ch. 9. Pp. 249–287

14. Kennedy 1999. P. 641

15. DuBoff, Richard. 1989: Accumulation and Power. Armonk, NY: M.E. Sharpe, Inc. pp. 150–155

16. Bowles 1989. P. 109

17. Inflationdata.com July 21, 2010

18. Yergin 1991 Ch. 22: 431–449

19. Cleveland, C., Costanza, R., Hall, C.A.S., and Kaufmann, R. 1984. "Energy and the U.S. economy: A biophysical approach." Science 211: 576–579.

20. Bowles 1990. P. 27

21. Collins 2000. P.21

22. Collins 2000: Ch 2. Pp. 40–67

23. Yergin 1991. Ch 25–26: 499–540

24. Yergin 1991 Ch 29–31: 588–652.

25. Bowles, 1991, Bureau of Labor Statistics, Inflationdata. com.

26. Wachtel, H. and Adelsheim, P. 1977. "How Recession Feeds Inflation: Price Mark-Ups In a Concentrated Economy." Challenge. 20 (4): pp. 6–13

27. Yergin 1991 Ch 33: 674–698

28. Bowles 1990 Ch. 8: 121–135

29. Yergin, 1991 Ch.36: 745–768

30. Bowles Ch 10: 146–169, Ch 13: 199–216

31. Economic Report of the President. 2011. "Capacity Utilization Rates, 1962–2010." Table B-54.

32. Davidson, Adam and Blumberg, Alex. 2009. "The giant pool of money." This American Life, Program # 355. National Public Radio.

33. James Quinn. 2008. "The Great Consumer Crash of 2009." Seeking Alpha

34. Quinn, 2008. p. 6

35. Foster, John Bellamy and Magdoff, Fred. 2010. The great financial crisis. New York: Monthly Review Press. pp. 30–31.

36. Keynes, John Maynard. 1964. The general theory of employment money and interest. New York: Harcourt Brace. p. 159

37. The economic report of the president. Table B-79. Congressional Budget Office. "Historical Budget Data. Tables F 10 and F 11.

38. U.S. Bureau of the Census. "Historical Income Data, and "Top Incomes Database. Paris School of Economics.

39. Yergin 1991. Ch. 19. Pp. 368–388

40. Yergin 1991. P. 381

41. Kennedy 1999. P. 624–626

42. (US Bureau of the Census. 1975. Historical statistics of the United States: Colonial times to the present. Bicentennial edition. Series G 416–469, N 156–169, P 68–73, Q 148–162)

43. Yergin 1991 Ch. 21: 409–430.

44. DuBoff. 1989. Pp 134–235.

Globalization, Neoliberalism and Energy

8

Young adults today have grown up in a world where globalization is a pervasive reality and where most of our politicians accept its supposed virtues. Sometimes there are discussions about how globalization is losing (or gaining) us jobs, or whether we have globalized too much or not enough, but for most people it is just a fact represented by the labels from all around the world on their clothes or electronic devices. This was not the case when the authors of this book were young; at that time nearly everything we ate, wore, or drove was made in America. Anything from overseas – except specialized luxury goods – was normally assumed to be cheap and inferior. Globalization, at least on the scale we see it today, is a relatively recent phenomenon so it is important to understand how globalization grew so large so fast, what are the perceived and actual gains and costs, and how these are related to energy use.

Before we consider this from a modern perspective, however, we think it important to emphasize that trade has been important since before written human history, as indicated by many important artifacts found in archeological digs going back tens of thousands of years. People have always wanted luxury goods from abroad, and have always sought various tools, amusements, foods, and experiences not found locally. One of the clearest examples of long-range trade is the "spice route" connecting Europe and the Middle East to all parts of Asia (Fig. 8.1). Spices were very important in ancient times for their own sake and also to hide the sometimes tainted smell and taste of rotting food before refrigeration. Spices were good items of trade because they were exotic, relative light and compact, and could be carried for thousands of miles by camel and donkey and still make a profit. At an archeological dig near Stockholm, Sweden, we watched the excitement of the excavators of an ancient Viking site when they found a coin from Constantinople. The Vikings, often more traders than plunderers, traveled thousands of miles on European rivers. Many archeological digs of Native American sites find, for example, arrow heads made from stone quarried hundreds or thousands of miles away. With the advent of European colonization and imperialism in Africa, Asia and the Americas trade took on a whole new dimension. An early economic "school," the mercantilists (the fifteenth through the eighteenth century), believed wealth was measured in gold or silver, and promoted trade and imperialism to obtain these metals. Nevertheless the day-to-day lives of most people, including Europeans, remained based on materials that rarely traveled more than a few tens or rarely hundreds of kilometers from their growth or extraction.

Ricardo and Comparative Advantage

A great debate raged over the British Corn Laws, which limited the import of cheap grains from the continent, in the early 1800s. The landed classes favored these laws as they kept the price of food

C.A.S. Hall and K. Klitgaard, *Energy and the Wealth of Nations: Understanding the Biophysical Economy*, DOI 10.1007/978-1-4419-9398-4_8, © Springer Science+Business Media, LLC 2012

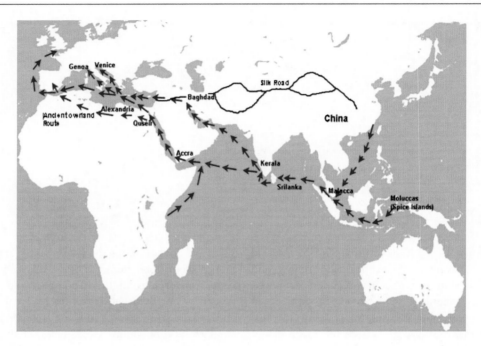

Fig. 8.1   The spice route (*Source*: Dyfed Lloyd Evans)

and rents high. The emerging capitalists wanted to see the laws repealed, as they raised wages (which were based on food prices) and reduced profits. Thomas Malthus, perhaps better known for his writings on population, defended the Corn Laws and the aristocracy. David Ricardo argued against them. His argument has become known to history as the doctrine of Comparative Advantage. In developing the principle of comparative advantage Ricardo extended to an international basis Adam Smiths idea that productivity and wealth will increase by the application of the division of labor. In short, the idea is that a county should produce what it does best and trade of the rest. For Ricardo "doing best" meant producing a commodity by means of the fewest number of labor hours.

Ricardo set up an entirely arbitrary and abstract example in which England and Portugal traded wine and cloth. It was arbitrary and abstract because Ricardo assigned Portugal and *absolute advantage* in both wine and cloth. In other words, Portugal could produce both commodities with fewer labor hours than could England. This was clearly different than actual historical conditions, in which England produced cloth at a much lower cost. After helping Portugal in their war with

Spain, England extracted many concessions from the Portuguese in return for their military aid, including the repeal of laws that banned Portuguese citizens from wearing English cloth. This destroyed the domestic textile trades and Portuguese capital moved into vineyards.

While Portugal was *absolutely* more efficient in the production of both wine and cloth, it was *relatively* more efficient in the production of wine. England was relatively more efficient in the production of cloth. If each country specialized in the commodity they were relatively most efficient in they could specialize and trade. The result would be that each country would be able to consume more wine and cloth without an increased expenditure of labor. His model, however, depended upon the fact that both capital and labor were immobile internationally. Only the finished products were exchanged. Ricardo's argument carried the day. On the basis of the fame he received from writing *Principles of Political Economy* Ricardo was elected to Parliament. There he argued tirelessly for the repeal of the Corn Laws. While his efforts were ultimately successful (the Corn Laws were abolished in 1846) Ricardo did not live to see their repeal, having died in 1823.

In subsequent years the Ricardian principle of comparative advantage was systematically sanitized and incorporated into the body of neoclassical economics. The historical context of the debate over the Corn Laws was removed, and advantage was determined by "resource endowments." Ricardo's use of labor hours was abandoned in favor of ratios of opportunity cost, and the need for the immobility of capital and labor was abandoned. The new argument, now cast within the confines of perfect competition, was now taught to generations of students as "the law of comparative advantage," devoid of any historical context or theoretical preconditions.

Trade and Imperialism

The advantages of trade were often conflated with those of raw exploitation of others and with imperialism. Beginning in the sixteenth century most European powers laid claim to territory in Africa and the Americas. We have already discussed the raw exploitation of natives in these areas by the Spanish as they sought gold and silver, the English for tea from India and Ceylon, and especially sugar from Barbados and so on. Much of the labor energy for the production of these products came from actual or virtual slaves.

Few consumers of cotton clothes in 1860 or of rubber tires in 1900 (or even of some diamonds or cell phone materials today) understood the human slavery that produced the products they purchased. We found a particularly chilling account in Adam Hochschild's book, *King Leopold's Ghost*, about how some estimated ten million Africans were savagely used and killed as Belgians and other Europeans "developed" the interior of Africa for ivory (much used before plastics were available for everything from false teeth to piano keys) and rubber (for tires and many other things). The lies used by Leopold to justify his horrendous abuse of the people living in the Congo basin are a reminder of how exploitive economic practices are sugar-coated by governments and in the press.

Whatever the virtues of globalization it is clear that it is a fact, and that the world has become enormously internationalized in the last decades (Fig. 8.2). All recent American presidents have called for more "free trade," implying a continuation of internationalization. Arguments against free trade tend to be about how U.S. factory jobs are moved overseas, resulting in economic hardship in the United States. An obvious example is automobiles, as the United States in 1950 produced some 99% of the automobiles it used, but now imports about half. As a consequence the

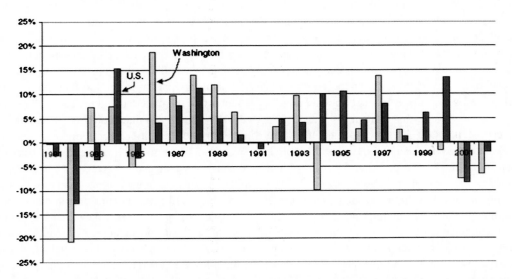

Fig. 8.2 Growth in international trade. International trade has been increasing from 5 to 10 percent for most years for both the United States and the State of Washington. Source: U.S. Census Bureau

city of Detroit and the state of Michigan, which once had the comparative advantage of relatively easy access to Minnesota iron ore and Pennsylvania coal, have suffered enormous economic impact. What is less obvious today, at least to the comfortable of the developed world, is that increased internationalization of trade means that the processes of exploitation of nature and of workers is also exported.

The Concept of Development and Its Relation to Trade

Most of the world today is quite poor. One billion of the Earth's seven billion people live on only one dollar a day, and another one to two billion live on between one and two dollars a day. There are very large pressures for the poorer countries to develop in order to become less poor. This development is often done in accordance within the concept of comparative advantage, that is, a search for some kind of product that might be produced well in that location. Ricardo originally devised his concept of comparative advantage around the idea of various labor efficiencies. However, one comparative advantage that poor developing nations almost always have is cheap labor. Some conditions of production are sufficiently irregular such that mechanization is not always profitable. This, along with a social structure of accumulation that degrades physical labor, can result in many dangerous and poorly paid jobs, many of them in poor countries. Many people who live there are driven from their traditional livelihoods by lower cost production wind up taking these low-wage and dangerous jobs, often because of the lack of alternatives. Of course development in fact requires the land, capital, other biophysical resources, transformations, and processes to occur if it is going to work. The pressure to develop comes from many sources including governments attempting to help or placate their constituents, idealistic foreign aid or NGOs (Non Governmental Organizations) from the developed world, and various business and economic interests who are interested in increasing their profit by exploiting the situation. What is rarely mentioned is that the real pressure behind

much, or perhaps most, development is simply an increase in population (a biophysical aspect), so that if some kind of development does not keep pace people will get poorer, which nobody wants.

The principal tool, or more accurately suite of tools, used to guide development is neoclassical (or free market or neoliberal or "University of Chicago") economics. (Curiously for much of history and much of the world today "liberal" means believing in maximum free trade, often the position of businessmen that in the United States might be more likely to be called "conservative.") The ascendancy of the neoclassical model occurred over the first half of the twentieth century as economists sought to generate a "scientific," "neutral" model that would focus on improving the welfare of the economy in general and leave the issue of the distribution of that wealth (properly in their view) to governments, hence absolving economists from any responsibility pertaining to that issue. The logic, summarized nicely in Palley [1] and Gowdy and Erickson [2], is that free markets will lead to "Pareto optimization" where, due to market pressures for lower prices from suppliers, the various factors of production (i.e., land, labor, capital, and so on) are being used so "efficiently" that they cannot be combined in any other way that would generate greater human satisfaction. The logic continues that if markets are completely "free" from government interference at each step of the production chain, each producer will be seeking the lowest possible prices, and each potential supplier will be seeking to cut his or her costs (ideally through "efficient" use of resources) so that the total net effect is that the final demand product will be generated as cheaply in that economy (which means, increasingly, the global economy) as possible. This should lead to the lowest possible prices to consumers, which is the objective of many economists. Most economists argue that this process works very well and generates substantial net benefits (e.g., Bhagwati [3]). Likewise most economists are enthusiastic about the free market system because, at least in theory, it is efficient; that is, economies are generating as much personal happiness as possible from their limited resources.

An important part of this is that there should be trade, and an important component of trade is that there should be more trading partners, including less-developed regions where there are resources that the developed world increasingly needs and where there is "unmet demand" for the products of the industrial countries [4]. This is an article of faith for most neoclassical economists and has guided how the United States undertakes trade and our relations with the less-developed world over the last half of the twentieth century. Development should lead to more wealth for both the nation becoming developed and for the developed country increasingly trading with it. In theory this should lead to efficiency, that is, that all parts of both economies are generating what consumers desire at a maximum rate given the resources at their disposal. The degree to which this in fact occurs is not at all clear from objective analyses of the behavior of real economies, and the converse is often true. (See Hall and Ko [5] and especially Bromely [6].)

The Leverage of Debt

In Latin America and Africa, especially, there have been pressures for development promoted by development agencies of the industrial nations, internal elites, foreign NGOs, and the World Bank for many decades. These efforts have been motivated by genuine humanitarian concerns. Less idealistically but perhaps more accurately they are characterized by Naomi Kline and others as, in reality, the self-serving efforts of powerful financial interests within the development agencies themselves, operating under the cover of humanitarian concerns. Certainly one net effect of either perspective on development has been an increase in debt of the poorer countries. There have been enormous pressures to repay debts associated with development and for revisions in how economics are undertaken, according to the neoliberal model imposed from outside entities, including especially the World Bank and the International Monetary Fund (IMF). The pressures have come from the leverage these institutions have because of outstanding international debt from many countries in Latin America and elsewhere. Given the nearly impossible demands

on governments due to poor and growing populations, and the difficulty in extracting taxes from rich elites who are often the same as those running governments, the easy solution has been and continues to be more debt, which is a tax on future citizens. When governments can no longer afford to pay their debt service, which often exceeds 25 or more percent of total GNP and perhaps all tax incomes, governments have sometimes defaulted. Default allows the banks and their agents to impose their sometimes Draconian "structural adjustment" programs which has meant, basically, reducing government services, eliminating tariffs that have protected home industries (such as agriculture), and basically opening countries to globalization. The basis for this is usually neoclassical economics as codified in the "Washington Consensus." The results are generally a disaster for the poorer countries and a continued bonanza for the banks. They are insightfully reviewed in Kroeger and Montanye [7].

Most structural adjustment programs also include policies and incentives for development, normally of industries that will generate foreign exchange (after all the bank's objective in structural adjustment is to get dollars or euros to repay the debt owed to them). For example, as part of the structural adjustment program implemented in Costa Rica in the mid-1990s there were large incentives to encourage the development of "nontraditional" agricultural crops from everything from macadamia nuts to cut flowers. These crops tend to be very dependent upon expensive imported agrochemicals, therefore it is not surprising that they did not have any significant effect on resolving debt. Meanwhile rising oil costs add greater balance-of-payment strains on most economies. In Costa Rica population growth has meant more food imports and the need for more agrochemicals for domestic crops, again making the resolution of debts more difficult [8]. The failure of many past development concerns, generally driven by neoclassical economic concepts of growth, to deal with the issue of population growth chains developing countries into seeking economic growth, whether real growth is possible or not.

The Logic for Liberalizing Economies

In the United States, especially, during the Reagan and Bush years, conservative leaders became adept at convincing many formerly apolitical or even labor union people that their own personal conservatism in issues such as family, society, religion, gun ownership, and so on could be best met through making an alliance with economic and political groups whose "conservative" agendas were quite different. These groups and their representatives in government were very much opposed to government in general and any interference with individual "freedom," especially intervention in the market. Thus they opposed, for example, government programs to generate energy alternatives (such as solar power or synthetic substitutes for oil), proclaiming that market forces were superior for guiding investments into energy and everything else. They also tended to be opposed to restrictions on economic activity based on environmental considerations and even mounted campaigns to discredit scientific investigation into environmental issues such as global warming.

We wish to point out that we use the term "liberal" and "conservative" as they tend to be used regularly and loosely in the United States to refer to the usually larger role of government by the Democratic party and smaller by the Republican party (at least in theory; the data are quite a bit more mixed). The terms themselves are often very misleading; e.g., many conservative people are extremely interested in conservation of nature, and that the concept of free trade is advocated by many liberals. In fact as we pointed out earlier in many countries such as Argentina "liberal" means liberal free trade, and is often associated with business interests.

These self-proclaimed conservative forces opposed government policies that restricted free trade. Their policy successes have contributed to the movement of many American companies or their production facilities overseas where labor is cheaper and pollution standards less strict. By 2000 the country seemingly had recovered from the stagnant 1970s and the recessions of the early 1980s and early 1990s. Stock values had increased

steadily and the general economic well-being of many Americans led to a general sense of satisfaction in market mechanisms. The end of communism in Eastern Europe and Russia effectively ended the Cold War, leaving the free market approach to economics as the only game in town with respect to economics. The presidential administrations of Republican George H. W. Bush and Democrat Bill Clinton alike pressed a free trade agenda. These programs included for many foreign lands reduced spending on social programs, the reduction of government ownership, and enhanced international trade. The terms of trade greatly improved for the United States as markets became "liberalized," and prices of basic commodities from coffee to cotton to oil declined by as much as half. Unfortunately poverty rates soared in Africa and Central America as a consequence. For example, the price paid to a farmer for a pound of coffee in Costa Rica (about a dollar per pound) barely changed from 1980 to 2005. These issues are discussed in depth by, for example, Annis [9] and Bello [10], and reviewed in Hall [8]. Fundamentally the arguments go back to Ricardo's concept of comparative advantage and to the concept that free trade will lead to efficiency. An implicit assumption of those who promote internationalism is that the players have equal power in the face of the supposedly neutral market. Of course this is patently absurd; a small coffee grower in Costa Rica does not have equal power in the face of some large national coffee buyer.

We Need to Test Our Economic Theories About Globalization, Development, and Efficiency

A recurrent theme of this book is that if economics is to be accepted as a real science we must expose the main ideas to empirical testing. For example, Gowdy has undertaken this by reviewing the work of those who have subjected the basic tenets of our dominant economic paradigms using the scientific method ("one cannot help but be impressed with the rigor of modern social scientists" [11]). There is a crying need to subject more of our economic theories to broad, unbiased, and thorough assessment of whether they deliver on what they promise [12–14]. There may be no trusted, or at least broadly accepted, concept within economics with a greater need of such testing than that of "efficiency." Efficiency is the principal argument used to promote the neoliberal model and of its application to international development and unrestricted international trade. Yet economists themselves have increasingly questioned the effectiveness of their development models. A particularly fine example of this is William Easterly's book, *The Elusive Quest forGrowth: Economists' Adventures and Misadventures in the Tropics*. Easterly reviews the use of economic theory (basically neoclassical) as applied to development, especially development in the tropics. Easterly did what few economists do: he actually tested whether the models of economists – that had been the backbone of billions of dollars of aid – had accomplished what they were supposed to do. In particular Easterly asked whether the main development model, the Harrod–Domar investment model, had, when applied, resulted in a measureable increase in GDP. His answer was that there was a perceptible increase in GDP for only for 4 of 88 cases where it had been tried. In other words, when tested these models were a disaster with respect to achieving their goals. LeClerc [15] arrived at a similar conclusion while testing a broader array of economic models as applied to development. Anyone involved in the world of investment economics should read these two studies.

It is often said that real economies are too complex, and the difficulty of undertaking proper tests and controls so daunting that you should not expect economic concepts to be explicitly testable. Hall's former student Dawn Montanye, however, asked whether the (neoliberal) structural adjustment model imposed upon Costa Rica by USAID (Agency for International Development) in the early 1990s had achieved its own clearly stated objectives when the subsequent behavior of the economy was examined [16]. This was a seemingly straightforward and reasonable thing to ask, based on readily obtainable data, although, curiously, it seems not to have been undertaken by USAID. Her results were yes for two and no for four out of their six principal objectives. In addition

there were a number of quite important but unanticipated negative consequences that occurred even for the cases where the objectives were met. If in fact there is such a large disconnect between theory and results then one wonders whether there should be so many routine pronouncements on how to run real national economies based on conventional theory and models [15].

If efficiency is the main reason that neoclassical economics is promoted, and if, to our knowledge, this efficiency has been tested barely or not at all by economists, how then might we go about testing efficiency? Countries such as Costa Rica have been subject to structural adjustment (a program often imposed upon debt-laden countries that had to turn to "the lender of last resort" (the World Bank and especially the International Monetary Fund). If as promised by theory structural adjustment does lead to efficiency this should be obvious from the data comparing pre- and poststructural adjustment. If this is not observed it seems to us hard to argue that structural adjustment and neoclassical economics do in fact lead to efficiency.

Definitions of Efficiency

The word "efficiency" is often confused with *efficacy*, which means "getting the job done," without regard to efficiency. *Efficiency* usually means output over input. There are two difficulties with defining efficiency: (1) it is hard to find a consistent definition of output of what? All GDP? Only those components you think are desirable? Or everything? And likewise (2) input of what? Economists usually think of efficiency (of, e.g., an economy) as the output of all desirable goods and service (i.e., capital, labor, and sometimes land) over the input of all resources available for production, usually referring to money or capital. Perhaps the best way to explain efficiency, as economists use the word, is by giving the counterexample of economics, that is, supposedly, not efficient. In the socialist states of Eastern Europe and the Soviet Union from roughly 1920 to 1990 the determination of how much of a good and service was produced (i.e., the allocation of productive resources) was decided in large part by central planning, that is, by government economists whose job it was to decide how many tractors, carrots, chickens, or other commodities were needed. There were some famous fiascos resulting from this (or at least good stories), so that, for example, in the 1950s in Russia and Poland too many tractors and too few refrigerators were ordered by the central planning committee, so that there were mountains of unused tractors and people were very unhappy because they needed refrigerators. To most Western economists this was a tragic example of how it was far better to leave the decisions of what to make up to markets, that is, Adam Smith's invisible hand of supply and demand. In other words in the centrally planned economy the productive resources of the nation, steel mills, labor, factories themselves, had been used *in*efficiently: they had produced too much of one thing that was not needed or wanted and not enough of another that was. In addition it had required a large, perhaps expensive, government bureaucracy to do the allocation decisions. It is this argument about efficiency that is used most commonly by neoclassical economists to argue for free markets and free trade. Two problems with estimating efficiency is that it is very difficult to decide just what output should be counted and what input is required for a particular economic activity. Despite the constant use of the word efficiency by economists you would be hard pressed to find where that has been measured or tested explicitly (except for some very general international comparisons often using rather arbitrarily defined quantifications of such terms as "level of financial development" and "improvements in efficiency," e.g., King and Levine [17]).

Engineers often use a very explicit measure of efficiency: simply the ratio of energy out of a process to the energy in. For example, coal is converted to electricity in a modern power plant at about 40% efficiency, and gasoline to road transport occurs at about 20% efficiency. Humans, too, generate work at roughly 20% efficiency. Some of the energy loss is inevitable losses to the second law of thermodynamics, some is related to needing to run the process at a more rapid rate than would generate maximum efficiency, and some is caused by poor design or poor housekeeping (i.e., not keeping the tires properly inflated).

A kind of combined ratio is often used to measure efficiency of economies within biophysical economics: the GDP output over the energy input, usually for a country. We call this the *biophysical economic efficiency*. The economic output must be corrected for inflation to compare different years. The ratio does not mean anything explicitly (as the engineering one does) but rather is one relatively unambiguous way that we can measure the efficiency of an economy and test explicitly the hypothesis that more free trade leads to greater efficiency. This, as we said above, is not possible to do with the more nebulous "productive resources" perspective usually given by economists. It is useful mostly for comparative purposes: either for different countries or for one country over time. Because the economies of many countries were explicitly or implicitly (via the general spread of neoclassical economic concepts) "converted" at least partially to more neoliberal models in the 1990s and early 2000s then it should be simple to test whether those economies subject to explicit structural adjustment consistent became more efficient during the 1990s. If their efficiencies are increasing then this would tend to support the hypothesis and the contrary.

Testing the Hypothesis that Freer Trade Leads to Economic Efficiency

The actualization of neoclassical economics in Latin America and elsewhere was carried out through a program sometimes called "The Washington Consensus" that was administered to, especially, countries that could not pay interest on their debts to The World Bank or to the International Monetary Fund (IMF) [18]. The program of increasing free trade and reducing governmental spending ("stabilize, liberalize, and privatize") were said to be good and tough medicine for the debtor nations. And they were supposed to lead to economic efficiency. Inasmuch as we could not find any studies by economists about whether economies had in fact become more efficient after "adjustment" we undertook this ourselves by examining time trends in biophysical economic efficiency.

Our methods were very simple: plot the biophysical efficiency (i.e., real GDP/energy used, agricultural output per unit of fertilizer and so on for various countries) and see if there is any trend toward increasing efficiency. Explicitly we tested the hypothesis that following the implementation of neoliberal policies (either in the country or more generally worldwide after 1990) there would subsequently have been an increase in the biophysical efficiency of nations. We undertook this explicitly for 3 countries on each "developing" continent [5] and for 133 countries more recently [19].

Results of Testing for Biophysical Efficiency Following Liberalization

We found in both studies that when the energy use and the GDP for all countries in the world are plotted on the same graph the results are basically linear (although with considerable scatter), indicating that energy is required, or at least associated with, increases in the production of GDP for essentially all nations (Fig. 8.3). We also found that for most countries which were increasing in per capita wealth

that per capita energy use increased at approximately the same rate as that wealth (Fig. 8.4).

Today there is an enormous variation in the wealth of different national economies, from the poorest, where people tend to live on but 38 cents a day (or 140 dollars a year) to the wealthiest in the developed world where according to the World Bank, annual incomes ranged from 50,000 to 87,070 dollars annually in 2008. Not surprisingly, from our perspective, the energy use by these different countries varies similarly from about 0.32 GigaJoule per capita to 694 GJ per capita for 2005 (Fig. 8.5). Additionally as countries have developed they have tended to use more energy roughly in proportion to their increase in wealth (Fig. 8.4). When we examine the relations of GDP and energy use for developing countries in Africa and Latin America (the region especially affected by "liberalization of markets") we find no evidence at all that biophysical efficiencies have increased in response to the liberalization trends of the 1990s and early 2000s. When all countries are considered, biophysical efficiency has tended to, if anything, remain the same or decrease, both since 1970 and also since 1990 (Fig. 8.6). Colombia, relatively unaffected by neoliberal policies, may be an exception. We found similar results for many

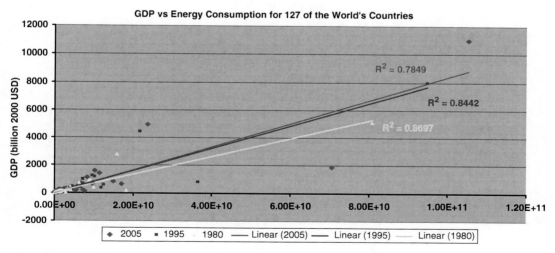

Fig. 8.3 The relation of energy use (Joules) and GDP for 127 countries in 1980, 1995, and 2005. The basically linear results indicate that energy is required for, or at least associated with, increases in the production of GDP for most nations and that whatever (small) increase in efficiency

that may have occurred (i.e., an increase in the slope of the line) tended to occur before the increased global liberalization of markets that began usually in the 1990s (*Source*: Ajay Gupta)

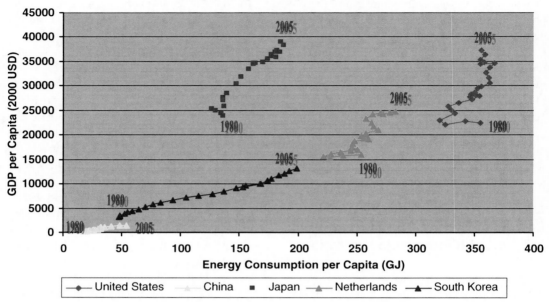

Fig. 8.4 The relation of GDP and energy use for five nations of the world. Developing nations including China, Korea, and Japan tend to have a strong correlation between their increase in energy use and their GDP, strongly implying that energy is needed for economic development. This is also true for the Netherlands, a developed nation, but appears less so for the United States which has increased its official GDP with little or no increase in energy use (*Source*: Ajay Gupta)

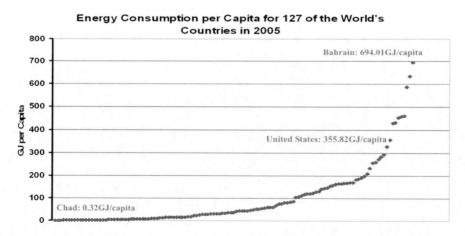

Fig. 8.5 All nations of the world ranked in order of increasing energy use per capita (*Source*: Ajay Gupta)

other countries [20–22]. Hence the hypothesis of this chapter, that the increasing use of the neoliberal, free market, neoclassical, or Washington Consensus approach to economics in the last decade of the twentieth century in the developing countries will necessarily bring increased efficiency of economies is not supported, and we must seek some other explanation for economic growth besides increased efficiency (as derived from neoclassical policies or anything else). These results

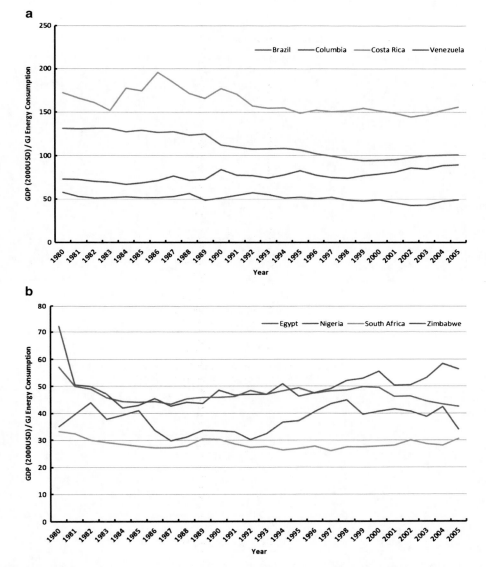

Fig. 8.6 (**a**) The ratio of GDP to energy use for four countries in Latin America from 1971 through 2005. The flat or decreasing lines for all countries are not consistent with the hypothesis that liberalizing markets increases efficiency. (**b**) The ratio of GDP to energy use for four countries in Africa from 1971 through 2005. The flat or decreasing lines for all countries after 1980 are not consistent with the hypothesis that liberalizing markets increases efficiency (*Source*: Ajay Gupta)

are consistent with the increasing view of many development economists themselves [23].

Our results do show, however, that efficiencies have increased in many developed nations (Fig. 8.7). Whether this is because highly developed countries are capable of becoming more efficient through pure technology, or rather have basically exported their frequently polluting and energy-intensive heavy industries to the rest of the world is another

question. For some countries, sometimes energy-exporting countries, efficiency declines (Fig. 8.8). Further analysis, in which the embodied energy associated with imports and exports is added to or subtracted from the energy use (denominator) of the equation suggests that the main reason for the disparity of the different countries may be the degree to which each country is associated with undertaking the "heavy lifting" for others [24].

a

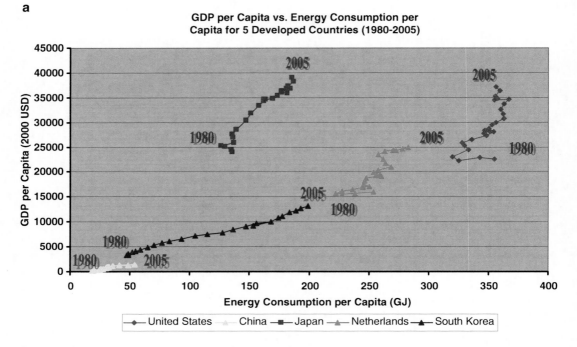

b

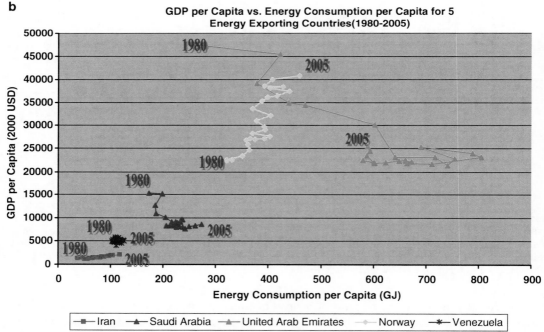

Fig. 8.7 (**a**) Our results do show, however, that efficiencies have increased in many developed nations (Source: Ajay Gupta). (**b**)The same relation as Fig. 8.7 but for oil exporting countries. This figure suggests that much of the reason for the apparent increase in efficiencies in Fig. 8.7 is due to the increasing import of energy-intensive components of the economy such as oil (*Source*: Ajay Gupta)

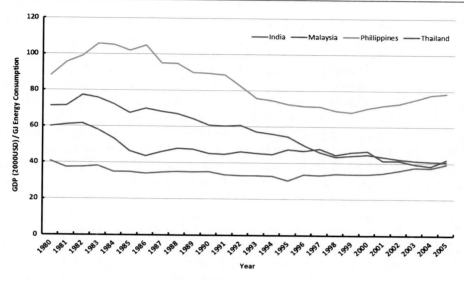

Fig. 8.8 The efficiency for some Asian countries, indicates little change or a slight decrease (*Source*: Ajay Gupta)

In our original paper we examined the efficiency of GDP versus water and forest products used as well as agricultural output versus input of fertilizer and found that there was always a strong relation between economic development, as indicated by GDP, and the increasing use of resources with no indication of an increase in efficiency over time.

What Should We Do Instead?

Our main conclusion from these and many other results is that for the majority of countries there has been no increase in efficiency since "liberalization" of the economy. Instead it is clear that whatever economic growth has occurred has occurred as a consequence of (or at least is highly correlated with) increasing the rate of the exploitation of energy and other resources. We conclude that neoclassical economics when applied does not increase wealth by increasing efficiency in any measurable way but only by increasing the rate of resource exploitation. These resources can be domestic or imported. If wealth comes from resource exploitation and not from efficiency then the concept of development must be very tightly tied to the soils, climate, agricultural potential, mineral resources, and other biophysi-

cal resources that to date have been given short shrift in conventional economic analysis. If the human condition is to be improved, it requires as much attention to the biophysical as well as the political and monetary environment, and it must be done within biophysical limits.

If neoliberal economics does not seem to be in agreement with empirical tests, violates the basic laws of physics, and is not consistent with its own assumptions, then what alternative do we have to guide development or to attempt to operate our economies by, at least in the macro sense?

Again We Return to Biophysical Economics

Our partial answer is *biophysical economics*, a rather imperfect but growing approach to economics that is based upon the recognition that wealth is fundamentally generated through exploitation of natural resources. Biophysical economics also recognizes that economic policies are mostly about directing how energy is invested in that exploitation. The fundamental approach to biophysical economics can be found in our books (Hall [8], Leclerc and Hall [22], and, of course, this one). This approach leads to some rather different views as to how we can improve the average economic plight of the poor of the world. In particular the biophysical model of development puts a real onus on the availability of affordable energy for successful development to occur.

There are many models of development including the Domar (focusing on the importance of savings), the Rostow (focusing on "stages of development"), and others. These are reviewed in LeClerc [15] who found that data that support their importance in generating or even explaining development is pretty thin. In all fairness it should be mentioned that it is not just the neoclassical model that seems to be having trouble generating economic growth. According to a Web-based review (cepa.newschool, 2004) there have been various models for encouraging development over time and each has basically been abandoned when it failed to generate much in the way of the desired development. This is in agreement with LeClerc's perspective. The cepa review, and our own, conclude that any rationale for the dominance of the neoliberal model today and the evidence for its effectiveness is ambivalent.

Among the theories of development (see review in LeClerc [23]) few of them are very powerful in predicting success or failure of development and not surprisingly, few of them connect development directly to energy use. We suggest another model, specifically the *biophysical model of development* which says that real material development (i.e., an increase in wealth) occurs only when the ratio:

energy resources / number of people

increases. This can be seen in Fig. 8.4, where per capita wealth increased only in those countries where per capita energy use increases (e.g., Korea, Mexico, Netherlands).

Hall [8] found many examples where development, at least as expressed as real GDP per person, is correlated very closely to the energy used per person. Where energy use has increased more rapidly than populations, people have become wealthier; where energy use has increased less rapidly than population, people get poorer (Figs. 8.9 and 8.8).

Although it is true that the United States and some other developed nations have become more efficient in turning energy into GDP, according to Robert Kaufmann about half of that is due to the increased use of higher quality input and much of the rest is due to the change in the economy from industrial production (much of which has been exported) to services (or even, strangely, consumption!). The degree to which this can occur for other countries is not clear. The GDP produced for the world as a whole has remained nearly constant or increased only slightly, suggesting that gains in the developed countries are matched by decreases in the less-developed countries that often are undertaking more of the heavy industrial work for the developed nations [19]. Our explanation for increases in economic activity is that the more resources, and explicitly more energy, can be developed the more economic activity can occur. This energy is used to fuel the productive process which in the contemporary world is more dependent upon energy than either capital or labor [8]. This is hardly news to most energy scientists, however, the degree to which it is a concept foreign to economists is quite remarkable [24]. Wealth comes from nature and the exploitation of nature, and much less so from markets or their manipulation.

Because it is clear that neoliberal policies have not resolved the persistent economic problems of the developing world, why then are they continually pursued? The cynical view is that they serve rather nicely the interests of those who impose them by maintaining cash flow to the banks and their shareholders. But that answer by itself is simplistic, for there are many well-meaning individuals who believe sincerely that these policies should be helping the countries where they are imposed. Whether these policies lead to net human welfare of all effected is a much more contentious issue within the broader world of those who think about these issues. If one can generate low prices by paying laborers as little as possible or by paying as little as possible for environmental cleanup then there will be strong pressures for this to happen, despite, or perhaps because, these benefits all go to the developed world. Such pressures are behind much of the move for globalization. We are unaware of much in the way of thorough systems-oriented case history research that examines whether this is true except perhaps for Brown et al. [25].

So, in summary, what must we do if we seek economic development that works?

1. Accept that neoclassical economics has failed for the most part.
2. Use the scientific method!
3. Build a biophysical model of the actual economic possibilities based on the real resources of a nation and its population trends.
4. Consider reducing demand through, for example, population control as an equally viable development strategy as opposed to increasing economic activity and, hence, the need for fossil fuels.

Specific Example: Assessment of Development and Sustainability in Costa Rica

We certainly recognize that the assessment above can be criticized for superficiality, although we believe the results are nevertheless basic and important. But we have, with our colleagues, undertaken such analyses in much more detail in the past. The most important of these studies has been for the economy of Costa Rica, which we have examined in great detail (e.g., Hall [22], a 761-page book published in 2000 with explicit, data-intensive chapters on each of the major segments of the economy) from a biophysical and conventional economics perspective. Our original purpose for undertaking this analysis was to determine how a sustainable society and economy might be developed. We view the book as a model for undertaking biophysical economics. But to our surprise our study (also given in Ref. [8]) found at least 19 reasons that Costa Rica (often the advertised model of sustainability) could not possibly be considered sustainable. Many of these reasons were based upon the interaction of energy and resource use to create a situation of decreasing efficiency (as defined in this chapter). Among these reasons:

1. Impossible debt loads that have been approximately constant since the 1970s, and which drain the government of substantial revenue each year.
2. There are too many people to feed, especially without imported fertilizers and other industrial input to agriculture. Even with these Costa Rica now imports about half its food.
3. This results in a need to generate foreign exchange for the necessary agricultural and food input.
4. Even with increasing input the yield per hectare for most crops has not increased since about 1985 due to erosion, depletion of nutrients, and a saturation of response to fertilizers.
5. Costa Rica, as a nation with no fossil fuels, has been, continues to be, and almost certainly will become even more dependent upon imported fossil fuels. This is true despite the very great efforts that Costa Rica has undertaken to exploit the natural advantage it has with many renewable energies: hydro power, wind, and geothermal, all a consequence of its extensive and high mountains.
6. Therefore Costa Rica is extremely vulnerable to an increase in oil prices and eventual oil depletion. Continued population growth makes this problem more severe every year. Attempts at a growth economy mostly have been negated by population growth, much of it from immigrants.
7. Despite enormous efforts there have been no "silver bullets" (i.e., magic solutions to problems) and, probably, the concept of sustainable development has no utility, at least so far, except, perhaps, to make the user feel good and to attract tourists.
8. Nevertheless, Costa Rica has generated an extremely good society on a relatively small resource base. There is a great deal that the rest of the world can learn about the efficiency by which Costa Rica generates good government services on a relatively small monetary and resource base.

Questions

1. Why do you think the world economy has been so globalized?
2. What early economist might be especially interested in seeing the degree of globalization that has taken place?

3. What was the "Spice Road?" What replaced, in part, its function? Can you give an energy argument for that?

4. What has been the relation between imperialism and foreign trade?

5. What does "development" mean? What are some of the groups that encourage development today (e.g., government foreign aid, NGOs, local investors)?

6. Many say that economic globalization is a two-edged sword with positive and negative aspects. What are some of the positive aspects? Negative?

7. Have most development models been tested? Why or why not? If so what results were found?

8. Do you think it is always difficult to test whether economic models work? Why or why not?

9. How is efficiency different from efficacy?

10. Define several uses of the word efficiency related to global issues.

References

1. Palley, T. I. 2004. From Keynsianism to Neoliberalism: Shifting paradigms in economics. in Johnston and Shad Filho (eds), Neoliberalsim – a critical reader. Pluto Press . tpalley@osi-dc.org).

2. Gowdy, J. and J. Erickson. 2005. The approach of ecological economics. Cambridge Journal of economics. 29 (2): 207–222.

3. Bhagwati. J. 2004. In Defense of Globalization. Oxford University Press. N.Y.

4. This perspective is a near mantra for many in both political parties in the US

5. Developed from : Hall, C.A.S. and J.Y. Ko. (2004). The myth of efficiency through market economics: A biophysical analysis of tropical economies, especially with respect to energy, forests and water. pp. 40–58. in M. Bonnell and L. A. Bruijnzeel (eds). Forests, water and people in the humid : Past, present and future hydrological research for integrated land and water management. UNESCO. Cambridge University Press.

6. Bromely, D. 1990. The ideology of efficiency. Journal of environmental economics and management 19: 86–107.

7. Kroeger T. and D. Montanye. 2000. Effectiveness of structural development policies. Pp. 665–694. In Hall, C. A. S. 2000. Quantifying sustainable development: the future of tropical economies. Academic Press San Diego.

8. Hall, C. A. S. 2000. Quantifying sustainable development: the future of tropical economies. Academic Press San Diego.

9. Annis, S. 1990. Debt and wrong way resource flow in Costa Rica. Ethics and International Affairs 4: 105–121.

10. Bello, W. 1994. Dark Victory: The U.S., structural adjustment and global poverty. Pluto Press, London.

11. Gowdy, J. (2005). Toward a new welfare foundation for sustainability Ecological Economics 53:211–222.

12. We cannot emphasize enough that anyone who wants to understand economic efficiency should read the paper by Bromley, 6 above.

13. Gintis, H., 2000. Beyond Homo economicus: evidence from experimental economics. Ecological Economics. 35: 311–322.

14. Hall, C.A.S., P.D. Matossian, C. Ghersa, J. Calvo and C. Olmeda. 2001b. Is the Argentine National Economy being destroyed by the department of economics of the University of Chicago? pp. 483–498 in S. Ulgaldi, M. Giampietro, R.A. Herendeen and K. Mayumi (eds.). Advances in Energy Studies, Padova, Italy.

15. LeClerc, G. 2008. Chapter 2 in LeClerc, G. and Charles Hall (Eds.) Making development work: A new Role for science. University of New Mexico Press. Albuquerque

16. Montanye, D. 1994. Examining sustainability: An evaluation USAID's agricultural export-led growth in Costa Rica. Master's Thesis. State University of New York, College of Environmental Science and Forestry.

17. Robert G. King, R. G. and R. Levine. 1993. Finance and growth: Schumpeter might be right. The Quarterly Journal of Economics, Vol. 108, No. 3 (Aug., 1993), pp. 717–737

18. Williamson, J. 1989. What Washington means by policy reform, in: Williamson, John (ed.): Latin American Readjustment: How much has happened, Washington: Institute for International Economics .

19. A. J. Gupta et al. Estimating biophysical economic efficiency for 134 countries (in preparation).

20. Ko, J.Y., C.A.S. Hall, and L.L. Lemus. 1998. Resource use rates and efficiency as indicators of regional sustainability: An examination of five countries. Environmental Monitoring and Assessment 51: 571–593.

21. Tharakan, P., T. Kroeger and C.A.S. Hall. 2001. 25 years of industrial development: A study of resource use rates and macro-efficiency indicators for five Asian countries. Environmental Science and Policy 4: 319–332.

22. LeClerc, G. and Charles Hall (Eds.) 2008. Making development work: A new Role for science. University of New Mexico Press. Albuquerque

23. Naím, Moisés. 2002. "Washington Consensus: A Damaged Brand." Financial Times, October 28.

24. Kunz, Hannes. 2010. Personal communication.

25. Brown, M.T. H.T. Odum, R.C. Murphy, R.A. Christianson, S.J. Doherty. T.R. McClanahan, and S.E. Tennenbaum. 1995. Rediscovery of the World: Developing an Interface of Ecology and Economics. In C.A.S Hall (ed) Maximum Power. University Press of Colorado Press. P.O. Box 849, Niwot, CO 80544, pp. 216–250

26. Klitgaard, Kent. 2005. Comparative advantage in the age of globalization. The international journal of environmental, cultural, economic & social sustainability. 1(#): 123–129

Are There Limits to Growth? Examining the Evidence

9

In recent decades there has been considerable discussion in academia and the media about the environmental impacts of human activities, especially those related to climate change and biodiversity loss [1]. Far less attention has been paid to the diminishing resource base for humans. Despite our inattention, resource depletion and population growth have been continuing relentlessly. The most immediate of these problems appears to be a decline in oil production, a phenomenon commonly referred to as peak oil, because global production appears to have reached a maximum and may now be declining. However, a set of related resource and economic issues are continuing to accumulate in ever greater numbers and impacts – water, wood, soil, fish, gold, copper – so much so that author Richard Heinberg [2] speaks of "peak everything." We believe that these issues were set out well and basically accurately by a series of scientists in the middle of the last century and that events are demonstrating that their original ideas were mostly sound. Many of these problems were spelled out explicitly in a landmark book called *The Limits to Growth,* published in 1972 [3]. In the 1960s and 1970s, during our formative years in graduate school, our curricula and our thoughts were strongly influenced by the writings of ecologists and computer scientists who spoke clearly and eloquently about the growing collision between increasing numbers of people, their enormously increasing material needs, and the finite resources of the planet. The oil-price shocks and long lines at gasoline stations in the 1970s confirmed in the minds of many

that humans were facing some sort of limits to growth. It was so clear to us then that the growth culture of the American economy had limits imposed by nature in 1970 that the first author made very conservative retirement plans in 1970 based on his estimate that we would be experiencing the effects of peak oil just about the time of his expected retirement in 2008.

These ideas have stayed with us, even though they largely disappeared for more than three decades from both public discussion and from college curricula. Although few people think about these issues today, most of those who do believe that technology and market economics will resolve the problems. The warning in *The Limits to Growth* – and even the more general notion of limits to growth – are dismissed as alarmist. Even ecologists have largely shifted their attention away from resources to focus on various threats to the biosphere and biodiversity. They rarely mention the basic resource/human numbers equation that was the focal point for earlier ecologists. For example, the February 2005 issue of the journal *Frontiers in Ecology and the Environment* was dedicated to "Visions for an Ecologically Sustainable Future," but the word "energy" appeared only for personal "creative energy." "Resources" and "human population" were barely mentioned.

But has the limits-to-growth theory failed? Even before the financial collapse in 2008, newspapers were brimming with stories about increases in the price of energy and food, widespread hunger and associated riots in many cities, and various

Fig. 9.1 The global population has doubled in the last four decades, as exemplified in this crowded market in India. Although some regions suffer from poverty, the world has avoided widespread famine mostly through the increased use of fossil fuels, which allows for greater food production. But what happens when we run out of cheap oil? Predictions made in the 1970s have been largely ignored because there have not been any serious fuel shortages up to this point. However, a re-examination of the models from 35 years ago finds that they are largely on track in their projections (*Source: American Scientist*)

material shortages. Subsequently, the headlines shifted to the collapse of banking systems, increasing unemployment and the evaporation of economic growth. A number of people blamed a substantial part of the current economic chaos on oil price increases earlier in 2008. Although many continue to dismiss what those researchers in the 1970s wrote, there is growing evidence that the original Cassandras were right on the mark. Their general assessments, if not always right in the details or exact timing, correctly warned about the dangers of the continued growth of human population and their increasing levels of consumption in a world approaching very real material constraints. It is time to reconsider those arguments in light of 30 years' experience and new information, especially about peak oil. Figures 9.1 through 9.6 give a vivid perspective on how some of these issues play out at the local level.

Early Warning Shots

A discussion of the resource/population issue always starts with Thomas Malthus and his 1798 publication *First Essay on Population*:

> I think I may fairly make two postulata. First, that food is necessary to the existence of man. Secondly, that the passion between the sexes is necessary, and will remain nearly in its present state. ...Assuming then, my postulata as granted, I say, that the power of population is indefinitely greater than the power in the earth to produce subsistence for man. Population, when unchecked, increases in a geometrical ratio. Subsistence increases only in an arithmetical ratio. Slight acquaintances with numbers will shew the immensity of the first power in comparison of the second.

Malthus continued with a very dismal assessment of the consequences of this situation and even more dismal and inhumane recommendations

Fig. 9.2 A village on one of Bangladesh's coastal islands was devastated by a cyclone in 1991, in which a total of more than 125,000 people were killed. Large storms had caused destruction in 1970, and would again in 2006. Although people in areas such as these are aware of the risk, overcrowding often prevents them from moving to safer regions (*Source*: American Scientist)

Fig. 9.3 In 1979 motorists were forced to line up for rationed gasoline during a period of oil price shocks and reduced production. Such events were compelling support for the argument that the world's population could be limited by a finite amount of natural resources (*Source*: American Scientist)

Fig. 9.4 In drought-stricken Southeast Ethiopia, displaced people wait for official distribution of donated water. Children who try to make off with the resource hours ahead of the appointed time are chased off with a cane. Such incidents demonstrate that water is another resource often available only in limited quantities (*Source*: American Scientist)

Fig. 9.5 Oil is not the only resource that may have peaked, with use outstripping the Earth's ability to support the level of consumption. In Sardinia, off the coast of Italy, commercial fishermen's catches are down by 80% compared to what their fathers used to haul in (*Source*: American Scientist)

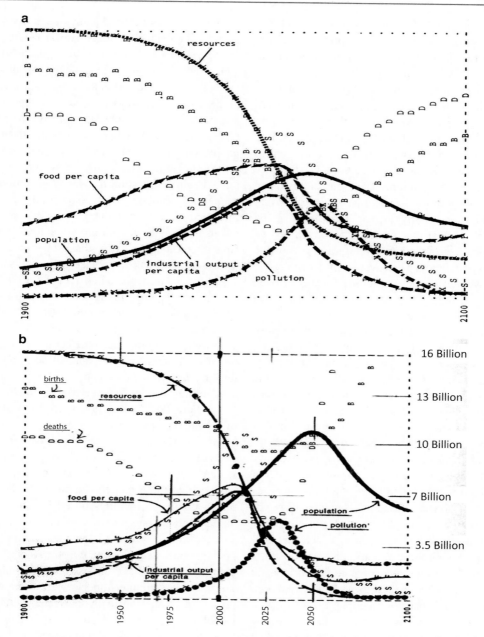

Fig. 9.6 (**a**) The original projections of the limits-to-growth model examined the relation of a growing population to resources and pollution, but did not include a timescale between 1900 and 2100 (Source: Hubbert 1968).

(**b**) If a halfway mark of 2000 is added, the projections up to the current time are largely accurate, although the future will tell about the wild oscillations predicted for upcoming years (*Source*: Hubbert 1968) as to what should be done about it: basically to let the poor starve. Malthus's premise has not held between 1800 and the present, as the human population has expanded by about seven times, with concomitant surges in nutrition and general affluence, albeit only recently. Paul Roberts, in *The End of Food* [4], reports that malnutrition was common throughout the nineteenth century. It was only in the twentieth century that cheap fossil energy allowed agricultural productivity sufficient to avert famine. The argument has been made many times before that our exponential escalation in energy use in agriculture is the principal reason that we have generated a food supply that grows geometrically along with the human population. Thus over the past century we have avoided wholesale famine for most of the Earth's people because fossil fuel use also expanded geometrically.

The first twentieth-century scientists to raise again Malthus's concern about population and resources were ecologists Garrett Hardin and Paul Ehrlich. Hardin's essays in the 1960s on the impacts of overpopulation included the famous *Tragedy of the Commons* [5], in which he discusses how individuals tend to overuse common property to their own benefit despite disadvantages to all involved. Hardin wrote other essays on population, coining such phrases as "freedom to breed brings ruin to all" and "nobody ever dies of overpopulation," the latter meaning that crowding is rarely a direct source of death, but rather results in disease or starvation, which then kill people. This phrase came up in an essay reflecting on the thousands of people in coastal Bangladesh who were drowned in a typhoon. Hardin argued that these people knew full well that this region would be inundated every few decades but stayed there anyway because they had no other place to live in that very crowded country. This pattern recurred in 1991 and 2006.

Ecologist Paul Ehrlich [6] argued in *The Population Bomb* that continued population growth would wreak havoc on food supplies, human health, and nature, and that Malthusian processes (war, famine, pestilence, and death) would sooner rather than later bring human populations "under control," down to the carrying

capacity of the Earth. Meanwhile agronomist David Pimentel [7], ecologist Howard Odum [8], and environmental scientist John Steinhart [9] quantified the energy dependence of modern agriculture and showed that technological development is almost always associated with increased use of fossil fuels. Other ecologists, including George Woodwell and Kenneth Watt, discussed people's negative impact on ecosystems. Kenneth Boulding, Herman Daly, and a few other economists begin to question the very foundations of economics, including its dissociation from the biosphere necessary to support it. Daly pointed out its blind faith in both growth and on infinite substitutability, the idea that something will always come along to replace a scarcer source. These writers were part and parcel of our graduate education in ecology in the late 1960s and 1970s. More recently Lester Brown [10] and others have provided convincing evidence that food security is declining, partly because of distributional issues and partly because of declining soil fertility, desertification, and a decrease in the availability of fossil-fuel derived fertilizer.

Jay Forrester, the inventor of a successful type of computer random-access memory (RAM), began to develop a series of interdisciplinary analyses and thought processes, which he called system dynamics. In the books and papers he wrote about these models, he put forth the idea of the coming difficulties posed by continuing human population growth in a finite world [11]. This view soon became known as the limits-to-growth model (or the "Club of Rome" model, after the organization that commissioned the publication). These computer models were refined and presented to the world by Forrester's students Donella and Dennis Meadows and their colleagues [3]. They showed that exponential population growth and resource use, combined with the finite nature of resources and pollution assimilation, would lead to serious instabilities in basic global economic conditions and eventually a large decline in the material quality of life and even in the numbers of human beings.

At the same time, geologist M. King Hubbert predicted in 1956 and again in 1968 that oil production from the coterminous United States would peak in 1970. Although his predictions were dismissed at the time, U.S. oil production in fact peaked in 1970 and natural gas in 1973.

These various perspectives on the limits to growth seemed to be fulfilled in 1973 when, during the first energy crisis, the price of oil increased from $3.50 to more than $12 a barrel. Gasoline increased from less than $0.30 to $0.65 per gallon in a few weeks and available supplies declined due to a temporary gap of only about 5% between supply and projected demand. Americans became subject for the first time to gasoline lines, large increases in the prices of other energy sources, and double-digit inflation with a simultaneous contraction in total economic activity. Such simultaneous inflation and economic stagnation was something that economists had thought impossible, as the two were supposed to be inversely related according to the Phillips curve. Home heating oil, electricity, food, and coal also became much more expensive. Then it happened again: oil increased to $35 a barrel and gasoline to $1.60 per gallon in 1979.

Some of the economic ills of 1974, such as the highest rates of unemployment since the Great Depression, high interest rates, and rising prices, returned in the early 1980s. Meanwhile, new scientific reports came out about all sorts of environmental problems: acid rain, global warming, pollution, loss of biodiversity, and the depletion of the Earth's protective ozone layer. The oil shortages, the gasoline lines, and even some electricity shortages in the 1970s and early 1980s all seemed to give credibility to the point of view that our population and our economy had in many ways exceeded the ability of the Earth to support them. For many, it seemed like the world was falling apart, and for those familiar with the limits to growth, it seemed as if the model's predictions were beginning to come true and that it was valid. Academia and the world at large were abuzz with discussions of energy and human population issues.

Our own contributions to this work centered on assessing the energy costs of many aspects of resource and environmental management, including food supply, river management, and, especially, obtaining energy itself [12]. A main focus of our papers was energy return on investment (EROI) for obtaining oil and gas within the United States, which had declined substantially from the 1930s to the 1970s. It soon became obvious that the EROI for most of the possible alternatives was even lower. Declining EROI meant that more and more energy output would have to be devoted simply to getting the energy needed to run an economy.

The Reversal

All of this interest began to fade, however, as enormous quantities of previously discovered but unused oil and gas from outside the United States were developed in response to the higher prices and then flooded into the country. Most mainstream economists, and a lot of other people too, did not like the concept that there might be limits to economic growth, or indeed human activity more generally, arising from nature's constraints. Mainstream (or neoclassical) economists presented, mostly from the perspective of "efficiency," the concept that unrestricted market forces, aided by technological innovation, seek the "most efficient" outcome, generally meaning the lowest prices, at each juncture, and the net effect should be a continued satisfaction of consumer demand at the lowest possible prices. This would also cause all productive forces, including technology, to be optimally deployed, at least in theory. They felt that their view was validated by this turn of events and by new gasoline resources.

Economists particularly disliked the idea of the absolute scarcity of resources, and they wrote a series of scathing reports directed at the scientists mentioned above, especially those most closely associated with the limits to growth. Nuclear fusion was cited as a contender for the next source of abundant cheap energy. They also found no evidence for scarcity, saying that output had been rising between 1.5% and 3% per year. Most important, they said that economies had built-in, market-related mechanisms (the invisible hand of Adam Smith) to deal with scarcities. An important empirical study by economists Harold J. Barnett and Chandler Morse in

1963 [13] seemed to show that, when corrected for inflation, the prices of all basic resources (except for forest products) had not increased over nine decades. Although there was little argument that the higher-quality resources were being depleted, it seemed that technical innovations and resource substitutions, driven by market incentives, would continue indefinitely to solve the longer-term issues. It was as if the market could increase the quantity of physical resources on the Earth.

The subsequent behavior of the general economy seemed to support their view. By the mid-1980s the price of gasoline had dropped substantially. The enormous new Prudhoe Bay field in Alaska came online and helped mitigate to some degree the decrease in production of oil elsewhere in the United States, even as an increasing proportion of the oil used in America was imported. Energy as a topic faded from the media and from the minds of most people. Unregulated markets were supposed to lead to efficiency, and a decline in energy used per unit of economic output in Japan and the United States seemed to provide evidence for that theory. We also shifted the production of electricity away from oil to coal, natural gas, and uranium.

In 1980 one of biology's most persistent and eloquent spokesmen for resource issues, Paul Ehrlich, was "trapped" (in his words) into making a bet about the future price of five minerals by actuary Julian Simon, a strong advocate of the power of human ingenuity and the market, and a disbeliever in any limits to growth. The price of all five went down over the next 10 years, so Ehrlich (and two colleagues) lost the bet and had to pay Simon $576. The incident was widely reported through important media outlets, including a disparaging article in the *New York Times Sunday Magazine* [14]. Those who advocated for resource constraints were essentially discredited and even humiliated.

Indeed it looked to many as though the economy had responded with the invisible hand of market forces through price signals and substitutions. The economists felt vindicated, and the resource pessimists beat a retreat, although some effects of the economic stagnation of the 1970s

lasted in most of the world until about 1990. (They live on still in places such as Costa Rica as unpaid debt from that period [15].) By the early 1990s, the world and U.S. economies basically had gone back to the pre-1973 pattern of growing by at least 2% or 3% a year with relatively low rates of inflation. Inflation-corrected gasoline prices, the most important barometer of energy scarcity for most people, stabilized and even decreased substantially in response to an influx of foreign oil. Discussions of scarcity simply disappeared.

The concept of the market as the ultimate arbitrator of value and the optimal means of generating virtually all decisions gained more and more credibility. This was partly in response to arguments about the subjectivity of decisions made by experts or legislative bodies. Decisions were increasingly justified or rejected by economic cost–benefit analysis where supposedly the democratic collective tastes of all people were reflected in their economic choices. For those few scientists who still cared about resource-scarcity issues, there was no place to apply for grants at the National Science Foundation or even the Department of Energy (except for studies to improve energy efficiency). Most of our best energy analysts worked on these issues on the weekend, after retirement or pro bono. With very few exceptions graduate training in energy analysis or limits to growth withered. The concept of limits did live on in various environmental issues such as disappearing rain forests and coral reefs, and global climate change. But these were normally treated as their own specific problems, rather than as a more general issue about the relationship between population and resources.

A Closer Look at the Arguments

For a distinct minority of scientists, including ourselves, there was never any doubt that the economists' debate victory was illusory at best, and generally based on incomplete information. Cutler Cleveland, an environmental scientist at Boston University, reanalyzed the Barnett and Morse study in 1991 and found that the only reason that the prices of commodities had not been increasing was that for the time period analyzed in the original study the real price of energy used to extract the commodities had been declining [16]. Hence, even as more and more energy was needed to win each unit of resource, the price of the resources did not increase because the price of energy was declining.

Likewise, when the oil shock induced a recession in the early 1980s, and Ehrlich and Simon made their bet, the relaxed demand for all resources led to lower prices and even some increase in the quality of the resources mined, as only the highest-grade mines were kept open. But in recent years energy prices have increased while demand for materials in Asia soared and the prices of most minerals increased dramatically. Had Ehrlich made his bet with Simon over the past decade, he would have made a small fortune, as the price of most raw materials, including the ones they bet on, had increased by two to ten times in response to increased demand from China and declining resource grades.

Another problem is that the economic definition of efficiency has not been consistent. Several researchers, including the authors, have found that energy use – a factor that had not been used in economists' production equations – is far more important than capital, labor, or technology in explaining the increase in industrial production of the United States, Japan, and Germany. Recent analysis by Vaclav Smil found that over the past decade the energy efficiency of the Japanese economy had actually decreased by 10%. A number of analyses have shown that most agricultural technology is extremely energy intensive [17]. In other words, when more detailed and systems-oriented analyses are undertaken, the arguments become much more complex and ambiguous, and show that technology rarely works by itself, but instead tends to demand high resource use.

Meanwhile oil production in the United States has declined by 50%, as predicted by Hubbert. The market did not solve this issue for U.S. oil because, despite the huge price increases and expanded drilling in the late 1970s and 1980s, there was a decrease in oil and gas production then, and there has been essentially no relation between drilling intensity and production rates for U.S. oil and gas since.

There is a common perception, even among knowledgeable environmental scientists, that the limits-to-growth model was a colossal failure, inasmuch as obviously its predictions of extreme pollution and population decline have not come true. But what is not well known is that the original output, based on the computer technology of the time, had a very misleading feature: there were no dates on the x-axis of the graph between the years 1900 and 2100 because the technology to do that had not been invented (Fig. 9.6a). If one draws a timeline along the bottom of the graph for the halfway point of 2000, then the model results are almost exactly on course some 35 years later in 2008 (with a few appropriate assumptions; Fig. 9.6b). Of course, how well it will perform in the future when the model behavior gets more dynamic is not yet known. Although we do not necessarily advocate that the existing structure of the limits-to-growth model is adequate for the task to which it is put, it is important to recognize that its predictions have not been invalidated and in fact seem quite on target. We are not aware of any model made by economists that is as accurate over such a long time span.

Avoiding Malthus

Clearly even the most rabid supporter of resource constraints has to accept that the Malthusian prediction has not come true for the Earth as a whole, as human population has increased some seven times since Malthus wrote his pamphlet, and in many parts of the world population continues to grow with only sporadic and widely dispersed starvation (although often with considerable malnutrition and poverty). How has this been possible? The most general answer is that technology, combined with market economics or other social-incentive systems, has enormously increased the carrying capacity of the Earth for humans.

Technology does not work for free, however, and it can be a double-edged sword whose benefits can be substantially blunted by Jevons's paradox – the concept that increases in efficiency

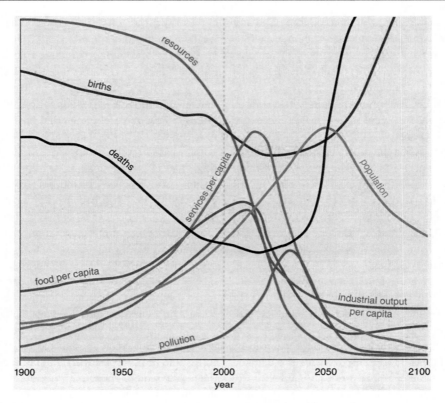

Fig. 9.7 The values predicted by the limits to-growth model in 1972 and actual data for 2008 are very close (See Figure 5 in [1] and Turner [20]. The model used general terms for resources and pollution, but current approximate values for several specific examples are given for comparison. Data for this long a time period are difficult to obtain; many pollutants such as sewage probably have increased more than the numbers suggest. On the other hand, pollutants such as sulfur have largely been controlled in many countries (*Source*: American Scientist)

often lead to lower prices (Jevons found in the middle of the nineteenth century that more efficient steam engines were cheaper to run so that people used them more, as today more fuel-efficient automobiles tend to be driven more miles in a year) and hence to greater consumption of resources [18]. Probably the more important problem with technology is its energy cost. As originally pointed out in the early 1970s by Odum and Pimentel, increased agricultural yield is achieved principally through the greater use of fossil fuel for cultivation, fertilizers, pesticides, drying, and so on, so that it takes some 10 cal of petroleum to generate each calorie of food that we eat. The fuel used is divided nearly equally among the farm, transport and processing, and preparation. The net effect is that roughly 19% of all of the energy used in the United States goes to our food system. Malthus could not have foreseen this enormous increase in food production through petroleum. In fact Malthus, who was associated with and supported by the landed gentry, tended to view machines not as a means of pushing back the collision between human food needs and agricultural production but rather as threatening the position of the landed class.

Similarly, fossil fuels were crucial to the growth of many national economies, as happened in the United States and Europe over the past two centuries, and as is happening in China and India today. The expansion of the economies of most developing countries is nearly linearly related to energy use, and when that energy is withdrawn, economies shrink accordingly, as happened with Cuba in 1988. There has been, however, some serious expansion of the U.S. economy since

Fig. 9.8 The annual rates of total drilling for oil and gas in the United States from 1949 to 2005 are shown with the rates of production for the same period. If all other factors are kept equal, EROI is lower when drilling rates are high, because oil exploration and drilling are energy-intensive activities. The EROI may now be approaching 1:1 for finding new oil fields (*Source*: American Scientist)

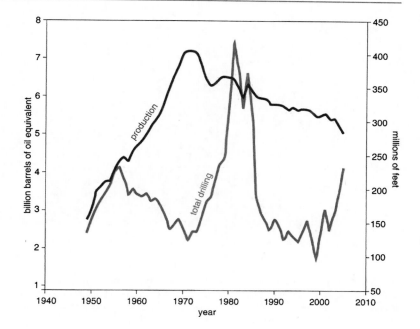

1980 without a concomitant expansion of energy use. This is an exception relative to most of the rest of the world, possibly due to the United States' outsourcing of much of its heavy industry, or to various accounting procedures, rather than a real increase in wealth. Thus, worldwide most wealth is generated through the use of increasing quantities of oil and other fuels.

A key issue for the future is the degree to which fossil and other fuels will continue to be abundant and cheap and how much can be extracted at a significant energy profit. The important remaining questions about peak oil are not about its existence, but rather, when it occurred or will occur for the world as a whole, what the shape of the peak will be, and how steep the slope of the curve will be as we go down the other side. Although we can still squeeze more oil out of known fields the prospects for large new finds remains bleak. According to geologist and peak-oil advocate Colin Campbell [19], "The whole world has now been seismically searched and picked over. Geological knowledge has improved enormously in the past 30 years and it is almost inconceivable now that major fields remain to be found." Increased drilling appears to not be a viable

approach to getting more petroleum as the finding and production of oil in the United States at least is not influenced by the amount of drilling above some very low rate (Fig. 9.8). Meanwhile the world uses two to four times more oil than it finds (Fig. 9.9), and the EROI for most alternatives are much less than what we have been used to with our fossil fuels (Fig. 9.10).

These Ideas Are Not New

These ideas were discussed intelligently, and for the most part accurately, in many papers from the middle of the last century. But then they largely disappeared from scientific and public discussion. This was in part because of an inaccurate understanding of both what those earlier papers said, the validity of many of their predictions, and in part because of a deliberate campaign by economists to negate the importance of resources and any concept of limits on economic growth. The failure to bring the potential reality and implications of peak oil, indeed of peak everything, into scientific discourse and teaching is a grave threat to industrial society. Instead the

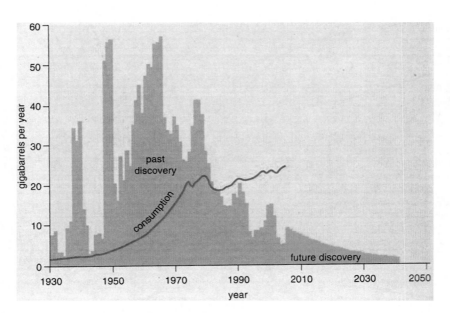

Fig. 9.9 The rate at which oil is discovered globally has been dropping for decades (*blue*), and is projected to drop off even more precipitously in future years (*green*). The rate of worldwide consumption, however, is still continuing to rise (*red line*). Thus, the gap between supply and demand of oil can be expected to widen (*Source*: American Scientist)

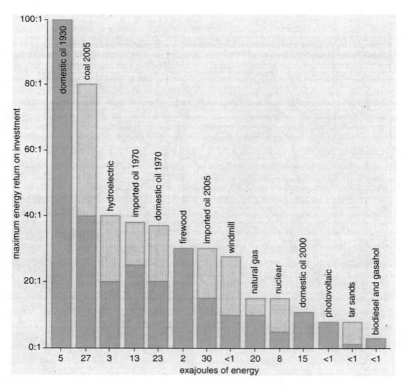

Fig. 9.10 Approximate values of energy return on investment is the energy gained from a given energy investment; one of the objectives is to get out far more that you put in. The size of the resource per year for the United States is given below the bar. Domestic oil production's EROI has decreased from about 100:1 in 1930 for discoveries, to 30:1 in 1970, to about 10:1 today, the latter two for production. The EROI of most *green* energy sources, such as photovoltaic, is presently low. (*Lighter* colors indicate a range of possible EROI due to varying conditions and uncertain data.) EROI does not necessarily correspond to the total amount of energy in exajoules produced by each resource (*Source*: American Scientist)

public is fed a pabulum of nonsolutions. For example, "green" energies are not displacing fossil fuels but rather just adding more output to the mix. If we are to resolve these issues in any meaningful way, we need to make them again central to education at all levels of our universities, and to debate and even stand up to those who question their importance, for we have few great intellectual leaders on these issues today. The possibility of a huge multifaceted failure of some substantial part of industrial civilization is so completely outside the understanding of our leaders that we are totally unprepared for it. Ugo Bardi has written a very fine new book [21] that determines just how the community of economists and others who disliked these results caused the very important and accurate results to be dismissed based on fallacious arguments and lies. They too must learn economics from a biophysical as well as a social perspective. Only then do we have any chance of understanding or solving these problems.

Questions

1. What are "the limits to growth"? Give two interpretations.
2. Who was Thomas Malthus? What, basically, did he say?
3. Have Malthus' predictions held? For what time period? Why or why not?
4. There was relatively little scholarly writing about Malthus' ideas until the 1960s, when some people again paid a lot of attention to them. Can you name any of the people who rediscovered Malthus' ideas? From what fields did they come? Why do you think this was the case?
5. What was the Club of Rome?
6. What happened in the 1970s that seemed to support the concept that there might be limits to growth?
7. How was this related to economic issues? Why did this have such a widespread impact?
8. What happened in the 1980s that completely changed the perspective of many on limits to growth?
9. Discuss the concept of markets in relation to this changing perceptions.
10. What additional insight to the influential work of Barnett and Morse was undertaken by Cutler Cleveland?
11. Why did many people think the limits to growth model failed? What do you think?
12. Why do many people think that technology can generate solutions to resource problems?
13. What is Jevons' paradox?
14. What is the general relation between energy availability and limits to growth? Do you think that technological advances change that relation?

References

1. Derived from: Hall, C. A. S. and J. W. Day, Jr. Revisiting the Limits to Growth After Peak Oil. American Scientist, Volume 97: 230–237.
2. Heinberg, R. 2007. Peak Everything: Waking Up to the Century of Declines. New Society Publishers. Gabriola Island B.C. Canada
3. Meadows, D. H., D. L. Meadows, J. Randers, and W. W. Behrens III. 1972. The Limits to Growth. Potomac Associates, Washington, D.C.
4. Roberts, P. 2008. The End of Food . Houghton Mifflin, N,Y.
5. Hardin, G. 1968. The tragedy of the commons. Science, 162: 1243–1248.
6. Ehrlich, P. The population bomb. Ballantine Books. N.Y.
7. Pimentel, D., L. E. Hurd, A. C. Bellotti, M. J. Forster, I. N. Oka, O. D. Sholes, and R. J. Whitman. Food production and the energy crisis. Science. 182: 443 – 449
8. Odum, H. T. 1973. Environment, Power and Society. New York: Wiley Interscience.
9. Steinhart, J. S. and C. Steinhart. Energy use in the US food System Science 184: 307–316.
10. Brown, Lester R. 2009. Could Food Shortages Bring Down Civilization? Scientific American, May, 2009
11. Forrester, J. W. 1971. World Dynamics. Cambridge: Wright-Allen Press.
12. Charles Hall webpage: http://www.esf.edu/efb/hall/
13. Barnett, H., and C. Morse. 1963. Scarcity and Growth: the Economics of Natural Resource Availability. Baltimore: Johns Hopkins University Press. See also Passell, P, M. Roberts, and L. Ross. 2 April 1972 "Review of Limits to Growth". New York Times Book Review.
14. Tierney, J. 1990. Betting the planet. *New York Times Magazine* December 2: 79–81.
15. Hall, C.A.S. Quantifying sustainable development: The future of tropical Economies. Academic Press, San Diego.

16. Cleveland, C. J. 1991. Natural resource scarcity and economic growth revisited: Economic and biophysical perspectives. In Ecological Economics: The Science and Management of Sustainability. Edited by R. Costanza. New York: Columbia University Press.

17. Smil, V. 2007. Light behind the fall: Japan's electricity consumption, the environment, and economic growth. *Japan Focus,* April 2.

18. Hall, C. 2004. The myth of sustainable development: Personal reflections on energy, its relation to neoclassical economics, and Stanley Jevons. *Journal of Energy Resources Technology* 126:86–89.

19. Campbell, C., and J. Laherrere. 1998. The end of cheap oil. Scientific American March: 78–83.

20. Note: when our original paper was in press (and an hour before we had to have it submitted) we became aware of a very similar analysis of Limits to growth in terms of the veracity of its predictions. This study concluded exactly as we did: the model was a remarkably successful predictor. See Turner, Graham M. 2008. A Comparison of The Limits to Growth with 30 years of reality. Global Environmental Change 18 397–411.

21. Bardi, U. 2011. The limits to growth revisited. Springer, N.Y.

Part III

Energy and Economics: The Basics

Economics is mostly studied and taught as a social science, with very little connection to the natural sciences except for (1) the frequent application of the mathematics and reductionist methods of physics, although usually in the absence of the constraints from conservation principles (2) attempts to measure the value of nature in economic terms. Part I of this book developed our basic belief that economies are in many ways completely dependent upon energy for their operation and indeed are basically about how energy is used to transform raw materials from nature into the products that are found in markets. Economics as a discipline should reflect this basic reality, but basically does not. One of the reasons for this is that the training of most economists is in the social sciences and a great deal of mathematics but very little natural sciences. But this problem is not entirely the fault of the economists themselves for we believe also that the teaching of the natural sciences is too often divorced from its application to the day-to-day issues of interest to economists. In an attempt to give those interested in pursuing, or learning about, the foundations for a biophysical approach to economics we provide in this section basic information about the natural world that we think is a sort of minimum course in what is needed, starting with energy and then going on to science more generally, mathematics, and then a final section considering the entire issue of whether economics should properly be a social science, a biophysical science, or some amalgamation.

What Is Energy and How Is It Related to Wealth Production?

10

Energy: The Unseen Facilitator

Energy is, at best, an abstract entity for most contemporary people. Only rarely does it enter our collective consciousness, generally in those relatively rare times when there are particular shortages or sharp price increases in electricity or gasoline. In fact, as this book demonstrates, energy and its effects are pervasive, relentless, all-encompassing, and responsible for each process and entity in nature and in our own economic life. Energy is also behind many aspects of the basic nature of our psyches and many of the ways that world history has unfolded. Few understand or acknowledge this role because the pervasive impact of energy shown in this book does not usually enter into our collective training and education, and it does not enter into our educational curricula. Why is this so? If energy is as important as we believe then why is that not more generally known and appreciated? The answers are complex. One important reason is that the energy that is used to support ourselves, our families, or our economic activity generally is used at some other location and by other people, or by quiet automatic machines whose fuel tends to be relatively cheap. After all, coal, oil, and gas, our principal sources of energy, are basically messy, smelly, dangerous, and unpleasant materials. The energy from food that we need to fuel ourselves surrounds most of us abundantly and is available readily and relatively cheaply. Society has gone to great lengths to isolate most of us physically and intellectually from the energy sources upon which our food, our comfort, our transportation, and our economy depend. It is convenient to ignore energy because many facts about it are uncomfortable to know.

Perhaps a more important reason for our failure to understand the pervasive role of energy is that most uses of energy are indirect. Humans are conditioned, both evolutionarily and in their social education, to want and need the goods or services rendered by energy, but not energy itself. In fact energy per se, with the exception of food energy and warmth in winter, is hardly ever desirable or useful directly. This conditioning, however, does not diminish the requirement for energy in virtually everything that we do, nor does it compensate for the fact that our use of energy has become enormous in contemporary life. Today each American has the equivalent of about 80 slaves toiling tirelessly to keep us at about 70°, well fed, mobile, entertained, and so on. Where are these slaves? We can see the car engine, the furnace, or the air conditioner, but who is aware of the electric pump supplying water or running the refrigerator, or the massive electrical and fossil-fueled devices digging up the Earth to bring us the energy to run these devices. Who thinks about the energy required to make the metals and plastics in our car, the timbers and concrete in our homes, offices, and schools, or the paper in this book? But they all require it, and a lot of it.

C.A.S. Hall and K. Klitgaard, *Energy and the Wealth of Nations: Understanding the Biophysical Economy*, DOI 10.1007/978-1-4419-9398-4_10, © Springer Science+Business Media, LLC 2012

Another reason that we do not think much about energy is that energy today remains enormously cheap relative to its value. If we want water delivered to our house we might hire a person to do the job. A very strong person can work at a rate of about 100 W, so in a 10-hour day could do 1,000-watt hours (one kilowatt hour) of work, say hauling water from a well to our sink or shower. If we paid that strong person at minimum wage he or she would charge about $80 for the 10 hours' work. But if we installed an electric pump we can get the same work done for about ten cents per kilowatt hour. Because humans work at about 20% efficiency but electric pumps perhaps 60%, the relation is tipped even more in favor of the electric pump. So to do the same physical work with a person that a pump could do would cost about 800 times 3 or 2,400 more with the worker compared to the electric motor! So this is the main reason that an average American or European today is far richer than the richest king of old: we have cheap energy to supply us with the necessities and luxuries in life. A problem is that we have become dependent upon this cheap energy and the goods and services it provides in many ways. The value of energy is far more expensive than we are used to paying, and its potential abundance potentially much more limited than our dependence would imply.

A History of Our Understanding of Energy

Two hundred years ago no one understood energy as a concept, although they certainly understood many practical consequences such as plants needing sunlight for growth and the need for wood to do many economic things such as cooking or making metals or cement. Any concept of energy was tied up in confusion, often mystical, about the actual results of energy use, because energy cannot be seen or felt, but only its effects. Fire was thought of as a basic substance (as in earth, air, fire, and water) rather than as the energy released from the destruction of chemical bonds generated earlier by photosynthesis and the formation of

new bonds with oxygen. How could people then possibly understand energy if they did not have any concept of chemical bonds, oxygen, or chemical transformations? How could they possibly understand that the growth of plants, the work of a horse, the erosion of water, and the heat generated by fire and their own exertions had some common something that tied them all together? To them they were independent entities.

As in most other things in their lives that they did not understand, earlier people attributed energy, or at least some aspects of it, to a god or gods; the sun, of course, was worshiped by many cultures who understood clearly its importance for their food and warmth, but there were many other energy gods: Promethius, Hephaestus, Pele, Vesta, Hestia, Brigid, Agni, and Vulcan to name a few. These people had no possible way to see that there were common concepts linking the sun and the fire resulting from burning wood, nor could they understand that so many other processes that they also attributed to different gods (wind, rain, agriculture, the existence of wild creatures, and so on) were also connected to the sun. The knowledge that a sharp sixth-grader today has about energy and science in general would be far beyond what the most learned person would understand 400 or even 200 years ago. We have learned an astonishing amount about how the world really works through science. Even today, however, we cannot measure energy directly but only its effects! But we have become much better at that and in understanding how all of this is related.

For 200 years, from roughly 1650–1850, a series of remarkable discoveries and experiments, mostly from French and English scientists, allowed us to understand in a comprehensive way the essentials of energy. First and foremost among these were the remarkable discoveries of Isaac Newton. Newton discovered the three laws of motion, and in the more than 350 years since then no fourth law has been discovered. He also derived the law of universal gravitation and wrote critically important books on optics. Nevertheless by his own admission he did not understand economics and he lost most of his money on an ill-advised investment scheme.

Newton's Laws of Motion

The first law says that a body in motion (and this includes no motion, i.e., rest) tends to stay in that state unless acted upon by an outside force. This is completely counterintuitive, as most moving things come to a stop! But Newton realized that it was an outside force, friction, that caused them to stop, and if there were no friction they would continue in their path indefinitely. The first law explains many things we experience – the momentum of an automobile when we put in the clutch, the path of a baseball (although we need to include gravity), and even centrifugal force – which actually is a misnomer for the tendency of a ball on a string or a planet "tied" to the sun by gravity to continue in a straight line. The inertial force of the moving ball is balanced by the centripetal force (i.e. the force of the string keeping the ball "tied" to the hand of the person rotating the ball or of gravity pulling the planet to the sun).

The second law says that the acceleration of an object, say a baseball being hit, equals the force applied to the object divided by the mass of the object. It is familiarly written as:

$$F = MA$$

which can be rewritten as

$$A = F / M$$

Thus a powerful baseball hitter, such as the legendary Babe Ruth, was capable of applying great force (F) to a baseball with his bat, accelerating it greatly (A), and giving it enough velocity (sometimes) to travel out of the ballpark. The force that he applied could be measured by the measuring the mass (M) of the baseball and the amount that the ball was accelerated. If one could make a baseball twice as heavy with a lead core it would, other things being equal, be accelerated only half as much.

Newton's third law of motion says that in an isolated system for every force there is an equal and opposite force, in other words all forces sum to zero. This is evident when you are in a small boat and move your body one way and the boat moves in the opposite direction. It is obvious to anyone who has fired a rifle that the gun moves back against your shoulder when the bullet is accelerated forward. It was also obvious to early designers of shipborne cannon that if proper arrangements were not made that the recoil of the cannon could do more damage to the ship shooting it than to the target!

Newton also determined the "universal law of gravitation," that two bodies would attract each other as the product of their masses and the inverse of their distance squared. This law is so powerful that it can be used to explain very precisely the orbits of planets around the sun, or the movement of a hit baseball. Chemical reactions take place according to a somewhat similar Coulomb force, which is electrical, and although completely different from gravitation physically, it has a similar dependence on the inverse of the squared distance between the two charges (i.e., the protons and the electron(s) involved in the reaction). Normal chemical reaction can be predicted, in principle, by quantum chemistry from the masses and charges of the nuclei and electrons involved.

Probably the most important result of Newton's work is that it showed that the physical world followed definite laws that appeared (and still appear) never to be broken no matter where or when applied, and that many of these laws could be expressed by simple mathematical equations. Although the concept of energy was not yet known to Newton we now understand that energy was related to matter by the relation of force to mass. In the hundreds of years of science that has followed many have tried to find simple, elegant, mathematical laws that were as powerful as Newton's, but with the exception of Albert Einstein and James Clerk Maxwell few succeeded.

The essence of what "force" (as in Newton's second law) was, where it came from, and how it changed over time remained elusive. The next important step in our understanding was the relation of physical energy to heat. It was obvious that over time a lot of fuel wood was needed to run any major production process. Also, it was certainly apparent to observers that many physical actions were associated with heat, as was obvious by the heating of turning wagon wheels or a hard working horse or person, or the drilling of the hole in a cannon. But why this should be or what it meant remained elusive.

The Mechanical Equivalent of Heat

Many early scientists and engineers, seeing and understanding the tremendous force made possible when water was heated to form steam, were interested in building engines to do mechanical work. According to Wikipedia, Thomas Savery built the first heat engine as early as 1697. Although his and other early engines were crude and inefficient, they attracted the attention of the leading scientists of the time. Classical thermodynamics as we know it now evolved in the early 1800s with concerns about the states and properties of everyday matter including energy, work, and heat. Sadi Carnot, the "father of thermodynamics," published in 1824 the paper that marked the start of thermodynamics as a modern science. It's title was "Reflections on the Motive Power of Fire, a Discourse on Heat, Power, and Engine Efficiency" which outlined the basic energetic relations among the Carnot engine, the Carnot cycle, and motive power. Only a few hundred copies were published and Carnot died thinking his work had had no impact. The term *thermodynamics* was coined by James Joule in 1849 to designate the science of relations between

heat and power. By 1858, "thermo-dynamics", as a functional term, was used in William Thomson's paper "An Account of Carnot's Theory of the Motive Power of Heat [1]." The first thermodynamic textbook was written in 1859 by William Rankine, originally trained as a physicist and a civil and mechanical engineering professor at the University of Glasgow.

The quantitative study of the relation between heat and mechanical work was undertaken further by Joule and Benjamin Thompson (also known as Count Rumford) who was astonished when he found that by immersing newly cast brass cannon into water while boring the hole in them using horsedrawn power he could actually make the water boil. He and other onlookers were astonished that they could generate heat without fire. The fact that the water would boil for as long as the horse kept turning the drill invalidated the earlier dominant "phlogiston" theory that heat was a substance that flowed from one object to another, because it never ran out! Great progress was made in understanding energy relations by Robert Mayer and James Joule who measured "the mechanical equivalent of heat" by taking a pulley and rope, attaching a weight to one end and wrapping the other end of the rope around a shaft that went into an insulated water chamber where it operated a paddle wheel (Fig. 10.1). As the weight dropped (doing so many kilogram meters of mechanical

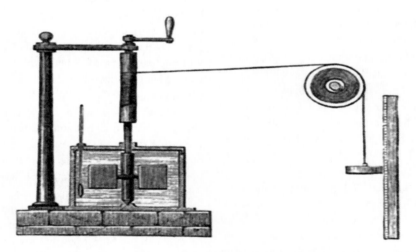

Fig. 10.1   Joule's machine for measuring the mechanical equivalent of heat, or perhaps better said as the quantity of heat released per unit of mechanical work done (*Source*: 2009 citizendia.org)

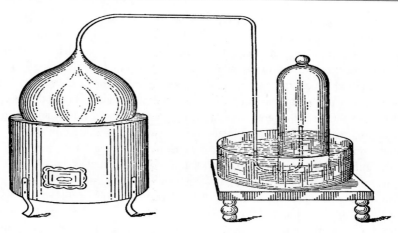

Fig. 10.2 Lavoisier's experimental approach to measuring the oxygen content of the atmosphere (*Source*: Florida Center for Instructional Technology)

work) the temperature increase inside the chamber could be measured. By doing so Joule found that one newton-meter of work (or 7.2 foot pounds) was equivalent to 1 J of heat energy. More commonly we use larger units. The kilojoule (kJ) is equal to 1,000 J. The average amount of solar energy received per second by one square meter of the Earth's surface is 239 J [i.e., the solar constant minus the albedo (reflectance) divided by 4, the ratio of Earth's surface to Earth's cross-section]. Thus one kilojoule is about the amount of solar radiation received by one square meter of the Earth in about 4 s. The megajoule (MJ) is equal to one million joules, or approximately the kinetic energy of a one-ton vehicle moving at 160 km/h (100 mph). The gigajoule (GJ) is equal to one billion joules. A gigajoule is about the amount of chemical energy in seven gallons of oil. A barrel of oil has about 6.1 GJ.

The French engineer Sadi Carnot found, while investigating why so much energy was required to drill the bore in a cannon, that the amount of work that could be done by a thermal engine was proportional to the difference between the temperature of the source (i.e., the furnace or steam at a turbine blade) and the temperature of the environment where the heat was dumped. His equation is:

$$W = (Ts - Te) / Ts$$

The Carnot efficiency of heat-to-work conversion of an ideal heat engine that receives heat of high absolute temperature, Ts, from a source (e.g., a furnace) and rejects heat of lower temperature Te < Ts to a sink (e.g., a river). By definition, it cannot exceed 1.

This equation explains why despite the vast amount of heart stored in, for example, the surface of the North Sea in summer, so little work can be done from it: the difference between the surface temperature (30°) and the deepest water (2°) is too small compared to, say the temperature difference in an oil-fired power plant, where temperatures at the turbine entrance may reach 817°C and the cooling water which might be from 6° (winter) to 17° (summer). It also explains why power plants are slightly more efficient in winter.

Also in England and France another very important discovery was made in the 1770s, that of oxygen. Probably Priestly in England discovered oxygen a little earlier than did Antoine Lavoisier in France, although the latter probably understood its significance better while quantifying its abundance and reactions (Fig. 10.2). Both derived oxygen by heating oxides of mercury. Lavoisier discovered that the atmosphere contained oxygen or "eminently breathable air" by showing that an animal lived longer in a container of pure oxygen than in a container of air. He also clarified the role of oxygen in combustion, the rusting of metal, and its role in animal respiration, recognizing that respiration was "slow burning." He also came up with the basis for the law of conservation of matter by showing that the elements after a chemical reaction always weighed the same as they did before the reaction.

These earlier investigators of energy turned what had been a completely mystery into a well-understood and quantifiable science, and we owe a great deal to their work. Except for Albert Einstein's discovery of the equation for turning mass into energy (and vice versa, as in the Big Bang) there have, arguably perhaps, not been any comparable discoveries of the basic physics of energy, especially that can be represented readily by simple equations. However, as we show, perhaps the most important discoveries came with applying basic energy laws and ideas to more complex systems, including ecology and economics.

What Is Energy?

A definition of energy turns out to be more difficult than what one might think. The high school physics definition, "The ability to do work," does not take us very far. Romer's physics text started from the concept that energy concepts can be used to understand all the conventional material of physics because "all physics is about energy." Yet even he admitted that he was unable to give a satisfactory definition of energy. He said we can see its effects, we can measure it, but we don't really know what it is. Usually we detect energy being used because something is moved, a car, a basketball player, chemicals against a gradient, and so on. For our day-to-day purposes energy is mostly either photons coming from the sun or chemically reduced (i.e., normally, hydrogen-rich) materials such as wood or oil that can be oxidized to generate work (i.e., move something) at some point in space and time. Most of our energy comes into the Earth originally in the form of photon flux from the sun. Some small part of this energy is captured by plants in chemical bonds and then passed through food chains. Thus we are able to use the energy in a hamburger by oxidizing the reduced matter in the animal tissue.

This energy initially was obtained by the cow when it ate grass that had in turn captured that energy from the photons, and then passed it as chemical bonds to the cow and then to us. Even when we are driving a car we are oxidizing formerly reduced plant material (oil) that is constructed of high-energy chemical bonds originally made with energy captured from the sun by algae. In general *reduced* means hydrogen-rich and oxygen-poor, so that a fuel is generally a hydrocarbon such as oil or occasionally a carbohydrate such as alcohol (the "ate" on the end refers to the presence of oxygen, so that a carbohydrate will have somewhat less energy than a hydrocarbon per gram but still enough to be used as a fuel). When a reduced fuel is oxidized energy is released, and the hydrogen released as water (H_2O) and the carbon as carbon dioxide (CO_2). The general equation for combustion of a hydrocarbon is:

$$CnH2n + O_2 \rightarrow H_2O + CO_2$$

the exact numbers required to balance the equation depending upon the exact form of the hydrocarbon burned but are for oxidation of common biological foods about:

$$C_6H_{12}O_6 + 6CO_2 \rightarrow 6CO_2 + 6H_2O$$

The equation for photosynthesis is the same but runs from right to left.

Power refers to the rate at which energy is used. For example, a light bulb is rated in kilowatts, a unit of power, so that a 100-W light bulb uses 360 kJ in an hour, equivalent to the energy in about 10 mL of oil. An automobile engine is rated in horsepower, roughly the rate at which a horse can do work, which was used to estimate the power of early steam engines. Automobiles today typically have 100–200-horsepower engines, thus one can see how much the increased ability to do work using fossil fueled-engines has given humans (Table 2.3). If we want to know the total energy used we multiply a measure of power (e.g., 100 W) times the time of use (say 10 h) to get the total energy use, in this case 1,000 W hours or one kilowatt hour).

The use of different terms to describe energy (e.g., calories, kcal, BTU, watt, joule, therm) may

Table 10.1 Energy conversions as well as the metric prefixes that establish magnitude (Kilo =1000, Mega= million, Giga = billion)

One calorie	4.18 J
One BTU	1.055 KJ
One kWh	3.6 MJ
One therm	105.5 MJ
One liter of oil	37.8 MJ
One gallon of oil	145.7 MJ
One barrel of oil	6.118 GJ
One ton of oil	41.87 GJ (6.84 barrels)

seem very confusing but they all measure one thing: the quantity or rate of heat produced when all of the energy has been converted to heat. Thus although electricity is usually measured in kilowatts it would be just as correct to say that a 100-watt electric light bulb uses 0.36 MJ per hour. The international organization that determines what are appropriate units for us to use (Systeme International) has settled on joules as the preferred unit of measure, although calories is much more familiar to us, so we try to always give our results in Mjoules (million joules) although occasionally in more familiar terms as well. Table 10.1 gives many energy conversions as well as the metric prefixes that establish magnitude. Although the metric system may be unfamiliar to American readers, trust us, with a little experience it is enormously easier than the English system. The metric system was designed in part by Lavoisier although he was guillotined in the French Revolution for being (also) a tax collector! If all energy was expressed in Joules then our ability to understand it would be increased enormously.

Quality of Energy

When considering energy as a resource in a general way there are several critical things to think about. First of all, there is the *quantity* of it, how much there is at the disposal of the species or human society using it. For example, there is roughly ten times more coal in the world compared to oil. Second is the *quality* of that energy: that is, the form that it is in, which has a great deal to say about that energy's utility. The most obvious

example is food. The energy in corn has obvious utility to us as food where the energy in wood or coal does not. There are many other aspects of quality. Corn, a grass, is a very productive crop so where the land is crowded people often eat nothing but corn (or other grasses such as wheat or rice) because it gives the most food production per area. But corn lacks a critical factor absolutely required for humans: the amino acid lysine. If the corn is fed to a cow then the energy bonds in the corn will be transferred to energy bonds in the flesh of the cow. This animal protein has a full complement of amino acids and hence is a higher-quality food, at least from that perspective. Many relatively poor humans in Latin America (and elsewhere) eat mostly rice and beans. This is actually a very good diet because the rice and beans are cheap and they complement each other: the amino acid lysine is missing in rice but found abundantly in beans, whereas rice is basically carbohydrates, a good energy source, and beans are protein-rich. Thus rice and beans provides an excellent diet for humans, although it is still missing one critical ingredient: vitamin C. Fortunately vitamin C is abundant in chili peppers, which is often used as a condiment by people who have a rice and bean diet. So cultural selection appears to be often associated with real dietary needs, all of which ensures that the energy that fuels humans has the required quality.

We often say that the energy in the protein-rich beans, or a chicken that is fed rice, is of a higher *quality* than the rice because the animal food contains more protein, a food type absolutely necessary for humans and most animals that is in insufficient supply in many plant foods. Many would say it tastes better too. Thus people may feed rice or other grain to an animal to get a smaller quantity of higher-quality chicken. Likewise coal or oil can be burned to generate a smaller quantity (as measured by heating ability) of electricity. But this electricity has a higher quality in that it can be used to do things such as light a light bulb or run a computer that one cannot do with the oil or coal. We are willing to take roughly three heat units of coal or oil and turn it into one heat unit of electricity because it is more useful to us; that is, it can do more work and hence is more economical, in that form. We say the quality of the electricity is higher, and a special term, called *emergy*, has been derived to represent quality of energy in a comprehensive fashion [2].

A related aspect of energy is its ability to do work defined by physicists in a very careful, specific way. The term used here is *exergy*, which is that component of energy that can actually do the work, as opposed to being transferred into heat due to the minimal second law requirements for some to be turned into heat. High quality (low entropy) energy is energy available to do work. Using it degrades most into low quality, high entropy heat that can not do more work. In formal second law analysis and technical thermodynamics in physics and certain engineering the terms exergy and enthaply are used to measure quality. There is a lot of discussion among energy purists about how exergy is a more proper term to use. In fact the efficiency of use of, for example, oil is determined by far more than just exergy but also a whole series of complex issues pertaining to the state of technological development of that country and the specific application, the age of the machinery, and so on. A number of analysts including Gaudreau et al. [3] believe that the exergy has been overused and that in general it is as just as useful to think about energy as joules of energy. Then we can concentrate on reserves versus actual empirical use rates.

There is a third component of energy, also related to its quality, which relates to the energy required to get that fuel. We normally measure this property as energy return on investment, and this issue is explored in much greater detail in Chap. 13. We often hear very bullish statements about the tremendous amounts of energy that are all around us just waiting for us to exploit. But there is a catch. The energy has to be of a high enough quality to make it worthwhile to exploit, and real fuels must have a very high EROI. For example, we normally can get only about a third of the energy out of an oil field simply because the remaining oil sticks tightly to the substrate. If we really wanted that oil we could get it; we could dig a 2-mile deep hole and shovel it out of the ground and heat it in a giant pot. But obviously that would require far more energy than one would get from the oil. In fact we use steam, water pressure, chemicals, and pumping, and to some degree it works.

But at some point getting more of the remaining oil out simply costs too much money for the energy to do it and the well is closed off. Reduced carbon, a potential fuel, is extremely abundant in shale rocks throughout the world, and as such it represents, some say, a tremendous energy source. In certain very carbon-rich rocks it is possible to get oil or gas out with a substantial energy profit. But for the majority of these rocks more energy would be required to get this dilute carbon out of the rocks than the energy contained within it, so that rock cannot be considered a fuel. Similarly, the oceans contain a tremendous energy potential in the hydrogen found in the water molecule. But that hydrogen is not a fuel, for it takes more energy to separate it from the oxygen it is combined with than can be recovered by later burning it.

Energy return on investment (EROI, the energy gained from energy used to find or extract it, comes into play more generally when we examine our commonly used fuels (See Chap. 14). For example, petroleum was discovered in the United States with an EROI of 1000:1 or greater in the 1920s, but it is 5:1 now. The EROI of production was initially low, about 20:1, then increased to about 30:1 in the 1950s, and then has declined to about 10:1 today. Finding and developing a brand new barrel of petroleum today (versus pumping out an existing stock) requires perhaps one barrel for each three to five barrels found. Similar patterns have held for other fuels, such as for coal, over time, although for coal the numbers, although decreasing, are much higher. Thus in general as time goes by the highest-quality fuels are used first and the EROI declines. Although it is true that occasionally brand new, very high-grade petroleum resources are found, the probability for most of our main resources is vanishingly small because, according to Colin Campbell, the whole world has been seismically and otherwise explored and picked over for many decades.

Similarly we have used up our highest-grade copper ores, so that the average grade mined fell from about 4% in 1900 to 0.4% in 2000. This lower-grade copper requires more energy to get a kg of pure copper out, and we can say that its RoE (material return on energy investment) is declining. Humans, usually being no economic fools, tend to use high-grade resources first, high-grade meaning more concentrated or easier to

Table 10.2 The energy cost of economic activity (on average)

Energy use per unit of economic activity average (in 2005) when GDP was 12.36 trillion dollars:

$$\text{Energy per \$GDP} = \frac{105 \text{ exajoules (e 18)}}{12.36 \text{ trillion (e 12) dollars}} = \frac{8.45 \text{ e6 J}}{\text{dollars}} = 8.45 \text{ mega Joules per dollar}$$

One dollar of economic activity requires	8.45 MJ	(0.0139 bands = 0.21 gal)
One thousand dollars requires	8.45 GJ	(1.3 barrels)
One million dollars requires	8.45 TJ	(1,380 barrels)
One billion dollars requires	8.45 PJ	(1.4 million barrels)
One trillion dollars requires	8.45 EJ	(1.4 billion barrels)
12.36 trillion dollars requires	8.45 EJ × 12.36 = 105 EJ	(17.3 e9 barrels)

Source for raw data: U.S. Dept. Commerce

access. This important concept is called the *best first principle*, and it is very important as we consider the possibilities before us. The principle was also derived very clearly in economics by David Ricardo two centuries ago. He talked about how the first farmers in a region would use the best soil (perhaps on flat land near a river) first, and that subsequent farmers were forced to use lower and lower-quality soils.

What Are Fuels?

Fuels are normally energy-rich, reduced compounds of hydrogen and carbon that we call carbohydrates if they also contain some oxygen or hydrocarbons if they do not. The key to the utility of energy is neither its availability nor its final end product (often oxidized carbon such as CO_2) but its ability to transfer electrons from one location on an energy *gradient* (i.e., from the fuels) to another location (usually the environment), doing work along the way. In an analogous way oil in the ground or even in the gas tank is not useful. Rather it becomes useful when it releases energy in the process of transfer from the reduced state to the oxidized state. A key to the way that organisms have evolved is that life has tended to break this process down into a series of tiny steps that captures some of this energy step by step. Thus electrons are passed through energy capture devices, such as a membrane or a whole plethora of oxidized-reduced chemical compounds, which cycle from energy-rich reduced forms to energy-poor oxidized states and, as appropriate, the converse.

In a way this flow of energy through biological food chains is not unlike the flow of electrons in a wire that we call electricity. Some energy source, the sun for biology or a generator fueled by falling water or the combustion of fossil fuel, gives the electrons a boost, a kick in the pants as it were. In electricity the wire provides a circuit for the electrons to travel along, and the energy represented by their excited state can be used by a device such as a motor or light bulb put in the path way of the electrons flowing from the source to what we call a sink, representing a place to which the low-energy electrons can return, generally to be kicked into a high-energy state again. The energy provided by the kick is simply moved to the place where it is utilized in a light, motor, or whatever. Similarly electrons that have received a "kick" from the sun in photosynthesis pass through the complex "wires" of biological circuits carrying the energy derived from the photon to reduced carbon compounds in a plant and then through food chains to various animals and decomposers. So when you eat your corn flakes or a hamburger remember that the energy that allows you to run, jump, or just exist came from the sun through the magic of photosynthesis (Tables 10.3 and 10.4).

Why Energy Is so Important: Fighting Entropy

When we think about energy it is normally from the perspective of going somewhere, or keeping warm in the winter, or some friend's high energy level. But the reach of energy is far more pervasive.

Table 10.3 Selected fuels and their approximate heat equivalents

Fuel	Heat equivalent (MJ)
Residual oil (1 barrel)	6,626
Crude oil (1 barrel)	6,164
Distillate oil (1 barrel)	6,140
Gasoline (1 gal)	132
Electricity (1 kW-hour)	3.6
Natural gas (1 cubic foot)	1.1

Source: State of Oregon DOE

Table 10.4 MJ used per 2005 dollar spent in select sectors of the economy

Sector	MJ
Oil and gas field machinery and equipment	7.36
Petroleum lubricating oil and grease manufacturing	61.30
Cement manufacturing	68.4
Rolled steel shape manufacturing	15.60
Fabricated pipe and pipe-fitting manufacturing	9.84
Water transportation	48.80
Other miscellaneous chemical product manufacturing	16.30
Other basic organic chemical manufacturing	21.70
Explosives manufacturing	22.70
Watch, clock, and other measuring device manufacturing	5.65
Oil and gas extraction	9.26
Drilling oil and gas wells	9.87
Support activities for oil and gas operations	6.98

Source: Economic Input–Output Life Cycle Assessment Model developed by the Green Design Institute at Carnegie-Mellon University (We do not know exactly how these were calculated so are simply passing them on. We also suspect that the nominal precision used does not reflect reality.)

The principal reason is due to what is normally called *entropy*. Entropy is often used inaccurately or vaguely. Physically it describes, essentially, the tendency of the components of a physical system to spread as evenly as possible in space and over all states of motion. Entropy is the physical measure of disorder, that is, randomness, like a group of molecules with no structure. The concept of molecules arranged in a definite pattern (such as in a building or an animal) is the opposite of those molecules being spread out randomly. The tendency is for molecules to be arranged

randomly, that is, to have high entropy. Some have called this property "the entropy law." Although this concept may seem far removed from our day-to-day existence where we live surrounded by ordered structures (such as in the computer I am using as I write this) it is in fact critical. Everything with which we deal is affected by entropy, and everything that we own tends to degrade (i.e., become more random) over time: our cars (that's why we need to take it to the shop), our homes (that's why we need fire insurance, repainting, plumbers, termite controllers, and so on), our food (that's why we need refrigerators), our closets and even ourselves (that's why we need to eat, and why most of us require medical intervention at various times in our lives). What all these things are – cars, houses, ourselves – is bits of negentropy, or negative entropy, an ordered structure of molecules, something that is by itself extremely unlikely. Life must be nonrandom to exist; that is, life consists of very specific aggregates of molecules that are completely different from the general environment within which it resides. But although by chance alone negentropy is extremely unlikely, in fact it is common around US. This is due principally to natural selection that has generated life plans that extract energy from the environment and invested that energy into creating, maintaining, and reproducing life forms.

The creation of negentropy requires energy to concentrate and organize molecules as well as a plan as to what reorganization will work. For life the plan is a specie's DNA, and analogously for a mechanic or plumber it is the wiring or piping diagrams, shop manuals, and so on, or her training and experience that allows the car or house to continue to exist. But without energy the plan is useless for it requires energy to take metals or other materials out of the ground and air to make new biomass or new brake drums or pads, cylinder blocks, pipes, faucets, and so on. It even takes our personal energy expenditure to reduce the daily entropy of our closets. More generally life, including civilization, is about very specific structures, or construction according to a plan, and then the maintenance of that structure. Both of these things are energy-demanding, the degree of which is a function of the complexity, size, and makeup of the plan. That is why we eat, why plants photosynthesize, and why modern civilization requires coal, gas, and oil: to get the energy necessary to maintain and in some cases build the very specific structures that we are and that characterizes all life and also our economies. An organism's DNA gives it the pattern or plan for the very specific structures, physiologies, and behaviors that have worked well in the environment in which it is found, or at least have worked well up to the present. Those patterns that did work in the past may or may not work in the future, depending upon whether there are environmental changes or whether some other species has figured out a new way to exploit that environment. But all organisms are in a sense betting that what they have will work well enough for what life is all about, propelling genes into the future. It is a wonderful and wondrous process and the results are magnificent!

A simple example will help to think about this. Both a ham sandwich and your own self are extremely unlikely, nonrandom structures of molecules of carbon, nitrogen, phosphorus, and so on that have been developed by taking the elements and materials of nature, initially scattered more or less at random over the surface of the Earth, and concentrating these elements and their compounds into structures that would be extremely unlikely (except for the investment of energy into a plan) a wheat plant, a pig, and ourselves for starters and then additionally all that goes into a ham sandwich. Once the structure is made energy must be continuously invested or the materials of which it is composed will go back on their own toward entropy, that is, a more random assemblage, and the structure will fall apart. A simple example is your ham sandwich. If that sandwich is put into a refrigerator, a device that uses energy to maintain the structures of its contents, the integrity of the sandwich will be maintained for some time. Pull the plug on the refrigerator (i.e., cut off the energy) and the sandwich begins to go into a more random assemblage, first smelly organic residues and then eventually carbon dioxide and simple nitrogen compounds such as ammonia. Pull the energy plug on yourself by not

eating and the same will happen to you, eventually. Likewise a car will not run without both fuel and the energy required for its repair, a city cannot run without its fuel supplies, power plants, and many kinds of repair personnel, or an entire civilization without all of these things, which must be supplied essentially daily. Most past civilizations that have lost their main energy supplies became extinct, as we develop later.

The practical meaning of this is that it is always necessary to find new energy resources to construct and maintain whatever structures we have, including houses, cars, civilizations, and ourselves. This is familiar to us in the shop costs, medical bills, and taxes we must pay to maintain our cars, ourselves, and our roads and bridges against the entropic forces of nature that would otherwise result in time in cracked and broken roads and bridges rusted to pieces. Curiously it is necessary to generate additional entropy to maintain areas of negentropy. The refrigerator must take high-grade electricity and turn it onto lower-grade (more entropic) heat in order to maintain the ham sandwich in its desired configuration, and each of us must take low-entropy food and turn it into high-entropy heat and waste products in order to maintain ourselves. Even the creation of this book, which we hope represents highly ordered information, requires the generation of excess entropy around us, as a look at either of our offices will confirm.

Laws of Thermodynamics

Thermo means heat (or energy) and *dynamics* means changes. *Thermodynamics* is the study of the transformations that take place as energy or fuels are used to do work. *Work* means that something is moved, including, for example, a rock or your leg lifted, a car driven, water evaporated or lifted up in the atmosphere, chemicals concentrated, or carbon dioxide transformed from the atmosphere into a green plant. There are two principal laws of thermodynamics, called the *first law of thermodynamics* and the *second law of thermodynamics*. Quite simply the first law says that energy (or for some particular considerations energy-matter) can never be created nor destroyed, but only changed in form. Thus the potential energy once found in a gallon of gasoline but then used to drive a car, say, 20 miles up a hill, is still found somewhere, as the momentum of the car, as heat dissipated by the radiator, exhaust or where the tires met the road, or in the increased potential energy of the car at the top of the hill. Most of the original energy will be found as heat dissipated into the environment, where it is essentially impossible to get any additional work out of it. (Technically you could capture that waste heat and use some of it, but it would require the use of even more energy to do so). But some fraction of the work done can be used again: for example, the automobile could be rolled back to its original downhill position using the force of gravity. The second law says that all real-life processes produce entropy. At every energy transformation some of the initial high-grade energy (i.e., energy that has potential to do work) will be changed into low-grade heat barely above the temperature of the surrounding environment. In other words the first law says that the *quantity* of energy always remains constant, but the second law says that the *quality* is degraded over time. The practical meaning is that with the exception of the reliable energy input from the sun it is always necessary to find new energy resources to construct and maintain whatever structures we have, including houses, cars, civilizations, and ourselves. The implications of this have had overwhelming effects upon all human enterprises and histories, and constitutes the remainder of this book.

To our knowledge there are no examples of any action occurring on earth, or anywhere else for that matter, that is not subject to the laws of thermodynamics. The only possible exception, given in the first part of this chapter, is that the law of conservation of energy needs to be expanded to the law of conservation of mass-energy when nuclear reactions (in a star, nuclear bomb, or nuclear power plant) are considered. This is because mass can be converted to energy (and the converse) according to Einstein's famous equation: $E = MC^2$, which says that under special circumstances energy created equals mass times the speed of light squared. In other words, in a nuclear conversion a small amount of mass can be converted to a huge amount of energy, although this can take place only under very special conditions. This is an example of how science often moves forward. The first law of thermodynamics might seem to have been violated when we learned about nuclear reactions, but with Einstein's help we only had to understand that although the first law works very well for everyday conditions we have to expand it to include mass for the very special circumstances of a star: we need to learn how to expand our law.

Types of Energy

Although energy is critical to all of our daily activities an actual definition, as we have said, is hard to come by. Energy is usually defined as the capacity to do work, where work implies something is moved (a rock or animal from here to there, chemicals concentrated, and so on). The most important routine work activities that take place within the human realm are driven by the sun (solar energies). These activities include: the evaporation and lifting of water from the sea to provide us with rains and rivers that flow from mountains, the concentration of low-energy carbon from the atmosphere into higher-energy tissues of a plant through photosynthesis, the passing of this energy through food chains (e.g., with a deer or a cow eating plants), the generation of winds that move atmospheric water from the ocean to the land and cleanse the local skies of pollutants, the generation of soils through complex processes of forests and grasslands, the running of many complex processes in natural ecosystems and so on (Fig. 10.3). Increasingly we also use fossil fuels such as coal, oil, and natural gas.

Energy that is being used at the time in question to undertake work is called *kinetic* energy, and energy that has the possibility to do work, but that is not doing it now is called *potential* energy. Examples include a rock at the top of a hill, the energy in a pile of firewood, the concentrated energy within a flashlight battery not being used, and the chemical energy in a gallon of gasoline sitting in a gas tank. When the gasoline

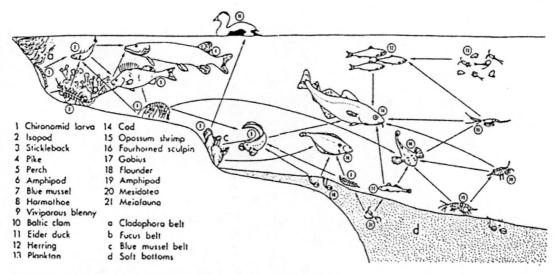

1	Chironomid larva	14	Cod
2	Isopod	15	Opossum shrimp
3	Stickleback	16	Fourhorned sculpin
4	Pike	17	Gobius
5	Perch	18	Flounder
6	Amphipod	19	Amphipod
7	Blue mussel	20	Mesidotea
8	Harmothoe	21	Meiofauna
9	Viviparous blenny		
10	Baltic clam	a	Cladophora belt
11	Eider duck	b	Fucus belt
12	Herring	c	Blue mussel belt
13	Plankton	d	Soft bottoms

Fig. 10.3 Energy flow through a Baltic ecosystem. The energy enters from the sun, is captured by green plants, and is passed to herbivores and then carnivores through food chains (sometimes called food webs) (*Source*: B-O Jannsen)

is used to move a car the potential energy of the gasoline is changed into the kinetic energy of the automobile in motion and into heat. Most energy that we use is derived directly from the sun either at present (i.e., wind, the power of dry air to evaporate, and so on) or in the past (i.e., the gasoline came from petroleum that was once solar energy captured by little floating plants (phytoplankton) in the sea. Other sources of energy besides the sun include the energy of planetary motions (which causes tides), geological processes such as volcanoes and crustal movements, and that of nuclear decay (which causes the interior of the earth to be hot).

Solar energy is especially important as it runs the whole "heat engine" of the earth (Fig. 10.4). In this process the more concentrated energy of the solar rays hitting the earth perpendicularly at the equator causes a warming of the Earth there relative to other regions. This in turn causes the air over the equator to rise. As this air rises it cools, and the associated loss of energy means that the atmosphere can hold fewer water molecules, which over time fall out as rain. Thus the equator is a very wet region, and it is here that tropical rain forests are found. The exact place of the greatest rain changes from north to south with

the seasons, but it is always directly "under" the sun, that is, at the location where the sun's rays are most nearly perpendicular. The rising air is eventually constrained by gravity and accumulates at about 5–10 miles high over the equator. This causes a high pressure zone there and, because the air masses have been moved upwards, a low pressure zone on the surface at the equator. This high pressure at altitude pushes the air north and south until it cools enough to descend at about 30° north and south. As the air descends it warms again and hence has the energy to hold more and more water, so that when it comes in contact with the earth's surface it literally sucks the moisture out of the soil and vegetation, generating the earth's great deserts. It also generates another high pressure area there (at the Earth's surface at 30° north and south) which pushes air back towards the low pressure air on the equator while being bent to the right in the northern hemisphere and left in the southern hemisphere by the Earth's rotation (the Coriolis force). This causes the steady *trade winds* characteristic of the tropics which become increasingly moisture-laden as they approach the equator. The high pressures at about 30° push air masses poleward, and these winds as affected by the Coriolis force

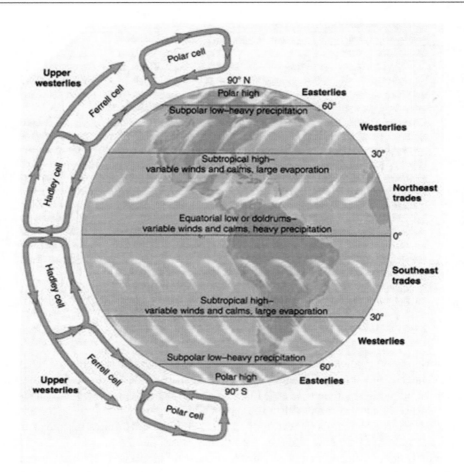

Fig. 10.4 The basic heat engine of the Earth. Electromagnetic radiation, usually considered as traveling in "packets" called photons, enter the Earth's atmosphere after traveling from the sun. Because the Earth's surface is more nearly perpendicular to their entrance path near the equator they tend to be more concentrated there and subsequently heat the Earth's surface especially well at the equator. This causes warm air masses to rise at the equator, and then disperse north and south as described in the text. As the air masses rise they cool, and as cooler air has less energy to keep the water molecules suspended, it rains a lot on the (thermal) equator, which moves north in summer and south in winter. The rising air masses create high pressure above the equator, pushing air masses north and south until they descend at 30° (*Source*: Kaufmann and Cleveland, 2008)

cause our familiar westerly winds in the temperate zones of the northern hemisphere.

When such steady winds are forced upward by a mountain in their path the air masses cool, generating a rainy region on the windward side (think Seattle, Washington) and a dry or even desert area on the leeward side (think Yakima, Washington). Thus the unequal interception of solar energy on different parts of the Earth's surface generates the world's winds, its wet and dry areas, and, more generally, its climatic zones. Solar energy also evaporates water from the surface of the ocean, lifting and purifying it in the process, moves it onto land masses while causing it to rain as solar-powered winds push the air masses up mountains, and in so doing generating the world's rains and rivers. We may not appreciate a particular rainy day, however, the rains are essential to our purified water supplies and the growth of plants upon which all animal life, including our own, depends. An understanding and appreciation of the world's hydrological cycle and the critical role of energy in it is perhaps one of the most fundamental things we can learn about how the Earth, and hence our economy, operates. Curiously this process is not considered part of most economics even though it is probably the most important step in the world economy, the purifying of water and the lifting of it to the land and to the mountains that supply most of the world with its water for agriculture, for all economic activity, and for life itself. It is not considered by conventional economics because it is free; that is, it does not enter into markets. But being free and indispensible makes it more, not less, valuable to our economy and we need to think of it that way especially as we have to pay more to compensate for the pollution and other abuse of water that is increasingly part of the hydrological cycle.

Energy and Life in More Detail

Life, in all of its manifestations, runs principally on contemporary sunlight that enters the top of our atmosphere at approximately 1400 W (1.4 kW or 5.04 MJ per hour) per square meter for a point perpendicular to the sun's rays. Roughly one quarter of that amount reaches the Earth's surface. This sunlight does the enormous amount of work that is the thermodynamic consequence of this energy input and that is necessary for all life, including human life, even when isolated from nature in cities and buildings. The principal work that this sunlight does on the Earth's surface is to evaporate water from that surface (evaporation) or from plant tissues (transpiration) which in turn generates elevated water that falls eventually back on the Earth's surface as rain, especially at higher elevations. The rain in turn generates rivers, lakes, and estuaries and provides water that nurtures plants and animals. Differential heating of the Earth's surface generates winds that cycle the evaporated water around the world, and sunlight of course maintains habitable temperatures and is the basis for photosynthesis in both natural and human-dominated ecosystems. These basic resources have barely changed since the evolution of humans (except for the impacts of the ice ages) so that preindustrial humans were essentially dependent upon this limited, or perhaps more accurately diffuse, although predictable, energy base.

In *photosynthesis* energy from the sun is captured by green plants using chlorophyll, a very special compound similar in structure to the hemoglobin in our blood. Chlorophyll appears green to our eyes because it uses (i.e., absorbs) the shorter red and longer blue wavelengths from the sun and reflects back the green wavelengths that it does not use. A thick layer of green plants covers the Earth wherever temperatures are moderate and water is abundant. The amount of energy trapped by photosynthesis is immense, roughly 3,000 exajoules) per year, which is about six times larger than the energy use of all human activities (488 exajoules per year). The first state occurs in the center of the chlorophyll molecule where electrons circling the magnesium–nitrogen compound are "hit" by a photon from the sun and "pushed" into a larger orbit, which allows them to store more energy and then pass it to special chemical compounds. This is similar to how a professional skater stores energy when she puts her arms out to the sides and accelerates her spin by pushing with her legs, and then uses that stored energy to speed up her spin by pulling her arms back to her sides. Free electrons are normally made available from reduced compounds and move through biological circuits to fuel biotic processes. That energy is first stored temporarily in reduced compounds in plants such as NADP, which are then used to split water to get hydrogen, and CO_2 to get carbon. Plants then combine the carbon and hydrogen to make reduced, energy-rich compounds such as a sugar. Eventually, the electron is passed to an electron acceptor, normally oxygen, but occasionally sulfur or some other element. These electrons are re-energized when green plants give a new kick to the electrons when a photon from the sun again drives photosynthesis. And hence the process continues, with the energy from solar-derived photons driving every biological activity including the movement of my fingertips on this keyboard. It is incredible!

Fig. 10.5   Herbivores grazing in Kenya (*Source*: Kathy Wooster)

The chemistry of photosynthesis is based on the energy from photons being used to split carbon dioxide and also water to get or *fix* reduced carbon and hydrogen, which is then used to generate sugars with oxygen as a waste product:

$$6CO_2 + 6H_2O \rightarrow C_6H_{12}O_6 + 6CO_2$$

The sugars are then synthesized into the more complex compounds of life. These include cellulose (the basic structural material of wood, which is just a lot of sugars attached one to another into a network of the same materials called a *polymer*) and, with the addition of nitrogen, the proteins of animal and many plant tissues. This same equation is "run backwards" by animals and decomposers that use the chemical compounds. When green plants first evolved some three billion years ago, and especially one billion years ago when plants colonized the land, they changed the atmosphere from an anaerobic one to an aerobic one. This can be seen in, for example, the rocks of Glacier National Park where there are green layers of iron-containing rock that were laid down before the oxygenation

of the atmosphere and similar "rusted" red rocks that were laid down later after the evolution of an oxygenated atmosphere.

What about animals? Take a look at most wild or domestic animals. Usually they are eating, that is, getting energy, or trying to position themselves to do so (Fig. 10.5). If they are not eating they tend to be resting, conserving energy. In the breeding season obviously things get a bit more complicated. Plants too spend most of their time dealing with energy: for example, they are photosynthesizing any time the sun is shining and in various ways attempting to protect themselves from energy losses by making natural pesticides, for example. Humans are a bit different because food energy is (at this time in our history) so abundant and cheap, at least for the richer half of humanity, that we have to invest relatively little time or personal energy to feed ourselves. We are also different now because our energy requirements are only about half of what they were when we were more active. For example, early New England farmers had to eat (and drink, especially ale!) about 7,000 kcal (30 MJ) each day to fuel their hard agricultural work, although many hard workers in poorer countries get by on less than half that. Any of us today who ate that many calories would become huge!

The study of biology, from biochemistry to ecosystem biology, is very much about the study of how energy is passed from one chemical entity to another within that level of inquiry. Biochemists often focus on the importance of the energy storage materials NADPH and ATP, scientists who study at the level of one organism often consider feeding behaviors and the physiology of energy transfer within and across the gut wall, whereas ecosystem biologists talk about the transfer of energy from plants to herbivores to predators. The importance of energy in biotic function has captured the attention of many of our great biological thinkers, including Alfred Lotka, Harold Morowitz, Max Kleiber, Howard Odum, and others. And what do they conclude? Basically that life, or more specifically the individual organisms and species that constitute the packages of life, is about capturing as much energy as possible per unit time with as little expenditure or investment as possible per unit gained, using that net energy gained to sequester more energy and other resources, and using them to create structures and fuel behaviors to propel their genes into the future. As far as we know this is entirely the result of the uncaring processes of natural selection; those organisms and, ultimately, genes that were successful at this pattern were those that tended to survive, prosper, and eventually be relatively dominant on the Earth's surface. Some people prefer to use the more general term "resources" rather than just "energy resources" when discussing these issues, and there is occasionally a good case to be made for that. Obviously water is a critical resource for plant growth, and all the solar energy a plant could ask for might be available in an Arizona desert although water is very much limiting. In other situations some specific nutrient, such as phosphorus or nitrogen may be limiting, but even these limitations can be mitigated by the plant investing more of its energy into growing longer roots to exploit more soil or transferring molecules across fine roots. Thus for most of the earth the critical issue is energy, and life seems to be very good at expanding to capture as much of the available energy around it as possible.

Two important concepts here are *energy investments* and *energy opportunity costs*. The former means that life must always invest energy into fighting entropy, getting other resources, and so on. The second means that because every organism has only a limited energy supply at any one time any particular investment in one process means that there is that much less energy to invest elsewhere. If a tree invests more energy in growing long roots to get more water or nutrients, then less will be available for growing tall, and it might be shaded by a competitor or eaten by an insect. If more energy is diverted into making natural pesticides (such as caffeine, mustard oils, or various alkaloids) then less is available for growing roots, and so on. Likewise if a civilization invests more energy in military activities, or expanding office space, or building fancy homes, or looking for oil, then less energy is left for repairing bridges or education. Politics is all about how to make energy investment decisions, although it is done through deciding where to spend money (more on that in the next chapter). In both trees and politics there is a tendency to invest in a way that can capture more energy (through plant or economic growth), but that can work only when there are additional energy resources that can be exploited by using less energy than that gained. If the energy resources become restricted, then investing in growth can be self-defeating, a situation that many of the world's economies now face.

Energy Storage

Life is of course about much more than simply gaining and using energy, for life must use energy when and where it needs it in order to help the organism adjust to a continuously changing environment. Just as a motor or light needs a switch to turn it off and on as needed, life too must have switches. As a simple example, if our muscles were firing all the time they would be useless, and in fact such a condition is a pathology called tetanus. Thus life has evolved a whole series of complex controls and switches that capture available energy from the sun or food a little at a time, stores it, and releases it as needed and as controlled by hormones and the nervous system operating through very complex biochemistry.

The general solution that has evolved for the storage and on/off problem has been through the use of various storage reservoirs. This allows for the capture, storage, transport, and release of the energy made available to the organism by photosynthesis or by ingesting food. The most common such compound for *short-term storage* is adenosine tri-phosphate (ATP) and its less energized form ADP. Whenever the body needs energy quickly it calls on ATP to deliver that. These compounds are ubiquitous to life, and are critical to all activities that an organism does. *Medium-term storage* is the glycogen in our liver, and *longer-term storage* is all too familiar to us as bodily fat. Very curious, although perfectly understandable from an evolutionary perspective, procedures to utilize energy have occurred in our remote evolutionary past that are very important to us today. For example, the area inside our cells where the energy is actually used (i.e., sugars or other fuels oxidized) are small organelles called *mitochondria*. These are the microfurnaces of our bodies, and indeed of most other organisms on Earth.

The curious thing about mitochondria is that they appear to have been free living single-celled organisms at one time, similar to a flagellated algae, and, as elegantly worked out by microbiologist Lynn Margulis, the microscopic structure of mitochondria and flagellates are extremely similar. Mitochondria appear to have evolved from food ingested by the host organism to symbiotic (i.e., co-occurring mutually dependent, mutually reinforcing organisms) to a fully incorporated component of the cells of both plants and animals because of a mutually beneficial arrangement. This must have happened at the dawn of life and was then passed on to nearly all life today. The host cell provides fuel to the mitochondria and the mitochondria process it efficiently and fully, returning some to the host as ATP. In other words their highly developed ability to release energy from food using oxygen became a great asset to the organism as the atmosphere went from anaerobic to aerobic that became incorporated in some

past organism in its own cellular structure. Aerobic versus anaerobic use of food gave such a powerful advantage to organisms that we now find mitochondria in essentially all cells of organisms living where there is oxygen.

Another problem facing life is how to get rid of the electron once its high-energy state is exhausted. This is analogous to a steam turbine that requires the venting or condensation of the steam behind the blades, otherwise the blades would not be pushed forward because of back pressure, and also to the ground-wire in a (direct) current circuit. We live in a world where oxygen is generally available and used as an electron receptor, and that is why we need oxygen. But this was not always the case. Food can be partially digested using fermentation where only part of a molecule is "sent" to an electron acceptor, leaving relatively energy-rich alcohol or vinegar as a by-product. Before the evolution of green plants a billion or so years ago free or uncombined oxygen essentially did not exist, and fermentation prevailed. Whenever oxygen did become available it was so reactive that it quickly combined with the many reduced compounds in the environment, such as ferric iron, reactive silica, and even the tissues of existing organisms. But upon the evolution of green plants so much more oxygen was produced as a by-product of photosynthesis that it accumulated in the atmosphere, where it now constitutes about 21% of the molecules there.

A Big Jump in the Earth's Energy Supplies for Life

Free oxygen increased with the first massive increase in land plants as a waste product of their photosynthesis that split water to gain the hydrogen needed to produce reduced carbohydrates such as sugars. For all the existing plants and animals and microbes on the Earth this free oxygen, itself extremely reactive, was initially a severe toxic threat, a widespread and dangerous pollutant, some say that the evolution of oxygen-releasing green plants was the greatest environmental impact the earth has ever faced! It has been argued that the mitochondria were initially evolved (or as we said above, "captured") to sequester the dangerous oxygen before it destroyed other parts of the organism, and only later developed the capacity to enhance the metabolic activity of the host cell. Over time natural selection created organisms (including humans) with protective skins that require oxygen to live. But even today there are many environments where oxygen is not present. They are normally obvious to us from their smell of hydrogen sulfide, characteristic of, for example, the mud of a marsh or the inside of our intestines, which would not be a good place to have oxygen for then the energy in our food would be used up before it got to do us any good. In these environments oxygen remains a poison for many of the organisms.

Thus it appears that evolution has operated in many complex ways, such as incorporating oxygen-using organelles (i.e., mitochondria) in all animals that live in an oxygen-rich environment, to derive means of using energy more powerfully. Apparently the main ways to do this were worked out very long ago in the evolution of life inasmuch as nearly all life has the same internal energy structures and uses the same basic phosphorus-based chemistry for storage and quick release. Biochemist Paul Falkowski makes an elegant argument that in many ways the biochemistry that life depends upon now is inappropriate for our existing oxidized environment and can be understood only as a "holdover" from life's anaerobic past: the anaerobic mechanisms that worked in the past were too deeply engrained into the processes of life for life to abandon, and so were retained and modified even if not perfectly suited for the new aerobic environment. Although complete oxidation of food using mitochondria allows the most complete use of food, many different approaches to utilizing food energy have evolved, and these different pathways are still used variously by different species and in response to different environmental conditions. If oxygen is not present the less thorough but quite adequate energy release process called fermentation still can be used, and this process generates energy-intermediate alcoholic residues which we have exploited to generate beer and wine. The partial transformation of the grain or

fruit into usable energy leaves as residues alcohol and CO_2, which generates the bubbles in beer.

A more general perspective is that energy is passed through and among organisms in a series of complex *redox* (reducing-oxidizing) reactions, until the full food value is extracted and some or all of the carbon base is turned into CO_2. Energy is passed from one organism to another through an ecosystem along food chains and food webs. Plants capture the energy from the sun and turn a portion of it into their own tissues, leaves, stems, roots, and so on. Then some of that energy is passed to herbivores (plant eaters) and then carnivores (meat eaters) and decomposers. The word trophic means food, and trophic dynamics is the study within ecology of how energy is passed along food chains within an ecosystem and what happens to that energy. An important thing that happens is that energy is lost (actually turned into heat) at every step as necessitated by the second law of thermodynamics. Most of the energy that is lost was actually used by the organism itself for its own maintenance metabolism. This is due to the necessity for each organism to "fight entropy" through energy investments and the necessary losses to heat arising from the second law of thermodynamics. Usually only a small proportion, very roughly 10%, is passed from one trophic level (such as plants) to the next (herbivores). This is one reason there are few top carnivores; if there are four or more trophic levels each passing on only 10% of the energy then only a very small amount of the original energy captured by photosynthesis makes it to the top carnivore.

Although it is obvious that an organism must get enough *quantity* of energy to maintain itself it is also necessary for it to get a sufficient *quality* of energy. Most generally the missing ingredient in vegetative material for humans or for other animals is sufficient protein. Kwashiorkor is a common disease of people on an insufficient protein diet, characterized by cinnamon-colored hair and a protruding belly, as well as many personal metabolic problems restricting the ability of people to work. Once in the 1950s well-meaning nutritionists made a large effort to increase protein production of certain groups of people who had this disease, for example, feeding existing grains to chickens and fish, to try to increase the protein available to these people. But the program backfired because the people were actually energy-starved, and their bodies were burning the proteins for fuel, not using them for structural development. In other words our bodies have an even greater need for energy than for structural building and repair. So feeding energy-rich grains to produce a smaller quantity of animals actually exacerbated the problem by reducing the energy available to the people. Even though they got more protein their desperate bodies had to use it for fuel, not maintenance or growth of new tissues! But where calories are sufficient, protein is critical for normal healthy development, thus the quality of energy is often as important as the quantity. Obviously in our food quality is a much more complex issue than simply protein or not.

Proteins are foods made of amino acids that are based upon nitrogen as well as carbon. One can think of a hamburger: the bun is a carbohydrate made of carbon, hydrogen, and oxygen, and the beef is protein, which has those elements and much nitrogen as well. We normally think of protein as meat, however, there are many other sources. For example, ecologists have found that many of the animals of an estuary or a forest are dependent upon detrital food chains, that is. food that has been dead a relatively long time before being consumed (as opposed to grazing or browsing food chains). Dead plant material is mostly carbon and as such contains little nitrogen, which is critical for the protein needs of the animals that feed upon it. But in estuaries and forest floors much of the decomposition of this material occurs by bacteria, and certain bacteria can do something that most other organisms cannot: they can fix nitrogen from the air and turn it into protein; thus the animals that eat microbially mediated food get much better nutrition because there tends to be more protein. This may sound repulsive to humans but maybe it is much less so if you think about the microbially mediated foods we eat: bread, cheese, beer, wine, salami, sour cream, and so on. In fact most of our party foods are microbially mediated!

More on Energy and Evolution

Plants and animals in nature have been subjected to fierce selective pressure to do the "right thing" energetically, to ensure that whatever major activity that they did, and do, gained more energy than it cost, and generally got a larger energy net return than alternative activities. Biology in the last century had, appropriately enough, focused mostly on fitness, that is, on the ability of organisms to survive and reproduce, in other words to propel their genes into the future. Although it is a no-brainer that a cheetah, for example, has to catch more energy in its prey than it takes to run it down, and considerably more to make it through lean times and also to reproduce, it took the development of double-labeled isotopes and the exquisite experimental procedures by the likes of Thomas and his colleagues [4] to show how powerfully net energy controlled fitness. Thomas et al. studied tits (chickadees) in France and Corsica. Found that those birds that timed their migrations, nest building, and births of their young to coincide with the seasonal availability of large caterpillars, which in turn were dependent upon the timing of the development of the oak leaves they fed upon, had a much greater surplus energy than their counterparts that missed the caterpillars. They fledged more, larger, and hence more likely to survive young while also greatly increasing their own probability to return the next year to breed again. Those of their offspring that inherited the proper "calendar" for migration and nesting were in turn far more likely to have successful mating and so on. Thomas et al. also showed how the natural evolutionary pattern was being disrupted by climate change, so that the tits tended to get to their nesting sites too late to capitalize upon the caterpillars, who were emerging earlier in response to earlier leaf-out. Presumably if and as climate warming continues, natural selection will favor those tits which happened to have genes that told them to move north a bit earlier.

Maximum Power

Howard Odum has taken the concepts one step further by arguing that it is not just the net energy obtained but the *power*, that is, the *useful energy per unit time*, that is critical. Odum argued that there is generally a tradeoff between the rate and the efficiency for any given process: the more rapidly a process occurs the lower its efficiency, and vice versa. Under a given set of environmental conditions it is not advantageous to be extremely efficient at the expense of the rate of exploitation, nor to be extremely rapid at the expense of efficiency. For example, in a series of elegant observations and experiments Smith and Li [5] found that a trout that feeds on drifting food in a rapidly flowing stream will acquire large amounts of food drifting by but at a low net efficiency; that is, much of the energy surplus created by the consumption of a large amount of food is spent in muscle contraction for the trout so that it can fight the faster current. Likewise a trout in slow water can be very efficient because its swimming costs are lower, but the slower water brings with it less food, and thus the overall energy surplus will be limited by the lower rate at which food is provided. Dominant trout will pick an optimum intermediate current speed, which will result in faster growth and more offspring. Subdominant trout will be found in water moving a little faster or a little slower. In some experiments trout with no competitive power will be found drifting aimlessly in still water slowly starving to death.

This kind of tradeoff can be found throughout the plant and animal kingdoms and even rates of power plant operation in industrial society [6]. It explains why we must shift gears to stay near the middle of each gear range in a stick-shift car when we want to accelerate, and why most businessmen once chose to take jumbo jets to cross the Atlantic rather than the Concorde or ocean liners. In fact it can be used to explain why the Concorde went extinct, and perhaps why the second Queen Mary was much smaller than the first.

Of course life in all of its diversity also has a diversity of energy life styles that have been selected for – sloths are just as evolutionarily successful as cheetahs, whereas warm-blooded animals pay for their superior ability to forage in cold weather with a higher energy cost to maintain an elevated body temperature – the list is endless. Yet there remains a rate-efficiency tradeoff within each lifestyle. Drift-feeding trout choose areas of intermediate current to maximize the energy surplus, however, suckers have "chosen" through natural selection (i.e., have been selected for) to maximize energy surplus by processing lower-quality food on the bottom, and probably have an optimum power output for that set of environmental conditions.

Nevertheless each life style must be able to turn in an energy profit sufficient to survive, reproduce, and make it through tough times. There are few, if any, examples of extant species that barely make an energy profit, for each has to pay for not only their maintenance metabolism but also their "depreciation" and "research and development" (i.e., evolution), just as a business must, out of current income. Thus their energy profit must be sufficient to mate, raise their young, "pay" the predators and the pathogens, and adjust to environmental change through sufficient surplus reproduction to allow evolution. Only those organisms with a sufficient net output and sufficient power (i.e., useful energy gained per time) are able to undertake this through evolutionary time, and indeed some 99% of all species that have ever lived on the planet are no longer with us; their "technology" was not adequate, or adequately flexible, to supply sufficient net energy to balance gains against losses as their environment changed. Given losses to predation, nesting failures, and the requirements of energy for many other things the energy surplus needs to be quite substantial for the species to survive in time.

Natural Economies

Of course in Nature plants and animals do not exist in isolation but combined in complex arrangements that we call ecosystems tied together by the movement of energy and materials from one species to another in what is often referred to as food chains or food webs. We know this because if we put a little radioactive phosphorus in the soil under some grasses that phosphorus will soon show up in leaves of the grasses, and then a few days later in the grasshoppers that eat the grass, and then a few days later in the spiders that eat the grasshoppers. The phosphorus, and other materials, and the energy originally captured from the sun by the plants is moved along from species to species by each one eating others. As given above we call the green plants that capture the solar energy *primary producers*, the animals that eat the grass *herbivores*, the animals that eat other animals *carnivores*, and so on. Eventually all plant and animal material ends up as dead organic material, often called detritus, and this material is then broken down into very simple materials or even elements by bacteria and other decomposers. We call the study of these relations *trophic* (meaning food) analysis and each successive step from the sun, trophic levels. It is rather amazing to think that all the energy necessary for all the animals and all the decomposers, and even the plants at night and in the nongrowing season, comes from the photosynthesis during the daylight hours during the growing season.

We can call all of these trophic interactions collectively *natural economies*, in other words Nature too, just like human economic systems, is all about production, exchange within and between species, and eventual degradation. Of course natural ecosystems are different from modern human economies in that there is no money, but the economy exists just fine without the money, as might conceivably ours (i.e., many economies are based on barter alone). This idea that Nature too has economies is a very powerful one for it allows us to focus on just what are the essential features of an economy when we strip it of the human additions of money, debt, credit, and so on.

Summary so Far: Surplus Energy and Biological Evolution

The interplay of biological evolution and surplus energy is extremely general, as emphasized a half century ago by Kleiber [7], Morowitz [8], Odum

[9], and others. Plants too must make an energy profit to supply net resources for growth and reproduction, as can be seen easily in most clearings in evergreen forests where living boughs on a tree in the clearing are usually lower down than they are in the more densely forested and hence shaded side of the tree. If the bough does not carry its weight energetically, that is, if its photosynthesis is not greater than the respiratory maintenance metabolism of supporting that bough, the bough will die (or perhaps even be sloughed off by the rest of the tree).

Every plant and every animal must conform to this "iron law of evolutionary energetics": if you are to survive you must produce or capture more energy than you use to obtain it, if you are to reproduce you must have a large surplus beyond metabolic needs, and if your species are to prosper over evolutionary time you must have a very large surplus for the average individual to compensate for the large losses that occur to the majority of the population. In other words every surviving individual and species needs to do things that gain more energy than they cost, and those species that are successful in an evolutionary sense are those that generate a great deal of surplus energy that allows them to become abundant and to spread. We are unaware of any official pronouncement of this idea as a law, however, it seems to us to be so self-obvious that we might as well call it a law, the law of minimum EROI, unless anyone can think of any objections.

Probably most biologists tacitly accept this law (if they have thought about the issue), but it is not particularly emphasized in biological teaching. Instead biology in the last century focused mostly on fitness, on the ability of organisms to propel their genes into the future through continuation and expansion of populations of species. But in fact energetics is an essential consideration as to what is and what is not fit, and many believe that the total energy balance of an organism is the key to understanding fitness. It took the development of double-labeled isotopes and the exquisite experimental procedures by the likes of Thomas et al. [4] to show how powerfully net energy controlled fitness.

Energy and Economics in Early and Contemporary Human Economies

Humans are no different from the rest of nature in being completely dependent upon sunlight and food chains for our own energy requirements and nutrition, and on being part of complex interactions among very complex food chains leading to ourselves. Human populations, like those of any other species, must capture sufficient net energy to survive, reproduce, and adapt to changing conditions in the area in which they live. Humans must first feed themselves before attending to other issues. For at least 98% of the million years that we have been recognizably human the principal technology by which we as humans have fed ourselves (obtained the energy we need for life) has been that of hunting and gathering. The study of contemporary hunter-gatherers such as the !Kung of the Kalahari desert in southern Africa that we introduced in Chap. 2, are probably as close to our long-term ancestors as we will be able to understand. Most hunter-gatherer humans were probably like other species in that their principal economic focus was on obtaining sufficient surplus energy as food gained directly from their environment. Studies by anthropologists such as Lee [10] and Rappaport [11] confirmed that indeed present-day (or at least recent) hunter-gatherers and shifting cultivators acted in ways that appeared to maximize their own energy return on investment, perhaps 10 J returned for each one invested. Angel found that agriculture actually decreased the average physical fitness of humans [12].

Human evolution, broadening the definition to include social evolution, is different, for the human brain, language, and the written word have allowed for much more rapid cultural evolution. The most important of these changes, as developed in Chap. 2, were energy-related: the development of energy-concentrating spear points and knife blades, agriculture as a means to concentrate solar energy for human use, and more recently the exploitation of wind and water power and, of course, fossil fuels. What is important from our perspective is that each of these cultural adaptations is part of a continuum in which humans invest some of their energy to increase the rate at which they exploit additional resources from nature, including both energy and nonenergy resources.

The development of agriculture allowed the redirection of the photosynthetic energy captured on the land from the many diverse species in a natural ecosystem to the few species of plants (called cultivars) that humans can and wish to eat, or to the grazing animals that humans controlled. It also allowed the development of cities, bureaucracies, hierarchies, the arts, more potent warfare, and so on, all that we call civilization, as nicely developed by Jared Diamond in his book *Guns, Germs and Steel*.

A human as a machine works at about 20% efficiency: the power output of a human (i.e., its muscular work) is about 20% of the food energy input to that machine. Thus over a 10-hour day a human can deliver about one half to one horsepower hours, or about 5–10% of what a horse can do (and on about 5–10% of the food) [13]. Put another way, the power output of a human at rest is about 60 W, and at peak performance a strong worker might generate about 300 usable watts, although that rate cannot be sustained for very long. A very strong person might be able to deliver 100 W, or one kilowatt hour (3.6 MJ) over a 10-hour day. The human machine cannot deliver this power if the temperature is above 20–25°C, so that other things being equal it is more difficult to generate surplus wealth in the hot tropics [14]. A horse can generate about 3 kW. By comparison a four-cylinder standard automobile engine generates about 1,000 kW, and a jet turbine engine about 1,000,000 kW. Clearly the world now has at its disposal a tremendous amount of power compared to the past (Table 2.1).

Anthropologist Leslie White once noted that a bomber flying over Europe during World War II consumed more energy in a single flight than had been consumed by all the people of Europe during the Paleolithic, or Old Stone Age, who existed when people lived entirely by hunting and gathering wild foods [15]. White estimated that such societies could produce only about 1/20 horsepower per person, an amount that today would not suffice for even a fleeting moment of industrial life. Over time humans increased their control of energy through technology, although for thousands of years most of the energy used was animate (people or draft animals) and derived from recent solar energy. A second very important source of energy was from wood, which has been recounted in fascinating detail in Ponting [16], Smil [17], and especially Perlin [18]. Perlin estimates that by 1880 about 140 million cords of wood were being used in the world per year. Massive areas of the Earth's surface – Peloponnesia, India, China, parts of England, and many others – have been deforested three or more times as civilizations have cut down the trees for fuel or materials, prospered from the newly cleared agricultural land, and then collapsed as fuel and soil become depleted. Archeologist Joseph Tainter [19] recounts the general tendency of humans to build up civilizations of increasing reach and infrastructure and complexity that, again and again, have eventually exceeded the energy available to that society.

People have understood how to get energy from winds or from steam for millennia, for example, to grind grain, but the technology and incentives to do so increased rapidly from about 1750 onward. Cottrell [13] gives a thorough review of the importance of the increased use of energy by civilization, to which he, rightly in our opinion, attributes most other advances in civilization. Water power was especially important, for example, in New England in the early years of this country. But the real push in the development of civilization came with learning how to burn coal to do many things, but especially to make iron and to run steam locomotives. With these inventions, mostly in England in the 1800s, industrial development really took off and this led to what most people call the industrial revolution. It was not simply the development of the use of coal but a whole suite of financial, chemical, metallurgical, and other developments that accelerated each other and led to the enormous production of wealth that took place in England, Scotland, and Germany during the 1800s. For example, James Watt could not develop his famous steam engine until his friend William

Wilkinson had perfected the iron refining and drilling technologies that allowed for the construction of the perfectly round cylinder needed for Watt's steam engine. Even their interactions required the social environment of the Scottish enlightenment for their ideas to evolve and to come to fruition as actual components of society. Most thinking people at that time believed that these were wonderful inventions that would finally free people from the drudgery of every day and allow them to build a better society through rational thinking. At the same time many of the English Romantic poets, notably William Wordsworth, were horrified by the smoke and grime and repetitive jobs of the industrial revolution and pined for bucolic preindustrial England. Our societies today need such vast amounts of energy that we provide it by mining stocks of solar energy accumulated eons ago, and converted into coal, natural gas, and petroleum. Without these stocks we could not live as we do. Clearly the world now has at its disposal a tremendous amount of power compared to the past.

In summary, it seems obvious that both natural biological systems subject to natural selection and the cultures and civilizations that preceded our own were highly dependent upon maintaining not just a bare energy surplus from organic sources but rather a substantial energy surplus that allowed for the support of the entire system in question, whether of an evolving natural population or a civilization. Most of the earlier civilizations that left artifacts that we now visit and marvel at, including pyramids, ancient cities, beautiful buildings and rooms, monuments, and the like, had to have had a huge energy surplus for this to happen, although we can hardly calculate what that was. Certainly massive works from the past represented small net surpluses from thousands or millions of people carefully organized or brutally forced to do this work. Archeologist and historian Joseph Tainter has written elegantly about the role of surplus energy in constructing and maintaining ancient empires, Mayan, Roman, and so on [19]. Tainter argues that as empires get larger they can spend more and more energy impressing potential adversaries, that the construction of impressive capital cities in itself shows potential competitors that the empire has so much surplus wealth that it makes much more sense for the competitor to knuckle under, become part, and pay tribute than to fight the empire. The ever-expanding frontiers, however, and the need for ever more surplus energy as the distance needed to bring in food and other resources from increasingly distant provinces, increasingly decrease the net energy delivered to the center and eventually the empire falls in on itself and collapses from its very need for the complexity required to generate the necessary surplus energy. This has happened again and again in antiquity, and more recently with the collapse of the German Third Reich, the British Empire, and the Soviet Union. An important question for today is to what degree does the critical importance of surplus energy in nature and in the past apply to contemporary civilization with its massive, although possibly threatened, energy surpluses. At what point have we developed so much infrastructure that it requires all the surplus energy we can get just for maintenance metabolism, so that growth is impossible?

Contemporary industrial civilizations are dependent upon the sun and in addition on fossil fuels. Today fossil fuels are mined around the world, refined, and sent to centers of consumption. For many industrial countries, the original sources of fossil fuels were from their own domestic resources. The United States, Mexico, and Canada are good examples. However, because many of these industrial nations have been in the energy extraction business for a long time they tend to have both the most sophisticated technology and the most depleted fuel resources, at least relative to many countries with more recently developed fuel resources. For example, as of 2010 the United States, originally endowed with some of the world's largest oil provinces, was producing only about 40% of the oil that it was in the peak year of 1970, Canada had begun a serious decline in the production of conventional oil, and Mexico in 2006 was startled to find that its giant Cantarell field, once the world's second largest, had begun a steep decline in production at least a decade ahead of schedule. We come back to these issues later but the important thing

for now is the extreme petroleum dependence of contemporary affluent society. Howard Odum's "maximum power" hypothesis is a very powerful and insightful way to think about the evolution of nature and of human society. It explains, for example, how oil-rich nations gained ascendancy over solar-based societies, at least as long as their oil lasts. But it also suggests that countries that waste their energy or are unable to come to grips with the finite nature of premium energy will not be selected for.

There have been many other important authors in the last century who also have emphasized the importance of energy and energy surplus, including chemists William Ostland and Frederick Soddy, railroad man Frederick Cottrell, anthropologist Leslie White, and economist Nicholas Georgescu-Roegen. Although each of these authors acknowledges that other issues, including human culture, soil-nutrient inventories, and investment capital (among many others) can be important, each is of the opinion that it is energy itself, and especially surplus energy, that is key. Survival, military efficacy, wealth, art, and even civilization itself were believed by all of the above investigators to be products of surplus energy. For these authors the issue is not simply whether there is surplus energy but how much, what kind (quality), and at what rate it is delivered. The interplay of those other factors determined the flow rate of net energy and hence the ability of a given society to divert attention from life-sustaining needs, such as agriculture or the attainment of water, towards luxuries such as military excursions, art, and scholarship. Indeed humans could not possibly have made it this far through evolutionary time, or even from one generation to the next, without there being some kind of significant net positive energy, and they could not have constructed such comprehensive cities, civilizations, or wasted so much in war. A scary thought is that it does not take an enormous amount of energy to generate horrific war: all of World War II, in which more than 50 million people lost their lives and a billion more were seriously compromised, was fought on seven billion barrels of oil, about the quantity that the United States uses in 1 year at relative peace.

Thus as we face the inevitable contraction in the availability of our most important fuels and as the difficulties of generating alternatives at the scale required seem to mount day by day, we must face the possibility that our own economy and indeed civilization, which is almost universally based on the concept of continual growth of just about everything, may need a massive rethinking of how to go about thinking about itself and planning for the future: in other words, a new economics. This book is meant to give you the conceptual tools to do that [20].

Questions

1. If energy is so important why are most people unaware of most of the energy that they use?
2. What is meant by "The mechanical equivalent of heat? How was this demonstrated?
3. Can you explain Carnot's equation: $W = (Ts-Te)/Ts$? What implications does this have for the limits with which we can turn fuel into work?
4. Why, if the amount of energy stored in the surface of the North Sea is so great, is it not possible to extract this energy for use by society?
5. What is oxygen? If oxygen is so reactive why do we have oxygen in the atmosphere?
6. What is the law of the conservation of matter?
7. What is energy? Do you think it has been defined adequately?
8. What is combustion? Can you give an example of an equation representing combustion?
9. What is the relation between energy and power?
10. Energy is often given in different units such as therms, kwatt hours, joules, calories, and so on. How are these units different? How are they the same? Which unit should you use? Why?
11. Define the relation between energy quantity and energy quality. Can you give an example where it is important?
12. Explain the differences among energy, exergy, and emergy.
13. What are fuels ?
14. What is entropy? What is negentropy? Can you give an example from everyday life? What is the relation between energy and entropy?
15. What is the relation between negentropy and a plan? Can you give several examples?
16. How does negentropy relate to biotic evolution?
17. Why does the maintenance of negentropy generate entropy?
18. What is the first law of thermodynamics?
19. What is the second law of thermodynamics?
20. Can you define the first and second laws of thermodynamics using the words quantity and quality?
21. What might be considered an exception to the laws of thermodynamics? In your opinion is this really an exception?
22. What is the difference between kinetic and potential energy? How are they related?
23. How is the surface of the Earth a heat engine? What are trade winds and how are they formed?
24. Give the basic equation for photosynthesis.
25. What is the relation between energy investments and energy opportunity costs?
26. What are some of the biotic chemical compounds in which energy is stored?
27. Discuss the terms aerobic and anaerobic in relation to the Earth's evolutionary history.
28. According to Paul Falkowski why do organisms carry within them inappropriate chemistry for today's environment?
29. If a metabolic process produced alcohol or vinegar, what does this tell you about the efficiency of the use of the original plant material?
30. What does redox mean?
31. Define trophic dynamics and give an example. What does this tell us about the efficiency of ecosystem processes?
32. What element characterizes proteins and makes them different from carbohydrates?
33. Relate energy to evolution.
34. What does the maximum power principle tell us about the efficiency of a biological or physical process?
35. Does nature have economies? How so? Do you think it is accurate to describe nature in that way?
36. What is the "iron law of evolutionary energetics"?

37. Relate the principles learned in the earlier part of this chapter to human societies.

38. How did wood use precede the industrial revolution?

39. Summarize your views on how natural and human societies use energy to survive and prosper.

40. Do you think that technology will make the end of the oil era of little concern? Why or why not?

References

1. Kelvin, W. T. (1849). "An account of carnot's theory of the motive power of heat – with numerical results deduced from regnault's experiments on steam." Transactions of the Edinburg Royal Society, XVI. January 2.

2. Odum, H. T. 1996. Environmental accounting: Emergy and environmental decision making. John Wiley, New York.

3. Gaudreau K., Fraser R.A., Murphy S. The tenuous use of exergy as a measure of resource value or waste impact. Sustainability. 2009; 1(4):1444–1463.

4. Thomas, D.W.; Blondel, J.; Perret, P.; Lambrechts, M.M.; Speakman, J.R. 2001. Energetic and fitness costs of mismatching resource supply and demand in seasonally breeding birds. *Science* 291, 2598–2600.

5. Smith, J. J. and H. W. Li 1983. Energetic factors influencing foraging tactics in juvenile steelhead trout, Salmo gairdneri. P. 173–180. In D. G. Linquist, G. S. Helfman and Ju. A. Ward. (eds). Predators and prey in fishes. Dr. W. Junk Publishers, The Hague Netherlands.

6. Curzon, F. L. and Ahlborn, B. 1975. Efficiency of a carnot engine at maximum power output. *Am.J.Phys.* *43*, 22–24.

7. Kleiber, M. 1962. The fire of life: An introduction to animal energetics. John Wiley, New York.

8. Morowitz, H. 1961. Energy flow in biology. John Wiley and Sons. New York.

9. Odum, H.T. 1972. Environment, power and society. Wiley-Interscience New York.

10. Lee, R.1969 !Kung bushmen subsistence: an input-output analysis. p 47–79 In Vayda, A. P., Ed. Environment and cultural behavior; ecological studies in cultural anthropology. Published for American Museum of Natural History [by] Natural History Press: Garden City, N.Y.

11. Rappaport, Roy A. 1967. Pigs for the ancestors. Yale University Press. New Haven.

12. Angel, J.L. 1975. Paleoecology, paleodemography and health. In: Polgar S, ed. Population Ecology and Social Evolution. The Hague: Mouton: 1975; p. 667–679.

13. Cottrell, F. 1955. Energy and society. McGraw Hill, N.Y.

14. Sundberg, U. and C. R. Silversides. 1988. Operational efficiency in forestry. Kluwer, Dordrecht, The Netherlands.

15. I thank Joe Tainter for bringing this quote to my attention. Leslie White was a great believer in the importance of energy for human affairs and is well worth reading today.

16. Ponting, C. 1991. A Green History of the World. The environment and the collapse of great civilizations. Sinclair Stevenson. London.

17. Smil, V. In Energy in world history, Westview Press: Boulder, 1994.

18. Perlin, J. 1989. A forest journey : the role of wood in the development of civilization, W.W. Norton: New York, 1989.

19. Tainter, J.A. In The collapse of complex societies, Cambridge University Press: Cambridge, Cambridgeshire. New York, 1988.

20. Kummel, R. 2011. The second law of economics. Energy, entropy and the origins of wealth. Springer, New York.

The Basic Science Needed to Understand the Relation of Energy to Economics

11

This chapter is designed to provide in a very basic way enough science so that it is possible for the reader who has not had an extensive background in science, or who simply wants a review, to do so relatively easily. The contents of this chapter are divided into five main sections: understanding nature, the scientific method, the physical world, the biological world, and the integrative science of ecology.

Understanding Nature

What Is Nature?

We start by considering what is nature and the natural world. In the most common view nature is all of the world that is not explicitly human or human-dominated. In the lovely Rocky Mountain rural environment where this is being written it seems obvious what nature is: go to one of the National Parks, get on a hiking trail, and hike until there seems to be little human influence. Nature clearly is the rocks, streams, clouds, and animals. But here too it may be difficult to find pure nature, as there is usually a trail under your feet maintained by other human hikers and the Park Service, many of the plants, including lovely flowers, are introduced pest species, all of the plants are growing in an environment influenced by carbon dioxide increased by human activity, the glaciers we may look at are shrinking, and the rainbow, brown, or brook trout you may see or catch were stocked from original populations in British Columbia, Europe, or the Eastern United States. On the other hand, humans are products of natural selection in natural environments, are animals just as much as deer or trout, and are limited in as many ways by their own genetic and physiological capacities as are the wild plants and animals. Humans, like other animals, can die from too much heat or cold, and they need water nearly daily and food regularly, or they die. But humans are different from most other animals in that they can modify their environment significantly. In addition humans can adapt rapidly through cultural evolution. For the purposes of this book we do not get very concerned about the nuances and usually say that although humans are derived from and still are a part of nature, *culture* is that which is human-dominated and *nature* is that which is not. Nature is also the natural forces that constrain all these things and allow them to operate. Humans of course have always sought to, indeed needed to, exploit nature for their own survival and, often, for the production of wealth. In order to do this it was necessary to understand nature to some degree. So how have humans gone about understanding nature?

C.A.S. Hall and K. Klitgaard, *Energy and the Wealth of Nations: Understanding the Biophysical Economy*,
DOI 10.1007/978-1-4419-9398-4_11, © Springer Science+Business Media, LLC 2012

Human Explanation of Nature

Human existence has always been fraught with uncertainty and with great difficulty in being able to understand and predict events. This has been especially true with respect to our economic lives. Early humans understood nature well enough to gather the plants and hunt the animals that were necessary for them to eat, to predict usual seasonal patterns of plant growth and animal migrations, and early farmers certainly understood a lot about plants, soils, water, manure, and so on. But humans have always sought more cosmic or at least comprehensive explanations for the natural events around them, and for more power in predicting or influencing whether a particular venture would be successful. Early Greeks and Romans, and indeed most prescientific peoples, believed that a god or whole series of gods controlled the day-to-day events in their lives, including the weather, how well their crops grew, and so on. Very often the ancients would make some sort of a sacrifice (frequently human) as an investment to please the gods and to help ensure the success of a planting, a military campaign, or whatever. Similar practices seemed to be characteristic of many other cultures around the world.

These practices gave humans a sense that there was something they can do to influence important events in their lives. But how do we know whether these various approaches, or any others, work at a rate any better than random? In other words, nearly any human endeavor will always have some chance of succeeding and some chance of failing, independent of any divine, governmental, or policy intervention, or even, perhaps, whether the endeavor itself is a particularly good idea. How can we increase our odds of getting something right? The answer is to use the scientific method. But first we need to think a little more about why prediction can be so difficult, even with the scientific method.

Most of us have had both good and bad things happen to us, and frequently these have been beyond our control. Why should the events of life be such a mixed bag of successes and failures? Is it just the random or at least unpredictable nature of the universe? Perhaps it is because natural selection itself must be based on both failures and

successes occurring. In other words evolution must have both successes to move genes forward in time and failures to help generate the most fit. This was obvious to Charles Darwin [1]. But how can we determine when something good happened as a result of our good decisions or actions versus by chance alone? This is where science comes in, for it can help us to determine whether something really works, or works just by chance alone. Certainly science cannot resolve all issues; for example, science may have little to say about what values should be pursued by a person or a nation (although it can help in understanding the effects of implementing certain values), but we do believe that the domain of science can and should be expanded, and this includes into economics and indeed the general understanding of our lives.

Cause and Effect

Normally in science we seek *cause* and an *effect* and reasons for their linkages. So if we observe an effect, such as an apple falling from a tree, we ask, as did the great early physicist Isaac Newton, "Why"? Newton determined that it was the attraction of the Earth to the apple, and the apple to the Earth, that caused this to happen, and expressed this idea in beautiful and elegantly simple mathematics: the force between two objects was proportional to the product of their masses divided by the square of the distance between them. This simple law, which works equally for molecules and for the sun and planets in our solar system, has been verified again and again by others.

We say that that the force is the *independent* variable, that it exists whether the apple falls or not, and the falling of the apple is the *dependent* variable, that it occurs when the force is applied in the right direction and at the right distance. Likewise in economies a dependent event (say the production of some corn) will occur only if the independent variable takes place, that is, the farmer plants the seeds. Of course the corn production will take place only if other things occur too: the sun must shine to provide energy, rain needs to fall or irrigation water be provided, there must be sufficient fertilizing elements in or applied to the soil, and so on. In this case we would say that the production of the corn is a multiparametered issue: the dependent variable occurs as a result of many independent variables. The various independent variables in turn may be a consequence of other independent variables, such as climate change or a farmer's economic ability to provide fertilizer or willingness to work hard. All of these factors operating together form a *system*, a series of interconnected causes and effects. Thus unraveling economic cause and effect is not always easy. This is why we advocate in later chapters a *systems approach* to understanding real energy and economic issues. This may seem impossibly complex to the reader now, but in fact with proper training is quite manageable.

The degree to which energy studies should be based in science has rarely been questioned, as energy analysis in many respects forms the basis of science. In addition most aspects of energy seem to follow known scientific laws. An important question, however, one to which we have no easy answer, is to what degree *economics* should be a science. Although economics is usually identified as a social science, the degree to which its basic assumptions are given using, and subjected to, the scientific method is not quite so clear. Introductory economics books rarely put forth their fundamental principles about human behavior as hypotheses to be tested but as truisms to be accepted without critical reflection or empirical testing. In addition there is usually no particular effort to ask, as we do here, whether or to what degree economic principles are consistent with the basic scientific laws. The reason that these issues are important is that real economic systems must operate in the real material world where the laws of science always apply.

The Scientific Method

Formalizing Our Search for Truth Amid Uncertainty

How do humans get to know things? How can we know things for sure? The answer is partly that there is no way that we can know anything absolutely for sure, and a common aphorism is that

those people who know little often know it with certitude, whereas those who know a great deal tend to approach that knowledge with great uncertainty and humility. Thus it is true that we cannot trust finally and forever even those things derived from good science, for there may be special cases or new information that causes us to change our minds or at least to understand how what we thought was true had some limitations. For example, we once thought that matter could not be created or destroyed, although changed in form. But the great physicist Albert Einstein found that under special conditions matter could be transformed to energy according to his famous equation $E = mc^2$. In this case the advance of science told us that the earlier law of conservation of energy worked under usual conditions, but that there are exceptions. This perspective enriches our understanding of the law of conservation of matter, which is now considered the law of the conservation of matter and energy. Angier [2] has written a useful book that summarizes much of what we have learned from the scientific method, and how we have learned it, in a very accessible style.

We believe very strongly that, even if there are many important exceptions to the power of science, if there is any knowledge we can trust it must be derived from, or at least be consistent with, science and the scientific method. We believe that because the economy must operate in the real world it cannot operate as if it were a perpetual motion machine, which in fact is the most common way of representing economic systems in introductory economics textbooks. More generally there is a great deal of information derived in the natural science disciplines that could be of great value for understanding actual economies, but that this information is rarely if ever put into economics textbooks. In addition, as we stated in the introduction, we do not believe that the education of our young people should be compartmentalized so that one learns natural sciences only in chemistry, physics, geology, or biology. At the same time human activity need not be missing from natural sciences.

But what is science? How does "scientific truth" agree with or differ from other kinds of truths, including logical truths, economic truths, religious truths, and so on. Before we give more economics we focus on more science, going beyond the basic energy needed to understand economics by developing some basic science needed to understand both energy and economics.

The Need for Science to Understand How Economies Work

The more we can increase our scientific understanding of the world the better we should be able to understand what good economics is, and should be. This follows in the same way that our ability to do medicine is improved as we better understand the human body, the environment of humans, the technology of disease prevention and control, and the social interactions between healthcare providers and sick people. In other words we believe in a comprehensive systems approach for all but the simplest problems. Our list of the most important things you need to learn about science includes especially the scientific method and the most basic concepts pertaining to nature including matter, energy, life, and the fundamental interactions of all of these within the biosphere. Economics, if it is to be a real science, must be consistent with, and constrained by, these scientific principles, for we know of no exceptions to them. Humans can want to do many things, but they are able to do only what is possible within the laws of nature and the resources actually available, and if these concepts are not understood human endeavors are apt to backfire (some might say continue to backfire).

Steps in the Scientific Method

The scientific method is usually taught to high school and college students as a series of experiments, with hypotheses, tests, and controls, that a scientist follows in the process of gaining new knowledge. The formalized procedures of the scientific method usually include observation of phenomena, the formation of hypotheses that are thought to explain those phenomena, and the rejection of those hypotheses that are not supported by appropriate experimentation. Usually the proce-

dure requires a "test" and a "control," identical in all respects except for the one factor that is being tested. Thus to test the hypothesis that phosphorus is needed for plant growth one might grow two plants in pots with the soil in the plots being identical except that one contains phosphorus and one does not. The use of a control is usually critically important to identify the causative agent.

In fact the process of science tends to be much more complex and messy, with many different pathways to new scientific understanding. Neither Isaac Newton nor Charles Darwin, probably the two most important and creative scientists who ever lived, particularly followed the scientific method as mentioned above. Rather they were extremely astute observers and thinkers about what might be behind what they observed. Today the fundamental criteria by which science is judged are that the mechanisms are consistent with known science and also that the results generated by the science *works*, "works" being defined as generating predictable results that are repeatable by others. For example, when we sent men to the moon we were able to aim the space capsule based simply on Newton's laws of motion, laws that worked so well that not even small mid-course corrections were necessary even though we had never tested them previously outside of the Earth's local environment. Surely we would like to have such predictive power in economics! A problem, however, as any good economist or scientist will tell you, is that it is difficult to make predictions in a multiparametered world, that is, in a situation where many factors in addition to the one you are interested in or have control over might influence the results. Because real economies have many inputs and many outputs determining which factor or factors may be most important can be quite difficult (but see the next chapter where we show how this can be done well, if not perfectly).

This is a problem faced by much of the rest of science too, and often it has been overcome with the help of statistical analysis designed for that purpose. But first we need to think more basically about how we seek and sometimes find truth using science. As defined by the scientific methodologist Glymore [3], "science" is that field of intellectual inquiry that is amenable to the scientific method. Exactly what constitutes the scientific method is certainly debatable. Most practicing scientists would agree that most good scientists, natural and social, strive for *rigor*. Generally rigor means intellectually defensible, using conceptual models that capture both reality itself and the mechanisms that determine the relation between cause and effect. It is often assumed that mathematical rigor means scientific rigor, but as we show, this is often not the case.

Within the natural sciences rigor generally means, at a minimum, that the concept, descriptor, or model used: (1) is explicitly and unambiguously defined, (2) is consistent with *first principles* (i.e., things we know to always hold), (3) have been tested with adequate controls using some form of the scientific method (where that is possible) and has survived that testing, (4) explains an appropriate and nontrivial set of observed phenomena well and, perhaps most important, and (5) is repeatable by others who also follow the above rules. If all of these criteria and, as appropriate, others, are not met then we have to consider the theory or approach in question as a theory, or an hypothesis, or a myth, or something else, but not yet in any sense a scientific law or even a scientifically supported concept or theory. Probably the strongest criterion that marks something as science is that the observation or experiment is repeatable by others who follow the appropriate directions of the person promoting the hypothesis, and who usually are trying to get it to fail. Although it is very hard to say something is unequivocally correct using the scientific method, and some philosophers of science (most notably Popper [4]) make the point that we can only fail to falsify an hypothesis, the true power of science comes from a theory's ability to withstand very explicit attempts to falsify it.

Many very exciting new concepts in natural science have fallen when they have failed to satisfy all the points given above. On the other hand there are some extremely powerful scientific theories, such as plate tectonics and natural selection, that explain a great many observations but that are amenable to experimentation in only a limited way. So it is not always required to meet all of the criteria listed above, but if they do not, then we have some very careful explaining to do. For example, Charles Darwin thought that we would never see natural selection in action, or be able to test it explicitly, because he thought the time scales were far too long and the experimental manipulation extremely difficult, in part due to the complex, multiparametered reality of nature. Nevertheless all other information – such as the fossil record – was so convincing that nearly all biologists came to accept Darwin's theory even without experimental verification. Recently, however, biologists such as Grant [5] and Schluter [6] have devised very clever observations and even experiments that have allowed us to observe and even manipulate natural selection, and it works essentially exactly as Darwin had hypothesized. So with very careful attention to scientific methodology and to the system in question it is possible to undertake experiments to test our hypotheses even when people originally thought it impossible. An amazing thing with all of our new molecular biology is that as we learn much more about the mechanisms of how life works we confirm that in fact nature behaves very much according to the basic principles that Darwin put down 150 years ago.

We next look at some fundamental physical and biological laws and principles that have been derived by using the scientific method that we believe are most solid and also most important for a good understanding of real economies and of biophysical economics. Most fundamentally we ask, "How does it work?" Whatever our answer, it must be consistent with science and derived by the scientific method, otherwise we cannot accept its validity.

The Physical World

The two fundamental divisions of the physical world are energy and materials. Thus we start our tour of scientific knowledge with a look at energy. Later we look at materials.

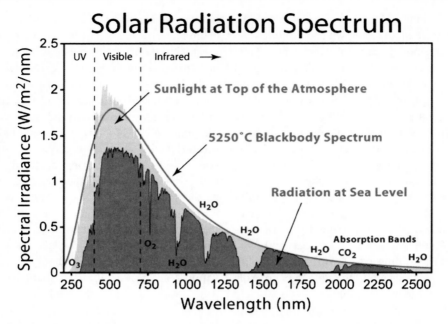

Fig. 11.1 Distribution of sunlight as a function of wavelength. (**a**) At top of atmosphere (**b**) on the Earth's surface. (*Source*: Wikimedia Commons)

Energy Sources

The principal sources of energy for the Earth are the sun, the movements of the sun and the moon relative to the Earth, and the radioactive decay within the Earth. The movements of the sun and the moon cause tides, and possibly some large-scale movements of portions of the Earth, The decay of radioactive elements within the Earth (plus residual heat from early Earth history) causes the interior of the Earth to be warmer than the surface. These factors also cause volcanoes and continental drift. Essentially all other energy, including wind, oil, gas, and coal, our food, and that of all nature, comes directly or indirectly from the sun, which is a thermonuclear furnace fed by hydrogen to make helium. We do not know exactly what the energy is that comes from the sun, but the effects are obvious. We know that it is electromagnetic energy and we know the spectrum of wavelengths, as per Fig. 11.1. Scientists have more or less settled on calling sunlight "photon flux" and the amount of it "photon flux density." Sunlight tends to be of relatively short wavelength (Fig. 11.1) and because of this has

very high energy and can do a great deal of work. The solar constant, the amount of sunlight received from the sun at the top of the atmosphere, is 1.3666 W/m^2 of cross-sectional surface area perpendicular to the sun, equal to 4.9 KJ/m^2/h. About one quarter of this, on average, gets transferred through the atmosphere to the Earth's surface (or about half directly under the sun). Figure 11.2 gives the disposition of this incoming solar energy. Some of it is immediately reradiated to space, some evaporates water, and the majority is turned into longer wavelength, less energetic waves that we call sensible (i.e., we can sense it) heat. This transformation is very obvious when you walk barefoot on a black surface when the sun is bright. Sunlight has a broad spectral distribution, meaning that when separated by a prism it has many different colors. Plants absorb and use for photosynthesis red and blue light but not green, and hence reflect green. The sky is blue because the small particles suspended in the atmosphere are at roughly the same size as the blue wavelength, so the other colors go straight through the atmosphere while some of the blue light is reflected (scattered) from the atmosphere to your eyes.

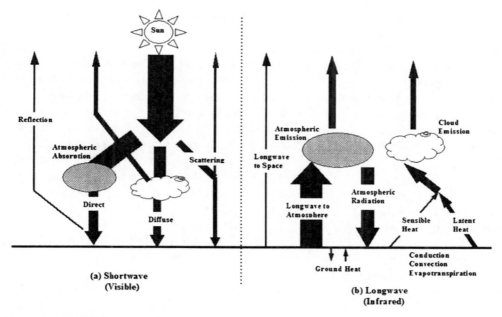

Fig. 11.2 Disposition of incoming solar radiation (Source: Amy Chen)

When the solar energy strikes the Earth's surface the portion that is not reflected does considerable work. We can feel the effects in the heating of dark surfaces. The largest amount of work that sunlight does on Earth is to evaporate water. Wind and more generally weather are caused by the uneven heating of the Earth's surface by the sun. Most important, the sun heats the Earth more at the equator than towards the poles because the land is perpendicular to the photon flux. The greater amount of heat at the equator is moved north and south in regular patterns by oceanic (e.g., Gulf Stream) and especially atmospheric (e.g., Hadley cell) systems. The intense heating of the Earth's surface on the equator creates strong upward movements of air masses there that cool and generate clouds and rain as they cool. These are readily seen in satellite pictures of the Earth as a band of clouds in the vicinity of the equator. The continual "piling up" of air at about 10 km above the equator pushes the air masses north and south near the top of the atmosphere. When these air masses reach about 30° north and south they descend. As they descend they generate a high-pressure mass of air that pushes surface air masses north and south towards

the low pressure area at ground level near the equator and the loop is finished. The Coriolis effect gives the winds the appearance of being shifted to the right in the northern hemisphere and to the left in the southern hemisphere (Fig. 10.4). The net result is the very steady trade winds of the tropics. A second loop, also driven by the high pressures of 30° north and south, cycles from about 30° to 60°, creating (with the help of the Coriolis force) the prevailing Westerlies that are familiar to those living in the temperate regions as they watch storm systems move across the land from west to east.

British meteorologist George Hadley figured out the first (equatorial) cell in 1735 which bears his name. What he did was to explain the wind patterns that savvy ship captains had known since the time of Columbus: use the aptly named trade winds for moving from Europe to the Americas and the Westerlies farther north to go from the Americas to Europe, while avoiding, where possible, the doldrums on the equator and the horse latitudes at 30° where air masses move vertically rather than horizontally. This is an early example of where scientific knowledge was of great assistance to the economic situation of those who understood it.

Basic Thermodynamics

The basic laws of thermodynamics were given in the previous chapter. Their importance includes the concept that although material can be recycled energy cannot. Once we have used energy it is, essentially, gone forever as a useful resource. This has enormous implications as civilization plows through its remaining resources of fossil fuels.

Entropy and Its Relation to Human Economies

When we think about energy it is normally from the perspective of our own personal ability to get something done, go somewhere, or keep warm in winter or cool in summer. But the reach and importance of energy is far, far more pervasive principally because of *entropy*, which we have covered in the previous chapter. The things bought and sold in economies, cars, houses, food, are bits of *negentropy*, or negative entropy, something that is highly organized or structures, something extremely unlikely by itself. A nation, civilization, or economy must constantly invest money and energy into maintenance, otherwise buildings, bridges, and even entire civilizations will collapse.

Most civilizations that have lost their main energy supplies have collapsed, as Tainter [7] and Diamond [8] have elegantly examined. Mexico is still rich in oil even as its main fields decline, and uses much of it to maintain the 20 million people concentrated in Mexico City. The need for a continual input of energy to that city was once made clear to us when we were caught in a 10-mile long traffic jam of bumper-to-bumper trucks that bring food and fuel into Mexico City every night. Mexico is filled with the ruins of enormous earlier cities and civilizations that, by some accounts, grew beyond their capacity to provide the energy resources that their large populations needed. Will the same fate befall Mexico City when oil becomes less abundant, as inevitably it will and which has already begun?

A Little Geology of Importance to Economics

We now shift our focus to materials. Economics is about goods and services. All goods are derived in some way from nature (including the soil and atmosphere), so it is useful to have information about their origins. Services too are generally derived from nature, for example, the fuel that runs a transportation service or the metals in a bus. Most of the materials that we use in our economic lives come from either plants (i.e., agriculture: food, some chemical feedstocks), forests (paper, lumber), or from the ground (rock, sand, cement, minerals such as iron, copper and aluminum, fossil fuels including coal, natural gas and oil, and groundwater). Most plastics are derived from fossil fuels, especially natural gas. The conditions under which these materials are found are normally considered the province of agronomy, forestry, or geology.

The first important geological fact about the Earth is that it is very old, roughly 4.5 billion years old. Over this very long time period mountains were thrown up by volcanic or tectonic activity, continents drifted across the ocean and life evolved, and in the process changed the Earth itself. Some kind of simple life has existed for about half to three quarters of that time, but fish, for example, and primitive life on land have existed for only about 500 million years [9]. Humans as a recognizable species have been around for about one million years, less than one thousandth of the time that the Earth has had life. It is usually thought that very large asteroids from outer space hit the Earth every few hundred million years and change things very much, for example, by eliminating dinosaurs and opening up the environment for the evolution of mammals.

There are three basic types of rocks: igneous (formed by volcanic activity), sedimentary (formed by deposition of sand, silt, or marine skeletons on the bottom of the sea or large lakes), and metamorphic, which are either of the former that have been transformed by crustal movements and pressures. Sedimentary rocks are further divided into sandstones, shales, and limestone, formed specifically from sand, silt, and marine organisms. In areas once covered by the ocean, such as central New York state, there are often alternating layers of sandstone and shale, representing successive geological eras. Why is there sometimes shale and sometimes sandstone and sometime limestone, sometimes in alternating bands? Once these sediments were found at differing distances from the source materials on the continental shelves. Because sand drops out of moving water relatively rapidly, the presence of sandstone implies that the source of the sediments was originally not very far. The finer silt that constitutes the shales could travel much farther from their continental origin before falling out, and limestone represents the remains of active populations of animals that made their shells out of calcium carbonate. Each of these materials can contain a certain amount of organic material (i.e., leftover plant and animal material) that can be the basis of the formation of fossil fuels.

The earth is a very dynamic place if you think in terms of geological time, with large crustal plates moving about its surface. For example, South America is separating increasingly from Africa to which it was once joined. Centers of activity where one plate smashes into another such as along the Andes of South America are characterized by mountain chains, volcanoes, and frequent earthquakes. The continents move about in response to geologic energies (deep "hot spots") that sometimes come up in the middle of the oceans, often causing volcanoes (as in Iceland and Hawaii) and continental drift. These hot spots generate island chains such as Hawaii, where a plate drifting over a single hot spot formed the islands from volcanic activity that is still continuing. At other locations the Earth pulls apart, causing rift valleys. Good examples are found in East Africa, where there is a series of large lakes formed in basins where the land is being pulled apart. Eventually the edges will move far enough apart so that the sea will tumble in and the lakes will become inland seas. This has already happened in the Red Sea where Egypt is separating from the Arabian peninsula, and where Madagascar has separated from Africa. Another example is where Scotland has drifted away from

Norway. As we show below, these rift areas are very important for the formation of oil.

Concentration, Depletion, and the "Best First" Principle

The most important geological issue relating to economics is that the materials that economies are based on, whether those of antiquity or of today, are not found distributed randomly (as we might expect from our discussion of entropy) about the Earth but rather in various concentrations of widely different purity and quality, the most concentrated being called ores or deposits. This is because past geologic energies, including volcanism, tectonic actions, river transport, microbial actions, and other processes have tended to concentrate the different elements (and certain compounds) in particular locations where they may be orders of magnitude (i.e., factors of 10) more abundant than the general background "crustal average." Such differences have been obvious for millennia to humans who have tended to exploit, and deplete, the highest-grade materials first. The initial copper and tin deposits in Crete, one place humans began the process of mining and smelting metals, were initially at such high concentrations that the metals abundantly flowed in a pure stream out of fireside rocks. When these rocks were all depleted, humans had to invent mining and much more complex metallurgy to supply the metals. Today the earth is a very well-explored place, and with a few exceptions there have been relatively few large discoveries of very important materials for many decades. Now rich mines are only a memory and we get most of our metals either from recycling (roughly half) or from huge, relatively low-grade deposits that require enormous machines and very large quantities of energy to extract the metals from the ores.

A good example is that of copper in the United States. Over time the best grades of copper were mined first because it takes less energy (and labor and equipment and hence money) to process these materials into forms that society finds useful. For example, if you go to the end of Main

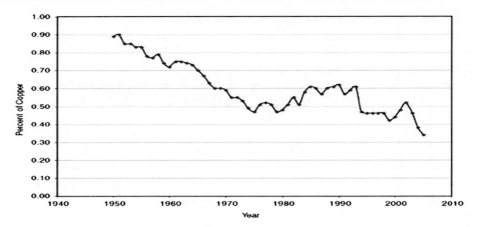

Fig. 11.3 Average grade of copper mined in the United States (Source: U.S. Bureau of Mines and the U.S. Geological Survey)

Street in Butte, Montana you will look into a hole several miles across and nearly half a mile deep. This was once a hill, and it had been called the "richest hill on earth." The hill contained copper ore that was up to 50% copper, and once the proper machinery was in place it was relatively easy and very profitable to mine that hill. Some 20 billion pounds of copper, plus gold, silver, zinc, and other minerals, were taken out of that hill. Some ancient geological processes (we are not sure exactly what, but it appears to have involved cooling of mineral-rich magma-heated water intrusions) concentrated copper there where it had lain until the miners dug it up. Now that rich copper ore is gone, and the huge hole has been slowly filling up with water which has turned to sulfuric acid because of the sulfur deposits that were associated with the copper. It is so acidic that if migrating waterfowl land on the lake in that hole they immediately die.

Today the average grade of copper ore extracted from U.S. copper mines contains about 0.4% copper [10], in other words only about 1% of what they were getting out of Butte at the turn of the last century (Fig. 11.3). Consequently some 100 times more ore has to be dug up, crushed and processed per kg of copper delivered to society than back then. Thus an important geological issue that affects economics is that, over time, the best deposits tend to be used first, so that the energy, dollar, and often environmental

cost of getting a purified product tends to increase. Of course technologies tend to improve over time, reducing costs and often energy use. Technology is in a race with depletion, sometimes one "winning," sometimes the other. In the case of copper it appears that at first the energy cost of getting a kilogram of pure copper decreased and subsequently it increased [10]. For most materials it appears that energy costs are increasing, but a much better case-by-case review is needed.

This "best first" principle is rarely mentioned in the economic literature (although it seems consistent with the law of diminishing returns). The concept also applies to many other aspects of human, and indeed other organismal, behavior. This principle has enormous economic implications as we deplete so many resources, especially as we can no longer count on more energy being available to mine ever lower-grade resources.

The Formation of Fossil Fuels

Because oil and gas are so important to our economic life and because there is so much controversy about how much is left to exploit, it is important to consider in some detail the very special circumstances that were required for their formation. Oil and gas are organic materials; they are plant and animal remains composed of mostly carbon (and also hydrogen) as is all life. (The word "organic" technically means carbon-based; organic chemistry is about the chemistry of carbon.) As life evolved a great deal of organic material was formed, most of which was oxidized relatively soon and turned back to carbon dioxide in the atmosphere, becoming available for new plant growth. But some of this organic material found its way to *anaerobic* (meaning without oxygen) basins. For example, coal was formed in great freshwater swamps in what is now Pennsylvania, Ohio, and Wyoming.

Oil was formed in two principal places: rift basins such as once existed between Scotland and Norway or Saudi Arabia, and river deltas such as off the Mississippi or Niger Rivers. Rift basins are formed when the land on one or both sides moves apart (as is the case today with East African lakes), generating deep basins called grabens, often with lakes or invading marine waters within (Fig. 11.4). Phytoplankton, tiny marine or fresh water plants, would grow in the water and fall to the bottom of some of the deep rift basins where there was no oxygen and hence little decomposition. This process is greatly assisted when the water cannot mix deeply (i.e., is *stratified* by temperature), a phenomenon familiar to many of you who have dived deeply into a lake. Hence a general requirement is that the oil-forming basins be located in the tropics or were active

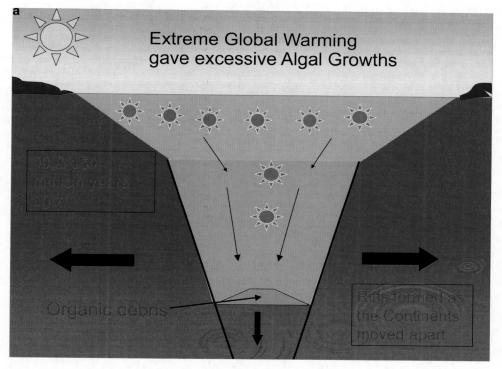

Fig. 11.4 The typical formation of oil. Oil is not formed often or in very many places, and requires very special conditions for formation. It was formed on the Earth in only two general geological times, about 90 and about 150 million years ago. In order for oil to form a series of steps must occur in sequence. (**a**) First a very deep lake or marine trench must be formed, such as when the crust moves apart forming a graben, during a period of climate warming. Phytoplankton, whose growth is encouraged by the warm conditions, sinks into the deep anaerobic waters. (**b**) Then it is necessary to have an extensive period of rains that wash sediments into the basin, covering the organic materials with thousands of meters of sediment. Then the organic material is pressure-cooked for many tens of millions of years, breaking down ("cracking") the complex molecules into simpler ones. (**c**) The relatively light hydrocarbons end up moving upwards from the source rocks. Most of it escapes to the atmosphere, but some small part is caught by impervious "trap rocks." This forms the oil and gas deposits we exploit (Source: Colin Campbell)

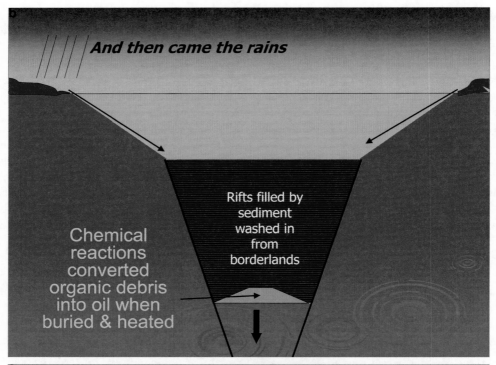

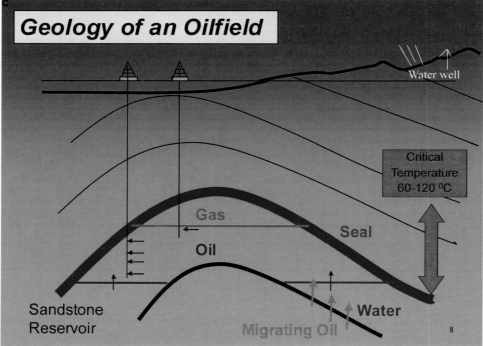

Fig. 11.4 (continued)

during periods of climate warming. Warm surface water can be mixed with deep water only with great difficulty (such as by a fierce wind). In the tropics both lakes and the ocean tend to be strongly stratified all year round, so that very often the deeper parts use up all of their oxygen and remain anaerobic.

Under extremely rare circumstances, often related to a warming climate with lots of evaporation, the sinking phytoplankton were protected from oxidation in the deep, nonmixing anaerobic bottom waters for long periods of time, thousands to many millions of years. As time went on and if the climate happened to change from dry to wet, sediments would wash down from the surrounding hills, covering the organic material with layers of sand and silt which, over time, became rock (Fig. 11.4b). If enough sedimentary rock (say 3,000–5,000 m) covered the basin the pressure would heat up the organic material, and over millions of years the ancient phytoplankton would be "pressure cooked" at about the temperature of boiling water, breaking the long plant molecules of typically hundreds of carbons tied together into shorter ones, thus forming oil and gas. The familiar word "octane" refers to oil with eight carbon atoms arranged in a ring which is the best formulation for gasoline as it does not combust too easily and hence cause preignition or knocking. Natural gas is what remains when the chains have been broken to lengths of only one or two carbon atoms. These very rare and special rocks are known in petroleum geology as *source rocks*. The oil and gas thus formed would then tend to rise upward over geological time as they are less dense than the earth's sediments within which they are found. Some small proportion, perhaps 1%, of the oil and gas migrating upward from the source rocks finds their way to particular rock formations impervious to their movement, such as salt domes or sandstone, where they are trapped. These rocks, which may be far above the source rocks, are known as *trap rocks* and are normally the locations that humans exploit (Fig. 11.4c). A good example of where all this took place was the rift valley where Scotland and England left Norway some 100–200 million years ago. The oil that we now exploit from the North Sea was created as a series of grabens were formed and flooded with water. Large phytoplankton growth in the productive water settled into deep basins, and was eventually covered by thick layers of sediments. Some of these layers, particularly those made of limestone but sometimes sandstone, formed both reservoirs and traps.

Similar burial of phytoplankton or other organic matter sometimes has taken place within and off river deltas where highly productive estuarine systems such as those associated with the Mississippi, Niger, and Orinoco Rivers generate a lot of organic material and where periodic sediment deposits covered over anaerobic basins. The general lesson from these descriptions is that the special conditions required for the creation of exploitable oil and gas fields have been quite rare in the geologic past (occurring mostly some 90 and 150 million years ago in very special and limited environments), and that the time to make oil and gas is extremely long. As a consequence significant commercially exploitable oil and gas are found in a relatively few regions of the Earth's surface. Coal, requiring similar but far less stringent conditions for its production, is much more common. Gas too is widely dispersed but the main reservoirs were relatively rare. On the other hand gas is found widely at low concentrations in "tight" shale and sandstone. Exploitation of these diffuse resources is becoming increasingly important as the large true gas fields found earlier face serious depletion. Whether these newer "unconventional" fields can maintain U.S. gas production at the present level for very long is unknown at this time.

As with copper (Fig. 11.3) and another example of the "best first" principle, humans have tended to exploit the large, high-quality and easy oil deposits first. They have exploited deeper, and deeper offshore regions, mainly off the Mississippi River, where there are more than 4,000 very expensive offshore platforms that are responsible for much of the United States' remaining oil and gas production. As this was being written there was considerable excitement about finding the new, possibly large Tiber oil field in the Gulf of Mexico. But the field is

35,000 ft (6 miles) under the Gulf of Mexico and would be extremely, perhaps prohibitively energy-intensive to develop. On the other hand the high pressures there may force the oil to the surface without expensive pumping or pressurizing. The United States found the most oil in the 1930s, and the world the most in the 1960s (Fig. 9.9). All of these factors have very important implications for EROI (Chap. 14).

A Little Chemistry of Importance to Economics

The world and everything in it, including yourself and your surroundings, is composed of chemicals. Economies generally mine or otherwise obtain source materials for chemicals (called *feedstocks*), refine or transform them, oftentimes combining them with other chemicals, and using or selling the products. The most fundamental chemicals, incapable of being transformed to other chemicals, are called elements; these include such familiar chemicals as hydrogen, oxygen, and carbon (Table 1). When two or more elements combine they generate *compounds*, which include most of the common materials of everyday life: hydrogen and oxygen combine to make water, hydrogen and carbon to natural gas or oil. The chemistry of the world and of our economy is extremely complicated, but usually it is based mostly on only about 20 or 30 elements and their compounds.

Conservation of Matter: Input Supplies

Perhaps the most important aspect of chemicals, or more generally all materials, for economics is the *law of the conservation of matter*. This law says that although matter (also called mass) can be transformed in many ways it can be neither created nor destroyed. Again there is the exception that under very special conditions (nuclear reactions) matter can be transformed to energy. There are two reasons that this law is of critical importance to economics. The first has to do with the supply of the materials required by the economy, the second (discussed below) relates to the

disposal of waste materials. In other words the goods that interest us as consumers or economists are derived from elements "borrowed" from nature. Cars are made from iron, copper, sand (for glass), natural gas (for plastics), and many other things; fish come from the sea; many houses, books and newspapers come from trees; clothes from plants, animals or, increasingly, petroleum; computers are made from plastics, copper, aluminum, gold, and silicon; and so on. Essentially every good starts as some material extracted from nature somewhere. Energy is required for the steps to make the final product.

Take, for example, plastics, a suite of materials made from hydrocarbons and chlorine. Plastics are ubiquitous and very useful, they can be formed into many shapes, and they are cheap. A common ketchup bottle may have seven layers of different kinds of plastics to protect the ketchup inside. Chemists have learned to be very clever at manipulating elements and molecules, but the carbon and hydrogen atoms in the plastic still have to start with raw materials from feedstocks, usually natural gas, or sometimes oil or coal. Increasingly plastics are recycled. As fossil fuels become more expensive the molecules can come from biomass such as from trees or crop residues.

Carbon Chemistry

Most of our food, fuels, plastics, and, many other things are carbon-based. Carbon can take many extraordinarily different forms, and can be transformed from one form to another relatively easily. Carbon may be found as carbon dioxide gas in the atmosphere; pure carbon in a pencil "lead" or a diamond; combined with hydrogen in hydrocarbons such as coal, gas, and oil; with hydrogen and oxygen in carbohydrates that includes most of the fuels that we eat; with the element calcium in limestone (from which we make cement); and so on. In general compounds with lots of hydrogen and little or no oxygen, such as hydrocarbons, are called *reduced* and serve as excellent fuels, and compounds that have a great deal of oxygen (such as CO_2) are called *oxidized* and are poor fuels. Carbohydrates are mostly reduced but slightly oxidized and hence do not make quite as good a fuel per gram as hydrocarbons. Combining oxygen with a hydrocarbon or carbohydrate releases energy that can be used to propel an athlete, an automobile, a chemical reaction, or a manufacturing operation.

You are made principally of carbon, with quite a bit of hydrogen, some oxygen and phosphorous, and more than a little nitrogen. Natural selection has chosen carbon as the basic skeleton for life because it has the possibility of combining with other atoms in four directions (i.e., it has four electrons in its outer or active ring), allowing the construction of the quite complex compounds that life requires, such as carbohydrates and fats. Because the element carbon is so closely associated with life the chemistry of carbon, living or not, is called *organic* chemistry. Carbohydrates, fats and protein are the basic biological compounds and also the basic food groups. Nitrogen too is an element of special importance. With five electrons in its outer shell and room for three bonds, it is also able to make very complex compounds that are often proteins. In its elemental form N_2, nitrogen forms about 78% of the atmosphere. In this state it is very inert, meaning that it does not react with most other elements except under very special conditions. But nitrogen can also be found combined with oxygen and with hydrogen, and in these states (nitrates and ammonia) it is extremely important for life because organisms can take the nitrogen from these compounds and (with carbon) make proteins. Proteins are important because they allow very great specificity, that is, very exact kinds of molecules. Nitrogen is critical for economies because it is the most important fertilizer used in agriculture, because plants need it to make their own proteins, and agriculture is usually one of the, or the, most important sectors in the economies of most nations.

Nitrogen Chemistry and the Haber Process

Although nitrogen is one of the most abundant elements on the Earth's surface (as N_2 in the air) it is relatively rare in its "fixed" form, that is,

combined with hydrogen or oxygen. Fixing is uncommon in nature because it takes a great deal of energy to break the three chemical bonds holding the two nitrogen atoms of N_2 together. This occurs only when great energy is applied to the atmosphere (as in a lightning bolt) or when special organisms (only certain bacteria and blue-green algae) invest a lot of their own photosynthetically derived energy into deliberately splitting the two nitrogen atoms apart so that they can get nitrogen for their own purposes, principally to make proteins. Until 1909 the major source of nitrogen for agricultural plants was from manure, and the first author's father remembers spending much of his childhood, as many did in 1920, hauling cow manure from the barn to the fields. Many of the readers of this book would be doing that too except for one great chemical discovery.

Ammonia (NH_3) is an extremely valuable chemical because of its long use in the dye industry and because it was the basis for explosives and fertilizer. Until 1908, however, ammonia was made only by natural process from certain bacteria and blue-green algae or from lightning in the atmosphere. As such its supply was limited. Although the principle by which synthetic ammonia might be made was simple and known for about 100 years, the actual process had eluded many important chemists. The equation is simply:

$$N_2 + 3H_2 \rightarrow 2NH_3$$

The N_2 is readily available as the major component of the atmosphere, although extraordinarily unreactive, and the hydrogen was readily available from coal or natural gas. After failing in several earlier attempts in 1909 the German chemist Fritz Haber discovered how to split the nitrogen molecules of the air industrially by adding a great deal of energy to N_2. He did this by heating a cylinder that was injected with air (the source of nitrogen) and natural gas (the source of hydrogen) while compressing the gases and using a special catalyst (initially osmium) [11]. The result was an output flow of ammonia (NH_3), a chemical very useful to plants and to industrial chemistry. None of Haber's university colleagues understood why he was so excitedly running around the campus shouting that he had done what no other person had done: to create "fixed" nitrogen from atmospheric nitrogen. Nor did they understand, as Haber did, why this was so important.

Haber had been assisted in this by a contract with the German Industrial firm BASF, which quickly scaled up Haber's mechanism to commercial scale and under the leadership of Carl Bosch eventually built very large factories. These factories required enormous amounts of energy to run the process. The early attempts produced some spectacular explosions however, once perfected, commercial ammonia freed all of us from carrying manure to the fields. It also had some rather different results, as industrially derived ammonium nitrate was and is the basis for gunpowder and other explosives. In 1914, at the start of World War I, The Germans had only 6 months of gunpowder, derived from Chilean guano (bird dung). Without the industrially produced gunpowder of the Haber–Bosch process the war would have ended quickly [12]. Thus the Haber–Bosch industrial fixation of gunpowder is credited with making World War I last for four additional miserable years, and, one might add, allowing World War II to be as devastating as it was. Even terrorists today blow up markets in Baghdad and buildings in Oklahoma City, and mining companies blow the tops off Appalachian mountains, using ammonium nitrate explosives.

Phosphorus

Plants need more than nitrogen fertilizer to survive and grow. Phosphorous and potassium, and in smaller quantities sulfur, molybdenum, and perhaps a dozen other chemicals are all essential plant nutrients. When the nuclear scientists Goeller and Weinberg [13] examined the entire periodic table they found that for all elements necessary to civilization at that time there was a substitute: aluminum wires could substitute for copper, energy could in effect substitute for nitrogen through the Haber process, and so on. But they found one exception: phosphorus. Phosphorus was completely necessary for plant

growth and life in general and there was no substitute. In the approximate words of geochemist Edward Deevey [14] some five decades ago, "[T]here is something peculiar about the geochemistry of the Earth today that life is so dependent upon phosphorus but it is now in such short supply." In other words it might seem that life evolved when phosphorus was more abundant. Today most phosphorus comes from mines in Florida and Morocco, and much of it goes in a one-way trip from mine to crops to animals to humans to toilets to waterways to the ocean. Thus the chemistry of phosphorus is of critical concern to modern economies because of its critical importance and nonsubstitutability for plant growth and because its main sources (in Florida and Morocco) are increasingly depleted. Thus more energy is required for fertilizer production, and because as a waste product, it causes very undesirable growths of algae in our water bodies.

Conservation of Matter: Wastes

A second implication of the law of conservation of matter is that all of the elements and materials that are extracted from the Earth must end up somewhere: as products or by-products, as recycled matter, or as wastes dumped into the environment. So if we manufacture a product, say a cleaning chemical, that material, or at least its elements, will be around indefinitely in some form or another. In the past (and still in many situations), whatever was left over after humans had used something was simply dumped into the river, into a landfill, or into the environment. When the economies of humans were based mostly on the products of nature directly, their wastes (e.g., food wastes, logging wastes) were normally the routine wastes that were part of ecosystems and could be processed by nature. Billions of years of natural selection had generated the dung beetles, bacteria, and so on to take advantage of these resources and in so doing keep things "cleaned up." Over the past few hundred years humans greatly increased the scale of everything, from agriculture to industry, through industrial and scientific processes. Humans also generated thousands of new chemicals that few organisms are able to process. The net effect was to overload many ecosystems that had previously been able to adapt to humans. For example, the large quantities of synthetic fertilizers that were generated from the Haber process and from mining phosphorus washed into rivers and lakes, where they often caused serious pollution even though these elements had always been part of nature. Although phosphorus and nitrogen are essential requirements for all plant life an excess amount generates undesirable algae growth and low oxygen conditions, a condition known as *eutrophication.*

Over time nature tends to process human-made chemicals into more innocuous forms, but there is often very serious production of pollution along the way. Humans have become much better at recycling materials in recent years and have reduced somewhat the amount of waste materials entering the environment. But recycling does not always reduce environmental impact as much as one might think as can be seen using a systems approach. For example, it would seem to be unequivocally good for the environment to recycle newspapers, to make new newspapers from old. But if newspapers are to be recycled first they need to be deinked, and then the fibers separated from the other materials. However, it takes more energy to make a ton of newspaper from recycled materials compared to virgin materials, and ultimately more wastes are produced, mostly from the old ink. This is a good example where understanding the law of conservation of mass (the materials in the ink) helps us to understand the implications of what might seem initially to be a good policy. It may still make sense to recycle newspapers, for example, to save space in landfills, and there are soy inks that are much easier to process, but it is not easy to make that judgment without undertaking a quite complex systems study. Probably the thing that makes the most sense is to reduce our use of paper as appears to be happening now as the Internet increasingly does the function once done by newspapers.

Natural and Man-Made Toxins

Toxic compounds do not come only from human activities. The natural world is full of complex chemical compounds, some of which are very toxic. Many natural compounds are designed to be toxic. All plant materials represent a potential food resource for a whole plethora of viruses, bacteria, insects, and grazers such as deer. One response has been for plants to develop over time various chemical defenses to make themselves unpalatable or even to kill their potential consumers. Familiar examples include mustard oils, caffeine, turpentine, and the alkaloid *Tetrahydracannabinol.* Although small amounts of these materials may make interesting dietary supplements, a diet composed of only one or more of them would kill us. That is a problem the insects face when they alight on, say, a mustard plant. Not surprisingly, most insects choose to go somewhere else for lunch and the plant is protected. Animals in turn have developed kidneys and livers to detoxify many of these chemicals, so that they can eat some of the material. Over evolutionary time there is a sort of cat and mouse game of defense and offense, as plants become more toxic and animals develop better ways to avoid or detoxify them.

Humans have changed the world around them in many ways through their understanding and application of industrial chemistry. One example is DDT, the first synthetic pesticide. DDT, developed in World War II, was considered a godsend to our soldiers, for it was cheap, nontoxic to humans, and eliminated many harmful and irritating pests such as body lice, with a single simple dusting. Soon it was used on agricultural crops with similar spectacular results in reducing the losses to insects. DDT seemed to be too good to be true, and it was. Rachel Carson, a marine biologist and gifted writer, published one of the most important books of all time, *Silent Spring* [15], which documented the very large impact that DDT had on bird reproduction. Her book launched the environmental movement and for the first time suspicion that not all new inventions, nor progress itself, were necessarily desirable.

DDT was especially a problem because it did not break down in nature: put it in the environment and it stays there, cycling through food chains and becoming concentrated as one organism ate another. The case against DDT was made further when it was discovered that the insects had become not only resistant to DDT but that some even required it for their survival! Natural selection can be that powerful and that fast! Chemists have responded to these problems by developing new pesticides that, although often more toxic directly to humans, break down to relatively harmless compounds in a matter of weeks. This appears to have solved the long-term toxicity problems, as long as these new chemicals are used! But the pests continue to evolve and over time the pesticides lose their effectiveness. Agronomist David Pimentel argues that even as we use far more pesticides we still lose about the same proportion of our food to pests that we did in the past, before pesticides.

Probably the most important pollutants worldwide, quantitatively, are the various carbonaceous waste products, including especially the fecal wastes of humans and domestic animals. This is a natural process but humans and their cities have completely changed the scale. Most natural bodies of water can handle moderate amounts of carbonaceous wastes through oxidation, changing the carbon materials into relatively harmless compounds such as water and CO_2. The problem occurs when too much polluting material is added. The oxidizing capacity of the water bodies are overwhelmed and all or most of the oxygen is used up, resulting in bad smells, fish deaths, and a generally degraded water. This is somewhat similar to what happens when too much phosphorus is added to water bodies. Sewage treatment plants and reformulating phosphorus-containing detergents can reduce impacts and are good examples as to how it is possible to successfully resolve serious externalities through good chemistry, good engineering, good economics, and especially good public policy implementation. But treating sewage uses a considerable amount of energy. At the base of it all is that the growing human populations and their growing use of materials continues to increase the pollution of the Earth as a whole.

Chemistry and Physics

Chemistry is usually considered independent from physics, but in fact the two interact in many many ways, and the disciplines were united by quantum mechanics. For example, essentially any chemical reaction is accelerated by increasing the temperature. Getting food particles to become unstuck from dishes requires work that occurs because you add physical energy by your scrubbing actions and by the chemical reactions in which you emulsify the food particles in a soap or detergent. Using hot water to clean the dishes adds additional energy to the process and accelerates the cleaning. Or leaving the dishes in the sink overnight after first filling them with clean water allows you to use the chemical energy of clean water (the molecules of water have positively and negatively charged ends that attract the materials stuck to your plate) to do work that you would otherwise have to do yourself. Polluted water has less energy to do that as the charged ends are already occupied. Thus clean water is economically much more valuable than soiled water because it can do more work, such as cleaning work, in industrial or other economic processes.

Another example is that chemical reactions are usually accelerated by increasing the surface to volume ratio, which is usually done by making the reactive particles smaller. This is familiar to most of us when we build a campfire: We must start with small dry twigs or even paper, with a very high surface-to-volume ratio and hence exposure to oxygen in the air, and then feed in progressively large twigs and the logs once we have a good hot bed of coals. In the process of industrial combustion some hydrocarbon (such as oil or coal) is combined with oxygen to produce energy that is then used in, say, some economic process. The efficiency with which oxygen combines with the carbon and hydrogen in the fuel depends upon how closely each oxygen molecule comes into contact with each fuel molecule.

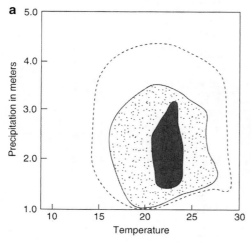
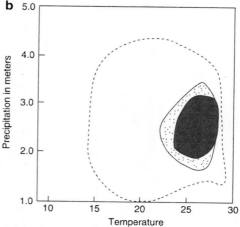

Fig. 11.5 The production of coffee and bananas in Costa Rica as a response to local variations in climate. The vertical axis is precipitation, the horizontal is temperature. Although coffee and bananas can grow essentially anywhere in Costa Rica sufficient yields of coffee (**a**) to make it economical are found only in the *central circle*, and likewise for bananas (**b**) (Source: Hall 2000)

Pulverizing coal before we burn it, so that the carbon is nearly completely oxidized, generates more energy and less pollution.

Climate and the Hydrological Cycle

A basic point of this book is that input to the economy is not simply labor and investments but also natural resources, especially energy, and a properly working environment including good soil and water, as well as a proper temperature and other attributes of climate. Climate refers to the average and normal range of temperature, rainfall, humidity, cloudiness, and so on that characterizes a spot or region of the Earth's surface. Each plant and animal species has a rather restricted range of temperatures and other environmental conditions they can withstand. There is a rather well-developed science that has undertaken considerable analysis of the response of organisms to gradients (i.e., ranges) of temperature and other factors. Figure 11.5, for example, shows the production of coffee and bananas in Costa Rica as a response to local variations in climate. What this means for economics is that each species of cultivated plant has an optimal place to

grow, and once these areas are planted it is much more difficult to make a good profit cultivating suboptimal lands. This basic idea was developed by the early classical economist David Ricardo as discussed earlier.

The climate of the Earth is extremely varied. Most obviously the tropics and subtropics are warmer, or at least they are at low elevations. More important, the temperatures there vary less over the year, and most of these areas do not freeze, a critical issue for many plant species. Less obviously there are many high elevation areas in the tropics that are very cold. In the troposphere temperatures decline by about 6.5 degrees Celsius per 1000 meters of elevation. For example, Mount Kenya and Kilimanjaro are nearly on the equator but they have, at least for now, permanent glaciers. Although the tropics at any one location tend to have very little temperature variation over the year they tend to have much greater rainfall variation, especially in the subtropics (Fig. 11.6). Temperate areas often have much more regular rainfall but greater extremes over the year in temperature. As you go toward the poles obviously it gets colder yet and the seasonal variation in temperature more extreme. Water, in general, and oceans, in particular, are much harder to heat than land. We say that water has a very large *thermal mass*. Oceans, in

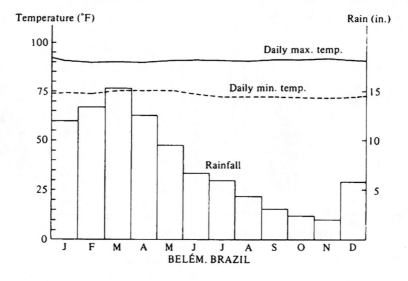

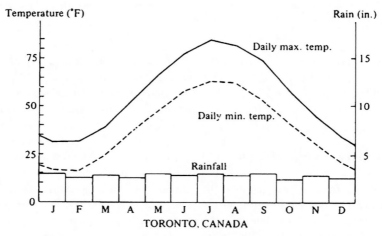

Fig. 11.6 Variability of temperature and rainfall in a typical tropical location (Belem, Brazil) and temperate location (Toronto, Canada) (Source: MacArthur 1972)

particular have a great deal of *thermal inertia* or resistance to change. Land areas that are far from oceans or great lakes tend to have *continental* climates: they get warmer in summer and colder in winter than land areas near the water which have what we call *maritime* climates. The west coast of the United States has less temperature variation than the east coast because the winds tend to blow from the west, bringing oceanic influences onto the land. Likewise areas downwind from or close to large lakes have less temperature extremes so that, for example, many wineries are associated with large lakes.

The Hydrological Cycle

The hydrological or water cycle is closely related to the climate and, like most things on this planet, it depends upon the sun. Solar energy strikes our atmosphere and about half of it reaches the Earth's surface where it is converted eventually to thermal energy. But first it does a great deal of work, the largest component of which is evaporating water. The proper functioning of the hydrological cycle is probably the most important component of the economy, although it is hardly mentioned in most economic analyses.

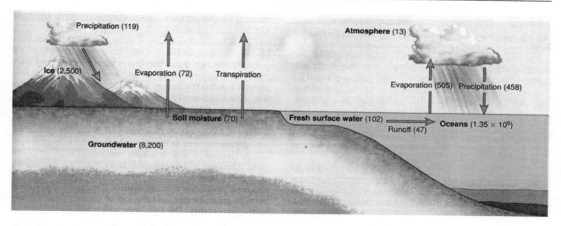

Fig. 11.7 The hydrological cycle. Water is evaporated from the land and (especially) the sea by solar energy, carried to land areas by winds where, if temperatures decline, it is deposited on land. From there it is evaporated or runs to the sea in rivers (Source: Kaufmann and Cleveland 2008)

The fundamentals of the water cycle can be seen in Fig. 11.7. Water evaporates principally from the sea, travels about the Earth as clouds and also invisibly in the atmosphere, propelled by winds, then falls onto the earth where it is held in the soil and then, if it does not evaporate, travels underground to rivers, and returns to the sea. Evaporation purifies water (because salt and pollutants essentially do not evaporate) and lifts the water into the atmosphere. This pure water then falls onto the Earth's surface, especially over land and most important, over mountains, providing soils, rivers, lakes, ecosystems, and people with clean water. The reader can get some idea as to how much work nature does for us through the hydrological cycle by considering what people living in New York City would have to do to get their clean water were it not done for them by the sun. They would have to go to the Atlantic Ocean, say at Jones Beach, dip out two pails of water, and somehow remove the salt. Probably the easiest way would be to build a fire and boil the water, collecting and condensing the steam that would be given off. Then that purified water would have to be put back into the buckets and you would have to start hiking to, and then up to the top of, the Catskill Mountains, where you would empty your buckets into the streams there. Even if all of the people that lived in New York City did this it would be but a trickle compared to what comes out of the actual rivers. Nature does a great deal of work for us

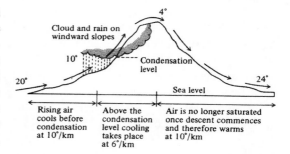

Fig. 11.8 The orographic effect. As air masses are pushed up mountains the air cools and loses energy so that it can no longer keep water molecules in suspension, and rain occurs. As the air descends on the leeward side of mountains deserts are created by the dry air (Source: MacArthur 1972)

and for our economy, however, this work rarely enters into the economist's calculations because there is no money involved.

Rain is more abundant near oceans, especially downwind from them, and at higher elevations due to the *orographic* effect. When air masses are lifted, or pushed up a mountain by winds, the air cools (Fig. 11.8). Cool air has less energy and so can do less work, including the work of keeping water molecules in suspension. The net result is that more water falls from the atmosphere in mountains, especially tall mountains. As the air masses move over the mountains and descend the other side they warm and can hold more water, especially as most of the water that once was in the air masses was lost on the windward side. Thus there tends to be a *rain shadow* downwind from mountains. Most of the rain, both in the mountains and elsewhere, falls onto the ground and filters slowly downhill underground. In general all of the soil under our feet contains water which is usually flowing very slowly towards the sea. Some soils can hold a considerable amount of water because there are considerable spaces between the soil particles. Gravel and sand hold water much better than silt or clay. Any water-holding below-ground substance is called an *aquifer*. The depth below which the soil contains water is called the *water table*, and if you dig a hole to a little below that depth you can have a *well*. When water flows naturally from where the ground level intercepts an aquifer we call it a *spring*. More generally where the surface of the ground is below the water table we find a *river*. Where rivers are dammed by some natural process, such as glacial debris, volcanic flow, or a beaver, we have a *pond* or *lake*. When this blockage is by human activity we call it a *reservoir*.

In as much as moving water has a great deal of energy, rivers can erode and hold in suspension many particles. Very fast water can move boulders, fast water can move gravel, and medium velocity water can move sand, but slowly moving water moves only silt. Rivers erode landscapes, making valleys, and depositing particles alongside the rivers when there are floods. When something is moved by a river it is called *alluvial*, and the general word for areas next to a river is called *riparian*. Steep upland areas are called *erosional* areas because the action of the river erodes away the rocks there. Where the river slows down in flatter sections, usually downstream, we find *depositional* environments. Hence riparian or streamside soil tends to be especially fertile for both natural vegetation and for agriculture because new soils are deposited frequently from floods. Small floods can occur yearly, moderate ones at decadal levels, and large ones less frequently. Rivers travel through *floodplains*, which are obvious when you look at a topographical map or a river from the right place. Rivers meander back and forth across the flood plain over time, and

often shift their position entirely. Some of our societal infrastructure is destroyed each year because people do not respect that eventually rivers will flood floodplains. Misguided federal flood insurance has encouraged people to live where they should not. An interesting comprehensive plan for reconsidering how we manage the floodplains of the Mississippi River is given in Mitsch et al. [16].

The economic value of the various parts of the hydrological cycle is immense, essentially incalculable. Most important, it provides rain for our agriculture and rivers to bring water to cities, to industries, and to irrigation. Most of the things that we make require very large amounts of water (Table 11.3). Rivers also build soil when they flood in the spring, and the reduced energy of slower flow allows suspended particles to fall out on the above-mentioned floodplains. It is no accident that human agriculture first started in such fertile riparian regions as the Tigris–Euphrates Rivers of present day Iraq, the Indus River of India, the Yangtze River of China, and the Nile River of Egypt. The yearly flooding of most natural rivers builds new fertile soil each year. When rivers are dammed there are many obvious gains (hydroelectricity, water for irrigation) but also many costs. The costs include the burying of fertile soils under the reservoir and the cessation of the soil-regenerating processes below the dam because the particles sink into the still, low-energy waters of the reservoir and are lost from the river.

Natural ecosystems, such as the forests that cover the Catskill Mountain watersheds that supply water for New York City, also do work for the human economy because they clean and purify water as well as regulate stream flow. Forests and grassland soils and aquifers absorb some of the excess of a heavy rainstorm and then release it slowly over time. Where forests are cut the water cycles are disrupted and humans must use more energy and more money to correct these problems. Rivers will always eventually go where the forces of nature dictate. Humans invest huge amounts of money and energy trying to keep rivers where they want them, but it will always be temporary. Sensible people understand what nature will do eventually and will build accordingly. Arrogant ones build where they should not, too often encouraged by flood insurance.

Humans exploit, and often overexploit, whatever water supplies they can find. When there are few people in a region, water is taken from streams or a well, but as, over time, more humans move in, rivers become polluted and wells are pumped dry. The city of Los Angeles is a great example. The early explorer John Fremont said of Southern California that it was a lovely spot but there never would be very many European-Americans living there because it was simply too dry. Nevertheless the people came. The larger rivers in Northern California were diverted through canals all the way to Southern California, allowing the great city of Los Angeles to be developed in a near-desert. Water also was diverted from the Owens River far to the East in California and eventually even the Colorado River. All of this water allowed not only the existence of Los Angeles but also much subsidized agriculture in Southern California. What is less talked about, however, is the costs of diverting that water, for example, destroying the once very large salmon fisheries of Northern California, causing San Francisco Bay to become much more saline with many adverse effects and completely drying up the Colorado River so it no longer flows to the ocean. How do we weigh the costs and the benefits? We talk about that later. What is clear is that often certain people get the benefits and others get the costs.

Humans have continued to exploit, develop, manage, pollute, and otherwise influence Earth's natural water supply. Presently water is an extremely serious issue for much of the world's population. Two especially difficult issues are human population growth where water is least available (such as in the Middle East) and the potentially disastrous effects of climate change. In general these extremely important issues, as with the economic benefits of a well-functioning hydrological cycle, are not in most economic analyses because we do not pay in our markets for the work of nature, but rather just for our cost of exploiting nature. Water is often considered a "free good" having little value for those who measure the value of things by their price.

Climate Change

Climate has a major impact on most economies, including the crops that can be grown profitably and the rate of their production, the amount of energy needed to keep people comfortable, the availability of water, and so on. Of increasing concern is the degree to which Earth's climate is changing. So the first question might be: will the climate change? That answer is easy, yes, certainly, it has always changed. The problem is that natural selection has prepared both humans and their important plants and animals for only a relatively small range from within the possible temperatures, soil moistures, water levels, and so on. Our second question is: is the Earth warming? Again here there is little disagreement among most environmental scientists: the Earth is indeed getting warmer, glaciers are melting, the polar ice appears to be shrinking, the temperature of the sea and probably the land is warming, and many areas seem to be getting drier. Our third question is: is the present climate change a function of human activities such as putting more and more carbon dioxide into the atmosphere? This turns out to be more difficult to answer than the first two questions. The initial lines are theoretical, and go back to the great Swedish chemist Svante Arrhenius who noted the property of CO_2 to absorb thermal energy in the laboratory in the 1880s. He reasoned that the burning of fossil fuel generated CO_2, therefore it would inevitably lead to a warming of the Earth's surface. But determining whether the present warming is due to greenhouse gases is more difficult, for the Earth warmed 15,000 years ago to end the ice age with no help from greenhouse gases. So the answer to this third question is probably, and in the minds of very many scientists, most certainly due to the greenhouse effect, the process where atmospheric gases, principally water vapor and carbon dioxide (CO_2), but also methane, nitrous oxides, and other gases, act as a one way "blanket." This blanket allows high-energy, short-wave radiation (i.e., photons from the sun) to penetrate the Earth's atmosphere to a greater degree than lower-energy, long-wave heat can leave. When the photons strike the Earth's surface they are transformed to heat (according to the second law of thermodynamics). Because this heat is trapped to some degree by the greenhouse blanket, the Earth warms.

There are at least four main lines of empirical argument that show the climate is changing: (1) the surface of the earth is getting warmer, as revealed by thermometers, satellite surveys of, for example, temperatures and polar icecaps, and most critically the temperatures of both deep wells and of the ocean itself (which are very hard to heat), (2) glaciers and tundra are melting all around the world, (3) many plants and animals are moving poleward and plants and rocks are appearing on the South Pole land mass that have never been previously observed by humans, and (4) the stratosphere, robbed of some of its heat from the Earth's surface, is cooling, something that was predicted by climate models before it was observed. Initially, real measurements of temperature change were difficult to interpret, and in the 1960s temperatures actually seemed to decline. What we understand now is that industrial fuel processes do at least two things to the atmosphere: they increase the CO_2 and they release dust, especially sulfate particles, which reflect sunlight and cause a cooling. But the dust settles out in roughly 2 weeks, whereas the CO_2 is cumulative: once it goes into the air it stays for a very long time. By the 1980s the CO_2 effect (in both models and reality) became more powerful than the dust cooling. Observed temperatures of the Earth have continued to set new records, more or less year after year and decade after decade.

The majority of scientists who work on this problem believe that it is the human-caused release of CO_2 and other "greenhouse gases" that are responsible for the global warming that we have observed. Starting in about the 1970s computers began to be large and fast enough to run global climate models and these showed again and again that if we kept increasing CO_2 that temperatures would rise. But because the Earth warmed considerably 12,000 years ago as we came out of the last ice age (with no help from human release of CO_2) there are some who say that the warming we are seeing today

is just a continuation of that process. Perhaps the Earth is still responding to whatever caused those changes. Important drivers in this long-term glacial cycling process are thought to be Milankovich cycles, relating to the distance and tilt of the Earth to the Sun, which tends to be repetitive on three very long time scales, changes in solar output (associated with sunspot activity), or something else. The arguments between these two groups are often extremely acrimonious and, to your somewhat neutral authors, at least often not argued in a way that allows for a point-by-point analysis of each perspective. Thus we come down on the side that the observed climate change is caused mostly by industrial activity, but acknowledge that the case is not airtight. However science is about probability and not certainty, and the evidence for human-induced climate change increases with each passing year.

How Climate Change Can Affect Human Economies

The effects of climate change are expected to be overwhelmingly negative for most economies around the world. For example, Rind [17] predicts that huge areas of the tropics will suffer from serious drying of soils. Considerable information exists that suggests that many tropical and warm-climate diseases and pests are moving northward in the United States. The Atlantic Ocean is measurably warming, and, because the heat in oceans is the source of energy that fuels hurricanes, more powerful hurricanes are thought probable. Because the winters are no longer as severe, bark beetles are moving north in the Rocky Mountains with devastating results on forests, many birds and ocean fish are moving northward, and Australia and Africa are seeing prolonged and unusual droughts. This climate change can affect entire regions and countries: entire cities and island nations, such as the Seychelles, may disappear under the waves as the sea level rises with glacial melt and thermal expansion. This would displace millions of people inland to regions already stressed by excess populations. Many of the world's great cities in South America and Asia are completely dependent upon the summer melt of glaciers to supply water during that part of the year, but glaciers and sometimes their flows are declining. For example, the glacier that supplied warm-weather water to the city of La Paz, Bolivia, finally disappeared in 2009.

These various impacts are clearly occurring now, with some severe economic impacts at this time. The economics of stopping or reversing global warming is overwhelmingly huge, but the consequences of not dealing with it are potentially more serious [18]. If the majority view is correct, then we must make enormous investments replacing carbonaceous fuels with solar or some other alternative, or suffer the consequences. If, on the other hand, the minority opinion is correct or, to further complicate matters, if there were a great increase in the number and severity of volcanoes that throw dust into the stratosphere, the climate could become cooler. The likelihood of this occurring is very small, but the impact potentially very important. Then it would be a poor use of our resources to change so quickly to expensive and intermittent solar energy sources. What a dilemma! Clearly climate is a very complex and important issue! Our view is that making our new energy investments in solar rather than fossil fuel is probably justified for other reasons too, including long-term energy availability, economic and national security issues, making jobs at home rather than abroad, making communities more self-reliant, and protecting the ocean from acidification and the land from the mercury that is released by burning coal. But we also believe that the full accounting has yet to be done, and this is a critical area for the application of biophysical economics (see Chap. 19). Of special importance is whether we can run our complex civilizations on low EROI alternative energy sources.

Natural Selection and Evolution

We now turn in our quest for the basic science needed to understand economics from the physical world to the biological world. We start with a further consideration of natural selection and evolution. All of life is the product of relentless natural selection operating over millions and even billions of years. Evolution has been a complex process that has resulted in the immense diversity of life as we know it and also our own genetic makeup. It has large elements of chance: will a meteor's path, set by some cosmic forces perhaps a million or billion years ago and light years away, intercept the Earth's orbit or not? Is that meteor large enough to cause a tsunami that wipes out half of Tokyo or an even larger one that might extinguish major components of life? This almost certainly happened some 55 million years ago when, apparently, a comet or asteroid struck the Earth, probably near Yucatan, and a large number of species went extinct. These elements of chance have operated in many many ways and often under what seem to be quite peculiar circumstances. The opposable thumb with which I carried the computer to the table and the stereoscopic vision I am reading the words on the screen are almost certainly an artifact of our ancestor's arboreal existence extending perhaps some 4–20 million years ago.

But although the events that evolution must adapt to can be random, the responses of organisms are less so. The process called adaptive convergence generates similar-appearing and similarly adapted plant and animal species in the different deserts of the world even when starting from completely different raw genetic materials. In each environment of the Earth there are problems that have to be solved, and only so many ways (e.g., thick cuticles, spiny defense of water reserves, and so on for deserts) in which that can be done well. Thus many different species "converge" in the ways that they solve the problems imposed by a particular environment, for example, the similar appearing desert plants in Southern Africa and Southern America were derived from very different genetic stocks, Euphorbs and Cactaceae, respectively. Thus part of the reason for evolutionary convergence is the relatively limited raw materials from which it makes sense to use for construction, and partly because the problems that all life must solve are similar for similar environments.

All around the world trees look basically the same: they have trunks, roots, and leaves to solve the above problems, although with somewhat different morphologies and physiologies for specific environments. This is because the problems that trees must solve in the environments where they are found are similar: they must stand up, capture sunlight, get nutrients and water, and so on. Where water is rarer, the approach of grass-like organisms works better, and so on. So although evolution is unpredictable, due to the importance of random environmental events and random mutations, to some degree it is comforting to the experienced biologist that there common ways to solve these problems. There is an ecological "theatre," that is the environmental milieu, within which the evolutionary "play" takes place. The theatre is a dynamic and changing place, requiring organisms to adapt to those changes, migrate, or die.

How does Natural Selection Work? The Ecological Theatre and the Evolutionary Play

Charles Darwin made the fundamental observation that populations of reproducing organisms tended to generate many, many more offspring than were necessary to replace the parents. There are three properties of the biotic world that necessarily lead to a world in which natural selection must operate. These three properties are: first, that there is variation among the genome of a given species; second, that these variations are to at least some degree passed from one generation to another; and third, that this variation leads to differential survival, that over time from among the variability some properties of organisms will be more likely, however slightly, to lead to organisms that are more successful at surviving and reproducing.

We examine some additional evidence for this third proposition below. But the logic of this argument is overwhelming: if these three properties of the biotic world are true then natural selection must occur. To our minds and those of most biologists the evidence is overwhelming and accumulates every year as we find more and more "missing links" in the fossil record. We watch natural selection work before our eyes as agricultural pests and human pathogens acquire resistance to our once-trusted tools of pesticides

and antibiotics in a way that is straightforwardly explained by simple Darwinian selection, and as scientists who study the design and behavior of organisms operating in nature see that those designs and behaviors consistently fit Darwinian predictions. The net result of natural selection has been evolution of life over time and the natural world as we observe it today, including ourselves.

Natural selection operates on three characteristics of an organism, its morphology (shape), physiology (function, chemical, and otherwise) and behavior. Characteristics of each of these are determined by the genetic plan inherited from the organism's parents and by the environmental conditions of its life. But the expression of genes is not perfectly straightforward, for, as Mendel showed, the expression of any characteristic may depend upon how genes from the mother and the father come together, including many issues related to dominance and recessiveness, and because many genes can determine any particular characteristic. We call the genetic makeup of an organism its *genotype* and its actual expression its *phenotype*. Phenotype, the outwardly expressed genetic makeup, is what we observe and is operated on by natural selection. Thus an important issue is that natural selection cannot operate simply and directly on genes but only more indirectly through their collective and environmentally contingent phenotypic expression. An important new discovery in biology is that we are finding that traits are not simply determined by genes for that trait, but also by other "regulator" genes that turn particular "expression" genes on and off. These genes are also subject to natural selection but the net effect is to make the possibility for more rapid evolution than we had previously thought.

Throughout evolutionary time evolution has finely tuned organisms to their environment by eliminating those genes that do not contribute to fitness: survival and reproduction. But what is fit is not a constant, for natural selection is chasing a moving target. For example, Jim Brown [19] and his students have unraveled the interaction of climate and the size of packrats in Colorado and Nevada and found that the size of the rats increased during cooler geological periods and decreased during the warmer periods as the climate cycled over long time periods. Although it is clear why it should be advantageous to be large (e.g., in competitive trials for mates) it is not so clear why it should be advantageous to be smaller. These investigators found that during warm periods a large surface-to-volume ratio, characteristic of smaller organisms, was important for dissipating heat, so large rats would get too warm when the climate was warm. This might not kill the rat directly but would, for example, make it more difficult to forage and hence to get enough food. Without a food energy surplus a female would have a much harder time getting enough energy to reproduce and provide lactation for her young.

Adaptation to Biotic Agents

Probably the biotic components of the environment, including predators, pathogens, and perhaps competitors are even more important than the biophysical components such as climate in determining the natural selection forces on an organism. These too are related to energy cost. The ultimate example is, of course, loss to predation. Other interactions are more subtle, and there is an ongoing cat and mouse game of energy losses and investments among different species throughout evolutionary time. Trees, for example, are great food for many insects. Trees can hardly hide from the insects that want to eat them, which of course would rob them of their energy reserves and of their ability to generate an energy profit that would allow for reproduction. The evolutionary response of trees has been to generate what are called secondary compounds, for example, tannins in oak leaves, that defend the trees against most insects. But there is an energetic cost for the tree to make most of these secondary compounds, so through evolutionary time there has been a tradeoff of more versus less natural pesticides. For oak trees the "correct" amount of tannins seems to be about 20% of the dry matter of the leaf.

Pathogens too impose an energy loss on organisms even when they do not kill them. A particularly nice study was done by Moret and Schmid-Hempel [20] who trained bumblebees to feed off small glass spheres, which the bees mistook for pollen. When the bees were fed this diet they would die from lack of energy in about 5 days. When the investigators infected the bees with a bumblebee pathogen the bumblebees would survive if they had real food but would die in only 3 days when fed the glass spheres. This shows that when challenged with the pathogens the bumblebees need to use their own energy reserves to fight them.

Finally competitors decrease the energy flow and sequestration in organisms either by forcing the organism to invest energy or lose exploitable resources. Commonly, they reduce the light, nutrients, or food available, or increase the energy cost of sequestering it. Examples are common in any forest. Among evergreen trees that grow next to a path or clearing those branches that are shaded die (or are thrown off) sooner than branches that are not shaded. If a branch does not pay the energy cost of its maintenance metabolism through sufficient photosynthesis it is sloughed off.

Ghosts of Natural Selection Past

Within each species there is a tradeoff between being well adapted to today's particular conditions and maintaining contingencies for more extreme but rarer events. An example is all around those who live in the more northerly latitudes. The trees that live in these locations obviously must be well adapted to the conditions that exist there today. Each adult tree produces hundreds to thousands of offspring annually of which far less than one can survive. These seedlings will tend to have some genetic variation among themselves, and if the region is a bit drier or wetter, warmer or colder, subject to more or less impact from a certain herbivore, then some genetic properties are likely to be a little more frequent in future years. There is also genetic selection for the tree to send well-equipped seeds into the world, for a young tree with large food reserves (think of an acorn or a beech nut) would, other things being equal, be more likely to "make it" in the world. But there is a cost too: heavy seeds tend not to travel far.

At the same time all of these trees genetically "remember" the ice age, when only those trees with long-range migratory capacity (e.g., smaller seeds that could travel better on the wind, or at least fall farther from the parent in a heavy wind) were able to migrate and hence survive better. This ability to migrate is well represented in present-day trees in New England, for the region was entirely under ice 12,000 years ago and no trees were found within thousands of miles. And because there were at least five major ice ages then there was a strong selection against those genetic groups that "forgot" how to migrate. There may be less selective pressure on organisms to be able to disperse their seeds widely today, but many trees retain that capacity, for once it was extremely valuable. Another example is the common salt marsh grass, *Spartina alterniflora,* found along most seacoasts in the temperate regions. Each fall this plant produces millions of seeds at great energy expense. Nevertheless the plant rarely reproduces through these seeds, but rather through the use of underground stems or rhizomes. Why then should the plant produce seeds? The answer is that the seeds are necessary to colonize new areas, and new areas were constantly being formed as the sea rose against the land following the cessation of the past glacial period. *Spartina* plants that did not produce seeds were drowned as the sea level rose, whereas those that did were able to colonize new areas as they occurred. With climate change again increasing the level of seas those "migratory" genes are likely to again be advantageous.

The Units of Selection

Natural selection works most obviously on individuals, for individuals are obviously the only ones that contribute to future generations. Perhaps it is more accurate to say that organisms that survive and leave the most surviving offspring are the ones that are more likely to be represented in the future, that is, to propel their genes into the future. But the situation is a bit more complicated for we have found increasingly that evolution works in complex ways. Richard Dawkins [21]

talks of "the selfish gene," arguing that what survives over a longer period of time is not the species (for after all most species that have been on this Earth are extinct) but rather genes. To Dawkins the genes are "selfish" in that they "use" organisms and species as their temporary receptacle to carry them forward in evolutionary time. Again it is not that they are deliberately doing this through some kind of cognitive process, but that the patterns that cause this to occur will be selected for. From this perspective genes are molecules capable of reproducing, and they exist in populations to the degree that they are successful in doing that.

There are others who argue that the units of selection are larger than the individual organism. The simplest and clearest example is that parents will often risk their lives for their offspring: this is obviously a behavior that has been strongly selected for. The late William D. Hamilton argued that there has been selection for organisms to look after relatives not their offspring, cousins, for example, because whereas an offspring has half the genes from a particular parent a cousin has one quarter, and so on. According to Hamilton an organism should be willing to take on average half the risk to help a nephew or niece than it would for its own offspring other things being equal. The idea is consistent with a Darwinian perspective of propelling one's genes into the future. A more complex situation has been argued by Robert Trivers [22]. "Reciprocal altruism" is the situation where an organism will do something that costs it energy or something else (hence reducing its own fitness) in order to assist an unrelated organism, but with the expectation that the one being helped will return the favor at some future time. A clear example of this is a herd ungulate defending the young of another unrelated animal from a predator. Again this seems to have a clear Darwinian genetic basis with direct recompense to the genes of the organism doing the activity, and in fact all may benefit with relatively small costs.

It gets more complex with interspecies interactions, but these are very common and are generally called coevolution. The idea is that a close interspecies interaction often benefits both species. The most common example is honeybees and apple trees: the bee gets its food and the apple tree gets pollination services. More complex examples exist where the role of a predator in regulating the numbers of a prey can keep the prey from overexploiting its food resources. The more we look, the more of these we find. However, this does not occur through conscious altruism on the part of an organism but apparently only via a tit for tat where the interaction, no matter how complex, is always of direct (or occasionally indirect) benefit to the organism engaging in the activity.

Finally the most complex issue is to what degree does coevolution occur at the level of an entire ecosystem. Anyone studying ecosystems is impressed with the apparent "harmony" of the system. Although there may be important fluctuations in populations or overall structure, one gets the sense that year after year the system continues to "keep itself together," adapt to, and bounce back from incoming stressors such as variable climates or storms, while maintaining and even strengthening its basic structure. Herbivores tend to keep plants in check, but not cause their extinction, dead material is degraded into soil increasing its utility for other species, nutrients are maintained within the system, predators and prey increase and decrease but not to the extremes they might be capable. To what extent is this "balance of nature" a case of many complex coevolutions versus simply "every organism for itself?"

Or, perhaps, are ecosystems regulated by the principle of Le Chatelier $(A+B \longleftrightarrow B+C)$? This principle, derived in chemistry, says simply that as a chemical (or other) reaction goes forward it will tend to be limited eventually by the depletion of the source materials that allowed it to occur in the first place. In an ecosystem plant biomass grows and grows until it has used up the nutrient inventory, and then further growth must await the death, decay, and mineralization of earlier plants. We cannot answer this question of regulation at the level of an ecosystem very well at this time but one thing is clear: a natural ecosystem is a wonderful and mostly self-regulating thing, whatever the mechanisms that control it might be. They run themselves for free off the energy of the sun. In contrast, human-dominated ecosystems, such as agriculture, require our constant intervention and management to be maintained in the form we wish.

Ecology

Both ecology and economics are derived from the Greek word *oikos,* which means "pertaining to the household." Conceptually in this book we are talking about managing both our immediate and also our larger household, and we believe that proper management makes both ecological and economic good sense in the long run. *Ecology* refers most specifically to an academic discipline, "the study of interactions among plants, animals and their physical environment within the natural or human dominated environment" or "the study of environmental systems" [23]. This is a very different from the popular definition of ecology that emphasizes the normative or value-laden "protect the environment" or "concerns for human health" perspective and includes the perspective of values. Most professional ecologists certainly do not mind the word ecology being used to refer to environmentalist issues (and they may in fact be focused professionally on protecting the environment), however, most would agree that the word "environmentalist" or "environmental" is probably a better word to use for the activist or protectionist or other values-associated perspective. This retains "ecology" for the more academic or technical one. Finally the words *environmental scientist* refers to many different people, hydrologists, atmospheric scientists, ecologists, economists, and others, who study the environment using the scientific method. It may refer to a person who is a pure scientist or one oriented towards advocacy or policy. We believe that all people involved in studying the environment and making policy judgments based on such studies should use, be conversant with, or at least be very aware of, the scientific method.

We love the concept of ecology as a basis for thinking about economics because ecology is about interactions among the many physical and biotic components of a section of the Earth's surface, often natural but also including all systems with varying degrees of human influence. Real ecosystems are constrained by the laws of nature and the energy input and material circumstances of their environment, as are, ultimately, economic systems. We believe that academic ecology has suffered somewhat by being taught too often as a biological science with a focus almost entirely on natural plants and animals. Humans tend to be ignored as a component of ecosystems except as a provider of insults. More accurately, ecology is about the science of all environmental relations and interactions, both biotic and abiotic, including human-dominated systems. Humans are dependent upon complicated interactions among many natural and economic energy and material flows. Economic systems are very similar to natural systems in that energy must be used to exploit resources from the Earth and atmosphere and to move and recycle materials through the systems to build structures and to sustain reproduction, cities, and all systems.

Economists could learn a great deal from the work of ecologists – mostly about the many ways that nature has learned to live within limits. Ecologists have to study ecology at many levels, at the level of the *individual* organism, or of a *population* of individuals, or of a *community* of different populations (i.e., of all of the species), and finally of *ecosystems*, which includes all of the living and nonliving components of a landscape or a waterscape whether natural or human-influenced. The ecosystem perspective is most useful for understanding economics. Within any of these levels ecologists tend to study the *structure* and the *function* and the *controls* of ecosystems. Structure might include the physical nature of the ecosystems (i.e., size of individual plants), the abundance of different species (or kinds of plants and animals, collectively known as the *diversity*), the number of individuals of a species (e.g., number of whitetail deer per square mile), or biomass, meaning the total living weight of a species, or of all species, again usually expressed per unit area. Function can mean the rate of energy capture from the sun, the use of energy by various components, the transfer of energy from one group to another, the decomposition rate, the way nutrients are recycled, and so on. The con-

trols can include external or climatic controls (temperature, rainfall, catastrophic events, etc.) and internal controls (self-regulating population control, nutrient limitations etc.). Ecologists have tended to focus on these four levels in studying their discipline.

An ecologist interested in individual organisms may look at how such organisms interact with their local environment, for example, at the effect of temperatures, sunlight, or plant nutrients on the growth of individual plants. In this way we find that each species tends to do more or less well (i.e., grow, be abundant, or some other factor) along gradients of conditions [24] (see Fig. 11.6). This climate-dependence has very large implications in limiting the types of organisms that can or cannot live in different regions; for example, different agricultural crops can be grown profitably only where climatic conditions are favorable. Another consequence is that each general region of the Earth has only a relatively few species (at least as a proportion of all species) that can live there. When ecosystems or microclimates are destroyed for economic gain often many species are lost because they are found nowhere else.

An ecologist using the second approach (called *population dynamics*) might look at how populations change over time and what controls might affect those changes. Such controls may be density dependent (i.e., influenced by the density of the population) and density independent (i.e., influenced principally by external factors), a debate that was important in the history of ecology. Ecologists interested in *community* ecology might examine the interactions among all the different species and populations of an ecosystem. The community approach often asks what determines the number of species in a given location, and how these different species control how that ecosystem operates. Finally ecologists interested in the *ecosystem* approach often focus on energy flow or *trophic* (i.e., food) relations. For example, they might follow the flow of energy from the sun through the food chain of an ecosystem. Primary producers (mostly green plants) are able to capture solar energy and use that to turn CO_2 and water (with a little help from mineral fertilizing elements) into biomass. Herbivores (such as deer or grasshoppers) eat plant material. Carnivores, such as wolves or an insect-eating bird, eat other animals, and top carnivores, such as tigers, eat other animals including carnivores. Detritus is dead plant or animal material, and *detritivores* eat detritus, meaning dead organic material and the microbes within it.

Ecosystems and their energy transfers are best examined from a systems perspective. At each transfer of energy from one trophic level to another about 80 or 90% of the energy is lost as heat, mostly for the energy that is required to support the living organisms and the growth of each trophic level. At least seven trophic levels are needed to concentrate the energy of tiny phytoplankton into packages such as sardines or flying fish large enough to be food for a tuna. The low efficiency of transfer from one trophic level to the next (10% or so) reflects the need for maintenance metabolism at each trophic level. Omnivores are animals such as bears and humans that eat both plant and animal material. The implications of this for economics is principally related to food chain length. Where human population densities are relatively small or agricultural production is high relative to the number of people then people can afford to eat meat at every meal. Where people are crowded and poor, or where agricultural production is low, then people must eat only plants. So, for example, although rich people in India or China may have a considerable amount of meat in their diet the many poor people there must eat principally rice or other plant materials. There would not be enough plant material to afford the 80% or 90% that would be lost as heat if the food were transformed into another trophic level, such as a goat. Energy is also often the basis for understanding more fully evolutionary issues, as it appears that essentially all aspects of natural selection are at least in part about energy costs and gains [24, 25].

Ecologists are often called upon to help understand and mitigate particular environmental problems by studying relations among the parts of an ecosystem. These have become important issues economically in many different ways. For example, too much phosphorus (from fertilizers or laundry detergent) tends to make many water bodies *eutrophic*, often a very expensive issue economi-

cally. Fossil fuels are roughly 1% sulfur, and the burning of fossil fuels creates sulfuric acid, which then creates a condition called acid rain that has killed many plants and fish. Another form of acid rain can also be generated from nitrogen from air when air is used to provide oxygen for combustion. Sometimes serious regional issues occur. For example, acid rain produced in power plants in Ohio has been implicated in fish kills, and economic losses associated with loss of tourism, and so on in the Adirondack mountains of New York State, so that the economic cost of the activity falls on others who do not take part in the economic gain from burning the fuel. This is called an *externality* in economics, a cost that is not included in the price. Fortunately it has been possible to stabilize and even reduce acid rain, but it is an expensive process.

An important applied area of ecology that we cover here is that of biodiversity losses and more generally what is called *conservation biology*. Almost all human economic activity destroys at least some natural ecosystems, and often the organisms and even species that live therein. In about 1980 a varied group of ecologists, conservationists, and naturalists came together and pooled their different approaches to what they viewed as a global crisis: the global loss of very many species or of what they called biodiversity. A great deal of effort has been put into attempting to understand and reduce this loss. Because many species are very important for humans (e.g., for food, for pollination of plants, for the many different medicines that come from tropical rain forests, and for regulatory aspects of many ecosystems), there have been many studies of the economic importance of these issues.

Ecological Stability

We end our discussion of ecology with a less precise but important aspect, that of stability and control. Undisturbed natural ecosystems tend to be broadly the same from year to year. When they are subjected to impacts from changing weather, landslides, invasions, and human impacts they have within them a resilience or ability to spring back once the impacts are relaxed. When we study the reappearance of vegetation on the slopes of Mount Saint Helen, Oregon, which was destroyed when the volcano exploded in 1980 we find that the forests are coming back relatively rapidly. Likewise, when humans cut tropical rain forests new forests will form within years or decades if given a chance (if the soil is not destroyed). Again and again we find a certain resilience in many ecosystems even as they are affected by natural or human directed processes. This is sometimes called the "balance of nature", although "balance" is not exactly right as there are many fluctuations. But the fluctuations tend to be within certain broad ranges, and ecosystems tend to return to their base conditions if they are left alone, at least on the scale of human lifetimes. One exception to this can be when new species are introduced that are very different from the original species, such as brown snakes on Guam or starlings on Hawaii. Because the original species have not encountered anything like these species the ecosystems can be heavily affected.

In contrast human societies appear much less resilient, and as Tainter [7] and Diamond [8] point out the historical and prehistorical record is full of the collapse of once proud and dominant cultures and economies. How are these different from the much more stable natural ecological systems? A great deal of this resilience, at least compared to human systems, is that the energy sources (mostly the sun but also input from other ecosystems) tend to be constant and predictable. The amount of primary productivity tends to be limited by the amount of sun and the climate, both of which tend to change little from one year or decade or even century to the next. Nutrients are potential limiters to plant growth, but inasmuch as they are tightly recycled in undisturbed ecosystems they rarely limit a natural ecosystem. Even floods and droughts tend to come and go within long-term ranges to which the ecosystems are adapted. Humans, however, tend to exploit and then overexploit the basic energy and other resources upon which they are dependent. This leads to the great question facing humanity

today: are we exploiting the Earth at a level beyond what the Earth can provide, and if so do we have the ability to be as resilient as natural systems tend to be?

Although this consideration of ecology, like the other sections in this chapter on science, has been rather brief, we think it will help the reader understand many contemporary economic issues and appreciate the need for an ecological basis for approaching economics.

Questions

1. How have humans explained and tried to predict events traditionally?
2. Are humans part of nature?
3. Explain the difference between an independent and a dependent variable.
4. What does multiparametered mean? Can you give an example?
5. Would you, or how would you, reformulate the question: "The scientific method leads to truth?"
6. Give the steps of the scientific method.
7. How do we know when science "works"?
8. What does scientific rigor mean? Can you give five characteristics of scientific rigor?
9. Is it possible to test the theory of natural selection?
10. What are the energy sources for the Earth?
11. What work does solar energy do on the Earth?
12. What is a Hadley cell? How does it work?
13. What is continental drift? Where is it occurring?
14. What is the "best first" principle?
15. What, technically, does "organic" mean?
16. Can you give the geological steps usually associated with the formation of oil?
17. What is the difference between source rocks and trap rocks?
18. What are the characteristics of the oil deposits that we have tended to find and exploit first?
19. What is the "law of the conservation of matter"?

20. What does "reduced" mean? How is that different from something that is called "oxidized"?
21. Why is it so difficult for plants to get nitrogen?
22. Who was Fritz Haber and what did he do?
23. Why is phosphorus important?
24. Explain eutrophication.
25. What is pollution?
26. Discuss some characteristics of an environment that increase the reaction rate of a chemical.
27. Why does the west coast of the United States have a more regular temperature than the east coast?
28. Draw the basics of the hydrological cycle.
29. Define and explain the reason for the orographic effect and rain shadow.
30. Give an example of how natural ecosystems provide services to cities.
31. Will the climate change? Why or why not?
32. Give four observations consistent with the idea that the world is getting warmer. What are some other processes that might cause the Earth to get cooler?
33. What are three observations that, if true, must lead to organic evolution? Do you think these apply to humans?
34. What are the three general characteristics of organisms that natural selection effects?
35. Discuss the units of selection.
36. What is the principle of Le Chatelier and how might it effect ecosystems?
37. What is the difference between the usual public use and the academic meaning of the word "ecology".
38. Why does ecology make a good basis for thinking about economics? Why might it not?
39. Discuss structure, function, and control with respect to an ecosystem.
40. What does the word trophic mean?
41. What is an externality? Can you give several examples?

References

1. Darwin, C. 1859. The origin of Species.
2. Angier, N. 2007. The canon. A whirligig tour of the beautiful basics of Science. Houghton Mifflin.
3. Glymore, C. 1980. Theory and evidence. Princeton, University Press, Princeton N.J.
4. Popper, K. (1934) Logik der Forschung, Springer. Vienna. Amplified English edition, Popper (1959)
5. Grant, P. 1986. Ecology and Evolution of Darwin's Finches. Princeton University Press.
6. Schluter, D. 2000. The ecology of adaptive radiation. Oxford University Press, Oxford, UK.
7. Tainter, J.A. 1988. The collapse of complex societies, Cambridge University Press: Cambridge, New York
8. Diamond, J. 2005. Collapse: How societies choose to fail or succeed. Viking N.Y.,
9. Falkowski, P. 2006. Evolution: Tracing oxygen's imprint on earth's metabolic evolution. Science. 2006: 311: 1724–1725
10. Hall, C.A.S., Cleveland, C.J., Kaufmann, R. 1986. Energy and resource quality : the ecology of the economic process, Wiley: New York, 1986.
11. Smil, V. 2001. Enriching the earth: Fritz haber, carl bosch and the transformation of World food production. The MIT Press, Cambridge, MA
12. Tuchman. B. 1962. The guns of August. MacMillen, N.Y.
13. Goeller, H. E. and A. M. Weinberg. 1976. The age of substitutability. Science 191: 683–689.
14. Deevey, E. S., Jr. 1970. Mineral cycles. Scientific American, September, pp. 148–158.
15. Carson, R. 1962b. Silent spring. Houghton Mifflin, Boston.
16. Mitsch, W. J., J. W. Day, J. W. Gilliam, P. M. Groffman, D. L. Hey, G. W. Randall, N. Wang. 2001. Reducing nitrogen loading to the Gulf Mexico from the Mississippi river basin: Strategies to counter a persistent ecological problem. Bioscience, 51:373–388.
17. Rind, D., R. Goldberg, J. Hansen, C. Rosenzweig, and R. Ruedy, 1990: Potential evapotranspiration and the likelihood of future drought. J. Geophys. Res., 95, 9983–10004,
18. Stern, N. 2007. The economics of climate change. The Stern review. Cambridge University Press. Cambridge
19. Smith, J., J. L. Betancourt, J. H. Brown. 1995. Evolution of body size in the woodrat over the past 25,000 years of climate change. science. 270: 2012 – 2014.
20. Moret, Y., and P. Schmid-Hempel. 2000. Survival for immunity: The price of immune system activation for bumblebee workers. Science. 290:1166 – 1168.
21. Dawkins, R. 1976. The selfish gene. Oxford University Press. N.Y.
22. Trivers, R. L. (1971) The evolution of reciprocal altruism. Quarterly Review of Biology, 46, 35–57.
23. Any basic textbook on ecology.
24. Hall C.A.S, Stanford J.A., Hauer F.R. 1992. The distribution and abundance of organisms as a consequence of energy balances along multiple environmental gradients. OIKOS 65: 377–390. 25.
25. Thomas, D.W.; Blondel, J.; Perret, P.; Lambrechts, M.M.; Speakman, J.R. 2001. Energetic and fitness costs of mismatching resource supply and demand in seasonally breeding birds. Science 291, 2598–2600.

12

The Basic Mathematics You Need to Know to Understand Economics

Most of the time economists do not "do science." Rather they tell stories dressed up in mathematics. Neoclassical economists mostly tell stories of the magic of market self regulation. Keynesian economists tell stories about how correct amounts of spending, taxing, and money creation can balance an otherwise unstable economy and lead to economic growth. If you want to understand the economist's story you should learn the requisite mathematics. Perhaps more importantly, if you want economists to listen to your story, you need to learn to present it in a language they understand and respect. If you can't express yourself mathematically then most economists will not even bother to listen to your story, no matter how compelling or well-supported by evidence. Even if presented with mathematical elegance mainstream economists may still reject your story if it conflicts too badly with theirs. But at least speaking the language of mathematics will give you a fighting chance of being listened to. Far too many economists arrogantly dismiss the analyses of other social scientists whose valuable insights are expressed primarily in words or oral histories.

Generally all scientists and economists agree that their analyses should be *rigorous*, meaning that it is thoroughly researched and done well according to the standards of those who usually undertake similar analyses. There are at least two very different types of rigor important here, however,

scientific rigor and *mathematical rigor*. There is often confusion between them. *Scientific* rigor refers to whether or not the formulation of a problem, such as in an equation, is consistent with the known laws and processes of nature, the problem is well understood, including which factors influence which other factors, and the degree to which the actual phenomenon are accurately represented by the equations used. *Mathematical* rigor usually means whether or not the equations are solved correctly and less frequently whether they are well formulated. It often also means that the problems are solved "elegantly" by the use of analytic (pencil and paper) means. While for many problems both scientific and mathematical rigor are required, we find too often that there has been too much attention paid to mathematical rigor and not enough to scientific rigor. Examples of this have been given for Ecology in Hall [1] and for economics in Chap. 5.

Economists are very committed to models, in fact often more committed to the model that to acquiring a broad-based knowledge of how the economy works. Nobel Prize winning economist Paul Krugman lamented this tendency in the profession when he said

> The economics profession went astray because economists, as a group, mistook beauty, clad in impressive-looking economics, for truth. …the central cause of the profession's failure was the desire for an all-encompassing, intellectually elegant approach that also gave the economists a chance to show off their mathematical prowess

After the financial collapse of 2008, former Federal Reserve Chair Alan Greenspan expressed

C.A.S. Hall and K. Klitgaard, *Energy and the Wealth of Nations: Understanding the Biophysical Economy*,
DOI 10.1007/978-1-4419-9398-4_12, © Springer Science+Business Media, LLC 2012

shock and dismay that the model in which he believed so strongly (market self-regulation) let him down. During the initial stages of the authors' collaboration Hall expressed to Klitgaard that "you can't get a PhD in economics without knowing how the economy works." Klitgaard responded that unfortunately you can. Much of graduate economics training now consists of graduate mathematics. Students are often awarded doctorates for developing elegant models that have little or nothing to do with how the actual economy works.

What keeps the story of mainstream economics from being scientific, despite the mathematics? We contend that in order to be considered scientific a discipline must not only follow the scientific method, but be consistent with the rest of known science. Here is our problem with most mainstream economics. Most natural sciences begin their studies with observations of nature. They then codify these observations into hypotheses, which they test statistically after gathering evidence. A scientific theory is only valid if others can reproduce the results using the same methods. Unfortunately economics suffers from two problems. To begin with, as we showed in chapters four and five the basic pre-analytical vision of the circular flow is inconsistent with the second law of thermodynamics and the law of conservation of matter. Since the economic system is isolated, without inputs and outputs, entropy must always increase. But the circular flow model leaves no room for material or thermal waste. If it did the value of the output could never equal the sum of factor prices. Secondly, beliefs about human behavior have not changed significantly from the 18th century. Students are told that humans are rational, self-interested, and acquisitive. Rarely do economists gather evidence on actual human behavior, or subject the belief to statistical testing. Rather these ideas are accepted without reservation as "maintained hypotheses." Interestingly enough, those economists that do approach human behavior as a science often win Nobel Prizes in Economics, for finding that most humans do not behave as the models say they do. We provided a brief review of Behavioral Economics in

chapter five. Unfortunately little of this cutting edge science has filtered down to the introductory level textbooks or shaped the consciousness of economics teachers. Despite these drawbacks economics is largely a matter of constructing testable hypotheses and subjecting them to the rigors of statistics. Will a tax cut increase job growth? Will deficit spending lead to inflation? Will technological change lead to lower-cost production? While these can be legitimate scientific approaches they may not be resolvable mathematically.

In order to make sense of these and countless other questions economists often construct models. We define a model as a formalization of our assumptions about a system. Formalize means to make mathematical. A system consists of inputs and outputs, boundaries, and feedbacks. Robert Costanza and other practitioners of systems dynamics say regularly that: "All models are wrong, but some models are useful." What is it that makes a model useful? Two of the most important benefits of modeling are *simplicity* and the *separation of cause from effect*. What does simplicity mean? A simple model is one in which there are few causes. If there is more than one, the causes do not interact strongly with one another. Finally simple models are linear—the relationships between cause and effect can be drawn as straight lines. Please be advised that simple does not mean easy. Neither does it mean immediately apparent by means of casual inspection. Even simple mathematical models can be quite difficult. The model for exponential population growth meets all the criteria for simplicity, yet it takes some sophisticated mathematics (logarithms and integral calculus) to solve it.

Fortunately there is a tailor-made mathematical device for separating cause from effect. It is called *the function*. Functions are relations of dependency. Changes in the effect depend upon changes in the cause. The effect is also known as the *dependent variable*, and is often symbolized by the letter y. The cause is called the *independent variable* and is often denoted by the letter x. So a typical function might read $y = f(x)$. This can be translated into English to read "y is a function of x" or "changes in x cause changes in y." Sometimes functions can have more than one cause, so a multivariate function could be written as $Z = f(x,y)$. For example, Keynesian economics posits that investment depends both on the cost of borrowing money and upon the expected profits that the investment might bring. In other words: $I = f(i, \pi_{exp})$. The greater the number of independent variables, the less simple the model is. Some models have a large number of independent variables, and these causes often interact strongly with one another. One example is the effect of weather and climate upon global financial markets. These models are known as *complex*, and are very difficult to solve. Indeed most cannot be solved without the aid of powerful computers. One problem with complex models is that they contain not only self-extinguishing negative feedbacks, but self-perpetuating positive feedbacks. Systems dominated by positive feedbacks tend to exhibit *Sensitive Dependence on Initial Conditions*. In other words, tiny differences at the beginning can turn into wild and unpredictable oscillations down the line. So most of the time economists attempt to construct simple systems that are often "solved" using analytic techniques.

Simple Models and Linear Functions

The simplest type of function is one in which there is one independent variable, and the relation is linear. Linear models are often comprised of constants, which do not change as the cause changes, and variables, which do change. Examples from economics include a demand curve. Here the willingness and ability of consumers to purchase various quantities of goods and services is hypothesized as a function of price. As prices go up the quantity demanded goes down. This could be specified as a function by saying:

$$Q_d = 100 - 2P$$

Translated into English this means that consumers would buy 100 units of a good if it were given away for free, but every dollar increase in price would lead consumers to buy two fewer

units. Keynesian macroeconomics is built upon the idea of the Consumption function.

$$C = a + bY$$

In this function the letter a represents autonomous consumption, or consumption when income equals zero. The letter b represents the Marginal Propensity to Consume, or the fraction of additional income that is spent, and Y stands for income. To begin with, the amount of consumption depends upon the amount of income, so the most general function would be: $C = f(Y)$. If the function were to be further specified as

$$C = 50 + .9Y.$$

This would mean that consumers would spend $50 even if they had no income, which they would accomplish by using up their savings or borrowing money, and when they received any extra money they would spend 90% of it. So if consumers were to receive an extra income of $1000, they would spend $950.

These are very simple models, but even the simplest model can get you a long ways towards understanding a difficult phenomenon. One way by which to understand the relation between cause and effect is to examine the rate of change in the effect with respect to the change in the cause. One can do this by calculating and interpreting the slope and the intercept of a function. The intercept is a constant value, the value of the function when the independent variable equals zero. As you probably recall the slope can be found by calculating the rise/run. Another way of expressing this would be to say a slope is the change in y divided by the change in x, or simply ($\Delta y / \Delta x$, where the Greek letter delta (Δ) represents change). The simplicity of the linear function is found in the fact that the *slope is constant* throughout the range of the function. The rate of change of the effect with respect to the change in the cause is the same at high levels of x as it is at low levels of x. To use the consumption function example, the linear slope (or marginal propensity to consume) says that the rich and the poor spend the same fraction of their additional income.

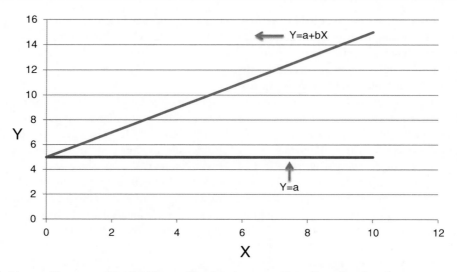

Fig. 12.1 Linear with y = a constant (in this case 5) and y = a constant plus a linear function of x

Often, these types of assumptions are not useful, because the rate of change varies as the independent variable changes. These more complicated phenomena require more difficult functions and mathematical techniques.

Economists have a special use for the slope. It is also called the *margin*. Students in introductory economics are told to associate margins with words like extra, additional, and one more. But margins are also mathematical concepts. Margins also mean "the change in the effect with respect to the change in the cause." So if consumption depends upon income then the *Marginal Propensity to Consume* is the change in consumption over the change in income. If a firm's output depends upon the number of workers they hire, given a fixed amount of equipment, then $\Delta Q/\Delta L$ is known as the *Marginal Product of Labor*. If a firm's total cost depends upon how much it produces then *Marginal Cost* = $\Delta C/\Delta Q$. Margins can always be expressed as slopes. In economics every analytical concept can be expressed by the slope of a line, and every line slope has an economic meaning! A simple linear function of one variable is seen in (Fig. 12.1)

Many economic phenomena have multiple causes. If you bought a sandwich for lunch you might have considered several variables: the price of the sandwich; the price of alternatives like soup; whether you had enough money; whether you liked what was on the menu. It is difficult to separate cause from effect when there are multiple causes. Yet the demand curve we specified above was a simple linear function of one variable. One method of simplifying is to pretend that things we know are variable are constant. These assumptions are known as *Ceteris Paribus* assumptions, where *Ceteris Paribus* is a Latin phrase that means "all other things remain constant." You might have spent many of your early days in introductory economics memorizing the lists of *Ceteris Paribus* assumptions for supply and demand. This method allows one to look for results by changing one variable at a time. Perhaps the most often made mistake in introductory economics is to confuse changes in quantity demanded (caused by a change in price) the changes in Demand caused by a change in one of the assumed constants.

Non-Linear Functions

While linear models are certainly most simple, many phenomena in nature and in the economy are simply not linear. In high school you probably learned the quadratic formula

$$Ax^2 + bx + c = 0.$$

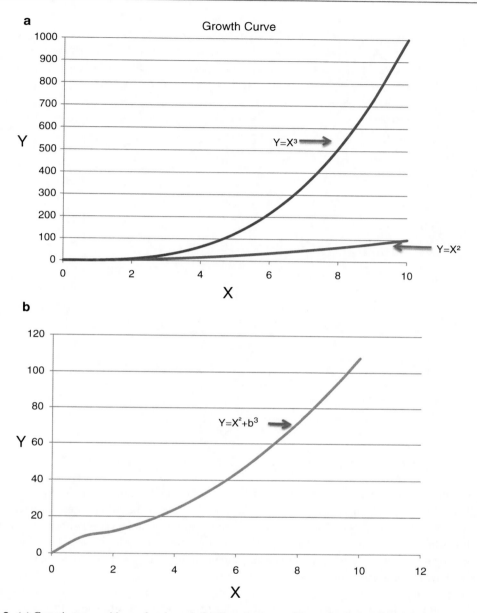

Fig. 12.2 (a) Growth curve, with $y = x^2$ and $y = x^3$. (b) Growth curve, with $y = x^2 +$ plus a cubed function

The graph of a quadratic equation is a parabola, as seen in (Fig. 12.3). Parabolas are especially useful for human communications, as the shape concentrates any rays to the center. Everything from satellite dishes for televisions to radio astronomy use the parabolic shape to their advantage.

Many curves depicting the behavior of economic firms are cubic, that is raised to the third power as shown in (Fig. 12.2b). Cubic equations in economics express mathematically the presence of diminishing marginal returns, the concept first enunciated by David Ricardo. Since the rate of change of effect with respect to cause, changes the slope will be different for different levels of x. Isaac Newton and Gottfried Leibnitz invented essentially simultaneously a special branch of mathematics

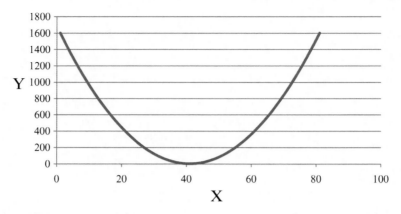

Fig. 12.3 Parabola generated from $y = ax^2 + bx + c$

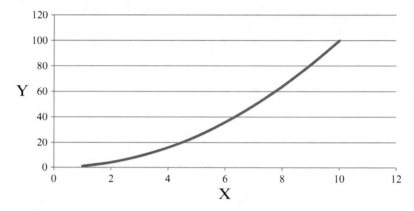

Fig. 12.4 Power curve, where $y = x^a + b$, $a = 2$, and $b = 0$

called calculus to model changing rates of change. We will come back to calculus later.

Newton, in particular, needed to understand how to do the mathematics to understand Kepler's laws about planetary motion, and invented calculus to integrate the motion of planets to show how the arcs intersected by planets during equal time intervals but at very different parts of their elliptical orbit intercepted the same area.

While there is a great deal that you can learn about calculus in many mathematics classes what you need to know about calculus for this book on economics is found in the next two paragraphs. How can that be, you say, when there are semester-long courses in calculus for economics, and in college there is calculus I, Calculus II, Calculus IV and more. Well that is true, and we do not want to discourage you from taking two or four semesters of calculus if you have not already. But

we have found again and again that even if our students have had two semesters of calculus they do not know, or at least remember, what calculus means essentially even if they were able to solve many homework questions when they took calculus. We know this by giving our upper division students who have taken calculus a simple calculus test, which is to draw the curve integrating a curve we draw on the blackboard, and then the first differential. We also ask them to write down the relation between the speedometer and odometer in a car in terms of calculus. The students get an average of about 25 percent on the test, the same as at an Ivy League University where one of us previously taught. Most of our science-based college seniors cannot answer these basic questions about calculus, although they have recently passed the course. Some of course can do that and far, far more, but they are not the average.

The students have been studying to the test, but in doing so did not learn the most fundamental aspects of calculus. So if you are in that category, here it is, in three minutes, how to think about what is most important in calculus.

Think of the speedometer and the odometer (the little mileage counter usually within the speedometer) in an automobile. In terms of calculus the odometer *integrates* the speedometer (Fig. 12.9), and the speedometer is the first *differential* of the odometer (Fig. 12.10). They are inverse functions of each other. So if you drive for one hour at 40 miles an hour and one hour at 60 miles an hour, after two hours the *integral* of your traveling will be 100 miles, that is, you will have traveled 100 miles. Likewise if you work for one hour at 10 dollars an hour and 3 hours at 12 dollars an hour at the end of 4 hours your integrated pay will be 46 dollars. The integral half of the relation is that if you have traveled 100 miles in two hours then by finding the first differential (and assuming a constant speed) your rate will have been 50 miles an hour. If you vary your speed then the first derivative of your speed, that is the rate of change of the speed, you will have a bit harder time deriving the integral, that is the rate of change. That, in a nutshell, is all that calculus is about, although the essence is that calculus does these calculations for "infinitely small" periods of time. This is not so hard to grasp, for a good odometer is integrating the speedometer at each second (or less) of time, and the speedometer is showing you the instantaneous rate of integration. Of course the math and the problems can get infinitely more complicated, but this is what is most important that you need to know about calculus for understanding the essence of biophysical economics.

Thus if you integrate your compound interest in the bank how much will your 100 dollars be worth in 5 years at ten percent interest? What will be the integrated cost of global-warming caused sea level rise over 100 years? We encourage you to learn much more about calculus, though, as the concept is really neat and useful. In practice the above examples can be solved easily in a computer using "finite difference" or time step arithmetic. But the answers should still be considered in terms of integrating something over time, and

that is what calculus is about. And remember that calculus was invented by Isaac Newton to solve a very practical question: how to understand and predict the motion of the planets around the sun. Calculus is important because it helps us focus not only on the present state of a system, but on how it is changing, and what the ultimate results of that change will be.

Exponential Functions

Some of the world's most important functions are exponential functions, and according to physicist Albert Bartlett, one of the primary problems of the American educational system is the failure to teach an appreciation of exponential growth. Exponential growth occurs when the next period's growth is added to the base, so a constant percentage is added to an ever expanding base and the growth rate increases with time. *Exponential* or geometrical growth means that the new value (Q_{new}) is added to the previously-determined independent quantity so that the independent value, and hence the dependent value, increases over time. This is the common situation of bank deposits growing, in theory, exponentially through compound interest. In that case even if the equation remains linear the solution, Q_{new}, will grow at an increasing rate over time as the amount added in to the quantity becomes more and more:

$$N_{new} = k * t * N_{t-1}$$

In this case N_{new} means the new quantity of something, for example number of people, is a variable that (usually) increases over time. k is a growth function or coefficient as before. t refers to time, and as before goes from one to two to three as the equation is solved for one, two or three (or more) years. N_{t-1} means the population number for the previous time it was solved, which is *not* the same as the original value after the first time the equation is solved. When this particular equation is solved over time, that is when we solve for many years, the results will look as in (Fig. 12.12), that is, it will be a curve increasing at an increasing rate line. In both cases we can solve these equations either analytically or more commonly numerically, that is with a computer. To do this we write an *algorithm* (a sequence of mathematical steps) and solve it *numerically*. A simple computer code to solve these equations is given as Table 12.1. Today and for several decades most complex mathematical equations are usually solved using computer models, which we introduce below.

Exponential growth catches those who expect linear growth unaware, unprepared, and too often incapable of making the necessary changes to fix the problem. The *Limits to Growth* study contains the ancient Persian legend of the Foolish Rajah. A bored potentate summoned his court magician to invent an amusement, and the result was the game of chess. The Rajah told the magician, who was also mathematically quite astute, that he could have anything he wanted. The magician asked for one grain of rice on the first square, with that quantity double on the next, and the next until the last square. How much rice was on the 64th square? This can be easily calculated using exponential growth from base 2.

Square	Grains	Exponent
1	1	2^0
2	2	2^1
3	4	2^2
4	8	2^3
5	16	2^4
-	-	-
-	-	-
64	?	2^{63}

Just how much rice is 2^{63} grains? A quick conversion to base 10 yields the approximate result of $1.8 * 10^{19}$ grains. This greatly exceeded the world rice harvest in 2010. Notice also that every doubling time includes all the increases that came before plus one: $4 = 2 + 1 + 1$; $8 = 4 + 2 + 1 + 1$. Using advanced techniques mathematicians have been able to calculate the concept of doubling time by a device known as the "Rule of 70s." The time for any quantity that is growing exponentially can be found by dividing the number 70 by the percentage growth rate (r %).

$$DT = 70/ \text{ r } \%$$

While a 3 percent annual increase in carbon dioxide emissions may not sound like much it adds up quickly. At this rate the amount of extra carbon we put in the atmosphere doubles every 20 years. And like our example of the rice, the next doubling time contains everything that came before plus one. So if the business as usual scenario continues we will have put more carbon dioxide into the atmosphere in the first two decades of the 21st century than all of humanity has done in all time. Exponential growth gives you a far different perspective. As it turns out, the growth rate of carbon emissions has decreased from about 3.5 to about 2.5 percent per year, and the primary driver has been the world recession. So what has the misery bought us? We now have an extra eight years to figure out how to move away from fossil fuels. If we do not, then climatologists predict that carbon dioxide will grow from the present level of about 390 parts per million of dry atmosphere to about 1200 ppm. Never in human history, and perhaps the history of the earth, has this occurred, and never has the increase in carbon concentration happened so rapidly. Exponential growth is powerful indeed!

The first political economist to make use of the concept of exponential or geometric growth was Thomas Robert Malthus. Malthus observed that while food production grew only arithmetically (what we call linearly) population, driven by the passion between the sexes, grew exponentially if left unchecked. While Malthus enunciated several "preventative checks" that would reduce the birth rate (moral restraint was his favorite) he thought, in the end that the "positive checks" of famine, plague, and war would be more effective once the food was divided among the smallest portions that would support life. Mathematically this would occur at the time period when the exponentially growing population curve intersected the linearly growing food production curve.

In fact since Malthus' time both the human population and food production have increased exponentially, with (arguably) food production even increasing somewhat more than the human population. The increase in food production is normally attributed to technology, which means plant breeding and better management but especially an increased use of fertilizers, tractors and so on. Essentially all of these inputs are based on an increasing use of petroleum, of course. Thus what Malthus' equations lacked was a factor for the invention and enormous expansion of petroleum-based agriculture. Of course if petroleum supplies becomes seriously constrained and good substitutes are not found then maybe in the long run Malthus' equations were right all along. While we believe that the constraints on food production Malthus envisioned may be a fact of life in the 21st century, we do not endorse Malthus' recommendation of increasing the death rate among the poor.

Exponential growth is very important in economics for at least two reasons. The first is the potential, and, in general, realized, exponential growth of the human population (and hence, in an approximate way, economic activity) increases sharply over time. The second is the exponential growth of money when invested. This concept excites many people who want to make a lot of money, for the potential is huge. A sobering reality check, however, can be found from the Bible. If we were to invest Judas' 30 pieces of silver (worth perhaps $500 today if they were the size of silver dollars) 2000 years ago at only 2 percent then they would be worth:

$$X = 500 \, e^{2000*0.02}$$

The answer to this simple equation is 208 quadrillion dollars, far more than all the money on earth now, which the World Bank estimates as 41 trillion dollars. A sobering conclusion from this is that on average investments on this Earth have yielded far less than two percent, which is less than the rate of inflation. That of course does not mean that you cannot do very well in the stock market, as long as the economy grows, anyway! But over the Earth's history investments have probably failed at least as often as not.

Another important set of functions are logarithmic functions. Logarithms are the inverses of exponential functions. If $10^2 = 100$ then $\log 10 = 2$. Scientists regularly use two forms of logarithms. Common logs are those to the base 10^2 as in the line above. However there is also a natural logarithm that is calculated to the base e, where is an irrational number approximately equal to 2.718282 (you can go on forever without a repeating pattern.) The number e possesses some very interesting mathematical properties (such as its rate of change is itself) that make it a very useful number. For example the natural logarithm (ln) of $2 = .693$. This is where the 70 in the "Rule of 70s" comes from. Logarithms are useful when one wants to compare absolute changes to relative changes, or wants to eliminate very big numbers by looking at percentage changes. Many natural processes can be modeled logarithmically. The saturation curve depicted in (Fig. 12.5) is a logarithmic curve.

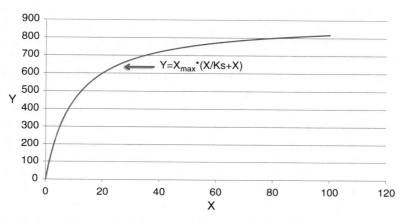

Fig. 12.5 Saturation curve (Michaelis Menten) where $y = x_{max} * (x/(Ks+x))$, where in this case x_{max} equals 900 and Ks equals 500

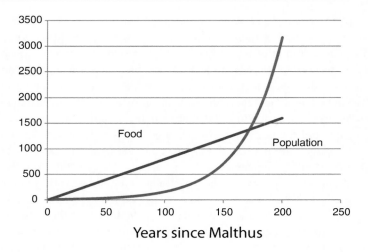

Fig. 12.6 Malthus: solutions solving Malthus' linear versus exponential equations, using approximate values for England in Malthus' time (1800)

The pH scale, which determines whether a compound is an acid or a base, is logarithmic, as is the Richter Scale which measures the intensity of the energy released from an earthquake.

Statistics

Perhaps the mathematical tool used most commonly in economics is statistics. Statistics is useful in many ways, but most important:

1. To help understand the degree of uncertainty associated with a number and
2. The degree to which different things are, or are not, related; that is, whether y is indeed a function of x and in what way.

Considering #2 above, we might want to know: is economic growth related to investments? To the number of workers? To the quantity of energy used? To technical innovations? To the exploitation of resources? Which resources? Obviously the answer is not simple. This is very difficult with economic relations. When one is trying to understand a solution of chemicals in the lab a chemist can usually undertake an experiment with and without a particular material added to the mix to get a pretty robust answer about what does or does not contribute to a particular end product. With economics it is normally a lot more difficult to undertake any such experiments because you are dealing with a system outside laboratory control and many things may be happening simultaneously. Nevertheless unraveling cause and effect

is not impossible, and is increasingly being done for some issues (see Chap. 12). So with experiments often difficult or impossible, economists often analyze existing economies over time, or compare many different economies, for example, between countries. To do this the most useful tool is generally some form of statistics.

Correlation

Probably the most basic statistical tool is *correlation*. Correlation examines whether when variable *a* gets larger does variable *b* also? Has economic growth depended upon increased energy use in the United States? In this case we might consider the economic growth the *dependent variable* and the energy use the *independent variable*, independent meaning that it changes without influence of the dependent variable. Plotting the data for 1900–1984 (Fig. 12.7) we would answer, "Yes, it appears that it does." The relatively high R^2, the most commonly used measure of goodness of fit. The high value of this coefficient of determination implies that the two are closely related, or at least tend strongly to co-occur. But if we think about it a little bit more we find at least two problems with what we have done. First of all we cannot say logically whether economic growth depends upon energy use or energy use depends upon economic growth. It is a chicken or egg question with no clear answer. What we can say is that the economic activity and the energy use are *correlated*, or co-related:

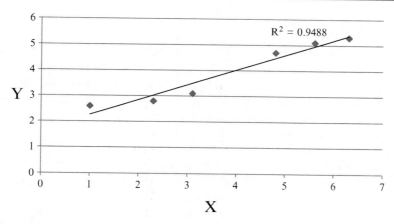

Fig. 12.7 Linear correlation

when one is high the other tends to be high and the converse. So that is a power (and a weakness) of statistical correlation. It does not tell you something that is not true, but it does not really help you as much as you would like either for determining which is the independent variable and which the dependent, or even if you are asking an appropriate question. Another problem is that if we look at the relation for 1984–2005 there appears to be considerable economic growth with relatively little increase in energy use (Fig. 1.1). This shows you another characteristic of statistics: what happened in the past may or may not continue into the future. (Or, as we believe, that we have not fully specified the problem, that is there are some indications that the inflation-corrected GDP has been exaggerated relative to the past (see "shadowstatistics.org") and, of course, the United States has outsourced a lot of its heavy industry since 1984).

A further problem dogging statistical analysis is *covariance*: two parameters may increase or decrease together but in fact have little or no relation to each other. The correlation would suggest that they are responding one to the other, but in fact both may be responding to a third. For example, both temperature and photosynthesis of plants in a field tend to increase during the first half of the day, and one might conclude that one causes the other. But in fact each is responding independently to an increase in sunlight.

The issue is further confused by multiparametered issues. Ideally we would like to be able to

study one independent variable and one dependent variable. If we are lucky we would find a straightforward relation, similar to what we see in Fig. 12.7. But what if some other factor were influencing the dependent variable? For example, we know that plants also need adequate water and nutrients. So if we want to understand or make a model of how plants grow we need to untangle the possible effects of each of these. If we are measuring the growth of a natural plant or one in an agricultural field we would need to collect considerable meteorological and soil data to unravel these effects, and we would then need to use multifactorial statistics to attempt to understand the influence of each one of them.

Econometrics

Econometrics is defined broadly as statistics on economics, but is increasingly associated with analyzing how variables change over time and also testing for causality. Most of these analyses attempt to account for statistical biases that arise when working with time-series variables. In addition several mathematical properties need to exist for the statistics to work properly. Error terms need to be distributed normally, and the error term in one time period is not supposed to be correlated to the error term in the next. Independent variables are supposed to be truly independent, and not interact with one another.

If any of these conditions hold then the statistician cannot make confident predictions or inferences because the measure of dispersion, called the variance will be too great. However, in nearly all applied work, these conditions do hold. Consequently a great deal of the econometrician's time is spent dealing heteroskedasticity, serial correlation, and multicollinearity. Today econometrics is a large academic field with its own textbooks, journals, and so on. These techniques are often very good ways of understanding what is really happening in real economies, as long as the proper factors are entered into the equations. For example, we have been very impressed with Robert Kaufmann's econometrics examining the degree to which the United States is or is not becoming less dependent upon fuels [5] and also where greenhouse gases are going [6].

Limits of Calculus

In fact most real problems cannot be solved through the use of complex mathematical analysis such as calculus [1, 2, 3, 4]. The reason is that economics is about, or should be about, many processes that are occurring simultaneously and analytic mathematics such as calculus can usually solve no more than one or two equations simultaneously (think back to your high school algebra when you were taught to solve one, then two, but not three, equations simultaneously). The problem becomes more difficult when the equations are non-linear (that is the basic factors are not represented by a straight line but rather a curved line) or when partial differential equations are required. In fact most real economies are about many non-linear things occurring and interacting simultaneously. If the price of one major commodity (say oil) changes it is likely to influence many other aspects of the economy, not just one or two. So a lot of the fancy-looking mathematics has to simplify these complex real problems into simpler "analytically-tractable" forms so that fancy solutions can be found through analytic means. The results

may look impressive (and indeed often are) but we have to ask very carefully whether the mathematical solution is in fact representative of the real world situation n or rather some simplified, and hence "analytically tractable", formulation. The answer is sometimes yes, sometimes no. The good news is that very powerful quantitative tools in computer models and even spreadsheets that allow people of good intuition but relatively modest mathematical skills to undertake extremely quantitative analysis of economies. But there are no spreadsheets that can test whether your concepts are accurately representing the phenomena analyzed. The power of mathematics (in its broad sense) is to make quantitative predictions from known (or hypothesized) relations of a system, which are usually called a model. The process of examining whether your model is a correct or at least adequate is called validation. This is the critical issue that seems to us to be lacking from most contemporary economic analysis. Then sensitivity analysis is the examination of the degree to which uncertainty in model formulation or parameterization allows one to determine how much to trust your results or reach certain conclusions. Economic models can be made much more in line with the procedures used by natural scientists by increasing the use of validation and sensitivity analysis, which are discussed further below.

An additional problem is that economic models tend to focus almost entirely on factors that are intrinsic to the system being modeled, such as interest rates, money supply and so forth and very little on what we call as modelers forcing functions, or factors that are outside the model that influence the behavior of the system. For example no economic models predicted the impacts of the oil price increases (and supply disruptions) of the 1970s or 2008 because they did not have the possibility of such "external forcing" built into the models. Similar problems occurred in modeling in ecology when populations were modeled as if only their own actions determined their abundance, rather than external factors such as climate. Thus it is important to

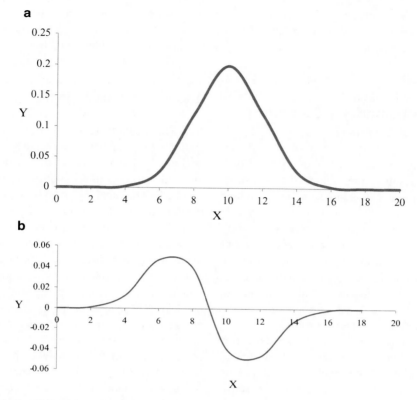

Fig. 12.8 Calculus: taking the first derivative

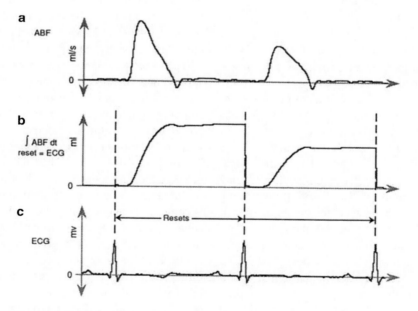

Fig. 12.9 Calculus: integrating a function

think of models as an attempt to reflect the entire workings of the economic system, that is to use a real systems approach.

What Is the Proper Use of Mathematics in Economics and Natural Science?

Part of what defines science to many people (including scientists themselves) is the use of mathematics, and mathematical models, to define and resolve problems. The power of mathematics (in its broad sense) is to make quantitative predictions from known (or hypothesized) relations of a system, which are usually called a model. The process of examining whether your model is correct or at least adequate is called *validation*. *Sensitivity analysis* is the examination of the degree to which uncertainty in model formulation (how it is structured) or parameterization (what numerical coefficients are assigned) allows one to trust your results or reach certain conclusions. It is through validation and sensitivity analysis that models generate their (sometimes) tremendous power in resolving and even predicting truth, such as that is possible and accessible to the human mind.

The reader by now has seen our distrust of many mathematical models. Even so we are strong advocates of modeling as a tool. What then is the proper role of mathematics in the scientific process if it is so frequently incorrect? First of all it is necessary to distinguish *mathematical* from *quantitative*. Quantitative means simply using numbers in an important way in your analysis (e.g., 3 salmon versus 7). This does not require any particular mathematical skill, although getting accurate numbers may require enormous skills of a different kind. *Mathematical* means using the complex tools of quantitative analysis to manipulate those numbers or to study relations among them. It includes algebra, geometry, calculus, and so on. It is our belief that it is much better to learn good quantitative methods that include understanding clearly the relation between the real world and the equations you are attempting to use than becoming a mathematical whiz at solving problems poorly connected to reality.

Analytic Versus Numeric

As we have said there are two principal means of manipulating numbers: *analytic* and *numeric*. In our opinion (but hardly everyone's) there are very few real problems in economics that can be represented adequately by analytic equations, and much of the economics that is done by complex analytic analysis is giving mathematical but not economic results. The use of analytical mathematics, however, does have one major benefit. Through the manipulation of equations you can transform a cause and effect relation that is stated in a way in which you cannot see the patterns you need to see into an understandable output and derive the patterns you need to understand. In other words sometimes analytic approaches can help you visualize clearly a concept you are trying to understand. In practice there are severe restrictions to the class of mathematical problems that can be solved analytically, often requiring a series of sometimes unrealistic assumptions to put the problem into a mathematically tractable format. In addition, the mathematical training required to undertake such analytic procedures precludes its use by many.

The second, *numerical*, technique gives approximate answers to an enormously broader set of possible equations using sometimes more complex equations often arranged in complex *algorithms* (or numerical recipes) solved stepwise in a computer. In theory either method can be used to solve many particular quantitative problems, and sometimes this is done. Fortunately if one learns computer programming or even becomes really good with a spreadsheet, one can solve complex multiple equations about quantitative relations that the best earlier mathematicians could not. More commonly equations are solved through the use of various spreadsheets, apps, and special programs, and the mathematics, usually numeric, are hidden from the user.

The use of analytic mathematics was especially important in the development of physics in the early part of the past century, and the creation of the atom bomb was tangible evidence to many of the power of pure mathematics combined with practical application. Even so the complex fluid dynamics equations required to build the bomb could not possibly be solved by analytic means, and as many of the nation's mathematicians spent the summer of 1944 in Los Alamos New Mexico, many solving the fluid dynamics equations numerically with hand-cranked calculators, something that a single good undergraduate computer student could solve now in an afternoon! [7] The success in physics of mathematics, both analytic and numeric, led many practitioners in other disciplines, including ecology and economics, to attempt to emulate (or at least give the appearance of) the mathematical rigor and sophistication of the physicists. This in turn led ecologist Mary Willson to decry many of their efforts which she said were undertaken for what she has called "physics envy" (Freudian pun intended) [8]. Nevertheless even Einstein preferred to solve his problems without mathematics when that was possible. Other sciences in which mathematical models have been especially important include astronomy, some aspects of chemistry, and some aspects of biology including demography and in some cases epidemiology.

A final problem with models is that there has been frequent confusion between *mathematical* and *scientific* proof. Mathematics can generate real proofs relatively easily because you are working in a defined universe (through the assumptions and the equations used) to which it applies. If you define a straight line as the shortest distance between two points then you can solve many problems requiring straight lines. But the world handed to us by nature is neither so straight nor so clearly defined, and we must constantly struggle to represent it with our equations. Hence a mathematical proof becomes a scientific proof only in the relatively rare circumstances when the equations do indeed capture the essence of the problem. We all have been seeking to follow in Newton's footsteps,

but Newton may have skimmed the cream from what nature has to offer. Meanwhile many economics ideas are more mathematical than real. At the extreme Krugman [9] has said that the main reason for the financial meltdown of 2008 was that Wall Street turned its analyses from people with financial acumen over to other people with extremely strong mathematical skills.

Despite all of the many problems of modeling we do not understand how one can use the scientific method (i.e., generate and test hypotheses) on complex issues *without* the use of formal modeling. This is because it allows one to apply the scientific method to complex real systems of nature and of humans and nature. *But it is critical that the right kind of models be used.* And the way to do that is quite simple: try to represent the real system that you are dealing with rather than some abstraction that happens to be analytically tractable or elegant.

Questions

1. What is the difference between mathematical and quantitative analysis?
2. Under what circumstances is scientific rigor the same as mathematical rigor?
3. Under what circumstances is analytical mathematics most useful?
4. What is the difference between constants and variables?
5. What does "is a function of" mean?
6. What does linear mean? Can you give an example of something that is linear?
7. Give three examples of nonlinear functions or relations.
8. What is an algorithm?
9. How is a correlation different from a function?
10. Define econometrics.
11. Define calculus in terms of something familiar in your everyday life.
12. How does "finite difference" relate to calculus?
13. What does validation mean? Sensitivity analysis?
14. Distinguish between analytical and numeric approaches to solving mathematical equations.

15. Analytical techniques are best suited to what kind of scientific problems?
16. If the equations of economics are often complex, why are they frequently described using analytical approaches?

References

1. Hall, C.A.S. 1988. An assessment of several of the historically most influential theoretical models used in ecology and of the data provided in their support. Ecological Modeling 43: 5–31.
2. Hall, C.A.S. and J.W. Day (eds). 1977. Ecosystem modeling in theory and practice. An introduction with case histories. Wiley Interscience, NY.
3. Hall, C. 1991. An idiosyncratic assessment of the role of mathematical models in environmental sciences. Environment International. 17: 507–517.
4. Gowdy, J., 2004. The revolution in welfare economics and its implications for environmental valuation. *Land economics* 80: 239–257.
5. Kaufmann, R. K. 1992. A biophysical analysis of the energy/real GDP ratio: implications for substitution and technical change. Ecological Economics 6: 35–56.
6. Kaufmann, R. K. and Stern, D. I. 1997. Evidence of human influence on climate from hemispheric temperature relations. Nature. 388: 40–44.
7. John G. Kemeny: BASIC and DTSS: Everyone a Programmer (Obituary). http://www.columbia.edu/~jrh29/kemeny.html
8. Williston, Mary. 1981. Physics envy. Bulletin of the Ecological Society of America.

Economics as Science: Physical or Biophysical?

13

Introduction

Economies exist independently of how we perceive or choose to study them. For more or less accidental reasons we have chosen over the past 100 years to consider and study economics as a *social science*. Before the late nineteenth century economists were more likely to ask "Where does wealth come from?" than are most mainstream economists today. In general, these earlier economists started their economic analysis with the natural biophysical world, probably simply because they had common sense but also because they deemed inadequate the perspective of earlier mercantilists who had emphasized sources of wealth as "treasure" (e.g., precious metals) derived from mining, trade, or plunder. In the first formal school of economics, the French Physiocrats looked to the biophysical world and especially land as the basis for generating wealth (e.g., Quesnay, 1758; see Christensen [1, 2]) focused on agriculture.

The biophysical perspective continued with Thomas Malthus, whose famous *Essays on Population* (there were six of them) assumed that human populations would grow exponentially. It seemed unlikely that anyone would control the "passion between the sexes," and that populations would continue to grow unless somehow "checked" by factors that either reduced the birth rate or increased the death rate. Inasmuch as Malthus had little faith in the "moral restraint" of the working classes, and believed that birth control was "vice," he recommended a rather

Draconian social policy to increase the death rates among the poor. In Malthus' view the agricultural production needed to feed this exponentially increasing human population could grow only linearly (i.e. less rapidly than the number of humans). He also opposed the importation of cheaper continental grains, as a limited food supply assured increasing rents for his patrons, the landed aristocracy, and squeezed the profits of the rival capitalists. It was this view that the human prospect was limited by inadequate food supplies, and that class conflict was inevitable, which led the Victorian philosopher Thomas Carlisle to give economics the label of "the dismal science." Adam Smith and other classical economists focused on both land and labor as means of transforming the resources generated by the natural world into materials that we perceive as having wealth. Later, David Ricardo made important observations about the general need to use land of increasingly poor quality as populations (and hence total agricultural production) expanded. Even Marx, although focused firmly on labor, was keenly interested in the long-term adverse effects of large-scale agriculture on soil quality. He firmly believed that capitalism exploits the land in the same way it does labor. Essentially all important early economists were focused at least as much on the biophysical basis of economics as on the social or human aspects of the economies they were trying to understand.

In the 1870s, these biophysically based perspectives in economics were displaced by the "marginalist revolution" of William Stanley Jevons,

C.A.S. Hall and K. Klitgaard, *Energy and the Wealth of Nations: Understanding the Biophysical Economy*,
DOI 10.1007/978-1-4419-9398-4_13, © Springer Science+Business Media, LLC 2012

Karl Menger, and Leon Walras. Their perspective was based on abstractions such as "subjective utility" that for the first time in economics ignored measurable physical input and output of material or energy. Economics became almost entirely a social science, focusing almost entirely on consumption rather than production. As we said above, this novel approach to economics was called neoclassical, and the ideas of the marginalists still dominate today. In the words of the early marginalist Frederick Bastiat: "Exchange is political economy." Hence production, a biophysical perspective requiring a knowledge of the natural sciences, became a less important, even nonexistent issue to economists compared to market-based human preferences. The common sense biophysical basis for economic analysis was snuffed out intellectually, although of course not in real economies. By the early twentieth century land, representing all of nature, was simply omitted, along with energy (which had never been considered), from neoclassical production functions. Generations of economists subsequently have been trained from a perspective that is divorced from biophysical reality except, occasionally, as it affects prices, within a worldview that is often extremely mathematical, theoretical, and even doctrinaire. On the other hand, one might say that neoclassical economics does a good job of reflecting the essential human characteristic of a desire for more of whatever and the reality that much of what happens within economics does indeed occur within what we may call markets. But the overall movement was away from economics being based on material reality, and hence amenable to the tools of natural sciences, to one focused on the human, social, or even psychological perspective. In short, the intellectual basis of economics changed from one that is quite comfortable with the natural sciences to one that is viewed and studied only as a social science.

At the turn of the last century, economists chose physics (and, more explicitly, the analytic mathematical format of classical mechanics) as a model for capturing the essence of their discipline. This is reflected in the familiar graphs and equations of commodity value and cost versus quantity, with price determined as the intersection of downward trending demand curves (derived from utility curves) and upward trending cost of supply curves. Although physics served as the model and its intellectual popularity as the motivation, the resulting economic model was physically unrealistic because it represented a dynamic, irreversible biophysical process with a static and reversible set of equations. In fact the conservation principles that constrained the equations of physics were incompatible with capital accumulation and, indeed, growth or even production in the economists' model [3].

Economics as a Social Science

So far we have focused on whether we should use the word "social" (versus "natural") in our consideration of the words "social science" as used as a descriptor of economics. Now we want to focus on the use of the word "science" in that descriptor. Banker DeLisle Worrall [4] has said recently (and we agree):

> There are no laws in economics. A law in the physical sciences, as Beinhocker reminds us is a *universal regularity with no known exceptions.* There is nothing in economics which meets that standard. What we have are theories: explanations for why regularities exist and explanations of how they work. We need to desist from writing papers that "prove" theories; they always turn out to be mathematical exercises of no practical relevance, yielding no insight about how the economy really works. In our empirical work we must accept the reality that the limitations of model specification, measurement error, choice of proxy variable, etc. are so formidable that we can never "prove" anything in economics by appealing to the numbers.

So if we are to take this position, and we do, we have to ask why, then, is economics called a social science, or indeed any kind of a science, if it has no ability to generate laws that we can count on? Why do so many important Wall Street financial institutions turn over their analyses to highly mathematical (but barely financially literate) "quants" when they universally led their institutions and their investors off the cliff? (See the quote by Krugman page 287).

This reintroduces the most basic message in our book. Should economics be principally about the social sciences, about human wants and desires and the ability of markets to fulfill these optimally, as most economists would say, or should it be about the biological and physical (i.e., biophysical) conditions that are behind the generation and even distribution of wealth. We believe that it is some mixture, but we also believe that by focusing almost entirely on the social science aspects of economics while essentially ignoring the biophysical aspects, conventional economics has failed to understand the processes that are in fact the essence of the economy. Consequently economics is essentially unable to deal with the new realities imposed upon the world by peak oil and climate change. The planet now is very, very crowded and depletion is looming for many, probably most, of our resources. Conventional economic theories and concepts can make only a small impact on mitigating the problems associated with absolute scarcity.

In summary, we cannot accept economics as presently practiced as any kind of science because it does not follow the rules of science as we summarized them in Chap. 11. This is true for both the behavioral aspects of humans (i.e., how they in fact interact with others) versus how the basic neoclassical model assumes that they do) as well as for the degree to which the model is inconsistent with the laws of nature as summarized in Chap. 5.

The Magnitude of the Problem

Economics as presently perceived may be the most widely, consistently, and incoherently taught course in American higher education, and probably in most other countries. By *widely* we mean that there may be more young people taking an introductory economics course than nearly any other single course in college except perhaps biology or English literature of some kind. By *consistently* we mean that in preparation for writing this book we reviewed about two dozen basic economics books and found them to be basically the same, and all build up a system of economics consistent with the basic neoclassical framework. This consists of a caricature of real economies as that of simply firms and households interacting through markets, with a focus on humans, their

wants and needs, and their independence in deciding what is good for them through their individual decisions in markets. In other words there is a consistent body of theory, known as neoclassical economics, that is accepted or promulgated by essentially all academic economists, at least as represented in their fundamental textbooks. By *incoherently* we mean that many of the assumptions that conventional economists must make to generate their world of theoretical economics, the associated equations, and their applications, defy logic to anyone trained in the natural sciences, the scientific method, or even possessing a reasonable degree of common sense. As some support for that point of view we note that as of 2010 ten of the last eight most recent recipients of the Nobel Prize in economics were people whose worked challenged, in various very fundamental ways, the basic existing neoclassical paradigm.

Most knowledgeable economists, when pressed, will acknowledge at least some of this, yet economics as a discipline rumbles onward year after year with little real change in the way that our young people are indoctrinated. This point of view – that much of what is taught in economics is quite divorced from biophysical reality – is apparent to most of our own students (especially those with a focus or at least reasonable experience with natural science). Although our students can indeed learn the principles of economics in their first course in that subject, and can pass and even do well on the tests, many generally do not, or barely, believe the concepts that they are taught there. Because many of the principles seem unrealistic to them they are often deeply bored. They sometimes use very harsh words to describe their disbelief in what they are being taught. In France economics students staged a revolution against what they were being taught and demanded better. They even founded their own journal, *The Post-Autistic Economics Review*. We agree with all of these students, and believe that collectively we have been teaching something like one million young people a year in the United States alone something that might reasonably be considered, at worse, complete fabrications, or at best a very simplistic and incomplete perspective on the reality and richness of thought that can be brought to bear on economic issues and problems.

The issue is not simply scientific or logical but also has a moral dimension. Most of our students, possibly more idealistic than the average, are also very much put off, for both scientific and moral reasons, by the essential selfishness that is accepted by and even celebrated in the basic economic theory found in introductory economics textbooks. This point perspective was made to us even more strongly by our colleague, Donald Adolphson, a very popular and thoughtful professor of economics and finance at Brigham Young University. He told us:

> The students at BYU are virtually all practicing Mormons. They are trained at home to think of their relation to God and then family first, community second and then the world community. Most travel to a foreign country as a late teenager as part of their preparation for life. When they take Introductory Economics, they are told in their textbooks that what the basic neoclassical model uses as a basis for that course, and economics in general, starts with the assumption that humans are "rational," rational meaning entirely selfish, or at least self serving, and principally materialistic. This just strikes them as wrong, and they reject their basic economics textbooks.

Well, it strikes us as wrong too. It also strikes most of our own students in upstate New York as wrong morally and with respect to their own motivation. In particular it seems wrong to the majority of our students because they have a high sense of idealism towards nature and towards other people, neither of which they wish to see sacrificed for mere self-serving and often superfluous economic goods and services. This is especially the case when they view the world around them as full of hyperaffluence bought at enormous expense to the environment, and the enormous disparities between rich and poor.

Other Economists Agree with Us

We find that there are many, many other scientists and economists who basically believe the same things as we do: that neoclassical economics is intellectually bankrupt at its core. For one of many possible examples we continue with the statement of Worrell given above:

Just recently I have read a provocative book entitled *The Origin of Wealth*, which challenges what its author, Eric Beinhocker calls "traditional economics," and proposes that it be replaced by a new paradigm, "complexity economics." His argument is fascinating, and I now propose to spend a little time talking about some of the things I have gleaned from it. **Economics uses the laws of physics as they were known in the *nineteenth* century... .** Early in the book Beinhocker describes a cross-disciplinary workshop on economics, arranged by Citicorp's CEO John Reed, that brought together ten leading economists (including Nobel Laureate Kenneth Arrow) and ten physicists, biologists and computer scientists. The physical scientists were "really shocked" to find that "economics was a throwback to another era" (page 47). Economists' mathematics [seemed] "like a blast from the past," and physical scientists were surprised by economists' assumptions, objecting particularly to the assumption of perfect rationality. Physical scientists craft their assumptions to ensure that they do not contradict reality, though they are designed to simplify it. The assumption of perfect rationality contradicts reality, and economists know that, but they still use the assumption, however modified. Beinhocker argues that what he calls Traditional Economics (TE) remains trapped in a time warp defined by the concepts that Leon Walras borrowed from the physics he knew at the time of the development of the marginalist theory of market economics which underpins the classical, Keynesian, neoclassical and new Keynesian views of the world. At that time only the first law of thermodynamics – the conservation of energy – was known. The notion of equilibrium [which is used in TE] is a form of expression of the first law. Physics subsequently discovered the second law – that entropy (disorder or randomness) is always increasing. The implication of the second law is that if the system were ever to reach equilibrium it would be dead. In effect, TE classifies economy as a closed equilibrium system, which violates the laws of physics, as they are now known to exist. Beinhocker proposes an alternative to TE, which he terms complexity economics, CE... .Beinhocker concludes that viewing the economy as a complex adaptive system provides us with a new set of tools, techniques and theories for explaining economic phenomena.

We agree with Beinhocker's perspective, however, we would like to take it a step further to the biophysical and empirical basis that forms the core of this book. This is even true for the social or behavioral basis of neoclassical economics. As summarized in Chap. 5, there has been a great increase in the degree to which basic human economic behavior has been tested using the scientific method, and through some clever experiments in behavioral economics. The results have tended to show that the (traditional neoclassical assumption of) *Homo economicus* view is false, or at least very poorly predictive. For example, Henrich et al. [5] after examining the results of behavioral experiments in 15 small-scale societies ranging from hunter-gatherers in Tanzania and Paraguay to nomadic herders in Mongolia conclude: "[T]he canonical model [i.e., *Homo economicus*] is not supported in any society studied."

Although "pattern explanation" models (such as the model of evolution by natural selection) are often as important to scientific methodology as quantitative predictive models, the ability to predict is a crucial criterion for any economic model that is to be used to influence policy and hence the lives of many people. But in fact we find that the core models used by economists (economic man and perfect competition) consistently fail the "good prediction" test. In addition the most basic models are not consistent with the laws of thermodynamics, nor do most economists even think about such laws [4]. This alone would be enough to disqualify any model in the natural sciences, but it has not seemed to bother economists. This failure to follow the laws of science is one of the reasons that we have looked for another framework to generate a theoretical basis for economics, while, however, attempting to retain that portion of conventional economics that is useful and consistent with the laws of science and empirical reality.

Is Economics Science?

This chapter has been a review of the scientific principles that we believe are important for understanding real economic systems. A question that may remain in the mind of many readers is, "To what degree does existing economics follow these rules of science?" The answer is rather discouraging. One can certainly find plenty of hypothesis generation and testing in learned economics journals. For example Hall [6] examined some 127 articles in the leading economics journal, *American Economic Review*, and found that for this subset of papers about 10% did test explicit hypotheses, which is good. Only 3%, however, could be construed as testing fundamental economic theories. These papers found more often than not that the basic economic theories tested in specific applications were generally not supported. So we might say based on this study that economics is at least sometimes a good science because ideas were being subjected to the scientific method, or perhaps bad science because such results have no impact on the center of gravity of conventional economics, as is clearly stated by leading economists themselves (e.g., Krugman).

An especially important issue is the degree to which humans behave "rationally" (meaning essentially materialistic and entirely self-regarding) as is assumed in the neoclassical model [7]. Although we all are aware that human behavior is almost infinitely variable, the neoclassical model is based on the assumption that human behavior is at all times "rational." For nearly 100 years it was simply accepted by most economists as the basis for how humans behaved. But the testing considered in Chap. 5 shows the assumption to be erroneous: that humans are as likely to be altruistic or vindictive as "rational." Another core belief of many economists is that good models make good predictions, and that this is more important than whether the model is consistent with known mechanisms [7]. This too is inconsistent with how natural scientists proceed.

So our answer to the question posed by the title of this chapter is that, no, the dominant economics at this time is not a science. Its basic models violate too many scientific principles including: the laws of thermodynamics, the law of the conservation of matter, the ways that people actually do behave according to empirical studies, and so on. In addition, even when economics appears to be "borrowing" equations from physics it is doing so incorrectly, even in violation of the physics it is trying to emulate. Instead of following these principles, principles that all natural science follows or risks rejection or humiliation from peers, neoclassical economics has generated its own world, a world that reflects the real world in only the most contrived ways. In theory there is a model of physics behind it, but the equilibrium model is just a copying of the equation form without any understanding of the actual physics; in fact it violates the second law of thermodynamics. In addition, the assumptions of "rational actors" required to make this model work are inconsistent with how real humans in fact interact with each other. The generation of theory based on a market concept of perfect information and equal power of interacting buyers and sellers that has not existed since agrarian England, if indeed they ever existed, combined with failure to make and test hypotheses makes an acceptance of the basic neoclassical model an article of faith, not rationality. Unfortunately the ascendance and the power of the ideas of the advocates of market theory and self-interest have spilled over to our public and political life, destroying many economies in the less-developed world [9, 10]. These ideas have completely changed the political perspective of many Americans from community, civic responsibility, fairness in distribution of wealth and care for others, to one of unbridled greed and self-focus. This market approach has turned, to some degree, universities from learning communities where highly trained and caring professors held students up to their own high standards of deep and skeptical thought to credentials markets where students perceive that they buy their education and expect As with little work. They have also given a green light to those who have enormous financial power to buy and to manipulate our political system while convincing many that "big government," which is in fact their only defense against much more pernicious big money corporations, is something to be avoided. It is a very large impact of a theory which is scientifically indefensible at its heart [11]. Moreover, neoclassical economics was conceived on the upslope of the Hubbert Curve. All the myriad problems of economics are likely to be exacerbated the coming post-peak world. To develop a rationale as to why, we turn to the concept of the Energy Return on Investment.

Questions

1. Why is economics usually considered a social rather than a biophysical science? What is your view?
2. Do you agree, from your own experience, that humans are essentially selfish or at least self-regarding? Or does it depend upon the circumstances?
3. With respect to the previous question is this pattern of basic selfishness found in all cultures around the world?
4. What are the characteristics of an endeavor that qualify it as a science? Do you think conventional economics qualifies as a science? Why or why not, or where and where not?
5. In the world of conventional economics what does rational mean? What does it mean to you?
6. Conventional economics is usually classified as a social science. In your opinion does economics qualify as a science? Why or why not?

References

1. Christensen, P. 1994. Fire, Motion and Productivity: the proto-energetics of nature and economy in francois quesney. Pp 249–288 in P. Mirowski. Natural images in economic thought. Cambridge University Press.
2. Christensen, P. 1984. Hobbes and the physiological origins of economic science. History of Political Economy 21:4
3. Mirowski, P. 1984. More heat than light. Cambridge: Cambridge University Press.
4. Worrell, DeLisle, 2010. Governor of the Central Bank of Barbados, in an address to the Barbados Economic Society (BES) AGM, Bridgetown, 30 June 2010.
5. Henrich, J. et al., 2001. Cooperation, reciprocity and punishment in fifteen small-scale societies. American Econ. Review, 91: 73–78.
6. Hall, C. 1991. An idiosyncratic assessment of the role of mathematical models in environmental sciences. Environment International. 17: 507–517.
7. Hall, C. and J. Gowdy, J. 2007. Does the emperor have any clothes? An overview of the scientific critiques of neoclassical economics. In G. LeClerc and pp. 3–12 C. A. S. Hall (Eds.) Making world development work: Scientific alternatives to neoclassical economic Albuquerque: University of New Mexico Press).
8. Friedman, M. 1955. Essays in positive economics. University of Chicago Press, Chicago.
9. Hall, C.A.S., P.D. Matossian, C. Ghersa, J. Calvo and C. Olmeda. 2001. Is the argentine national economy being destroyed by the department of economics of the University of Chicago? pp. 483–498 in S. Ulgaldi, M. Giampietro, R.A. Herendeen and K. Mayumi (eds.). Advances in Energy Studies, Padova, Italy.
10. Easterly, W. 2001. The elusive quest for growth: economists' adventures and misadventures in the tropics: Economists' Adventures and Misadventures in the Tropics. The MIT Press, Cambridge.
11. Hall, C.A.S. and K. Klitgaard. (2006) The Need for a New, Biophysical-Based Paradigm in Economics for the Second Half of the Age of Oil. Journal of Transdisciplinary Research Volume: 1, Issue 1, 4–22.

Part IV

The Science Behind How Real Economies Work

Our book so far has reviewed how economics progressed historically and how many important trends and issues in real economies can be understood much better through an appreciation of the role that energy plays in real economies. We also developed our perspective on the extreme limitations of the approach used by economics today. We developed in some detail the basic science needed to understand real economies, a kind of training that is missing from the education of most economists. We, however, are fundamentally committed to the idea that economics can be approached scientifically. In this section we apply the concepts of science to further understand in some new and important ways how real economies have operated and are likely to operate in the future.

Energy Return on Investment

14

Introduction

Many important earlier writers, including sociologists Leslie White and Fred Cottrell and ecologist Howard Odum, as well as economist Nicolas Georgescu-Roegen have emphasized the importance of net energy and energy surplus as a determinant of human culture. Human farmers or other food gatherers must have an energy surplus for there to be specialists, military campaigns, and cities, and substantially more for there to be art, culture, and other amenities. Net energy analysis is a general term for the examination of how much energy is left over from an energy-gaining process after correcting for how much of that energy (or its equivalent from some other source) is required to generate (extract, grow, or whatever) a unit of the energy in question. Net energy analysis is sometimes called the assessment of energy surplus, energy balance, or, as we prefer, energy return on investment, depending upon the specific procedures used. To do this we start with the more familiar monetary assessment and then develop how this relates to the energy behind economic processes [1].

Economic Cost of Energy

In real economies, energy comes from many sources: from imported and domestic sources of oil, coal, and natural gas, as well as hydropower and nuclear, and from a little renewable energy, mostly that of firewood but increasingly from wind and photovoltaics. Some of these are cheaper per unit of energy delivered than oil and some are considerably more expensive. It is possible to examine the ratio of the cost of energy (from all sources, weighed by their importance) relative to the benefits of using it to generate wealth:

$$\text{Economic cost of energy} = \frac{\text{Dollars to buy energy}}{\text{GDP}}$$

In 2007 1 trillion dollars of the U.S. GDP of 12 trillion dollars, roughly 9%, was spent to purchase the energy of all kinds used by the U.S. economy to produce the goods and services that comprised the GDP (Fig. 14.1). Over recent decades that ratio has varied between 5% and 14%. The abrupt rise and subsequent decline in the proportion of the GDP spent for energy has been seen before during the "oil shocks" of the 1970s, in mid-2008, and again in 2011. Each of these increases in the price of oil relative to GDP had large impacts on discretionary spending, that is, on the amount of income that people can spend on what they want versus what they need. An increase in energy cost from 5 to 10 or even 14% of GDP would come mainly out of the 25 or so percentage of the economy that usually goes to discretionary spending. Thus changes in energy prices, much of which goes overseas, have very large economic impacts on the economy which is very sensitive to discretionary spending changes, which is mostly domestic. Each increase in the

C.A.S. Hall and K. Klitgaard, *Energy and the Wealth of Nations: Understanding the Biophysical Economy*, DOI 10.1007/978-1-4419-9398-4_14, © Springer Science+Business Media, LLC 2012

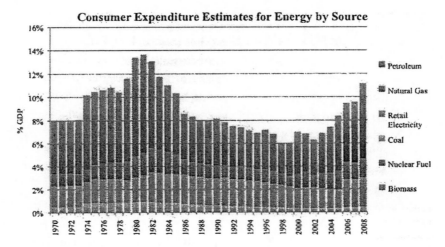

Fig. 14.1 Percentage of GDP that is spent on energy by final consumers

price of oil (and of energy generally) has been associated with an economic recession. What future energy prices will be is anyone's guess. There is a great deal of information implying that dollar costs of fuels will continue to increase substantially. Our guess is that this is in large part due to declining EROI and that it will take a substantial economic toll in the future. This chapter develops that argument.

What Is EROI?

Energy return on investment (EROI or sometimes EROEI, with the second E used to refer to the use of energy in the denominator) is the ratio of energy returned from an energy-gathering activity compared to the energy invested in that process. EROI is calculated from the following simple equation, although the devil is in the details.

$$\text{EROI} = \frac{\text{Energy returned to society}}{\text{Energy required to get that energy}}$$

The numerator and denominator are necessarily assessed in the same units so that the ratio so derived is dimensionless, for example, 30:1, which can be expressed as "30 to 1". This means that a particular process yields 30 Joules on an

investment of 1 J (or kcal per kcal or barrels per barrel). EROI is usually applied at the mine-mouth, wellhead, farm gate, and so on, that is, at the point that it leaves the extraction or production facility. We denote this more explicitly as EROI_{mm}. Sometimes corrections are made for the quality of the energy obtained or used. EROI should not be confused with *conversion efficiency*, that is, going from one form of energy already obtained to another, such as upgrading petroleum in a refinery or converting diesel to electricity.

The authors of this book and advocates of EROI believe that net energy analysis is a very useful approach for assessing the advantages and disadvantages of a given fuel or energy source, and offers the possibility of looking into the future in a way that markets are unable to do. The advocates of EROI also believe that in time market prices must approximately reflect comprehensive EROIs, at least if appropriate corrections for quality are made and subsidies removed. This perspective is supported in a paper by Carey King and Charles Hall that is in press. EROI by itself is not necessarily a sufficient criterion by which judgments about one fuel or another, can be made, although it is the one we favor, especially when it indicates that one fuel has a much higher or lower EROI than others. In addition it is important to consider the present and future potential magnitude of the fuel, and how EROI might change if a fuel is expanded.

History

The concept of EROI was derived conceptually from Howard Odum's teachings on net energy [2, 3], some earlier work by anthropologists and sociologists, and explicitly from Hall's PhD dissertation on the energy costs and gains of migrating fish [4]. The concept was implicit in Hall and Cleveland's 1981 paper [5] on petroleum yield per effort although the term net energy was used there. The first publication using the name EROI was in 1982 [6], and it received much more attention in a paper in *Science* [7]. More detailed and comprehensive summaries of the literature on EROI then available were put together in a large book that reflected the many high-quality energy data and studies available then [8]. The use of the concept lagged during the "energy lull" 1984–2005 but has picked up post-2008 with a new summary [9] and a flurry of new EROI-related papers [10–13]. An entire issue on EROI of the online journal *Sustainability* is being published in 2011.

There is very little quantitative information about actual EROIs for energy-producing systems for the medium or distant past, which is not surprising because we did not even understand the concept of energy until about 1850. But there are exceptions. Sundberg made a quite detailed assessment of the energy cost of energy in earlier Sweden [14]. From 1560 until 1720 Sweden was among the most powerful countries in northern Europe, based mostly on its very productive metal mines, but also an aggressive foreign policy backed up by high-quality weapons. The production of these mines required enormous amounts of energy for mining and especially smelting. The source of this energy was wood cut from Swedish forests and especially charcoal made from that wood and needed to get the high temperatures steel required. Sundberg gives a detailed calculation of how a typical forester and his family, self-sufficient on 2 hectares of farmland, 8 hectares of pastures, and 40 hectares of forest (collectively intercepting 1,500 Terajoules (or 10 exp 12 joules) of sunlight) generated some 760 Gigajoules (or 10 exp 9 joules) of charcoal in a year for the metal industry. To do that required about half a GJ of human energy or 3.5 GJ if we include the draft animal labor. This yields a rough EROI of the human investment as high as 1,500:1, or some 250:1 if we include the animals. But that is just the direct energy, as it took 105 GJ to feed, warm, and support the farmer and his family (which includes his replacement) and probably at least that to support the animals. So if we include both direct and indirect energy the EROI is down to roughly 4:1. The system was sustainable as long as the forests were not overharvested. By the middle of the nineteenth century, however, the forests were severely overharvested and many cold starving Swedes left for America, where they settled in the cold and forested Midwest.

Much of the still relatively sparse current literature on net energy analysis tends to be about whether a given project is or is not a net surplus, for example, whether there is a gain or a loss in energy from making ethanol from corn [15]. The criteria used in much of the current debate are focused on the "energy breakeven" issue, that is, whether the energy returned as fuel is greater than the energy invested in growing or otherwise obtaining it (i.e., if the EROI is greater than 1:1). If the energy returned is greater than the energy invested, then the argument is that the process should be done and the converse.

Several of the participants in the current debate about corn-derived ethanol believe that corn-based ethanol has an EROI less than 1:1, whereas others (summarized in Ref. [15]) argue that ethanol from corn is a clear energy surplus, with from 1.2 to 1.6 units of energy delivered for each unit invested. Further aspects of this argument center around the boundaries of the numerator: whether one should include some energy credit for nonfuel coproducts (such as residual animal feed; i.e., soybean husks or dry distiller's grains), the quality of the fuels used and produced, and the boundaries of the denominator (i.e., whether to include the energy required to compensate for environmental impacts in the future such as for the fertilizer needed to restore soil fertility for the significant soil erosion occasioned by corn production).

Such arguments are likely to be much more important in the future as other relatively low quality fuels (e.g., oil sands or shale oil) are increasingly considered or developed to replace conventional oil and gas, both of which are likely to be more expensive and probably less available in the not so distant future. If, of course, the alternatives require much oil and or gas for their production, not to mention water, which is often the case, then an increase in the price of petroleum could make these new alternatives relatively cheaper, or they could increase the cost of the input to negate that advantage. We believe that for most fuels, especially alternative fuels, the energy gains are reasonably well understood. Unfortunately the boundaries of the denominator, especially with respect to environmental issues, are poorly understood and even more poorly quantified. Thus we think that most calculated and published EROIs, including those we consider here, are higher (i.e., more favorable) than they would be if we had complete information.

Seeking an Acceptable EROI Protocol

Given the rather different quantitative responses from analyses (such as the corn-based ethanol example given above) we need some good and consistent way of thinking about the meaning of the magnitude of the EROIs of various fuels. It is our opinion that many of the EROI arguments so far are simplistic, or at least incomplete, because the "energy breakeven" point, although usually sufficient to discredit a candidate fuel, should not be the only criterion used. In addition it seems to us that many EROI analyses are generated from the perspective of defeating or defending a particular fuel rather than objectively assessing potential alternatives. Perhaps we need some way to understand the magnitude, and the meaning, of the overall EROI we might eventually derive for all of a nation or society's fuels collectively by summing all gains from fuels and all costs from obtaining them (i.e. *societal* EROI; E = energy):

$$\text{EROI}_{\text{soc}} = \frac{\Sigma \text{ of the E content of fuels delivered}}{\Sigma \text{ of all the E costs of getting those fuels}}$$

To our knowledge this has never been done.

We need to have a straightforward and universally accepted approach to EROI even while accommodating different approaches or philosophies. Of greatest concern is the boundaries of the analysis: should coproducts be included (such as hulls left from generating biodiesel from sunflower seeds that can be fed to animals, reducing energy needed to make the animal feed). Or should we include the costs of the energy to support a laborer's paycheck? Inasmuch as there are no clear and unambiguous answers to those questions, Murphy et al. [16] have advocated a basic EROI approach using simple standardized energy output divided by the direct (i.e., on site) and indirect (energy used to make the products used on site) energy used to generate that output to generate a standard EROI, EROIs. This approach allows the comparison of different fuels even when the analysts do not agree on the methodology that should be used. Murphy et al. also allow for the use of other EROIs, including new approaches that allow for special consideration of other aspects of that EROI, as long as the standard approach is also undertaken. We believe this allows for both standardization and flexibility.

$$\text{EROI}_{\text{st}} = \frac{\text{E returned to society}}{\text{Direct and indirect E required to get that E}}$$

The Best Analyses of the Costs

Determining the energy content of the numerator of the EROI equation is usually straightforward: multiply the quantity produced by the energy content per unit. Determining the energy content of the denominator is usually more difficult. The energy used directly, that is, on site, includes the energy used to rotate the drilling bit, operate the farm tractor, and so on. Sometimes governments keep these data. One also should include the energy used indirectly, that is, the energy to make the drilling bit and associated materials, the tractor, and so on. Unfortunately companies generally do not keep track of their energy expenditures, but only their dollar expenditures. But it is possible to convert dollars spent to energy spent using "energy intensities" for a dollar spent in different parts of the economy. Forty years ago this was easy and accurate. A remarkable group at the University of Illinois, including Bullard et al. [17], Hannon [18], Herendeen and Bullard [19], and Costanza [20] undertook such calculations for every sector of the U.S. economy, so that it was possible to have good estimates of the energy needed to generate each dollar's worth used. Using Leontief input-output matrices (who buys how much of what from other sectors) and very comprehensive energy use information they were able to generate very detailed determinations of how much energy it took to make all

products of the U.S. economy. These allowed us at the time to gain very detailed assessments of where and how energy was used in the U.S. economy and also the energy costs of getting energy (Table 14.1).

These analyses also showed that (except for energy itself) it does not matter where money is spent within final demand due to the complex interdependency of our economy (i.e., the energy cost per dollar of the final products that consumers buy are similar in energy intensities because there are so many interdependencies). Each sector purchases from many others within our economy. (This does not apply, however, to the intermediate

Table 14.1 Some estimates of EROI from the literature (Summarized, with methods, in The Oil Drum, 2007). TOD = The Oil Drum. Many updates in Hall (ed) Special issue of Sustainability on EROI (in press).

Resource	Year	Magnitude (EJ/year) in 2005 etc	EROI	Reference	Approach Appendix
I. Fossil Fuels					
Oil and gas (finding)	1930	5	>100:1	[10, 11]	TD
Oil and gas (production)	1970	28	30:1	[7, 8]	TD
Discoveries	1970		8:1	[7, 8]	
Production	1970	10	20:1	[7, 8]	
Oil and gas (production)	2005	9	11-8:1	[10]	TD
World oil production	1999	200	35:1	[13]	EI/A
World oil production	2006	200	18:1	Gagnon et al. 2007	EI/A
Imported oil	1990	20	35:1	Herweyer and Palcher (TOD)	EI/B
Imported oil	2005	27	18:1	Herweyer and Palcher (TOD)	
Imported oil	2007	28	12:1	Extrapolated from above.	
Natural gas	2005	30	10–80:1	Button and Sell (TOD;Sell in press)	BU
Coal (mine mouth)	1930		80:1	[7]	EI
Coal (mine mouth)	1970		30:1	[7, 8]	
		5	>100:1	[10]	
Bitumen from Tar sands		1	2–4:1	Gupta et al. (TOD)	BU/D
Shale Oil		0	5:1	Gupta et al. (TOD)	BU/E
II. Other nonrenewable					
Nuclear		9	15:1 (2–50:1)	Powers (below)	LR/F
III. Renewables					
Hydropower		9	>100:1	Schoenberg (TOD)	LR/G
Wind turbines		5	18:1	Kubiszewski & Cleveland (2007)	MA
Geotherrnal		<1		Hallorin (TOD)	LR/H
Wave Energy		<<1	?	Hallorin (TOD)	LR/I
Solar collectors					
Flat plate		<1	2–8:1	[8, Kubiszewski pers.]	BU
Concentrating collector		0	1.6:1	[8]	BU
Photovoltaic		<1	6.8:1	Cleveland (pers.; Battisti et al 2004)	LR/J
Passive solar		?	Prob. high	Giermek (TOD)	LR/J
Biomass					
Ethanol (sugarcane)		0	0.8–1.7:1	[8]	LR
Corn-based ethanol		<1	0.8–1.6:1	[15]	LR
Biodiesel		<1	1.3:1	[1]	LR/K

products purchased by manufacturers.) According to Costanza [20], the market selects for generating a similar amount of wealth per unit of energy used within the whole economic "food chain" leading to final demand. Although this is not exactly true it is close enough for our present purposes and it is certainly true for the average of all economic activity, with the exception of purchases for energy itself, which includes a similar amount of embodied energy per dollar spent as anything else but in addition the chemical energy of the fuel. Unfortunately there has been little such analysis of "sector interdependencies" since the pioneering works at the University of Illinois, so that it is harder, or rather less precise, to make such assessments today. The closest assessment that is available (to our knowledge) are the analyses undertaken by various people at Carnegie-Mellon University and available on their website.

We next show how an EROI analysis can generate some quite interesting results that can help us understand the importance of EROI for running an actual economy.

How Much Energy Is Needed to "Get the Job Done": Calculating EROI at the Point of Use

The EROI that is needed to undertake some activity, such as driving a truck, is far more than just what is needed to get the fuel out of the ground. This was assessed in 2008 by Hall et al. [1]. We introduce here new concepts that start with $EROI_{mm}$, the standard EROI at the mine mouth (or farm gate, etc.), and then take it farther along the use "food chain". We call the next step EROI at the "point of use," or $EROI_{pou}$, and it helps to understand the total energy required to get and use energy:

$$EROI_{pou} = \frac{E \text{ returned to society at point of use}}{E \text{ required to get and deliver that } E}$$

To do this we need to generate a more comprehensive EROI including the costs associated with refining, transporting, and using a fuel. As we do this the energy delivered goes down and the

energy cost of getting it to that point goes up, both reducing the EROI. This begins the analysis of what might be the minimum EROI required in society. We do this by taking the standard EROI (i.e., $EROI_{mm}$) and then including in addition in the denominator first the energy requirements to get the fuel to the point of use (i.e., $EROI_{pou}$) and then the energy required to use it to generate $EROI_{ext}$, (i.e., extended EROI).

Refinery losses and costs: Oil refineries use roughly 10% of the energy of the fuel to refine it to the form that we use and to generate other products. In addition about 17% of the material in a barrel of crude oil ends up as other petroleum products, not fuel. So for every 100 barrels coming into a refinery only about 73 barrels leaves as usable fuel. Natural gas does not need such extensive refining although an unknown amount needs to be used to separate the gas into its various components and a great deal, perhaps as much as 25%, is lost through pipeline leaks and to maintain pipeline pressure. Coal is usually burned to make electricity at an average efficiency of 35–40%. However, the product, electricity, has at least a factor of three higher quality, so that we do not count as costs the inefficiency of the process of upgrading quality. One thing this means is that oil resources that have an EROI of 1.1 MJ returned per MJ invested at the wellhead cannot provide an energy profit to society because at least 1.37 MJ (1/0.73) of fuel is required to deliver that 1 MJ to society.

Transportation costs: Oil weighs roughly 0.136 tons/barrel. Transportation by truck uses about 3,400 BTU/ton-mile or 3.58 MJ/ton-mile. Transportation by fuel pipeline requires 500 BTU/ton-mile or 0.52 MJ/ton-mile. We assume that the average distance that oil moves from port or oil field to market is about 600 miles. Thus a barrel of oil, with about 6.2 GJ of contained chemical energy, requires on average about 600 miles of travel × 0.136 tons/barrel × 3.58 MJ/ton-mile = 292 MJ/barrel spent on transport, or about 5% of the total energy content of a barrel of oil to move it to where it is used (Table 14.2). If the oil is moved by pipeline (the more usual case), this

Table 14.2 The energy cost of transporting oil and coal

	Energies cost (MJ/ton-mile)	Miles traveled	Energy cost (MJ)	Energy cost of energy unit delivered (%)[a]
Oil truck	3.58[b]	600	292	5
Pipeline	0.52[b]	600	42	1
Coal train	1.81[b]	1,500	2,715	8

[a] Energy unit delivered: oil = 1 barrel = 6.2 GJ/barrel; coal = 1 ton = 32 GJ/ton
[b] Ref. [22]

percentage becomes about 1%. We assume that coal moves an average of 1,500 miles, mostly by train at roughly 1,720 BTU/ton-mile or about 1.81 MJ/ton-mile, so that the energy cost to move a ton of bituminous coal with about 32 MJ/kilogram (kg) (32 GJ/ton) to its average destination is 1,500 miles × 1.81 MJ/ton-mile = 2,715 MJ/ton, or 2.715 GJ/ton of coal, which is about 8% (Table 14.2). Line losses if shipped as electricity are roughly similar. So adding between 1% and 8% of the energy value of fuels for delivery costs does not seem unreasonable. Perhaps 25% of the energy in natural gas is used to move the gas down the pipeline, and an unknown but significant amount to build and maintain the pipeline. We assume that these costs would decrease all EROIs by a conservative 5% to get it to the user, in other words the fuel must have an EROI of at least 1.05: 1 to account for delivery of that fuel.

Thus we find that our EROI$_{pou}$ is about 32% (17% nonfuel loss plus 10% to run the refinery plus about 5% transportation loss) less than the EROI$_{mm}$ indicating that at least for oil one needs an EROI at the mine mouth of roughly 1.5 (i.e., 1.0/0.68) to get that energy to the point of final use.

Extended EROI: Calculating EROImm Required at the Point of Use, Correcting for the Energy Required for Creating and Maintaining Infrastructure

We must remember that usually what we want is energy services, not energy itself, which usually has little intrinsic economic utility, for example, we want kilometers driven, not just the fuel that does that. That means that we need to count in our equation not just the "upstream" energy cost of finding and producing the fuels themselves but all of the "downstream" energy required to deliver the service (in this case transportation) and that for: (1) building and maintaining vehicles, (2) making and maintaining the roads used, (3) incorporating the depreciation of vehicles, (4) incorporating the cost of insurance, (5) and the like. All of these things are as necessary to drive that mile as the gasoline itself, at least in modern society. For the same reason businesses pay some 45 or 50 cents per mile when a personal car is used for business, not just the 10 cents or so per mile that the gasoline costs. So in some sense the money required for delivering the service (a mile driven) is some 4–5 times the direct fuel costs, and this does not include the taxes used to maintain most of the roads and bridges.

Many of these costs, especially insurance, use less energy per dollar spent than fuel itself and also less than that for constructing or repairing automobiles or roads, and certainly this not the case with the money used to deliver the fuel itself used in these operations. Thus we may wish to determine the energy required not only to get but also to use a unit of energy. We call this an "extended EROI" or EROI$_{ext}$. We define it formally as:

$$EROI_{ext} = \frac{E \text{ returned to society}}{E \text{ required to } get, \ deliver, \ and \ use \text{ that energy}}$$

More accurately, perhaps, this is the required EROI energy at the mine mouth for that energy to be minimally useful.

The energy intensity (embodied plus chemical) of one dollar's worth of fuel is some 8–10 times greater than that for one dollar's worth of

Table 14.3 Estimates of energy and dollar expenditures within the total U.S. transportation sector [b, c]

Category	Dollars (10^9)	As percent of total dollar Expenditures (%)	Conversion factor (MJ/$)	Total (EJ)	Total energy (%)
Expenditures					
Federal highway administration spending (2005)[a]	30	3.45	14	0.420	3.86
State highway spending (2005)[a]	11	1.26	14	0.158	1.45
Local disbursements for highway spending (2005)[a]	57	6.55	14	0.804	7.38
Motor vehicles and parts (2005)[d]	443	50.92	14	6.203	56.94
Automobile maintenance (2005)[d]	143	16.44	14	2.008	18.43
Automobile insurance spending (2007)[e]	162	18.62	7	1.134	10.41
Automotive service technicians and mechanics (2007)[f]	24	2.76	7	0.166	1.52
Total cost of transportation infrastructure	870	100.00		10.893	100.00

[a] FHWA: Highway Statistics 2005
[b] FHWA: Motor-Fuel Use 2008
[c] EIA: Retail Motor Gasoline and On-Highway Diesel Fuel Prices 1949–2007
[d] BEA: Personal Consumption Expenditures by Type of Product
[e] Statement Database
[f] Bureau of Labor Statistics: Occupational Employment and Wages, May 2007

infrastructure costs. Table 14.3 gives our estimates of the energy cost of creating and maintaining the entire infrastructure necessary to use all of the transportation fuel consumed in the United States. The energy intensities are rough estimates of the energy used to undertake any economic activity derived from the national mean ratio of GDP to energy (about 8.7 MJ/dollar; the Carnegie-Mellon energy calculator website and from Robert Herendeen, personal communication). Specifically one can use the earlier data from the Illinois group, corrected for inflation, and the Carnegie-Mellon numbers for heavy industry to generate an educated guess that in 2005 heavy construction used about 14 MJ/dollar. Because in the 1970s insurance and other financial services had about half the energy intensities as heavy industry, our estimate of the energy required for infrastructure replacement and maintenance for the entire United States for 2005 is equal to about 38% of the energy used as fuel itself.

Our calculation, then, of adding in the energy costs of getting the fuel to the consumer in a usable form plus the energy cost of the infrastructure necessary to use the fuel is equal to about 0.32 plus 0.375, respectively, or about 0.695 in total [1]. Thus the

$EROI_{mm}$ necessary to provide transportation from crude oil is 3.3–1 (1/0.305). To deliver the transportation services associated with 1 gallon of fuel to the consumer requires close to 3 gallons produced at the wellhead, and probably similar ratios for other types of fuels. Thus the minimum $EROI_{mm}$ required for society to drive a car or a truck would be at least 3:1. If we are to put something in that truck, say grain, the $EROI_{mm}$ required would be more. If we want to replace the worn-out workers as well as the worn-out trucks then you need a much higher $EROI_{mm}$ to support households, health care, education, and so on. Thus we need fuels with a very positive, not a bare positive, EROI. Future research might further "extend" our "$EROI_{ext}$" by including the energy of all of the people and economic activity included directly and indirectly to deliver the energy. Because, as we have indicated, roughly 5% of the economy (money spent) is associated with getting oil (this includes even those farmers who grow the grain for workers or the laborers who build the airplanes used indirectly to feed laborers or to get engineers to the site), and about 10% for all energy, we might say that as a nation that part of the denominator for the $EROI_{ext}$ would be 10% of all of the energy used in the country.

EROI for U.S. and North American Domestic Resources and Its Implications for the "Minimum EROI"

We start with historical, ecological, and evolutionary considerations, both because they have helped us a great deal to clarify our own perspectives on these issues and because in the unsubsidized world where evolution operates there are no bailouts or explicit subsidies, a very different situation from the one in which we operate in human society today.

In the past Charles Hall worked with Cutler Cleveland and Robert Kaufmann to define and calculate the energy return on investment of the most important fuels for the United States' economy [7, 8]. Since that time Cleveland and Hall have undertaken additional and updated analyses for the U.S. oil and gas industry [10, 11] and Nate Gagnon and Hall have done that for the world average [13]. Our results indicate that there is still a very large energy surplus from fossil fuels: variously estimated as an EROI (i.e., $EROI_{mm}$) from perhaps 80–1 (domestic coal and perhaps gas) to 10–18 to 1 (U.S. oil and gas) and 18 to 1 for contemporary oil and gas globally (Table 14.1). In other words, for every barrel of oil, or its equivalent, invested globally in seeking and producing more oil some 10–20 barrels are delivered to society. The ratio is higher, but also declining, for coal and gas. Thus fossil fuels still provide a very large energy surplus, obviously enough to run and expand the human population and the very large and complex industrial societies around the world. This surplus energy of roughly 20 or more units of energy returned per unit invested in getting it, plus the large agricultural yields generated by fossil-fueled agriculture, allows a huge surplus quantity of energy, including food energy, to be delivered to society. This in turn allows most people and capital to be employed somewhere else other than in the energy industry. In other words these huge energy surpluses have allowed the development of all aspects of our civilization, both good and bad.

But the problem with substitutes to fossil fuels is that, of the alternatives available, none appear to have the desirable traits of fossil fuels. These include: (1) sufficient energy density (Table 3.1), (2) transportability, (3) relatively low environmental impact per net unit delivered to society, (4) relatively high EROI, and (5) are obtainable on a scale and timing that society presently demands. Thus it would seem that society, both the United States and the world, is likely to be facing a decline in both the quantity and EROI of its principal fuels. The principal benefit of alternative fuels is that they emit less carbon into the atmosphere.

The Energy Used to Run the Economy

In 2005 the U.S. gross domestic product was about 12.4 trillion dollars, and the economy used about 105 Exajoules (or 10 exp 18 joules). Dividing the two we find that we used an average of about 8.5 MJ to generate one dollar's worth of goods and services in 2005. The amount of energy used to generate a unit of real GDP barely changes for most countries of the world over time and for the United States before 1984. Cleveland et al. [8] found a very high correlation between quality-corrected energy use and GDP from 1904 to 1984. Since then, however, the economy has increased much faster than energy use. Whether this is a true efficiency increase is debatable. If one uses inflation rates calculated using the pre-Clinton era equation for CPI (such as that provided by www.shadowstatistics.com), the GDP does not increase as rapidly over time and a relatively tight relation between GDP and energy use returns.

Gasoline at $3 per gallon delivers about 44 MJ/dollar (at 131 MJ/gallon of gasoline), plus roughly another 10% to get that gasoline (refinery cost = 4 MJ), so if you spend one dollar on energy directly versus one dollar on general economic activity you would consume about 48/8.4 or 5.7 times more energy. If energy costs were to increase indefinitely then at some point all the economic activity would be required just to pay for the oil. In fact when oil prices increase above about 5%, and energy prices increase above about 10% of the total GDP, the economy has invariably gone into recession (see Fig. 18.6).

EROI of Obtaining Energy Through Trade

An economy without enough domestic fuels of the type it needs must import the fuels and pay for them with some kind of surplus economic activity. The ability to purchase critically required energy depends upon what else it can generate to sell to the world as well as the fuel required to grow or produce that material. The EROI for the imported fuel is the relation between the amount of fuel bought with a dollar relative to the amount of dollar profits gained by selling the goods and services for export. The quantity of the goods or services that need to be exported to attain a barrel of oil depends upon the relative prices of the fuel versus the exported commodities.

Kaufmann [21] estimated from roughly 1950 through the early 1980s the energy cost of generating a dollar's worth of our major U.S. exports, including wheat, commercial jetliners, and the like, and also the chemical energy found in one dollar's worth of imported oil. The concept was that the EROI for imported oil depended upon what proportion of an imported dollars' worth of oil you needed to use to generate the money from overseas sales that you traded, in a net sense, for that oil. He concluded that before the oil price increases of the 1970s the EROI for imported oil was about 25:1, very favorable for the United States, but that dropped to about 9:1 after the first oil price hike in 1973 and then down to about 3:1 following the second oil price hike in 1979. The ratio returned to a more favorable level (from the perspective of the United States) from 1985 to about 2000 as the price of exported goods increased through inflation more rapidly than the price of oil. As oil prices increase again in this decade, however, as more of the remaining conventional oil is concentrated in fewer and fewer countries, and with the future supply of abundant conventional oil in question, that ratio has again declined to roughly 10:1 [9]. Estimating the EROI of obtaining energy through trade may be very useful in predicting economic vulnerability in the near future.

To some degree we have managed to continue to purchase foreign oil through debt, which gives us a temporarily higher EROI. Were we to pay off this debt in the future, and those who got the dollars wished to turn them into real goods and services (which seems a reasonable assumption), then we will have to take some substantial part of our remaining energy reserves out of the ground and convert it into fish, rice, beef, Fords, and so on that those people would be able to buy from us.

The Tradeoff Between EROI and Total Energy Used in Generating "Civilization"

The basic goods and services that we desire and require to have what we call modern civilization are highly dependent upon the delivery of net energy to society. This is a point made again and again by the authors quoted in the introduction to this chapter, although they do not provide any quantification. The total net energy that we have at our disposal in the United States, say roughly 95% of the 105 or so EJ, would decrease to 80 if the cost of energy were to double (as happened in the first part of 2008), or down to 60 if it were to double again and so forth, all of which is very possible. Thus as the EROI declines over time, the surplus wealth available to do important things in society such as healthcare, higher education, the arts, and so on, declines. We believe that the issues of peak oil and declining EROI are at least partly behind the increasing situation today where pensions, state and university budgets, healthcare plans, and the like become impossible to maintain at the level people have become used to expect. From this perspective we think it very likely that declining EROI is likely to dominate our future economy and quality of life.

Conclusion

Our educated guess is that the minimum $EROI_{mm}$ for a fuel to deliver a given service (i.e., miles driven, house heated) to the consumer is about 3.3:1 when properly accounting for all of the additional energy required to deliver and use that fuel. This ratio would increase substantially if the energy cost of supporting labor (generally considered consumption by economists although definitely part of production here), business services, or compensating for environmental destruction were included. (See Henshaw et al. [23] for a comprehensive view of this.) It is possible to imagine that one might use a great deal of fuel with an EROI of 1.1: 1 to pay for the use of one barrel by the consumption of another one or two or ten, but we believe it more appropriate to include the cost of using the fuel in the fuel itself. Thus we introduced the concept of "extended EROI" which includes not just the energy of getting the fuel, but also of transporting and using it. This process approximately triples the $EROI_{mm}$ required for fuel to make a contribution to the economy. Any fuel with an EROI less than the mean for society (about 20–30 to 1) may in fact be subsidized by the general petroleum economy. For instance, fuels such as corn-based ethanol that have marginally positive EROIs (1.3: 1) are subsidized by the infrastructure support (i.e., construction and maintenance of roads and vehicles) undertaken by the main economy which is two thirds based on oil and gas. These may be more important questions than the exact math for the fuel itself, although all are important.

Finally future analyses might even go so far as to include the money/energy to support and replace the oil worker or truck driver. We believe this is important as there is little argument about the need to amortize the maintenance and depreciation of the oil derrick, so why not some prorated portion of medical care for the worker or education of his or her children for eventual replacement of the worn-out worker? Economists have some serious problems with this line of reasoning because they say that, for example, medical care of workers or their children is consumption, not production. But, as with energy itself, a certain amount of consumption is essential for production and maybe we need to rethink when and how we draw the line between them. Perhaps it is best considered from the perspective of the two paragraphs above: as the EROI of fuels continues to decline in the future then the rest of us will be supporting more and more workers in the energy industry, and there will be fewer and fewer dollars and energy delivered to the rest of society.

The 3.3:1 minimum "extended EROI" that we calculate here is only a bare minimum for civilization. It would allow only for energy to run transportation systems, but would leave little discretionary surplus for all the things we value within civilization: art, medicine, education, and so on; things that use energy but do not directly contribute to getting more energy or other resources. If we are to support all the infrastructure needed to train engineers, physicians, and skilled laborers needed by society we would need a far higher EROI from our primary fuels. The calculation of this is beyond the scope of this chapter but our guess is that we would need something like a 5:1, and probably a 10:1 EROI from our main fuels to maintain anything like what we call civilization. Perhaps a future paper could undertake these difficult calculations.

Questions

1. Define net energy, energy surplus, and EROI. Are they just different ways to say the same thing?
2. Who were some of the pioneering thinkers about net energy?
3. Why do they think of net energy as a "determinant of human culture"? Do you agree?
4. Define the economic cost of energy.
5. What are some of the precedents that led to the development of the concept of EROI? What was the role of fish?
6. EROI can be calculated at various points, starting at the wellhead or the farm gate. Give some additional places in the energy use "food chain" where it might be useful to calculate EROI.
7. Do you think that including the energy to maintain the infrastructure required to use a fuel should be included in an EROI assessment? Why or why not?
8. What are some typical EROIs for various fuels in the United States?

9. Why would environmental considerations change an estimate for EROI?
10. What is the approximate proportion of our economy attributable to energy costs?
11. Explain how a nation with no energy resources invests energy to get energy.
12. What is the relation between EROI and the amount of, for example, education, medical care, and culture that a society can sustain?

References

1. Modified from Hall, Balogh and Murphy 2008, which contains supporting references. An update is available in Murphy, Hall and Cleveland (in press).
2. Odum, H.T. (1972). Environment, power and society. New York: Wiley-Interscience.
3. Odum, H. T. 1973. Energy, ecology and economics. Royal Swedish Academy of Science. AMBIO 2(6):220–227.
4. Hall, C.A.S. 1972. Migration and metabolism in a temperate Stream Ecosystem. Ecology. 53: 585–604.
5. Hall, C.A.S. and C.J. Cleveland. 1981. Petroleum drilling and production in the United States: Yield per effort and net energy analysis. Science 211: 576–579.
6. Hall, C.A.S., C. Cleveland and M. Berger. 1981. Energy return on investment for United States petroleum, coal and uranium, p. 715–724. in W. Mitsch (ed.), Energy and Ecological Modeling. Symp. Proc., Elsevier Publishing Co.
7. Cleveland, C.J., R. Costanza, C.A.S. Hall and R. Kaufmann. 1984. Energy and the United States economy: a biophysical perspective. Science 225: 890–897.
8. Hall, C.A.S.; Cleveland, C.J.; Kaufmann, R. 1986. Energy and resource quality: the ecology of the economic process, Wiley: New York, 1986
9. The Oil Drum: search for "Hall EROI". A summary is in: Heinberg, R. 2010 Searching for a Miracle: 'Net Energy' Limits & the Fate of Industrial Society. Post Carbon Institute and International Forum on Globalization
10. Cleveland, C. J. (2005). Net energy from the extraction of oil and gas in the United States. Energy: The International Journal. 30(5): 769–782.
11. Guilford, M.C., P. O'Connor, C. A.S Hall, and C. J. Cleveland. A new long term assessment of EROI for U.S. oil and gas production and discovery. Sustainability (in press).
12. Murphy, D., Hall, C. and R. Powers. 2010. New perspectives on the energy return on (energy) investment (EROI) of corn ethanol. Environment, Development, Sustainability. 13: 179–202.
13. Gagnon, N. C. A.S. Hall, and L. Brinker. 2009. A preliminary investigation of energy return on energy investment for global oil and gas production. Energies 2009, 2(3), 490–503.
14. Sundberg, U. 1992. Ecological economics of the Swedish Baltic Empire: An essay on energy and power, 1560–1720. Ecological Economics. 5: 51–72.
15. see review by Farrell et al. 2006, [Farrell, A.E.; Plevin, R.J.; Turner, B.T.; Jones, A.D.; O'Hare, M.; Kammen, D.M. Ethanol can contribute to energy and environmental goals. Science 2006, 311, 506–508.] as well as the many responses in the June 23, 2006 issue of Science Magazine for a fairly thorough discussion of this issue.
16. Murphy, D.J., C.A.S. Hall and C.J. Cleveland. (In preparation). Developing an EROI protocol. Sustainability.
17. Bullard, C.W.; Hannon, B.; Herendeen, R.A. 1975. Energy flow through the US economy. University of Illinois Press: Urbana, 1975.
18. Hannon B. (1981). Analysis of the energy cost of economic activities: 1963 2000. Energy Research Group Doc. No. 316. Urbana: University of Illinois.
19. Herendeen, R. and Bullard, C. 1975. The energy costs of goods and services. 1963 and 1967, Energy Policy: 3: 268.
20. Costanza R. 1980. Embodied energy and economic valuation. Science 210:1219–1224.
21. Kauffman 1986. Energy return on investment for imported petroleum, p. 697–702. in W. Mitsch (ed.), Global Dynamics of Biospheric Carbon. U.S. Department of Energy CO2 Research Series 19. Washington, D.C. (see also chapter 8 of reference 8).
22. Hall, C. A. S., S. Balogh and D.J. Murphy. 2009. What is the Minimum EROI that a Sustainable Society Must Have? Energies. 2: 25–47.
23. Henshaw, P.,C. King, and J. Zarnikau.(in press). System Energy Assessment (SEA), Defining a physical measure of EROI for energy businesses as whole systems. Sustainability Energy.

15

Peak Oil, EROI, Investments, and Our Financial Future[1]

Introduction

The expansion of the human population and the economies of the United States and many other nations in the past 100 years have been facilitated by a commensurate expansion in the use of fossil fuels. To many energy analysts that expansion of cheap fuel energy has been far more important than business acumen, economic policy, or ideology, although they too may be important [1–15]. Although we are used to thinking about the economy in monetary terms, those of us trained in the natural sciences consider it equally valid to think about the economy and economics from the perspective of the energy required to make it run. When one spends a dollar, we do not think just about the dollar bill leaving our wallet and passing to someone else's. Rather, we think that to enable that transaction, to generate the good or service being purchased, an average of about 8,000 kilo-Joules of energy (roughly the amount of oil that would fill a standard coffee cup) must be extracted and turned into roughly a half kilogram of carbon dioxide. Take the money out of the economy and it could continue to function through barter, albeit in an extremely awkward, limited, and inefficient way. Take the energy out and the economy would immediately contract. Cuba found this out in 1991 when the Soviet Union, facing its own oil production and political problems, cut off Cuba's subsidized oil supply. Both Cuba's energy use and its GDP declined immediately by about one third, groceries disappeared from market shelves within

a week, and soon the average Cuban lost 20 lb [16]. Cuba subsequently learned to live, in some ways well, on about half the oil as previously, but the impacts were significant and the transition was difficult. Yet Cuba moved away from monocrop agriculture to food production. There are more rooftop gardens per capita in Havana than in any other city. The United States has become more efficient in using energy in recent decades, however, most of this is due to using higher-quality fuels, exporting heavy industry, and switching what we call economic activity (e.g., [15]), and many other countries, including efficiency leader Japan, are becoming substantially less efficient [17–20].

The Age of Petroleum

The economy of the United States and the world is still based principally on "conventional" petroleum, meaning oil, gas, and natural gas liquids (Fig. 15.1). *Conventional* means those fuels derived from geological deposits, usually found and exploited using drill-bit technology. Conventional oil and gas flows to the surface because of its own pressure or with pumping or additional pressure supplied by injecting natural gas, water, or occasionally other substances into the reservoir. *Unconventional* petroleum includes shale oil, tar sands, and other bitumens usually mined as solids and converted to liquids and also natural gas from coal beds or "tight" deposits where the gas is found in low concentrations in rock. For the economies of both the United States

C.A.S. Hall and K. Klitgaard, *Energy and the Wealth of Nations: Understanding the Biophysical Economy*, DOI 10.1007/978-1-4419-9398-4_15, © Springer Science+Business Media, LLC 2012

Fig. 15.1 Pattern of
energy use for the world
(*Source*: Jean Laherrere).
Gtoe means gigatons oil
equivalent, with one
Gigaton equal to 41900
GigaJoules. The US
pattern looks broadly
similar, but at about 20
percent

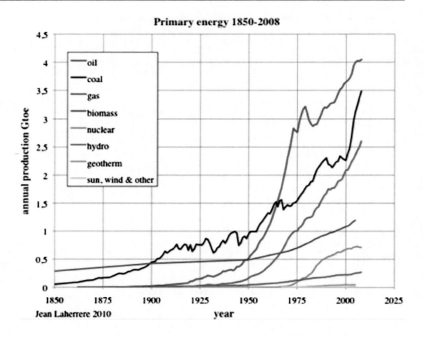

Primary energy 1850-2008

and the world, nearly two thirds of our energy
comes from conventional petroleum, about 40%
from liquid petroleum and another 20–25% from
gaseous petroleum [21] (Fig. 15.1). Coal and natural gas provide most of the rest of the energy that
we use. Hydroelectric power and wood together
are renewable energies generated from current
solar input and provide about 5% of the energy
that the United States uses. "New renewables"
include windmills and photovoltaics. In recent
years the annual increase in oil and gas use is
much greater than the new quantities coming from
the new renewables, or indeed their total production, so that their are not displacing fossil fuels but
just adding to the mix. All of these proportions
have not changed very much since the 1970s in
the United States or the world. We believe it most
accurate to consider the times that we live in as *the
age of petroleum*, for petroleum is the foundation
of our economies and our lives. Just look around.

Petroleum is especially important because of
its unique attributes leading to high economic
utility including very high energy density and
transportability [21], massive availability, and
relatively low price. Its future supply, however, is
worrisome. The issue is not the point at which oil
actually runs out but rather the relation between

supply and potential demand. Barring a massive
worldwide recession, demand will continue to
increase as human populations, petroleum-based
agriculture, and economies (especially Asian)
continue to grow. Petroleum supplies have been
growing since 1900 at roughly four or five, but
later two or three, percent per year, a trend that
most observers think cannot continue [22, 23].
Peak oil refers to the time at which an oil field, a
nation, or the entire world reaches its maximum
oil production and then declines. It is not some
abstract issue debated by theoretical scientists or
worried citizens but an actuality that occurred in
the United States in 1970 and in some 60 (of 95)
other oil-producing nations since [24–27]. Several
prominent geologists have suggested that it may
have occurred already for the world, although
that is not clear yet [28–30]. With global demand
showing no sign of abating at some time it will
not be possible to continue to increase petroleum supplies, or even to maintain current levels
of supply, regardless of technology or price. At
this point we have or will enter the second half
of the age of oil [30]. The first half was one of
year-by-year growth; the second half will be of
year-by-year decline in supply, with possibly an
"undulating plateau" around the peak. Some help

from still-abundant natural gas will separate the two halves and buffer the impact somewhat for a decade or so. We are of the opinion that it will not be possible to fill in the growing gap between supply and demand of conventional oil with alternatives on the scale required [31], and even were that possible the investments in money, energy, and time required would mean that we needed to start some decades ago [32]. When or as the decline in global oil production begins we will see the "end of cheap oil" and a very different economic climate, one that appears to have started in 2008.

The very large use of fossil fuels in the United States means that each of us has the equivalent of 60–80 hard working laborers to "hew our wood and haul our water" as well as to grow, transport, and cook our food; make, transport, and import our consumer goods; provide sophisticated medical and health services; visit our relatives; and take vacations in faraway or even relatively nearby places. Simply to grow our food requires the energy of about a gallon of oil per person per day, and if a North American takes a hot shower in the morning he or she will have already used far more energy than probably two thirds of the Earth's human population use in an entire day.

How Much Oil Will We Be Able to Extract?

So the next important question is how much oil and gas are left in the world? The answer is a lot, although probably not a lot relative to our increasing needs, and maybe not a lot that we can afford economically or energetically. Although we will probably always have enough oil to lubricate our bicycle chains, the question is whether we will have anything like the quantity that we use now at the prices that allow the things we are used to having, and whether growth is possible. Worldwide we have consumed a little over 1.1 trillion barrels of oil, mostly in the past 25 years. The current debate is fundamentally about whether there are 1, 2, or even 3.5 trillion barrels of economically extractable oil left. Fundamental to this debate, yet mostly ignored, is an understanding of the

capital, operating, and environmental costs, in terms of both money and energy, necessary to find, extract, and use whatever new sources of oil remain to be discovered. We also need to consider whatever alternatives we might be able to develop. These investment issues, in terms of both money and energy, will become ever more important.

There are two distinct camps for this issue. One camp, the "technological cornucopians," led principally by economists such as Michael Lynch [33, 34], believes that market forces and technology will continue to supply (at a price) whatever oil we need for the indefinite future. They argue that we now are able to extract only some 35% of the oil from a field, that large areas of the world (deep ocean, Greenland, Antarctica) have not been explored and may have substantial supplies of oil, and that substitutes, such as oil shale and tar sands, abound. They are buoyed by the failure of many earlier predictions of the demise or peak of oil, two recent and prestigious analyses by the U.S. Geological Survey and the Cambridge Energy Research Associates that tend to suggest that remaining extractable oil is near the high end given above, the recent discovery of the deepwater Jack 2 well in the Gulf of Mexico and the development of the Alberta Tar Sands.

A second camp, the "peak oilers," is composed of scientists from diverse fields inspired by the pioneering work of M. King Hubbert [24], a few very knowledgeable politicians such as U.S. Congressman Roscoe Bartlett of Maryland, private citizens from all walks of life and, increasingly, members of the investment community. All believe that there remains only about one additional trillion barrels of extractable conventional oil and that the global peak, a "bumpy plateau," will occur soon, or, perhaps, has already occurred. The arguments of these people and their organization, the Association for the Study of Peak Oil (ASPO), was spearheaded by the analyses and writings of geologists Colin Campbell and Jean Laherrere. They are supported by the many other geologists who agree with them, the many peaks that have already occurred for many dozens of oil-producing countries, the recent collapse of production from some of our most important oil fields, and that we now extract and use two to four barrels of oil for each new barrel discovered (Fig. 3.3). They also believe that essentially all regions of the Earth favorable for oil production have been well explored for oil, and there are few surprises left except perhaps in regions that will be nearly impossible to exploit.

There are several issues that tend to add confusion to the issue of peak oil. First, some people do, and some do not, include natural gas liquids or condensate (liquid hydrocarbons that condense out of natural gas). These can be refined readily into motor fuel and other uses so that many investigators think they should simply be lumped with oil, which most usually they are. Because a peak in global natural gas production is thought to be likely one or two decades after a peak in global oil, inclusion of natural gas liquids extends the time or duration of whatever oil peak has occurred or may be occurring (Fig. 3.7). The second is what characteristics of the peak will cause the largest economic impact. Is it the peak itself, or the ratio between the declining production rate and the potential consumption rate. Both the production and the consumption of oil and also natural gas had been growing at roughly 4% a year before 1930 declining gradually to 2% by 2005 and 1% or not at all since. The great expansion of the economies of China and India have recently more than compensated for some reduced use in other parts of the world. Meanwhile the growth rate of the human population has continued, so that "per capita peak oil" probably occurred in 1978 [35]. What the future holds may have more to do with the desired consumption rate than the production rate. If and when peak petroleum extraction occurs prices will rise. Any economic slowdown would decrease oil use which might decrease prices and . . . the chickens and eggs can keep going for some time. That is why many peak oilers speak of a bumpy plateau.

The rates of oil and gas production (more accurately extraction) and the onset of peak oil are dependent upon interacting geological, economic, and political factors. The usual economic argument is that if supply is reduced relative to demand then the price will increase which will then signal oil companies to drill more, leading to the discovery of more oil and then additional supply. Although that sounds logical the results

from the oil industry might not be in accordance to that logic as the empirical record shows that the rate at which oil and gas are found has little to do with the rate of drilling (Fig. 3.3). Recent experience may be changing that for "tight gas," where limited gas can be obtained by drilling many low yielding wells.

Finally, output can be limited or (at least in the past) enhanced for political reasons, which are even more difficult to predict than the geological restrictions. Certainly the events of the "Arab spring" of 2011 were completely unpredictable. Empirically there is a fair amount of evidence from postpeak countries, such as the United States, that the physical limitations become important when about half of the ultimately recoverable oil has been extracted. But why should that be? In the United States it certainly was not due to a lack of investment, inasmuch as most geologists believe that the United States had been overdrilled. We probably will not know until we have much more data, and much of the data are closely guarded industry or state secrets. According to one analyst if one looks at all of the 60 or so postpeak oil-producing countries the peak occurs on average when about 54% of the total extractable oil in place has been extracted [36]. Finally oil-producing nations often have high population and economic growth, and are using an increasing proportion of their own production [37].

The United States clearly has experienced "peak oil." As the price of oil increased by a factor of ten, from 3.50 to 35 dollars a barrel during the 1970s, a huge amount of capital was invested in U.S. oil discovery and production efforts. The drilling rate increased from 95 million feet per year in 1970 to 250 million feet in 1985. Nevertheless the production of crude oil decreased during the same period from the peak of 3.52 billion barrels a year in 1970 to 3.27 in 1985 and has continued to decline to 1.89 in 2005 even with the addition of Alaskan production. Natural gas production has also peaked and declined, although less regularly and with a possible new peak (Fig. 3.7). Thus despite the enormous advancement of petroleum discovery and production technology, and despite very significant investment, U.S. oil production has continued its downward trend nearly every year since

1970. When drilling rates are high apparently poorer prospects, on average, tend to be drilled. The technological optimists are correct in saying that advancing technology is important. But there are two fundamental and contradictory forces operating here: technological advances and depletion. In the U.S. oil industry it is clear that depletion is trumping technological progress, as oil production is declining and oil is becoming much more expensive to produce. Contrary to market theory increases in prices do not necessarily lead to increased production, and in fact because oil exploration is very energy intensive it can lead to less oil being delivered to society.

Decreasing Energy Return on Investment

Energy return on investment is simply the energy that one obtains from an activity compared to the energy it took to generate that energy. The calculations are generally straightforward, although the data may be difficult to get and the boundaries uncertain (see previous chapter). When the numerator and denominator are derived in the same units, as they should be, the units can be barrels per barrel, kcals per kcal, or Mjoules per Mjoule, the results are in a unitless ratio. The running average EROI for the finding U.S. domestic oil has dropped from greater than 1000 kJ returned per kilojoule invested in 1919 to about 5 to 1 today. The value for producing that oil has declined from 30 to 1 in the 1970s to around 10 to 1 today. This illustrates the decreasing energy returns as oil reservoirs are increasingly depleted and as there are increases in the energy costs as exploration and development are increasingly deeper and offshore [13, 21, 38]. Even that ratio reflects mostly pumping out oil fields that are half a century or more old because we are finding few significant new fields. The increasing energy cost of a marginal barrel of oil or gas is one of the factors behind their increasing dollar cost, although if one corrects for general inflation the price of oil has increased only a moderate amount.

The same pattern of declining energy return on energy investment appears to be true for global petroleum production, but getting such information is very difficult. With help from the extensive financial database on "upstream" (i.e., preproduction) maintained by the John H. Herold Company, Gagnon and colleagues [38] were able to generate an approximate value for global EROI for finding new oil and natural gas (considered together). Our results indicate that the EROI for global oil and gas (at least for that which was publicly traded) was roughly 23:1 in 1992, increased to about 33:1 in 1999, and since then has fallen to approximately 18:1 in 2005. The apparent increase in EROI during the late 1990s reflects the effects of reduced drilling effort, as was seen for oil and gas in the United States (e.g., Fig. 3.3). If the rate of decline continues linearly for several decades eventually it would take the energy in a barrel of oil to get a new barrel of oil. Although we do not know whether that extrapolation is accurate, essentially all EROI studies of our principal fossil fuels do indicate that their EROI is declining over time, and that EROI declines especially rapidly with increased exploitation (e.g., drilling) rates. This decline appears to be reflected in economic results. In November of 2004 *The New York Times* reported that for the previous 3 years oil exploration companies worldwide had spent more money in exploration than they had recovered in the dollar value of reserves found. This illustrates that even though the EROI for producing oil and gas globally is still about 18:1, it is possible that the energy breakeven point has been approached for finding new oil. Whether we have reached this point or not the concept of EROI declining toward 1:1 makes irrelevant the reports of several oil analysts who believe that we may have substantially more oil left in the world. It simply does not make sense to extract oil, at least for fuel, when it requires more energy for the extraction than is found in the oil extracted.

How we weather this coming storm will depend in large part on how we manage our investments now. There are three general types of investments that we make in society. The first is investments in getting energy itself, the second is investments for maintenance of, and replacing, existing infrastructure, and the third is for discretionary expansion. In other words before we can think about expanding the economy we must first make the investments in getting the energy necessary to operate the existing economy, and also in maintaining the infrastructure that we have to compensate for the entropy-driven degradation of what we already have. The required investments in the second and especially the first category are likely to increasingly limit what is available for the third. The dollar and energy investments needed to get the energy needed to allow the rest of the economy to operate and grow have been very small historically, but this is likely to change dramatically. This is true whether we seek to continue our reliance on ever-scarcer petroleum or whether we attempt to develop some alternative. Technological improvements, if indeed they are possible, are extremely unlikely to bring back the low investments in energy to which we have grown accustomed.

The main problem that we face is a consequence of the "best first" principle. This is, quite simply, the characteristic of humans to use the highest quality resources first, be they timber, fish, soil, copper ore, or fossil fuels. The economic incentives are to exploit the highest-quality, least-cost (both in terms of energy and dollars) resources first (as was noted by economist David Ricardo in 1962 [39]). We have been exploiting fossil fuels for a long time. The peak in finding new oil was in the 1930s for the United States and in the 1960s for the rest of the world. Both have declined enormously since then. An even greater decline has taken place in the efficiency with which we find oil, that is, the amount of energy that we find relative to the energy we invest in seeking and exploiting it.

That pattern of exploiting and depleting the best resources first also is occurring for natural gas. Natural gas was once considered a dangerous waste product of oil development and was flared at the well head. But during the middle years of the last century large gas pipeline systems were developed in the United States and Europe that enabled gas to be sent to myriad users who appreciated its ease of use and cleanliness, including its relatively low carbon dioxide emissions, at least relative to coal. Originally U.S. natural gas came from large fields, often associated with oil fields, in Louisiana, Texas, and Oklahoma. Its production has moved increasingly to smaller fields distributed throughout Appalachia and the Rockies. A national peak in production occurred in 1973 as the largest fields that traditionally supplied the country peaked and declined. Later as "unconventional" fields were developed a second, somewhat smaller peak occurred in the 2000s. Gas production has fallen by about 6% from that peak, and some investigators predict a "natural gas cliff" as conventional gas fields are increasingly exhausted and as it is increasingly difficult to bring smaller unconventional fields on line to replace the depleted giants. However, this "cliff" appears unlikely to occur for at least several decades because of the new technologies of horizontal drilling and hydrofracturing, which as of this writing are bringing in new "unconventional" gas at just about the rate that the conventional supplies are declining. It is quite difficult to predict the future of natural gas because of the many environmental and social issues associated with horizontal drilling and fracking.

The Balloon Graph

All sources of energy used in the economy, except the free solar energy that drives ecosystem processes, have an energy cost, and all of them have different magnitudes of importance to society. The energy cost of obtaining coal or oil or photovoltaic electricity is straightforward even if difficult to calculate, but there are other sources and other ways payment is needed. For example, we pay for imported oil in energy as well as dollars, for it takes energy to grow, manufacture, or harvest what we sell abroad to gain the dollars with which we buy the oil, (or we must in the future if we pay with debt today). In 1970 we gained roughly 30 megajoules for each megajoule used to make the crops, jet airplanes, and so on, that we exported [40]. But as the price of imported oil increased, the EROI of the imported oil declined. By 1974 that ratio had dropped to 9:1, and by 1980 to 3:1. The subsequent decline in the price of oil, aided by the inflation of the export products traded, eventually returned the energy terms

Fig. 15.2 "Balloon graph" representing quality (*y* graph) and quantity (*x* graph) of the United States economy for various fuels at various times. Arrows connect fuels from various times (i.e., domestic oil in 1930, 1970, 2005), and the size of the "balloon" represents part of the uncertainty associated with EROI estimates (*Source*: US EIA, Cutler Cleveland, and C. Hall's own EROI work. Note 1930 value is for finding, not producing, oil)

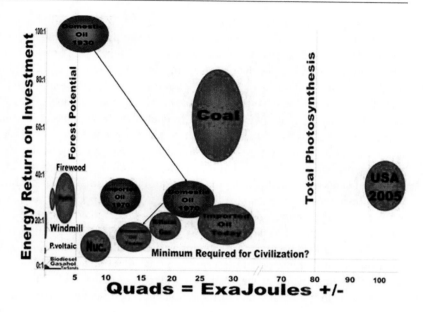

of trade to something like it was in 1970, at least until the price of oil started to increase again after 2000, again lowering the EROI of imported oil. A rough estimate of the quantity used each year and the EROI of various major fuels in the United States, including possible alternatives, is given in Fig. 15.2. An obvious aspect of that graph is that qualitatively and quantitatively alternatives to fossil fuel have a very long way to go to fill the pumps of fossil fuels. This is especially true when one considers the additional qualities of oil and gas, including energy density, ease of transport, and ease of use. The alternatives to oil available to us today are characterized by even lower EROIs, limiting their economic effectiveness. It is critical for CEOs and government officials to understand that the best oil and gas are simply gone, and there is no easy replacement.

If we are to supply into the future petroleum at the rate that the United States consumed in recent decades, let alone an increase, it will require enormous investments in either additional unconventional sources or payments to foreign suppliers. That will mean a diversion of the output of our economy from other uses into getting the same amount of energy just to run the existing economy. In other words from a national perspective investments will be needed just to run what

we have, not to generate new real growth. If we do not make these investments our energy supplies will falter, and if we do the returns to the nation may be small, although the returns to the individual investor may be large. Furthermore, if this issue is as important as we believe it is, then we must pay much more attention to the quality of the data we are getting about the energy costs of all things we do, including getting energy. Finally the failure of increased drilling to return more fuel (Fig. 3.3) calls into question the basic economic assumption that scarcity-generated higher prices will resolve that scarcity by encouraging more production. Indeed scarcity encourages more exploration and development activity, but that activity does not necessarily generate more resources. Oil scarcity will also encourage the development of alternative liquid fuels, but their EROIs are generally very low (Fig. 15.2).

Economic Impacts of Peak Oil and Decreasing EROI

Whether global peak oil has occurred already or will not occur for some years or, conceivably, decades, its economic implications will be large because we have no feasible substitute on the scale required and at the EROI that is needed. Any alternatives will require enormous investments in money and energy when both are likely to be in short supply. Despite the projected impact on our economic and business life within a relatively few years, neither government nor the business community is in any way prepared to deal with either the impacts of these changes or the new thinking needed for investment strategies. There are many reasons for this but they include: the role of economists in downplaying the importance of resources in the economy, the disinterest of the media, the failure of government to fund good analytic work on the various energy options, the erosion of good energy record-keeping at the Departments of Commerce and Energy, and the focus of the media on trivial "silver bullets" despite the inability of any one of them (except economic contraction and in some few cases conservation) to contribute anything like 1% to the total energy mix.

Of perhaps greater concern is that none of the top ten or so energy analysts that we are familiar with are supported by government, or generally any, funding. There are not even targeted programs in NSF or the Department of Energy where one might apply if one wishes to undertake good objective, peer-reviewed EROI analyses to see what options might actually be able to contribute significantly. Consequently much of what is written about energy is woefully misinformed or simply advocacy funded by various groups that hope to look good or profit from various perceived alternatives. Issues pertaining to the end of cheap petroleum will be the among most important challenges that Western society has ever faced, especially when considered within the context of our need to deal simultaneously with climate change and other environmental issues related to energy. Any business or political leaders who do not understand the inevitability, seriousness, and implications of the end of cheap oil, or who make poor decisions in an attempt to alleviate its impact, are likely to be tremendously and negatively affected. At the same time the investment decisions we will make in the next decade or two will determine whether civilization is to make it through the transition away from petroleum.

What would be the effects of a large increase in the energy and dollar cost of getting our petroleum, or of any restriction in its availability? Although it is extremely difficult to make any hard predictions, we do have the record of the impacts of the large oil price increases of the 1970s as a possible guide. These supply restrictions or "oil shocks" had very serious impacts on our economy which we have examined empirically in past publications [13]. At the time many economists did not think that even large increases in the price of energy would affect the economy dramatically because energy costs were but 3–6% of GDP. But by 1980, following the two "oil price shocks" of the 1970s, energy costs had increased dramatically until they were 14% of GDP. Actual shortages had additional impacts, when sufficient petroleum to run our industries or businesses were not available at any price. Other impacts included an exacerbation of our trade imbalances as more income was diverted overseas, adding to the foreign holdings of our debt and a decrease in discretionary disposable income as more money was diverted to access energy, whether via higher prices for imports, more petroleum exploration, or the development of low EROI alternative fuels. As EROI inevitably declines in the future more and more of the economy's output will have to be diverted into getting the energy to run the economy. This in turn will affect those sectors of the economy that are not essential. Consumer discretionary spending will probably fall dramatically, greatly affecting businesses such as tourism, housing and higher education.

The "Cheese Slicer" Model

We have attempted to put together a conceptual computer model to help us understand what might be the most basic implications of changing EROI on the economic activity of the United States. The model was conceptualized when we examined how the U.S. economy responded to the "oil shocks" of the 1970s. The underlying foundation is the reality that the economy as a whole requires energy (and other natural resources derived from nature) to run, and without these most basic components it will cease to function. The other premise of this model is that the economy as a whole is faced with choices in how to allocate its output in order to maintain itself and to do other things. Essentially the economy (and the collective decision makers in that economy) has opportunity costs associated with each decision it makes. Figure 15.3 shows our basic conceptual model parameterized for 1949 and 1970, before the oil shocks of that decade. The large square represents the structure of the economy as a whole, which we put inside a symbol of the Earth biosphere/geosphere to reflect the fact that the economy must operate within the biosphere [41]. In addition, of course, the economy must get energy and raw materials from outside the

economy, that is, from nature (the biosphere/geosphere). The output of the economy, measured as GDP, is represented by the large arrow coming out of the right side, where the depth of the arrow represents 100% of GDP. For the sake of developing our concept we think of the economy, for the moment, as an enormous dairy industry and cheese as the product coming out of the right-hand side, moving towards the right. This output (i.e., the entire arrow) could be represented as either money or embodied energy. We use money in this analysis but the results are probably not terribly different from using energy outputs. So, our most important question is, "How do we slice the cheese;" that is, how do we, and how will we, divide up the output of the economy with the

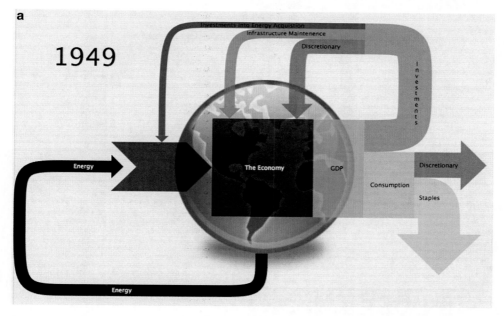

Fig. 15.3 The "cheese slicer" diagrammatic model, which is a basic representation of the fate of the output of the U.S. economy. (Source: Hall et al. 2008) (**a**) 1949 and (**b**) 1970. The box in the middle represents the U.S. economy, the *input arrow* from the *left* represents the energy needed to run the economy, the *large arrow* on the *left* of the box represents the output of the model (i.e. GDP) which is then subdivided as represented by the *output arrow* going to the *right*, first into investments (into getting energy, maintenance, and then discretionary) and then into consumption (either the basic required for minimal food, shelter, and clothing or discretionary). In other words the economic output is "sliced" into dif-

ferent uses according to the requirements and desires of that economy/society. (**c**) Same as (**a**) but for 1981, following large increases in the price of oil. Note change in discretionary investments. (**d**) Same as (**a**) but for 1990, following large decreases in the price of oil. Note change in discretionary investments. (**e**) Same as (**a**) but for 2007, following large decreases in the price of oil. Note change in discretionary investments. (**f**) Same as (**a**) but projected for 2030, with a projection into the future with the assumption that the EROI declines from 20:1 (on average) to 10:1. (**g**) Same as (**a**) but projected for 2050, with a projection into the future with the assumption that the EROI declines to 5:1

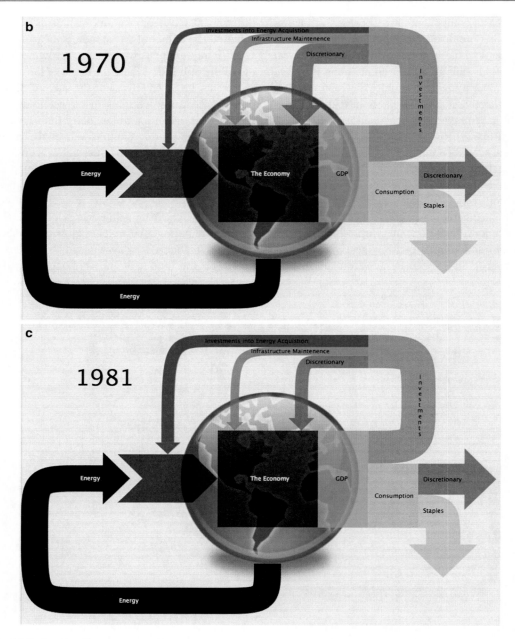

Fig. 15.3 (continued)

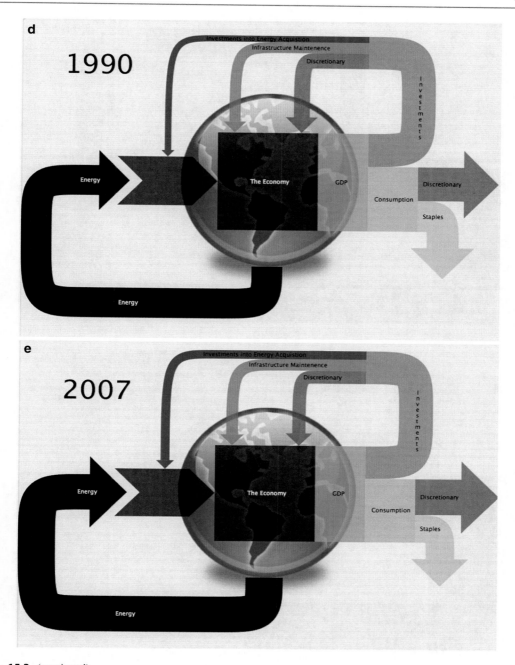

Fig. 15.3 (continued)

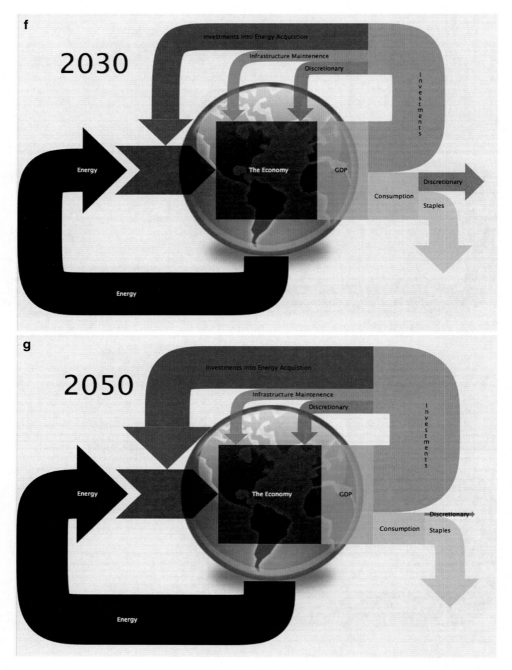

Fig. 15.3 (continued)

least objectionable opportunity cost. Most economists might answer "according to what the market decides," meaning according to consumer tastes and buying habits. But we want to think about it a little differently because we think things might be profoundly different in the future.

Most generally the output of the model (and the economy) has two destinations: *investment* or *consumption*. *Required* expenditures (without which the economy would cease to function) include (1) investments in maintaining societal infrastructure (i.e., repairing and rebuilding bridges, roads, machines, factories, and vehicles, represented by the middle arrow feeding back from output of the economy back to the economy itself), (2) some kind of minimal food, shelter, and clothing for the population (represented by the bottom rightward pointing arrow) required to maintain all individuals in society at the level of the federal minimum standard of living, and (3) the investments in, or payments for, energy (i.e., the amount of economic output used to secure and purchase the domestic and imported energy needed for the economy). This energy is absolutely critical for the economy to operate and must be paid for through proper payments and investments, which we consider together as investments to get energy: no investment in energy, no economic output. This "energy investment" feedback is represented by the topmost arrow from the output of the economy back upstream to the "workgate" symbol [42]. The width of this line represents the investment of energy into getting more energy. Of critical interest here is that as the EROI of our economy's total combined fuel source declines then more and more of the output of the economy must be shunted back to getting the energy required to run the economy if the economy is to remain the same size.

Once these necessities are taken care of what is left is considered the discretionary output of the economy. This can be either discretionary consumption (a vacation or a fancier meal, car, or house than needed, represented by the upper right pointing arrow in the diagrams) or discretionary investment (i.e., building a new tourist destination in Florida or the Caribbean, represented as the

lower of the arrows feeding back into the economy). During the last 100 years the vast wealth generated by the United States economy has meant that we have had an enormous amount of discretionary income. This is in large part because the expenditures for energy represented in Fig. 15.3 have been relatively small in the past.

The information needed to construct the above division of the economy is reasonably easy to come by for the U.S. economy, at least if we are willing to make a few major assumptions and accept a fairly large margin of error. Inflation-corrected GDP, that is, the size of the output of the economy, is published routinely by the U.S. Bureau of Commerce. The total investments for maintenance in the U.S. economy are available as "Depreciation of Fixed Capital" (U.S. Department of Commerce, various years). The minimum needed for food, shelter, and clothing is available as "Personal Consumption Expenditures" (or the minimum of that required to be above poverty), which we selected from the U.S. Department of Commerce for various years. The investment in energy acquisition is the sum of all of the capital costs in all of the energy-producing sectors of the United States plus expenditures for purchased foreign fuel. Empirical values for these components of the economy are plotted in Fig. 15.3a–g. When these three requirements for maintaining the economy – investments and payments for energy, maintenance of infrastructure, and maintenance of people – are subtracted from the total GDP then what is left is discretionary income.

We simulated two basic data streams: the U.S. economy from 1949 to 2005 (representing the growth prior to the "oil crises" of 1973 and 1979) and the impact of the oil crisis and the recovery from that, which had occurred by the mid-1990s. Then we projected these data streams into the future by extrapolating the data used prior to 2005 along with the assumption that the EROI for society declined from an average of roughly 20:1 in 2005 to 5:1 in 2050. This is an arbitrary scenario but may represent what we have in store for us as we enter the "second half of the age of oil," a time of declining availability and rising price when more and more of society's output needs to be diverted into the top arrow of Fig. 15.3.

Results of Simulation

The results of our simulation suggest that discretionary income, including both discretionary investments and discretionary consumption, will move from the present 50 or so percent in 2005 to about 10 percent by 2050, or whenever (or if) the composite EROI of all of our fuels reaches about 5:1 (Fig. 15.3e, f).

Discussion

Individual businesses would be affected by increasing fuel costs and, for many, a reduction in demand for their products as people's income goes increasingly for energy. This simultaneous inflation and recession happened in the 1970s and is projected to happen in the future as EROI for primary fuels declines. According to Keynesian economic theory called the Phillips curve the "stagflation" that occurred in the 1970s was not supposed to happen. But an energy-based explanation is easy [43]. As more money was diverted to getting the energy necessary to run the rest of the economy, disposable income, and hence demand for many nonessentials declined, leading to economic stagnation. Meanwhile the increased cost for energy led to cost-push inflation, as no additional production occurred from higher prices. Unemployment increased during the 1970s but not as much as demand decreased, for at the margin labor became relatively useful compared to increasingly expensive energy. Individual sectors might be much more affected as happened in 2005, for example, with many Louisiana petrochemical companies forced to close or move overseas when the price of natural gas increased. On the other hand alternate energy businesses, from forestry operations and woodcutting to solar devices, might do very well.

When the price of oil increases it does not seem to be in the national or corporate interest to invest in more energy-intensive consumption, as Ford Motor Company found out in 2008 with its former emphasis on large SUVs and pickup trucks. When oil was cheap we overinvested in remote second homes, cruise ships, and Caribbean semiluxury hotels, so that we had a massive loss of the value of real estate. This was called the "Cancun effect"; such hotels require the existence of large amounts of disposable income from the U.S. middle class and cheap energy, even though that disposable income may have to be shifted into the energy sector with less of an opportunity cost to the economy as a whole. Investors who understand the changing rules of the investment game are likely to do much better in the long run, but the consequence of having the "rug" of cheap oil pulled out from the economy will affect us for a long time.

So what can the scientist say to the investor? The options are not easy. As noted above, worldwide investments in seeking oil have had very low returns in recent years. Investments in many alternatives have not fared much better. Ethanol from corn projects may be financially profitable to individual investors because they are highly subsidized by the government, but they are a very poor investment for the nation. It is not clear that ethanol makes much of an energy profit, with an EROI of 1.6 at best, and less than one for one at worst, depending upon the study used for analysis [31, 44]. Biodiesel may have an EROI of about 3:1. Is that a good investment? Clearly it is not relative to remaining petroleum. However real fuels must have EROIs of 5 or 10 or more returned on one invested to not be subsidized by petroleum or coal in many ways, such as the construction of the vehicles and roads that use them. Other biomass, such as wood, can have good EROIs when used as solid fuel but face real difficulties when converted to liquid fuels, and the technology is barely developed. The scale of the problem can be seen by the fact that we presently use several times more fossil energy in the United States than is fixed by all green plant production, including all of our croplands and all of our forests (Pimentel, D., personal communication). Biomass fuels may make more sense in nations where biomass is very plentiful and, more important, where present use of petroleum is much less than in the United States. Alternatively one might argue that if we could bring the use of liquid fuels in the United States down to, say, 20% of the present then liquid fuels from biomass could fill in a substantial portion of that demand. We should remember that historically we in the United States have used energy to produce food and fiber, not the converse, because we have valued food and fiber more highly. Is this about to change?

Energy return on investment from coal and possibly gas is presently quite large compared to alternatives (ranging from perhaps 50:1 to 100:1 at the point of extraction), but there is a large energy premium, perhaps enough to halve the EROI by the time they are delivered to society in a form that society finds acceptable. The environmental costs may be unacceptable, as may be the case for global warming and pollutants derived from coal burning. Injecting carbon dioxide into some underground reservoir seems infeasible for all the coal plants we might build, but it is being pushed hard by many who promote coal. Nuclear has a debatable moderate energy return on investment (5–15:1, some unpublished studies say more). Newer analyses need to be made. Nuclear has a relatively small impact on the atmosphere, but there are large problems with public acceptance and perhaps safety in our increasingly difficult political world.

Wind turbines have an EROI of 15–20 return on one invested, but this does not include the energy cost of backup or electricity "storage" for periods when the wind is not blowing. They make sense if they can be associated with nearby hydroelectric dams that can store water when the wind is blowing and release water when it is not, but the intermittent release of water can cause environmental problems. Photovoltaics are expensive in dollars and energy relative to their return, but the technology is improving. One must be careful about accepting all claims for efficiency improvements because many require very expensive "rare-earth" doping materials, and some may become prohibitively expensive if their use expands greatly [45, 46]. According to one savvy contractor the efficiency in energy returned per square foot of collector has been increasing, but the energy returned per dollar invested has been constant as the price of the high-end units has increased. In addition, although photovoltaics have caught the public's eye the return on dollar investment is about double for solar hot water installations. Wind turbines, photovoltaics, and some other forms of solar do seem to be a good choice if we are to protect the environment, but the investment costs up front will be enormous compared to fossil fuels and the backup issue will be immense. Meanwhile the use of fossil fuel in the past decade has increased relative to all of these solar technologies.

Energy and money are not the only critical aspects of development of energy alternatives. Recent work by Hirsch and colleagues [32] have focused on the investments in time that might be needed to generate some kind of replacement for oil. They examined what they thought might be the leading alternatives to provide the United States with liquid fuel or lower liquid fuel use alternatives, including tar sands, oil shales, deep water petroleum, biodiesel, high MPG automobiles and trucks, and so on. They assumed that these technologies would work (a bold assumption) and that an amount of investment capital equal to "many Manhattan projects" (the project that built the first atomic bomb) would be available. They found that the critical resource was time. Once we decided to make up for the decline in oil availability these projects would need to be started one or preferably two decades in advance of the peak to avoid severe dislocations to the U.S. economy. Given our current petroleum dependence, the rather unattractive aspects of many of the available alternatives, and the long lead time required to change our energy strategy the investment options are not obvious. This, we believe, may be the most important issue facing the United States at this time: where should we invest our remaining high quality petroleum (and coal) with an eye toward ensuring that we can meet the energy needs of the future. We do not believe that markets can solve this problem alone or perhaps at all. Research money for good energy analysis unconnected to this or that "solution" is simply not available.

Human history has been about the progressive development and use of ever higher quality fuels, from human muscle power to draft animals to water power to coal to petroleum. Nuclear at one time seemed to be a continuation of that trend, but that is a hard argument to make today. Perhaps our major question is whether petroleum represents but a step in this continuing process of higher quality fuel sources or rather is the highest quality fuel we will ever have on a large scale. There are many possible candidates for the next main fuel, but few are both quantitatively and qualitatively attractive (Fig. 15.2). In our view we cannot leave these decisions up to the market if we are to solve our future climate or peak oil problems. One possible way to look at the problem, probably not a very popular one with investors or governments, is to pass legislation that would limit energy investments to only "carbon-neutral" ones, remove subsidies from low EROI fuels such as corn-based ethanol, and then perhaps allow the market to sort from those possibilities that remain. Or should we generate a massive scientific effort, as objectively as possible, to evaluate all fuels and make recommendations?

A difficult decision would be whether we should subsidize certain "green" fuels. At the moment alcohol from corn is subsidized four times: in the natural gas for fertilizers, the corn itself through the Department of Agriculture's 100 or so billion dollar general program of farm subsidies, the additional 50 cents per liter subsidy for the alcohol itself and a 50 cents per gallon tariff on imported alcohol. It seems pretty clear that the corn-based alcohol would not make it economically without these subsidies as it has only a marginal (if that) energy return. Are we in effect simply subsidizing the depletion of oil and natural gas (and soil) to generate an approximately equal amount of energy in the alcohol? We think so [31]. Wind energy appears to have about an 18:1 EROI, enough to make it a reasonable candidate, although there are additional costs relative to backup technologies for when the wind is not blowing. So should wind be subsidized, or allowed to compete with other "zero emission" energy sources? A question might be the degree to which the eventual market price would be determined by, or at least be consistent with, the EROI, as all the energy inputs (including that to support labor's paychecks) must be part of the costs. Otherwise that energy is being subsidized by the dominant fuels used by society.

Conclusion

It seems obvious to us that the U.S. economy is very vulnerable to a decreasing EROI for its principal fuels. Increasing effects will come from an increase in expenditures overseas as the price of imported oil increases more rapidly than that of the things that we trade for it, from increased costs for domestic oil and gas as reserves are exhausted and new reservoirs become increasingly difficult to find, and as we turn to lower EROI alternatives such as biodiesel or photovoltaics. Our cheese slicer model suggests that as economic requirements for getting energy increases a principal effect will be a decline in disposable income as a proportion of GDP. Because more fuel will be required to run the same amount of economic activity the potential for increased environmental impacts is very strong. On the other hand protecting the environment, which we support strongly, may mean turning away from some higher EROI fuels to some lower ones. We think all of these issues are very important yet are hardly discussed objectively in our society or even in economic or scientific circles.

Acknowledgments: We thank our great teacher, Howard Odum, many students over the years, colleagues and friends including Andrea Bassi, John Gowdy, Andy Groat, Jean Laherrere, and many others who have helped us to try to understand these issues. Jessica Lambright produced Fig. 15.2 and Nate Hagens made many useful comments. The Santa Barbara Family Foundation, ASPO-USA, The Interfaith Center on Corporate Responsibility and several individuals who wish not to be named provided much appreciated financial help.

Questions

1. What was the experience of Cuba that allows us to understand better the role of energy in an economy?
2. What is meant by the phrase "the second half of the age of oil"?

3. Argue for or against the following question. The important issue is "When will we run out of petroleum?"
4. How much oil do we discover for each barrel that we burn?
5. What happens to pressure as an oil field matures? Why?
6. What is the "cheese slicer" model?
7. Explain the difference between investment and consumption?
8. What is discretionary consumption?
9. What is the "Cancun effect"?
10. What resource do Hirsch and his colleagues think is especially important to adjust to a postpeak oil society?

References

1. Derived from Hall, Charles A. S., Powers, Robert C. and William. Schoenberg. 2008. Peak Oil, EROI, investments and the economy in an uncertain future. pp. 113–136. In Renewable Energy Systems: Environmental and Energetic Issues. Pimentel, D., Ed.; Elsevier: London, 2008.
2. Soddy, F. (1926). Wealth, virtual wealth and debt. New York: E.P. Dutton and Co.
3. Tryon FG. 1927. An index of consumption of fuels and water power. Journal of the American Statistical Association 22: 271–282.
4. Cottrell, F. 1955. *Energy and society*. Dutton, NY: reprinted by Greenwood Press.
5. Georgescu-Roegen, N. 1971. The entropy law and the economic process. Cambridge, MA: Harvard University Press.
6. Odum, H.T. 1972. Environment, power and society. New York: Wiley-Interscience.
7. Kümmel R. (1982). The impact of energy on industrial growth. Energy - The International Journal 7, 189–203.
8. Kümmel, R. 1989. Energy as a factor of production and entropy as a pollution indicator in macroeconomic modelling. Ecological Economics 1, 161–180.
9. Jorgenson D.W. (1984). The role of energy in productivity growth. The American Economic Review 74(2), 26–30.
10. Jorgenson D.W. 1988. Productivity and economic growth in Japan and the United States. The American Economic Review 78: 217–222.
11. Daly, H. E. 1977. Steady-state economics. San Francisco: W. H. Freeman.
12. Dung, T.H. 1992. Consumption, production and technological progress: A unified entropic approach. Ecological Economics, 6: 195–210.
13. Hall, C.A.S., Cleveland, C. J. and Kaufmann R. K. 1986. Energy and resource quality: The ecology of the economic process. New York: Wiley Interscience.
14. Ayres, R.U. 1996. Limits to the growth paradigm. Ecological Economics, 19, 117–134.
15. Ayres, R. U., and Benjamin Warr. 2009. The economic growth engine: how energy and work drive material prosperity. Edward Elger, Northhampton Mass.
16. Quinn, M. 2006. The power of community: How Cuba survived peak oil. Text and film. Published on 25 Feb 2002. Archived on 25 Feb 2006. Can be reached at megan@communitysolution.org
17. Kaufmann, R. 2004. The mechanisms for autonomous energy efficiency increases: A cointegration analysis of the US Energy/GDP Ratio. The Energy Journal 25, 63–86.
18. Hall, C.A.S. & Ko, J.Y. 2006. The myth of efficiency through market economics: A biophysical analysis of tropical economies, especially with respect to energy, forests and water. (pp.40–58). In G. LeClerc & C. A. S. Hall (Eds.) Making world development work: Scientific alternatives to neoclassical economic Albuquerque: University of New Mexico Press)
19. LeClerc & C. A. S. Hall (Eds.) 2006. Making world development work: Scientific alternatives to neoclassical economic theory: University of New Mexico Press. Albuquerque
20. Smil, V. 2007. Light behind the fall: Japan's electricity consumption, the environment, and economic growth. Japan Focus, April 2.
21. EIA 2009. U.S. Energy Information Agency website, accessed June 2009.
22. Campbell, C. and J. Laherrere. 1998. The end of cheap oil. Scientific American (March), 78–83.
23. Heinberg, R. 2003. The Party's Over: Oil, War and the Fate of Industrial Societies. (Gabriella Island, B.C. Canada: New Society Publishers)
24. Hubbert, M. K. 1969. Energy Resources. In Resources and Man. National Academy of Sciences. pp. 157–242 San Francisco: W.H. Freeman.
25. Strahan, D. 2007, The Last Oil Shock: A survival guide to the imminent extinction of petroleum man. Hachette Publisher, London.
26. Energyfiles.com Accessed August 2007. www.energyfiles.com
27. Deffeyes, K. 2005. Beyond oil: The view from Hubbert's Peak. New York: Farrar, Straus and Giroux.
28. EIA. 2007. U.S. Energy Information Agency website, accessed June 2007.
29. IEA 2007 IEA. 2007. European Energy Agency, web page, accessed August 2007.
30. Campbell, C. (2005). The 2nd half of the age of oil. Paper presented at the 5th ASPO Conference, Lisbon Portugal
31. Murphy, D., C. Hall and R. Powers. 2010. New perspectives on the energy return on (energy) investment (EROI) of corn ethanol. Environment, Development, Sustainability 13: 179–202.

32. Hirsch, R., Bezdec, R. and Wending, R. 2005. Peaking of world oil production: impacts, mitigation and risk management. U.S. Department of Energy. National Energy Technology Laboratory. Unpublished Report.

33. Lynch, M. C. 1996. The analysis and forecasting of petroleum supply: sources of error and bias. In D. H. E. Mallakh (Ed.) Energy Watchers VII. International Research Center for Energy and Economic Development.

34. Adelman, M. A. and Lynch, M. C. 1997. Fixed view of resource limits creates undue pessimism. Oil and Gas Journal, 95, 56–60.

35. Duncan, R. C. 2000. Peak oil production and the road to the Olduvai Gorge. Keynote paper presented at the Pardee Keynote Symposia. Geological Society of America, Summit 2000.

36. Brandt, A. R. 2007. Testing Hubbert. Energy Policy 35: 3074–3088. Also energyfiles.com 2007)

37. Hallock, J., Tharkan, P., Hall, C., Jefferson, M. and Wu, W. 2004. Forecasting the limits to the availability and diversity of global conventional oil supplies. Energy 29,1673–1696.

38. Gagnon, N., C.A.S. Hall, and L. Brinker. 2009. A Preliminary Investigation of Energy Return on Energy Investment for Global Oil and Gas Production. Energies 2009, 2(3), 490–503

39. Ricardo, David. 1891. The principles of political economy and taxation. London: G. Bell and Sons. Reprint of 3rd edition, originally pub 1821.

40. Cleveland C. J., Costanza, R., Hall, C.A.S. & Kaufmann, R.K. 1984. Energy and the US economy: A biophysical perspective. Science, 225, 890–897.

41. Hall, C., Lindenberger, D., Kummel, R., Kroeger, T. and Eichhorn, W., 2001. The need to reintegrate the natural sciences with economics. BioScience, 51: 663–673.

42. Odum, H.T. 1994, Ecological and general systems: an introduction to systems ecology. University Press of Coloirado, Niwot, CO, 1994.

43. Hall, C. A. S. 1992 . Economic development or developing economics? (pp. 101–126). In M. Wali (Ed.) Ecosystem rehabilitation in theory and practice, Vol I. Policy Issues The Hague, Netherlands: SPB Publishing.

44. See review in Farrell, A. E., Plevin, R. J., Turner, B. T., Jones, A. D., O'Hare, M. and Kammen, D. M. 2006. Ethanol can contribute to energy and environmental goals. Science Jan. 27 2006 and also the many letters on that article in Science Magazine, June 23, 2006).

45. Andersson, B. A., Azar, C., Holmerg, J. & Karlsson, S. 1998. Material constraints for thin-film solar cells. Energy, 23, 407–411.

46. Gupta, A. In preparation

47. With thanks to : Millennium Institute. Data principally from the U.S. Department of Commerce. Extrapolations via the Millennium Institute's T-21 model courtesy of Andrea Bassi.

The Role of Models for Good and Evil

16

The words "model" and "modeling" are found frequently in economics and in science in general. Therefore it is important that we consider here some of the most important characteristics of these words and concepts and introduce the reader to how they are used in energy studies and in economics. Some of what follows is a deliberate repeat of material in Chap. 12 because the points are important in both places.

Definitions: Models and Analytic Versus Simulation Models

There are many definitions of the word *model*. One is a purposive simplification, such as a model airplane. A second is "a device for predicting a complex whole from the operation of parts that are thought to be known." The definition that we like the most, from Hall and Day [1], is "a formalization of our assumptions about a system." Whether we do so formally or not, most of us use models constantly: models of scientific outcome, models of economic decisions, or models of your own behavior or that of others. Despite the many problems of modeling we do not understand how one can use the scientific method (i.e., to generate and test hypotheses) for any reasonably complicated system without the use of formal modeling. This is as true for management and policy-related issues as for theoretical ones.

The power of models derives from making our assumptions explicit, and hence testable. The power of mathematics (in its broad sense) and of mathematical models is to make the results of a prediction quantitatively explicit and hence quantitatively testable. Generally we are seeking a solution, a quantitative prediction for the value of some variable at some different place or time. The process of examining whether your model is correct or at least adequate is called *validation*. The examination of the degree to which uncertainty in model formulation or parameterization allows you to trust your results, or reach certain conclusions, is called *sensitivity analysis*. It is through validation and sensitivity analysis that models generate their (occasional) tremendous power in determining truth, such as that is accessible to the human mind.

What then is the role of mathematics in this process? First of all it is necessary to distinguish *mathematical* from *quantitative*. Quantitative means simply using numbers in your analysis: three salmon versus seven salmon. This does not require any particular skill (although getting accurate numbers may require enormous skills of a different kind). Mathematical means using the complex tools of quantitative analysis to manipulate those numbers, often to make a prediction of the value of something modeled (such as oil production or GDP). Mathematics allows one to manipulate relations among data so they can be put in recognizable patterns. These tools include algebra, geometry, calculus, simulation, and so on. There are two principal means of manipulating or solving quantitative problems: *analytic* (or closed form) and *numeric* (or simulation). The numeric approach gives sometimes more approximate answers to a broader set of possible problems using (generally) simpler equations put together in complex patterns and solved stepwise on a computer. In theory, either method can be used to solve any quantitative problem, and sometimes both approaches are used. It is, however, often difficult to solve analytically equations with a large number of independent variables. In practice the mathematical training required to undertake analytic approaches precludes its use by many. In addition there are severe restrictions to the class of mathematical problems that can be solved analytically, often requiring a series of sometimes unrealistic assumptions to put the problem into a mathematically tractable format. On the other hand simulation allows one to solve very complex problems using relatively simple mathematics. Hence there is a curious paradox: the most complex mathematics actually requires the simplest basic equations for starters.

A final problem is that there has been frequent confusion between *mathematical* and *scientific rigor* or *proof*. Mathematics can generate real proofs relatively easily because you are working in a defined universe (through the assumptions and the equations used) to which it applies. If you define a straight line as the shortest distance between two points, then you can solve many problems requiring straight lines. But the world handed to us by nature is not so cleanly defined, and we must constantly struggle to represent it with our equations. Hence a mathematic proof becomes a scientific proof only in the relatively rare circumstances when the equations do indeed capture the essence of the problem.

Models Can Get in the Way of Understanding Reality

Humans have often found that models of reality are much easier to deal with conceptually and operationally than reality itself, which tends to be very messy. There is a very long history of models getting in the way of truth, and this continues today. Perhaps the clearest and oldest example of both the strength and the potential fallacies of models are those associated with our understanding

of astronomy. Many educated ancients were extremely interested in what we now call astronomy because of their belief that the movements of heavenly bodies had great importance to their day-to-day affairs (this lives on today as astrology). Sometimes the reasons were clear and scientific. As agriculture became more and more important it became obvious that an understanding of the apparent movement of the sun north and south with the seasons was a much more reliable index of when to plant than was temperature, which could not be measured anyway and which varied much more than the stately daily progression of the sun. Thus the ancients built entire buildings and even cities to help measure the movement of the sun and other heavenly bodies, as told by various archeoastronomers such as Anthony Aveni [2]. These ancient astronomers needed very large instruments before the invention of brass instruments so that the relation of inaccuracies in construction was not too large compared to the size of the instruments. They built entire cities that would track the movement of the sun and other heavenly bodies through the seasons. The story goes that those who planted according to the schedules of astronomer-priests tended to get rewarded with larger and more reliable crops, and political power flowed to the priests accordingly. Stonehenge and the pyramids of Mexico as well as many lesser-known ancient cities appear to be built in part as giant celestial observatories.

Many of these ancient astronomers thought that the sun, the moon, and the planets went around the Earth (after all, it appeared obvious) and many thought that all heavenly bodies traveled in perfect circles, because that would reflect the perfection of God, as well as God putting humans at the center of all things. Probably the greatest of these ancient astronomers was the Greek–Egyptian Ptolemy, and today we must understand him as a person with tremendous mathematical and modeling skills. Ptolemy could predict the seasons and the movement of the planets with great precision, and even predict when the Nile would flood even though the rain that caused this to occur was thousands of miles to the south. They were able to do this with relatively simple mathematics, with one exception. In order to explain the observations of the interior planets (Venus and Mercury), Ptolemy and his colleagues had to come up with a series of circular "epicycles" in which these planets circled the sun, which was circling the Earth. This was a remarkably successful approach to astronomy, and could explain the observed data to within a few percent.

We now know that Ptolemy, even though a gifted mathematician, was dead wrong. It took more than a 1000 years for the Polish astronomer Copernicus to come along and, with the extremely accurate observations of Tycho Brahe, show that the Earth revolved around the Sun. When Johannes Kepler placed the Earth in an elliptical orbit around the sun, the need for epicycles disappeared. Newton provided the formal mathematics for Kepler's system. The search for a perfect model (a circular orbit) had got in the way of understanding elliptical reality. Or we might say putting too much faith in religion got in the way of science. Today there are many scientists who do not wish to abandon mathematically "perfect" solutions for a more accurate but less elegant explanation.

Seeking perfect and simplified models in other disciplines has sometimes interfered with understanding truth and reality. In fact it is probably more often the rule than the exception. The creation "model" from the Bible of the story of creation probably made as much sense as any other explanation until Charles Darwin came along and gave us a real model that was much more consistent with our observations and the fossil record. God may certainly exist (that question is well outside the aegis of science) but so does evolution. Just ask any hospital administrator or agricultural pest manager who has to deal with the routine evolution of hospital and agricultural pests.

For another example, fisheries science lost decades of understanding and ultimately contributed to the destruction of many of the world's most important fisheries because fisheries scientists chose to believe a model, the Ricker curve, rather than to look at their own data. For a long time fisheries scientists believed that the number of fish in a population (e.g., sockeye salmon in British Columbia) depended principally upon the number of parents, with there being the possibility of both too few and too many parents for maximum production of young. This idea "allowed" managers to let fishermen take large numbers of salmon that had been considered "excess." More recent work has shown that the main determinants of the salmon populations are climate and other environmental factors. The number of parents may also be important although not necessarily in the way initially proposed [3, 4]. More generally in population biology it has become clear that the simple, elegant mathematical models that once dominated thinking about populations were often misleadingly incomplete when not entirely wrong [5]. But we have learned, and now it is usually the case that we use ongoing data from the fisheries itself to set the seasons and otherwise manage the fish (where politics or too much greed does not get in the way). Likewise ecologists and game managers believed for too long in the simplistic, "perfect" and almost always wrong logistic and Lotka–Volterra (simple predator–prey models) models of population dynamics rather than to concentrate on the environmental forcing factors that were generally far more powerful predictors of actual populations. In all of these issues there is a huge tendency for people to want to believe in clever and neat models rather than in the messy reality that surrounds us.

The Importance of Paradigms

Models are far more than mathematical entities that live in mathematics books or in computers. More generally models are *conceptual*, mental pictures of the structure or function of a system, or of how something operates. Such conceptual models are often called *paradigms* when they become expansive and general. Nearly all disciplines, including economics, work from what is usually called a paradigm, or a set of paradigms. Paradigms are conceptual overviews that synthesize the main ideas of a discipline, explain a wide set of observations, and allow for the positioning of new ideas into the existing intellectual structure. Examples include evolution in biology and plate tectonics in geology. Before Charles Darwin's synthesis in 1859 there were many observations of nature that simply did not make sense, and were not related to each other. These observations include the fact that organisms tended to have far more offspring than were needed to replace themselves, that animal breeders were able to select certain characteristics of their animals and these characteristics were passed on to their offspring, and that there existed a vast record of past life in rocks that in some cases showed a regular progression of change. At that time the principal idea as to where life had come from for most Europeans was the story of creation in the Bible. Darwin was himself religious and initially believed like most educated people of his time that the Biblical explanation for creation was all that one needed to know. But Darwin also knew from the work of the earlier geologists Hutton and Lyell that the earth was very old, and that processes that had shaped the earth in the past were often still occurring at the time. Finally he knew that these processes (such as erosion of landscapes) could be very powerful even though they were very slow because they played out over such a large amount of time. Darwin brilliantly synthesized all of these different observations, and many more, in his book *The Origin of Species*. His concept of evolution through natural selection has become a paradigm for all of the biological world since then. His particular genius was to come up with the mechanism, *natural selection*, that could explain the process, which was evolution. As we gain new information we have made additions and revisions to his basic idea, but the idea itself has withstood the test of time very well. For example, in the past several decades we have made astonishing progress in understanding the nature of DNA and the many ways it works at the cellular and the molecular level. Nevertheless all of this exceedingly detailed and powerful new information has not changed the basic Darwinian way that we understand how evolution works and in fact adds considerable additional insight and support, although sometimes it adds a new understanding

of mechanism. Evolution through natural selection is the paradigm for all of biology.

In the 1950s geology was a rather sleepy science that had a whole series of unrelated observations about the earth: that volcanoes appear in specific regions and that earthquakes were associated with these regions as were mountain chains. As probably every school child has thought about when staring at a map of the world during a boring class, the shape of the African West Coast snuggles up very nicely against South America and so on. In addition, they knew that biologists had found that a particular type of tree, the Southern beech (the genus *Nothofagus*) was found in very similar, but not exactly the same, forms in Southern South America, Australia, New Zealand, and South Africa. Although physical scientists such as Wegener and biologists such as Darlington had considered for a long time that the continents must have moved, geologists were not buying it, or more usually not even thinking about it, because they had no idea of a *mechanism* to move the continents. Remember *theory* is the device that explains the mechanism that underlies our observations. The continents were just too large, there was no concept of where the energy might come from to do that much work, and the concept was too weird. But in the 1950s a group of geologists, many of them at Princeton University, began to connect the dots [6]. The most important knowledge was coming from, surprisingly, the bottom of the oceans. Oceanographers had begun to map the bottom of the ocean with powerful new sonar, and they found a very surprising thing: the middle of the Atlantic (and other) oceans had a series of underwater volcanoes that stretched from Iceland in the north (Iceland is itself a series of volcanoes) to below the tip of South America. Further studies showed that some of these volcanoes were actually active, spewing forth lava and heat underwater, and that the sea bottom on either side of the volcanoes was spreading apart. Here was the needed mechanism to explain continental drift! It was energy from deep inside the earth, moving up in these oceanic rift zones, that was pushing the continents apart. Soon geology was abuzz with excitement and many new concepts tumbled out, all aided by this continental drift paradigm. For example, we could now see and even measure with lasers, that the Red Sea was hinging apart. In addition, the beautiful rift lakes of East Africa could be seen as the first stage in land masses splitting apart. In time lakes such as Tanganyika and Malawi will split apart entirely and the sea will pour into what is now the middle of Africa, as it has with the Red Sea and the area between Africa and Madagascar.

Paradigms and Models

Scientists are usually most satisfied when they can formalize their paradigm, or some derivative of it, as a *model*. As we stated earlier the definition for a model that we like best is "a formalization of our assumptions about a system." The beauty of models from that perspective is that it says essentially that a model is a working hypothesis about how the world works, and as such it can be tested explicitly. It allows one to put reasonably complex issues such as continental drift into a format where they can be tested quantitatively. There are five major types or classes of models: conceptual, physical, diagrammatic (or graphical), mathematical, and computer. Each of them in some way attempts to capture the essence of a problem or situation, in a formalized although simplified way. Of course models can be good or bad, correct or incorrect, and complete or incomplete. But they should be consistent with the general principles of science outlined at the start of Chap. 11. They should contain appropriate mechanisms, and they should explain considerable empirical observations. A good paradigm meets all those criteria and can be considered a sort of supermodel that cements knowledge in an entire discipline. Both of the paradigms given above, natural selection and continental drift, meet those criteria. But many other models, and even some paradigms, have been found to be sadly lacking.

Clearly we have to build our models and our paradigms very carefully. The beauty of science and the scientific method is that it allows one to construct tentative models of how the world might work. Subsequent research may find through empirical (i.e., related to observation and data) observation and testing that the assumptions used to construct that model were good or poor. Then the model can be accepted, adjusted, or abandoned. There is no disgrace in constructing a model that turns out to be incorrect. That is how science moves forward. In science when one model or paradigm is shown to be false there is often another to take its place, or sometimes we have to conclude that a good model just is not possible yet, or maybe ever.

Models are great devices for bringing problems that are otherwise too large or too small into a scale humans can understand and conceptualize. Trying to imagine something as large as the atmosphere or the world economy, or as small as a hydrogen atom, can be a pretty daunting task. But by using models we can render them into terms and a scale that is more easily understood. A model must be simpler than the world it is attempting to explain. What we mean by simplification is that the model contains fewer variables than the world we are trying to explain by means of the model. In addition, the independent variables should be as independent as possible. If they are not, it becomes very difficult to separate cause from effect. Finally, if the models starts out with its variables arrayed linearly, the model is simpler. We would like to include a word of warning for the student who is not yet accomplished in modeling. Please do not confuse simple with easy. Simple means there are but a few independent variables which (usually) are linear and do not strongly interact. Simple does not mean immediately apparent by casual observation. Even simple models often require a great deal of work to get their structure correct and their predictive output to represent the real world reasonably.

A further conundrum is that just because a model is simple, or even just because it might make considerable intuitive sense, that does not mean that the model has correctly captured the essence of the system or question being asked at the time. It is amazing how infrequently this question has been asked. In our extensive and very different experience with modeling we do not believe that 10% of the models that we have seen are sufficiently well constructed to be appropriate for the questions they are supposedly representing. Is the Hubbert model of oil production sufficient to predict the future? If so what data do we need to parameterize it? Certainly the simple "firms and households" model of economics (Fig. 5.1) is completely inappropriate to resolve questions about national debts, pollution, climate change, or a host of other issues for which it or its various permutations have been used. Why is this so? Is it because economists have no other place to turn? One gets that impression from the review of development models by LeClerc [7].

So, Then, Why Is Economics, Which Is So Complex, So Analytical?

There remains within academic economics a great deal of what has been called physics envy, a desire to emulate the power and prestige of successful applications of simple equations in physics. Mathematical rigor is often very important for impressing colleagues and deans whether the analysis has a secure connection with reality or not. In some few cases it has led to the most brilliant and important advances in all of human knowledge. Mathematical rigor, however, although useful in its own right and in some applications, is hardly by itself a criterion of acceptable science, although it is often promoted as such. Thus the advanced economist is often reduced to simplifying quite complex economic questions into a format that is analytically tractable: that can be solved using analytic means, and sometimes using essentially ideological assumptions such as "free markets generate the optimal use of resources." Such impressive analytical modeling requires enormous skills and concentration, and sometimes generates very useful results. Very often, however, we believe it generates results that represent only the mathematics and not the real system. We give some examples throughout this book but especially in Chaps. 5 and 8.

To make *analytical* mathematical models work one requires very simple systems, often described as *two-body* systems. Real atmospheric or real economic systems are not so simple, and

pushing those real systems "kicking and screaming" into a small enough box (i.e., few enough equations) to be analytically tractable is not science. In our opinion there are very few real problems in economics that can be represented adequately by such simple relations, and much of the economics that is done by complex analytic analysis is deriving mathematical and not economic results. But the use of analytical mathematics does have one major benefit. Through the manipulation of equations you can transform a cause and effect relation that is obtuse into something wherein you can sometimes see the patterns you need to see, and derive the patterns you need.

How Have Models in Fact Been Used in Economics?

We believe that models have rarely been used in economics in their proper role, that is, as a formalization of assumptions to allow the testing of the hypotheses that are represented by the equations therein. Rather, models have been used mostly as conceptual shortcuts that take the very complex biophysical and social entities that comprise real economies and represent them as caricatures that demand acceptance (or dismissal), but not testing. Any model needs to be a simplification, therefore that simplification must represent the basic reality modeled. But the most important models in economics, such as the firms–household model, do not represent the essential biophysical reality that constitutes real economies.

It is true that within economics there are complex empirically based models. An example is the University of Pennsylvania Wharton model, a huge, data-rich computer simulation of linked economic transactions throughout the economy. It gives very detailed predictions about each section of the economy, although it failed to predict the 2008 market crash [8]. As such it is a useful predictive device if nothing unusual such as an oil price increase happens, and as such it *could* be used to generate and test hypotheses. But it was not generated upon a series of hypotheses about how the economy works, nor, to our knowledge, was it asked to test the basic maintained hypotheses of economics. Instead the structure of the economy is specified (given) and then a massive amount of empirical information on the economy is fed into the calibration phase of the model. The computer cleverly fits all of the actual data collectively to all of the equations in a process known as parameterization. The net effect is that the model can predict well small changes, say from 1 year to the next, because of the "can't fail structure" of the model, which is in some ways a tautology. But in no way that we are aware of, does this model test the underlying conceptual base of the neoclassical model of economic reality.

Some Other Problems with the Standard Neoclassical Model

Although there are some good attributes to the basic neoclassical supply–demand–market model there are also some extreme problems, as the reader probably has guessed by now. The first problem, well understood by any economist, is that of *externalities*. Externalities are a loss in utility imposed upon someone who is not a party to the transaction. Furthermore, these costs are not captured in the market price, and markets cannot allocate goods (or in this case bads) that do not command a price. A commonly used example is sulfur dioxide pollution. When a mill produces steel it purchases market inputs and combines them to produce steel. The output, like the inputs, has a market price. But the sulfur dioxide from burning coal to fuel the furnaces, goes into the atmosphere where it blows towards the East on the prevailing winds. This decreases the pH of the rain as sulfur dioxide is converted into sulfuric acid when it combines with water vapor. The costs of the acid rain are borne by those who purchased neither inputs nor outputs. If a factory were to dump toxins in a river the cost would be borne, and the externality felt, by downstream fly-fishers who see their endeavors ruined by the polluted waters. One might also argue the externality also falls upon the fish.

Can we now internalize into the basic economic model the much larger problems of climate change, depletion of highest quality fuels, the displacement of workers by still cheap fossil fuels, and so on? Even were we somehow able to determine the monetary value of, for example, the depletion of oil, would valuing these things in dollars be the appropriate way to value these things? And how would we incorporate these numerical values into price? Who would get the money? Some government entity? How would the government spend that money without causing further depletion? We do not have an answer to these questions but the issue seems daunting.

If the Basic Neoclassical Model Is Unrealistic, Why Do Economists Continue to Use It?

All cultures live at least partly by myths; a set of deeply held and sometimes true beliefs that validate everyday experiences and propagate patterns of behavior. Today when we study ancient cultures we often marvel at what we perceive to be the strange and foolish myths that guided their activities. Contemporary Western society also operates according to a number of sometimes contradictory myths embodied in various established conventions, religious tenets, folk wisdoms, and even economic "myths." The basic tenets of market capitalism, include the primacy and/or virtue of individual initiative, survival of the economic fittest, the need for economic growth, the indefinite possibilities of exploitation of resources, material consumption as the road to happiness, unlimited substitution, technology as the unfailing resolution to any economic shortage, that humans are meant to subdue and have dominion over nature, and so on. These are myths just as much as the meaning of the Rapanui (Easter Island) statues. So far the application of conventional economics, based on a series of myths or partial myths has given the residents of the wealthy North an unprecedented material standard of living and tremendous technological achievements. Yet we believe that continued adherence to these myths now threatens to undermine the affluent society they helped build while, clearly, generating enormous misery to people everywhere. One of our freshman asked us, "Do you mean that capitalism actually encourages its own destruction by encouraging the destruction of its resource base as rapidly as possible?" To this we say, "Look around." We also encourage them to read the works of Karl Polanyi, who first advanced this proposition clearly.

We believe that what separates myths from reality is the judicious use of the scientific method. Of course science itself has hardly been, and remains, immune to the prevailing myths. In the natural sciences we are familiar with large-scale "paradigm shifts," where fundamental scientific ideas that have been widely accepted and well developed are suddenly found to be quite wrong, leading to the replacement of the entire conceptual basis of a discipline. The main questions are: do we need, and are we ready for, such a paradigm shift in economics? And if we are prepared intellectually for that task, can we possibly implement it given the enormous intellectual, political, and financial investment in neoclassical economics as it is applied around the world? Our answer to the first question is that it is probably too late for most of the older economists steeped in neoclassical theory, but there are many younger economists, economists to be, and certainly a vast army of environmental, geological, and physical scientists ready to learn and help create a new economics more consistent with their own empirically based view of the world. The answer to the second is that it would be an extremely difficult and demanding task to actually implement a new policy approach to economics, even if it could be agreed as to what that should be. It might be much worse not to do so, however, especially if the plight of the world degrades substantially, which we perceive as not unlikely as oil and other critical materials become increasingly scarce over the coming decades.

There exist procedures by which we can do this. It is called the scientific method as it is applied in the biophysical sciences. Within this framework one is able to come up with whatever hypothesis you might want as to what is the truth. But if you are inconsistent with known reality you will get nowhere. For example, when in the early 1950s several scientists were closing in on the structure of DNA (where the "A" stands for acid) the great biochemist Linus Pauling proposed a chemical model that he thought represented DNA. But Watson and Crick, who later came up with the correct structure, noticed that Pauling's structure was not an acid, and so immediately shot down Pauling's model; it was not consistent with known science. Likewise all kinds of mechanical devices for generating energy have been shown false because they are thermodynamically incorrect. We think that there has to be a lot more of this reality-based kind of analysis applied to all economic models.

We expect a great deal of resistance from established economists to what we develop here. Past criticisms of neoclassical welfare economics are almost invariably dismissed by economists as attacks on a "straw man." This response by economists is so prevalent it is worth addressing in some detail. In one sense economists are correct to point out that the theory of many present-day economists has gone too far, such as the restrictive and unscientific assumptions of *Homo economicus* and perfect competition. A small but growing number of economists, particularly the many respected theorists, have already abandoned the models of human behavior we criticized earlier. The applied work and policy recommendations of most economists, however, remain grounded in these models. Many economists believe that their contemporary work can be integrated into the standard model, even though that model may be false. But we say this is wishful thinking, for if the restrictive assumptions of *Homo economicus* are relaxed to incorporate current knowledge about actual human behavior, the conditions for efficient resource allocation by markets (Pareto efficiency) cannot be met.

For example, according to the economist Gintis a definition of the "rational actor model" is that it: "holds that individual choice can be modeled as maximization of an objective function subject to informational and material constraints." In other words people try to do the best they can with the limited means at their disposal. Their objective is said to be "utility" or "well-being" broadly defined. These seem to be reasonable and harmless assumptions. But in economic texts and applied work "well-being" is equated only with the consumption of market goods chosen in a manner that conforms to the mathematical requirements of constrained optimization. We ask the reader to think what are the most important factors in your own life. For most of us family; friends; health; justice; fairness; a clean, undegraded, and uncrowded environment; spiritual issues; and good associates are all ahead of issues that could be bought or sold in the market. But every leading text in economic theory follows the pattern based only on consumption. For example, the respected economists Pyndyck and Rubenfeld [9] write: "In everyday language, the word *utility* has rather broad connotations, meaning, roughly, 'benefit' or 'well-being.' Indeed, people obtain 'utility' by getting things that give them pleasure and by avoiding things that give them pain. In the language of economics, the concept of **utility** refers to *the numerical score representing the satisfaction a consumer gets from a market basket*." (italics and bold in the original).

Thus the complex issue of individual utility becomes reduced to only the consumption of collections of market goods. The analysis of market choice proceeds by making the three basic assumption of completeness, transitivity, and that more is always preferred to less. These are the kinds of assumptions economists have refused to test empirically until recently. Without these assumptions Walrasian (neoclassical) analysis cannot work. As a leading microeconomic text points out regarding just one of these assumptions: "...substantial portions of economic theory would not survive if economic agents could not be assumed to have transitive preferences." Therefore we believe that attempting to "fix" the NCE model through, for example, internalizing externalities is missing the point of what should be (in our opinion) our major undertaking, which is to start our economic conceptualization from scratch in a way that represents what actually occurs in a real economy. In essence we must put our conceptual economic models inside nature where an economy must exist (e.g., Fig. 16.1), rather than attempt through internalizing externalities to put nature inside the economic framework (Fig. 16.2).

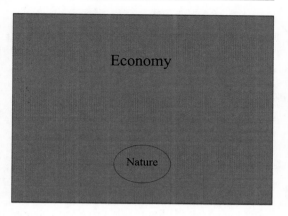

Fig. 16.1 Too often in ecological economics nature is placed "within" the economy where functions of nature are given monetary values that were originally evaluated in the economy

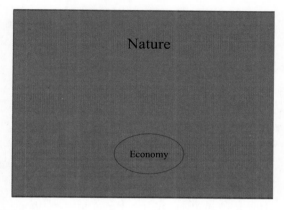

Fig. 16.2 The economy must exist within nature for it cannot exist any other way

A Final Thought on the Proper Use of Mathematics

The use of mathematics was especially important in the development of physics for centuries, and the creation of the atom bomb was tangible evidence to many of the power of mathematics combined with practical application. Nevertheless even Einstein preferred to solve his problems without mathematics when that was possible. Other sciences in which mathematical models have been especially important include astronomy, some aspects of chemistry, and some aspects of biology such as demography and, in some cases, epidemiology. Genetics, from

Mendel to contemporary population genetics, has been heavily influenced by, and sometimes tends to lend itself well to, mathematics. The importance of mathematics for most of biology is a little harder to pin down. Certainly the most important discovery in biology was that of Charles Darwin, who used essentially no mathematics beyond the concept of the potential of organisms for exponential growth in the development of the theory of natural selection. Likewise mathematics by itself had little to do with the development of the cell theory, the structure and nature of DNA, and most modern molecular biology. Hall [5] found that there had been an uncritical acceptance of some basic mathematical models in ecology, and this had in some cases led to misunderstanding and even damage to natural populations. All of this has been related, sometimes, to a confusion between mathematical and scientific proof. A mathematical proof becomes a scientific proof only in the relatively rare circumstances when the equations do indeed capture the essence of the problem.

Nevertheless despite all of the many problems of modeling we do not understand how one can use the scientific method, that is, generate and test hypotheses, on complex issues without the use of formal modeling. This is as true for management and policy-related issues as for theoretical ones. The reason is that models are an explicit formalization of our assumptions about a system, and as such allow for explicit testing of how you think the world works. In our view quantitative (or occasionally nonquantitative) models of at least sufficient complexity are necessary in the complex world of economics (and of environmental sciences) because it allows one to apply the scientific method to complex real systems of nature and of humans and nature. *But it is critical that the right kind of models be used.* And the way to do that is quite simple: Try to represent the real system that you are dealing with rather than some abstraction that happens to be analytically tractable. Quite simply most real problems require computer modeling, not analytic modeling. The power of models is to make our assumptions explicit, generally quantitative and hence testable.

Questions

1. What is a model? Where do you find models?
2. When speaking of a model what do we mean by "solution?"
3. What is the difference between something that is mathematical and something that is quantitative? Can something be both?
4. What is the difference between an analytical solution and a numerical one?
5. Explain how some cities were astronomical instruments.
6. Give one or more examples of a model that is conceptually incorrect but that nevertheless gives good predictions.
7. Give one or more examples of models that are very commonly used but that are probably incorrect.
8. What is a paradigm? Give several examples.
9. Can you give five general types of models? (Hint: One of them is computer.)
10. Can you explain the apparent paradox that one can use complex mathematics only on a rather simple model?
11. What is an externality?
12. How can we separate myth from reality?
13. What are some of the ways in which some economists today have criticized the basic models of contemporary economics?
14. Discuss the conceptual advantage of putting our economic models inside our models of nature versus the opposite.
15. What is validation? Sensitivity analysis? How would you use them in economics?

References

1. Hall, C.A.S. and J.W. Day (eds). 1977. Ecosystem modeling in theory and practice. An introduction with case histories. Wiley Interscience, NY.
2. Anthony F. Aveni (Editor). 2008. People and the sky: our ancestors and the cosmos. Thames and Hudson, London.
3. Pyper, B. J., F. J. Mueter, and R. M. Peterman. 2005. Across species comparisons of spatial scales of environmental effects on survival rates of Northeast Pacific

salmon. Transactions of the American Fisheries Society 134:86–104.

4. McAllister, M.K. and R.M. Peterman. 1992. Experimental design in the management of fisheries: a review. N. Amer. J. Fish. Management 12:1–18.

5. Hall, C.A.S. 1988. An assessment of several of the historically most influential theoretical models used in ecology and of the data provided in their support. Ecological Modeling 43: 5–31.

6. Hess, H.H. 1962. History Of Ocean Basins. IN: Petrologic studies: a volume in honor of A. F. Buddington. A. E. J. Engel, Harold L. James, and B. F. Leonard, editors. New York: Geological Society of America, 1962. pp. 599–620.

7. LeClerc, G. 2008. Pp. 13–38 in LeClerc, G. and Charles Hall (Eds.) Making development work: A new Role for science. University of New Mexico Press. Albuquerque.

8. http://knowledge.wharton.upenn.edu/article.cfm?articleid=2234.

9. Pyndyck, R. S. and D. L. Rubinfeld. © 2005 Microeconomics. Prentice Hall, Inc. Saddle River N.J.

How to Do Biophysical Economics

17

It seems imperative that we as individuals who care about the human condition in the poorer parts of the word and about nature must create a new way to undertake what is usually called developmental economics, usually seen as the application of economic principles to less-developed nations. Our reasons include: dissatisfaction with the intellectual foundations of conventional economic models used in development and with the results that have occurred with their use, the general sense of many development economists themselves that conventional economics has failed, the need to do something that will work, the concern that most knowledgeable people have that the future, and especially the future of most developing nations, will be much more constrained by the "end of cheap oil," and the need to protect whatever nature is left. We generate the "alpha version" of such a model in this chapter, summarizing certain useful approaches and successes of the past, and using a biophysical basis try to generate a synthesis to help the reader. We are not foolish enough to believe that we can in one fell swoop cure all the economic problems that generations of traditional economists have not been able to, but we believe that we do provide a useful basis here for beginning that process and for generating useful results now for field workers. Although our focus here is on developing nations the concepts are applicable anywhere.

We undertake this analysis with the full understanding that conventional (e.g., neoclassical) economics, for whatever its limitations, is an well-developed and integrated approach where, in general, the players are well entrenched and agree upon the rules. And we acknowledge that their influence is increasing in the applied world, even as many academic economists step back from the pure model. For example, "computable general equilibrium" models (CGE), which are applications of pure neoclassical economics, are increasingly used in World Trade Organization negotiation rounds that affect billions of lives. In addition, conventional economics has been developed in such a way (e.g., by emphasizing money rather than energy, demography, and other resources as we do) as to appear to be a logical extension of the day-to-day economics with which we are all familiar. These are significant hurdles to overcome for those of us who believe that a more useful and accurate economics can and must be developed. Nevertheless we perceive the importance of this to be so great as to require our best efforts. We know that we are not alone in challenging neoclassical economics, and that our best allies may be some of the economists themselves, especially those who spend their time in the realities of the developing world.

Hall, along with Gregoire Leclerc, have spent considerable time in the past developing a biophysical assessment for the country of Costa Rica and much of what follows is based on our experience in that assessment [1]. The Costa Rican book has 26 chapters with detailed assessments of essentially all important aspects of the Costa

C.A.S. Hall and K. Klitgaard, *Energy and the Wealth of Nations: Understanding the Biophysical Economy*, DOI 10.1007/978-1-4419-9398-4_17, © Springer Science+Business Media, LLC 2012

Rican economy and some novel visual procedures for examining the effects of development on economics over time.

One characteristic of traditional economic analyses is that usually the many varied and complex consequences of a given policy are reduced to a single scalar (such as increased GDP, as is usually the objective in, for example, most money-based economic cost–benefit analyses). In the Costa Rican work, in contrast, we developed procedures to put all of the dynamic information, including land use, demographic, environmental, economic, and so on, on the screen simultaneously, then let the user or decision maker (or the people affected) decide whether they prefer the existing path of development (by whatever criteria they chose) or rather something else. This approach can be particularly effective when integrated with historical patterns of land use, for example. Most people living in Costa Rica today are too young to understand how much their country has changed in one human lifetime, but they can see that clearly – and are often amazed – when they see this as a multidimensional visualization. So most of the rest of this chapter is a discussion of what kind of information you might want to include in such a visualization, or perhaps in some simpler analytic structure such as a spreadsheet.

A rough guess as to the cost of developing this kind of overall biophysical analysis for a small to medium-sized developing country is on the order of one to ten million dollars, assuming that you are undertaking this analysis with competent and not greedy investigators and that the biophysical and economic database is well developed, as was the case for Costa Rica. Our very thorough assessment of Costa Rica was done on a small fraction of that, although the work was subsidized in various ways, including the data, interest, skills, and good will of numerous Costa Ricans. Most of the examples we give here are done at a national level, although the biophysical approach that we are advocating is in theory applicable at any regional level that the investigator might choose. An important scale issue is that data are generally more readily available at the national level.

Other Approaches to Evaluating Resources and Nature

Before we give our own approach we think it useful to review a number of other economic approaches that have been developed to assess specific environmental impacts of an economic or other activity. These approaches do not give the full and comprehensive environmental and economic analysis, however, we believe in some cases they can be very useful supplements to the analysis we give below.

Our attempts to build procedures for biophysical assessment are related only marginally to most of what is being done under the aegis of "environmental economics," or the bulk of the activity in "ecological economics." The goal of environmental economics (and a substantial part of ecological economics) is to integrate the environment into economic analyses; in fact it has been mostly about putting a dollar price tag on all kinds of environmental objects and services. Much of ecological economics analyzes the interdependence and coevolution between human economies and their natural ecosystems (i.e., economics being a strict subsystem of human ecology [2]), but in practice much of the work is still about putting monetary values on nature. In fact dollar values often give extremely poor information about basic resources: for example, as wild salmon increasingly are disappearing and are hence of less and less value to our society their price goes up indicating they are becoming more valuable than when they were cheap and abundant!

Hence we believe that giving a dollar value to many things is often a rather poor estimate of the value of our most prized things, including our relations to those people close to us, justice before the law, the maintenance of natural environments, and the milieu of Earth that allows us to exist here in the first place. All of these are under assault by dollar-based aspects of our economy, and hence in our opinion dollar-based criteria are not appropriate for making assessments of the value of nature or our most essential resources. That said, we of course realize that we live in a monetary-based world where many things must be valued in monetary units for routine day-to-day transactions. So we sometimes have to walk an appropriate tightrope between using and not using monetary estimates.

The first biophysical assessment we review is that of *ecological footprint analysis* which examines the environmental requirements for a given region (for our purposes a social and economic unit such as a country or city) in terms of the quantity of land required to support the activities on that area considered. Comprehensive and thorough analysis of the ecological footprint is run by Mathis Wackernagel [3]. His team's footprint analysis found, for example, that the land area required to support the needs of the city of Vancouver, Canada was about 18 times the land area of the city itself. This included land areas needed for growing crops and producing cows, fish, and other animals consumed, growing timber, mining minerals, and so on (about half the area required) as well as assimilating the sewage, toxins, CO_2, and other wastes produced (the other half). Such assessments always show that the areas required to support people are much greater than the areas the humans actually occupy, and give lie to those who say that the Earth can support a much larger human population (or even the present level) indefinitely. They conclude that more than one Earth would be needed to support the world population if wealth were equally distributed. If everyone lived at the affluence level of the average North American we would need five planets. However, we only have one! Over time the authors have developed and refined their methodology impressively, and made its use on their website very straightforward and easy. Because they trace back virtually all the major material substances used by different groups of people their complete list of materials used constitutes a ready-made list of the biophysical materials required to support an economy. What they have not done yet is to relate the materials required to the level of monetary activity or asked these questions of developing countries. Once this is done we will have one rather good biophysical assessment at our fingertips.

The second biophysical approach we examine is *energy analysis*, which in its many variants means essentially the amount of energy required by various economic activities. These methods were developed mainly at the University of Illinois in the 1970s by Bruce Hannon, Clark Bullard, and Robert Herendeen, and were applied to most aspects of our economy including agriculture, manufacturing, provision of services, and so on [4–6]. These studies calculated not only the direct energy used (such as the energy used in a tractor factory to make the tractor) but the indirect energies as well (i.e., the energy to mine and refine the iron, plastics, and so on used by the tractor factory). As a rough estimate about half the energy used to make some product sold in "final demand" occurs in obtaining and refining the raw materials. Summaries of the results of such studies are given in Hall et al. [7, 8] and Cleveland [7]. One problem with this approach is that the numbers are old, and there has been little federal or other funding of such energy research for decades as energy analysis has fallen into political disfavor, or more accurately, indifference because in the minds of many (but not us) the market has resolved the energy issues of the 1970s. A recent study by Carnegie-Mellon has updated these analyses to 2004 (by methods that seem moderately defensible according to Robert Herendeen), and these estimates are readily available on their website [9]. Sergio Ulgaldi and his students at the University of Sienna, Italy, are putting together a Web-based system for calculating the material costs for many different commodities (e.g., a new building) including the associated environmental costs.

Howard Odum, Mark Brown, and others have argued that although the above energy analysis is useful it is incomplete because it does not take into account either the environmental energies required to manufacture something or correct for the fact that different types of energies have different qualities. For example, a kilojoule of electricity has value to society beyond its ability to simply heat water, and hence more value than a kjoule of coal. This is because of its special properties and because it takes about three heat units of coal in a power plant to produce one unit of electricity, the rest more or less of necessity being released into the air and water. Likewise a kjoule of sugar fixed by a plant has more value than a kjoule of the sunlight that made it and so on. Howard Odum generated the idea of embodied energy, or *emergy analysis*, to deal with these issues of environmental inclusiveness and of differential quality. Explicitly emergy (with an m, as in energy memory) is the total energy that had been required in the past to make a manufactured item. It is a concept analogous to *embodied labor*.

Odum and his student Mark Brown developed an extensive accounting scheme to measure this and to compute the quantities of emergy required to make, or cause to happen, many things [10–13]. An advantage of this approach is that it is obvious that if we want to account for the energy used to manufacture something we are missing the large quantities of environmental energies that are just as much needed to make it. These energies include, for example, the energy used to provide the water used, that is, to distill freshwater from the sea and lift it to mountain-tops which allows it to form rivers and hence become available to plants and to humans. Likewise the sun runs photosynthesis and everything that derives from that even though we do not pay Mother Nature for either the water or many of the products of photosynthesis. In addition it includes in the analysis an emergy assessment of the environmental services foregone because of the activity in question. Although the idea is tremendously appealing to us, and the comprehensiveness essential in our view, the difficulty in weighting one energy against another makes its use difficult. In addition, environmental energies are usually regenerated, whereas fossil energies are not, so their use is different.

All of these techniques are quite similar in that they are trying to get at real costs of economic activity, and that their utility may converge. Their use has not been compared often, but for example, Brown, Wackernagel and Hall [14] compared the carrying capacity of Costa Rica for humans using a comprehensive economic approach that went well beyond market costs, as well as two biophysical assessments: ecological footprints and emergy analysis. The results of the three approaches were very similar, giving hope that we are approaching a true cost using both biophysical

and comprehensive economic analysis. Although each of these procedures is helpful in assessing a biophysical economic analysis, we still feel that it is useful to generate a more explicit summary as to how we can undertake biophysical economics, which we do below. We look forward to the day when scientists and policy makers agree on a set of assessment procedures that are integrated into one useful package. One could go to a website, maintained by skilled professionals, and type in the quantity (in tons or dollars of a particular year) to get all of the material, energy, emergy, footprint, environmental degradation, and so on associated with that economic activity. A step in that direction is the triple bottom line approach (economic, energy, and environmental) of Barney Foran in Australia, who provides free software to help with the assessment [15]. Later this can be done also for different countries or international corporate entities to give more explicit values. Perhaps someday there will be a label on your breakfast cereal that gives, in addition to calories and sodium per serving, an assessment of the fuel and solar energy required to make it as well as the soil and biodiversity loss, maybe all summarized in terms of emergy. While we wait for this future Web-based synthesis there is a great deal of quantitative analysis we can do that can help provide the basis for this Web synthesis.

Explicit Procedures for Creating a Biophysical Economic Analysis for a Country or Region

We base what follows on living and working in the developing tropics for much of our lives (especially LeClerc) as summarized in our previous book *Quantifying Sustainable Development: The Future of Tropical Economies* [1]. This assessment included extensive discussion of our (and others') biophysical approaches including assessing land use change and its impacts [16–19]. From this we developed procedures for routine biophysical economic analysis, including rapid assessment of development schemes. We will be the first to recognize that undertaking biophysical economics analysis is a very imperfect activity, that we are just learning how to undertake such analyses, and that there are probably many changes that will be developed over time. Nevertheless this approach has served us and our colleagues and students well for analyzing many basic characteristics of a country or a region while dealing explicitly with many issues left out of conventional economic analysis.

The methodology unfolds in six steps that can be put simply as follows.

Step 1: State your objectives (while including the right people).

Step 2: Assemble a database of critical biophysical parameters.

Step 3: Make an assessment of critical economic parameters with as much data as possible from the past and present.

Step 4: Assess the relations between economic activity and biophysical requirements.

Step 5: Construct a comprehensive and accessible simulation of the future, including the possible impact of resource limitations.

Step 6. Make decisions consistent with biophysical possibilities.

We assume that after these steps are taken into account for devising a development scheme, money will flow in the right directions; schools will be built, equipped. and populated; and institutions will improve. Nevertheless we are also quite aware of the potential for, among other things, corruption of leaders to undermine our efforts. Does the use of explicit and open science make corruption less possible? We think so but do not really know! Part of what must be done is the professionalization of all government institutions and personnel, including accountants.

Step 1. State Your Objectives (While Including the Right People)

It is not possible to undertake a journey, no matter how sophisticated your vehicle, if you do not know where you are going. So the first thing to do in undertaking a biophysical assessment is to ponder, discuss, and then state explicitly your objectives. Often people confound problems and objectives. An objective should not be a series of problem-solving activities; it should be seen as long-term desired future conditions or outcomes. For the Costa Rica study the main objective was to determine to what degree, and in what ways, the country was or could become sustainable. This led logically to the next set of objectives which was then to determine what we meant by sustainability, which in turn led to some interesting literature that showed that very different people had very different perspectives on what sustainability meant, most of which were antithetical one to another.

A second part of this analysis is to examine what other objectives people had in the past for related issues and how well these were achieved. In other words, a review of pertinent literature both for the region being analyzed and also of past public and private development projects, their objectives, procedures, and successes and failures. Many of these analyses use (or should use) time series data of economics, agriculture, and so on. It simply is not possible to understand whether whatever plan you are undertaking is successful unless you have a yardstick of the past trends in time to which to compare it.

Very often the objectives will be stated in social, economic, or environmental terms. Given that we agree with that broad perspective, the reader might be curious as to why we then focus so much on the biophysical aspects of analysis. The answer is simple: we believe that social, economic, and environmental issues must be addressed and, where possible, resolved within the context of the possibilities of the biophysical environment. It is very easy to list the various things that you would like to have: higher incomes, less pollution, better healthcare, and so on. These and other objectives are very often not met because there are serious biophysical constraints. Some of course are social, and including especially corruption and the very unequal distribution of wealth. But much of what gets in the way of achieving one's social or economic objectives is biophysical, including resource availability, climatic constraints, or biophysical mismanagement, including, among others, overfishing, soil erosion, fuel limitations, and the ability to generate foreign exchange. It is important to understand what these limitations are or might be.

These biophysical aspects of development have been neglected during decades of neoclassical economic policies. Therefore the biophysical context must be restored in mainstream thinking as the framework within which the social and economic possibilities are considered. Most of our own research papers try to integrate the biophysical and the social sciences while attempting to meet social objectives.

If we are interested not only in the progress of science but also in its impact on the development of the country studied, then we have to find the right people to help develop the models. These people will help at many levels: to clarify the objectives, to obtain the data (not easy in many developing countries), to provide key insights to interpreting the data and for prospective analysis, and to make the connection with policy so that we can extend its use beyond the level of scientific research. If we are all involved from the start in developing an integrated model (i.e., "companion modeling") [15] there is a good chance that we will learn from each other and end up with a model (or a family of models) that is not only more relevant but one that will continue to be used for policy making. Allan and Holland and Beaulieu in Ref. [16] give several hints about how to identify who you should work with and how to connect them to the development process. A good starting point is to do a stakeholders' analysis and work, with the right people, on a shared vision for the country or region. This is where genuine objectives will appear more clearly to all, and when the collective learning process will begin.

Step 2. Assemble a Database of Critical Biophysical Parameters

The first step in undertaking biophysical analyses is to derive past time trends of pertinent scientific data and to determine the physical characteristics of the country or region being analyzed. Such analyses are far easier than in the past due to the increased availability of good digital summaries compared to 20 years ago. An example of how such a database has been developed is given in

Barreteau et al. [20]. The best way to do this is to generate an assessment of the physical resources of the region in question.

An essential requirement is a summary of energy resources including any known oil, gas, and coal deposits; assessments of what might be found in the future; developed and potential hydroelectric, solar, and wind potential (for which you need meteorological information); biomass possibilities; and so on. In all of these assessments it is important to realize that in general the better resources were developed first and that increased exploitation may require more energy and be monetarily expensive. For all of these generate a time series of their use. Different types of energy have different properties or qualities, and often it is useful to take that into account. Generally the data available will be in the form of heat units (i.e., therms, BTUs, kilowatthours, kcal, or the most commonly accepted units used today which are joules). Because these units are all intraconvertible, there is no real difference among them, but the use of different units for different fuels gets in the way of clear analyses. When electricity generated from hydro or nuclear power is compared to fossil fuels it is generally best to multiply it by a factor of about 2.6 to account for the difference in their ability to do work and also their opportunity (or conversion) cost if they are made from fossil fuels. Additionally we need to undertake an assessment of the various environmental energies that must be supplied for the economy to work properly. As stated above this can be done most comprehensively using an emergy analysis.

Similar assessments are required for natural resources that are not energy sources, such as the following.

1. Nonfuel mineral resources, such as metal ores. The important components of this are the size of the reserve (in tons), the quality (i.e., percentage of metal in the ore, both at present and as exploitation proceeds), the depth and ease of extraction, the energy cost of extraction of different amounts, and so on. Because in general the best grades were used first the remaining resources may not be as cheaply or profitably exploited as before. Often the

exploitation of minerals occasions significant pollution, therefore any such impacts, and a social and monetary estimate of that damage, must be made before the project begins. These issues must be considered in addition to expected market prices and other routine economic factors.

2. Water resources, both quantitatively and qualitatively, first in overview and then spatially. Some of the information that needs to be generated or summarized includes: rainfall and flow of major rivers (both as a mean and for drought and wet years), ground water resources and their vulnerability to depletion/salinization, evapotranspiration and soil moisture over space and time, water bodies that are significantly polluted, and so on.

3. Land resources for examining agricultural (and other) potential, that is,
   • A soil map, ideally with the soil units related to crop productivity, and including the location of possible potential and actual erosion
   • A digital elevation map
   • A land use map

Taking Demography into Account

Often our overall objective is to simulate how future land use, economic, and food security scenarios might be, as influenced by demography, erosion, policy, climate change, and so on. Thus a proper representation of human demography is fundamental to what one is trying to achieve with almost any biophysical model. Whatever economic growth might be, or be predicted, then that should be divided by the population level to get per capita wealth. Fortunately excellent datasets exists for LDCs, from nationwide census data every 5 or 10 years to yearly estimates based on samples in between. (Note that because NCE is based on the behavior of individual firms, it is insensitive to demography!)

For prospective analysis it is necessary to generate a demographic model based on actual demographic data. One simple model is:

$$P_t = P_0 e^{rt}$$

where P is the population level (normally in millions), P_t is the population at time t years into the future, P_0 is the population at some initial time t, e is the natural logarithm ($\cong 2.718$), and r is the intrinsic rate of growth, the rate at which the population is growing or, better, is expected to grow. The value r (in units of proportion of the existing population per year) is the birth rate (b) minus the death rate (d). Hence the term e^{rt} is a number that will usually be greater than 1.0 and will be the factor by which the population is larger (relative to the initial population) over time. The doubling time of a population can be calculated by dividing the number 70 by the growth rate expressed as a percentage so, for example, a population with a 2% per year growth rate will double in 35 years. This simple model is often reasonably accurate, at least within the restrictions of knowing the value of r, for a few decades.

Some analysts believe that to continue to use an exponentially growing model is seriously flawed, as populations cannot grow exponentially indefinitely as they would run out of food, resources, and/or space (i.e., carrying capacity). Some models, attempting to represent that fact, will assume or simulate some sort of empirical plateau (in other words, r diminishes) or saturation of growth. A logistic, or S-shaped curve, is used often to simulate that saturation effect. Although the logistic equation is simple and has some perhaps good logic behind it, in fact few populations in nature follow that pattern and attempts to use that model to predict human populations in the past failed miserably. The debate between "implosionists" and "explosionists" is still alive (because the data support either view equally well), and although the S-curve is still the most widely used distribution for making human population projections in LDCs (see www.prb.org), the beginning of the plateau could be put at any time after 2050. Others predict that populations cannot possibly continue to grow in an increasingly resource-constrained world, and may even shrink, perhaps catastrophically.

Both the exponential and the logistic model have a number of liabilities including that they are not sensitive to changing values of r over time and are insensitive to the more detailed

demographics such as the number of prerepro-
ductive versus postreproductive females, and of
course it is for only one geographical unit. More
complex and accurate, or at least sensitive, models can be made using what is known as a Leslie
matrix, which is usually solved in a spreadsheet
or a computer program. A simple example in
FORTRAN is given in Table 17.1. Data for all of
the world's countries can be obtained from FAO
or the CIA database. Sometimes the growth and
death rates are given for 5-year intervals when
annual values are needed. To use these data it is
necessary to enter the data into a spreadsheet
such as Excel and fit, for example, a second- or
third-order polynomial to the data to get a

relation from which you can generate values
for each year as well as predictions into the
future. Additional demographic information can
be developed including: poverty assessments,
health, and labor productivity.

Additional geographical information needs to be
developed on the location and extent of built infrastructure including cities, villages, transportation,
industries, ports, airports, protected areas, land ten-
ure (private and public), and so on. These can be built
into additional Geographic Information Systems
(GIS) data layers as is well understood from conventional GIS analyses. This information is useful in
understanding the accessibility of resources to populations and as drivers for predicting land use change.

Table 17.1   A simple Leslie matrix in FORTRAN

```
PROGRAM LESMATRIX
!***********************************************************************
! Dictionary:
!***********************************************************************
! ACLS              = Age class of the human population. 1 equals all people before
!                       their first birthday, 2 = all people between their first and second
!                       birthday and so on.
! PopNum(YR,ACLS)  = Population number for each age class for each year
!                       This state variable is updated each year.
! DRate(ACLS)       = Age-specific death rate
! Births (ACLS)     = Number of births per year per female by age class (this may be
!                       known only on average)
!***********************************************************************
!***********************************************************************
! Define variable type:
!***********************************************************************
INTEGER PopNum(100,100), YR, ACLS
REAL DRate (100), BRate(100)
!***********************************************************************
!***********************************************************************
! Open read and write files:
!***********************************************************************
OPEN (1,FILENAME = "LeslieMat.DAT", Status = "OLD")
OPEN (2,FILENAME = "LeslieMat.OUT", Status = "UNKNOWN")

! Read in initial population numbers (in thousands or millions) & age-specific death and birth rates
!***********************************************************************
READ (1,900) (PopNum(1, ACL), ACL = 1,80)
READ (1,900) (DRate (ACL), ACL = 1,80)
READ (1,901) BRate (ACL), ACL = 1,80)

! Write output headers:
!***********************************************************************
```

(continued)

Table 17.1 (continued)

WRITE (2,902) "Table 1, Population levels by age class"
WRITE (2,903) "Year Age Class >", (ACLS(I), I = 1,80)

```
! Solve equations annually for 50 years starting in year 2000
!*********************************************************************
DO YR = 1, 50
          Ryr = 2000 + YR              ! Real Year
          PopNum(Yr,1) = BirTot        ! Births from end of last year considered age class one
          BirTot = 0                   ! Initialize this year's total births
          ! Do for 80 year classes (assume 80 is oldest year people live or at least reproduce
     DO ACLS = 2,80                    ! New members of first age class already added in as
                                         births
     IF (ACLS.GT.15.AND.ACLS.LT.50) RepPop = RepPop + Pop(YR,ACLS)
     Births = RepPop * BRate (ACLS)   ! Sum up number of potentially reproducing
                                        females
                                      ! (here age 15 to 50)
     BirTot = BirTot + Births
                                      ! Move each year class forward, reduced by their
                                      ! death rate
     PopNum(YR,ACLS) = PopNum(YR-1,ACLS-1) – (1.0 * DRate(ACLS))
     END DO
     WRITE (1,904) YR, (PopNum(YR,ACLS), ACLS = 1,80)
END DO
!*********************************************************************
!Format:
!*********************************************************************
900 FORMAT (80I6)
901 FORMAT (F8.2)
902 FORMAT (A20)
903 FORMAT (A15,80I6)
904 FORMAT (15X,80I6)
!*********************************************************************
END PROGRAM LESMATRIX
```

Source: Charles A. S. Hall, with the assistance of Athena Palmer

---

Step 3. Make an Assessment of Critical Economic Parameters over Time

The first step is to undertake an assessment of the current economy and its recent history. There are a number of locations to find empirical information for this, but probably the easiest is to get the data from the Web. Good sources are the large multilateral organizations (FAO, UNDP, WTO, etc.), NGOs (WRI), and the unavoidable World Bank. Several organizations provide country fact sheets (The *U.S. Central Intelligence Agency Fact Book* http:// www.cia.gov/cia/publications/factbook/index. html/), and *The Economist*, (www.economist. com/countries/). As the digital divide between rich and poorer countries gets narrower there are more and more data from Less Developed Countries (LDC) government sites available. These government sites often contain key documents on policies, feasibility studies, law texts, economic summaries, and the like. Travel books are quite useful to grasp a country's idiosyncrasies. Many sites do not provide time series data, which makes the FAO data (Food and Agricultural Organization of the United Nations) probably the most useful, as they have consistently organized data back to 1961.

From this information a time series of economic activity can be derived. Some data that we suggest might be considered include a time series of basic monetary economic information, including GDP over time. Although any analysis of any raw GDP data almost always shows a rapid increase over time this is often misleading as much of the increase is due to inflation. So the first thing to do is to correct the data for inflation, normally by expressing all data in terms of monetary units for 1 year, for example in "2010 dollars" or "2005 pesos." This is done by using "implicit price deflators" (the easiest ones can be found in the *Statistical Abstracts of the United States*). The inflation-corrected result is called "real GDP" versus the "nominal GDP" which is not so corrected. This is especially useful when dealing in U.S. dollars, however, it is necessary to use corrections implicit for the country in question. In the United States and many other countries there are also more specific correctors for different sectors of the economy, for example, for energy and for food.

A second correction is sometimes required, which is to make an additional correction for purchasing price parity (PPP). If a nation's GDP is corrected for inflation relative to the U.S. dollar, as it often is, it is also necessary to correct for the fact that the increase in prices expressed in dollars does not reflect the fact that there is often far less inflation for local products such as food than for imported products or fuel paid for in dollars. On the other hand, if you are interested in the issue of how much it costs for importing oil (which must be paid for in dollars or euros) then correcting for PPP is not useful. For many developing countries the inflation rate applied to dollars is considerably greater than the rate applied to local items, therefore this can be an important issue.

To express the meaning of the real GDP changes in terms of how it affects the average person's ability to purchase goods and services the total national GDP, corrected as above, needs to be converted to per capita values. The total national GDP tells you little about how well individuals in that country are doing in terms of their own economic welfare or purchasing power.

Dividing the total wealth production by the number of people gives you per capita wealth, which is roughly proportional to important aspects of the average person's material well-being. To do this one simply divides the total GDP (corrected as above) by the number of people in the country for that year to get the per capita GDP. This results in a decrease in the apparent effect of GDP increases and in many cases where the population increases more rapidly than the GDP, people, on average, get poorer.

Even per capita changes do not tell the whole story, for most of the GDP may go to only a relatively few people. One way to examine this issue is to use or compute the Gini index, which is an overall measure of income inequality. The higher the value of the Gini Coefficient the greater the degree of inequality. Median income is more useful in this respect than mean income, and more complex such indices are also available.

One important aspect of sustainability is whether a nation is able to do whatever economic activity it does without going into international debt. It is often debt that leads otherwise excellent development schemes into failure. Because foreign products, both those essential for the development itself but also luxury items, require payment in foreign exchange, that is, in dollars or euros, it is essential for a country to export enough to pay for these items. The alternative is foreign debt, which in many countries is more or less the largest impediment to making an economy that works. Costa Rica, for example, needs to use about 15% of the foreign exchange it generates through the sales of bananas, coffee, and tourist services simply to pay for interest on its foreign debt. It uses perhaps another 20% of the foreign exchange it earns to pay for the generation of the exports, such as the fertilizers, plastics, and fuels required to make bananas. There are enormous demands in Costa Rica for imported items (from cars, buses, and trucks to fuel to run them to computers to apples) and a rather limited international demand (or more properly a huge oversupply) for bananas and coffee, so it is common for countries like Costa Rica to get into debt. On top of this governments often borrow from external banks to make payrolls or provide health services.

Although Costa Rica has done much better than many countries (including the United States) in not running up external debt it is a very difficult issue. Hence it is useful to plot imports, exports, and their difference, as well as debt and its accumulation or decrease over time, and to ask how much foreign exchange a development project requires.

Developing countries tend to be desperate for development capital, and that capital is rarely available internally. Costa Rica needs more electric power as its economy grows, and that can be supplied by developing more hydropower. But the Costa Rican government lacks investment capital. So Japanese power companies are more than happy to build the hydropower plants that are needed because they are happy to collect the revenue from those plants. The electricity problem is solved, but there is a new revenue flow out of the country. The point is that development projects need to be examined not only from the perspective of their promised gains but also their costs including, of course, their costs and gains to whom.

Step 4: Assess the Relations Between Economic Activity and Biophysical Requirements

The next major step is to look at the biophysical resources needed to make the economy do what it does, and presumably, to do more of the same in the future. Inasmuch as we also have developed time series of economic activity and also time series of energy used, we can quite easily develop the energy intensity, which is the energy used per unit of economic activity, either for the economy as a whole or for some aspect of interest. This is the first step required to understand the biophysical resources needed for the operation of the economy. A similar concept (actually the inverse) is assessing the efficiency of an economy, something we have discussed in the previous chapter. In general, efficiency is the output of a process divided by the input. One straightforward measure of efficiency that we might want to calculate then is the output of the economy divided by its

energy input, which if we have the information derived above, we can do very easily in a spreadsheet or computer program. The efficiency of the economy can be seen by the ratio of the two, and the changing efficiency by the changing slope of that metric over time. We have found that there is usually relatively little change in the energy required per unit of real GDP for nearly all developing countries (Gupta in preparation).

Depending upon the objectives of the study other indices can be used such as GNP per unit of imported or domestic energy or water, or agricultural production per unit of energy or fertilizer used, or GNP per unit foreign exchange gained or lost, or many other ratios of output vs input. When we have done these analyses in the past we have often found that GDP increases more or less in step with energy, water, fertilizer use, and so on, indicating that efficiency does not change much over time. This has important implications for the economic aspect of efficiency for if efficiency is not increasing the only way to generate wealth is through the further exploitation of resources, something that has serious environmental and supply implications. Much more detailed analyses can be undertaken through the use of input–output analyses.

Step 5: Construct a Comprehensive and Accessible Simulation of the Future, Including the Possible Impact of Resource Limitations

If there is an economic plan for development then the next step is to assess the energy, material, and other resource requirements for such a project. This can be an extremely difficult and comprehensive issue and there is not yet a clear-cut formula for how to undertake it. We had developed a computer simulation program to undertake this for Costa Rica which was certainly comprehensive and also a reasonably effective predictor of the future (except for forest cover; Hall and Leon [1]). Another recent computer program derived to examine the material costs of any development project is being developed by Sergio Ulgaldi and others at Parthenope University of Naples, Italy.

Thus if we have a list of materials required for a development project then we can assess the most important aspects of their use rather straightforwardly. The user simply puts in dollar amounts to be spent for different development categories according to the spreadsheet provided and the results are then printed out.

Presumably any such economic–biophysical analysis will show that the economy of the region is moderately to very energy-intensive and that any past expansion of the economy was based on oil (or at least energy). Thus future expansion of the economy presupposes the physical and economic availability of oil. Can this relation simply be extrapolated? At present there are about 38 oil-exporting nations. The economies of most of the smaller and medium-sized exporters are becoming themselves much more energy-intensive over time, and most will become net importers themselves within decades as their own domestic use intercepts their production [22]. If imported oil-dependent economies are to be expanded, that might be done in a way that makes them dependent upon perhaps unreliable or at least very expensive future oil supplies. This is an issue not normally considered within conventional economics as the recent market price of oil made it a seemingly attractive choice. But the price of oil has again increased substantially and the price increases that we have observed recently are likely only a small sign of what lies ahead as the world truly approaches the end of cheap oil. What this will mean for the world can only be guessed at, but for the non-oil-producing nations of the developing world the impact is likely to be enormous as populations and economies that had expanded based on cheap oil have the rug pulled out from under them. It is unlikely to be a pretty sight.

Predicting Land Use Change

An important part of many assessments of the future capacity of a nation or a region for providing economic or environmental services is an assessment of how much land is available in different categories. (This is loosely related to the concept of ecological footprint). The principle

tools for doing this are several computer models that start with one map of land use for a given year and then make assessments of what the land use might be in the future based on rates and patterns of development. Both rates and patterns tend to be derived from existing patterns that can be extracted digitally from one or more existing maps of land use. One of our favorite models for doing this, not surprisingly, is one that we derived ourselves. This model, called GEOMOD, is bundled with the most recent version of IDRISI, a commercial software package with powerful modules for assessing and predicting land use patterns [17, 18]. Farmers usually develop land in a way that represents the least effort or energy investment on their part (hence adjacent properties on the flattest land available) with the highest potential for agricultural production (usually near a river on flatter land), in other words land with a high EROI for themselves. This model searches for the highest EROI available for farmers over time and gives a reasonably accurate picture of how development is likely to occur.

Predicting Net Economic Output as a Function of Land Type

All land does not have the same capacity for economic production, and this is especially true when specific uses are examined. For example only about 19% of the total land area of Costa Rica is flat and fertile enough to be utilized for row crop agriculture. Such agriculture would be likely to cause irreparable erosional damage if attempted in other land categories. Another 9% of the land was suitable for pastures, and another 16% for tree crops such as coffee, which causes less erosion because of its continuous cover. The rest of the country, more than 56%, should have no human use at all except for sustainable forestry. In fact far more than 56% of the country has been developed for agriculture, pastures, or urban areas.

Farmers and many other humans are well aware of what land is best to use for various purposes. Thus over time the land available for development tends to be of poorer and poorer quality, as represented in the pioneering work of the early economist, David Ricardo. What this means for development is that average values of, crop production, for example, cannot be used to project what the yields might be for some development project. For example, coffee can be grown anywhere in Costa Rica. But high-quality coffee, of which Costa Rica has some of the best, requires very explicit environmental conditions (precipitation, temperature, soil, and so on) to get high yields, which tends also to mean best quality coffee beans. As of 1990 nearly all of the land that was best for growing coffee already had it growing there (or was covered by urbanized areas) and that if there were to be increased coffee production yields would probably be less or else more energy intensive than the average of what was occurring already. This is an example of what has been called, variously, diminishing returns to investment or declining energy and financial return on energy investment as the best resources are used. We found that over time for most crops any increase in area of land cultivated reduced the yield per hectare in that year, as the land added to production was, on average, of lower quality [17–22]. There are too many farmers to use only the best land, and hence more low quality land is used over time. Well-meaning foreign aid money designed to increase agricultural yields might be better spent on programs designed to improve the educational, health, and employment situation of women, generally the most effective way to decrease population growth rates and hence increase per capita wealth (p. 206).

Include Social Assessments

As we stated in Step one many of the issues that are most important to people interested in development are social and economic in nature. There is no easy formula for integrating the biophysical and the socioeconomic approaches, although much can be undertaken with an open mind, a willingness to work outside one's own discipline, and, perhaps most useful, an ability to find and work with others from other disciplines. Many of the chapters in Ref. [16], especially Beaulieu,

attempt to integrate biophysical and socioeconomic approaches. An important aspect of a biophysical (or any) assessment is that often there are no clear ways to achieve several goals at once, and one is left with tradeoffs, such as between GDP and dependence on imported energy or economic and environmental well-being. One of the most important conclusions from biophysical assessments is that many good and desirable social objectives are severely constrained, or impossible, given biophysical limits. Finally, development projects that were once very good often crash over time, as is classically illustrated with wild fisheries and aquaculture. These crashes are sometimes predicted through fisheries science but not ever, to our knowledge, through market assessments alone.

Step 6. Make Decisions Consistent with Biophysical Possibilities

A final step in undertaking a thorough assessment of the biophysical possibilities and constraints of a region is to examine alternative future environments in which one's decisions might be played out. Prospective analysis plays a fundamental role in shaping the development of a country. However, it is poorly done at best, policy makers having to juggle with too many parameters and being forced to use shortcuts, which opens the door to misconceptions and prejudice, wrong interpretation of the data, and shortsighted emergency measures. In *The Art Of The Long View* Swartz [21] describes the critical role of scenario analysis for positioning ourselves properly in the future. Scenarios are not predictions or forecasting: they are "vehicles for helping people to learn, alternative images of the future, to change the managerial view of reality."

At the core of this prospective analysis, one can easily imagine an environment to run and discuss comprehensive simulations of the future, for example, based on the previous five steps. It can contain some or all of the entities included above plus whatever other elements the user feels appropriate, including elements of neoclassical economic analysis, and the results can be compared or even integrated! Again our example of this approach is given in the CD that is included in the first of our previous books [1]. We believe especially in the development of good graphics and appropriate simulations for communication to stakeholders, and hold up the above CD as an example. Although many people are extremely suspicious of any such simulation models we think that formalizing one's knowledge and assumptions through modeling is a critical approach that needs to be undertaken much more with the decision makers of the developing countries in the future.

We must also face the fact that whatever good we might be able to do with the approach that we advocate, it can be undermined, like anything else, by the corruption and unresponsiveness of government in much of the world. We have no magic solution to this, although we are confident in the positive impact of a neutral and transparent scientific approach. But the main problem that we scientists face is that we are not very good at communicating our results to the public and therefore we have limited influence on the decisions that affect our society. This is where good computer graphics showing to the general populace the past and projected future aspects of their economies and environments as a function of whatever policies are implemented can be key. In fact we believe that if well done, political debates about the future might be carried out with the aid of good computer simulations and visualizations shown on national television. We often think while watching political debates that it would be very interesting if the promises of the candidates were subject to reality checks to see what was in fact possible, and at what energy cost.

Most people who are involved with such a comprehensive analysis are interested in implementing the results in what is normally called policy. Of course that can be an extremely difficult process but if you have worked with the right people from the start it will be possible to actually make better decisions. So it is important to involve decision makers from the beginning. From them (and ideally from the general affected populace as well) the scientist or economist can get a much clearer idea of desired ends (which might be quite different from what the scientist or economist assumes). In turn the decision

maker can learn to have a systemic, longer-term perspective for their country.

"Hybrid" forums where scientist and citizens meet and exchange views are ideal for social–technical debates and the education of each. Again the use of dynamic graphs that can convey to the user possible futures and tradeoffs as a function of policy today can be very useful. Finally with the new insights gained from the entire process given above, re-examine if and where conventional economics has failed and propose amendments to neoclassical economics-based policy or develop an entirely new perspective based on the analyses we have given above. It is a big charge to develop an entirely new economics but we think it critical, and what we have here is a formal start. And of course throughout the entire process of undertaking biophysical economic assessments and plans the scientific method must be used, theories need to be advanced in a way consistent with first principles, hypotheses need to be generated and tested, and so on. The final arbitrator of the correctness of our analyses is not whether this or that theory is the basis for our efforts but whether predictions based on policy prescriptions come to pass. This closes the loop on what is our basic wish: to bring the scientific method to developmental economics.

Questions

1. Explain some virtues of the process of visualization of model output (as was done, for example, for Costa Rica).
2. Distinguish among "environmental economics", "ecological economics" and "biophysical economics".
3. What is an ecological footprint? How does that relate to biophysical economics?
4. What is emergy analysis? How does it differ from energy analysis?
5. Give one example where biophysical economic, footprint, and emergy analyses give substantially the same answer.
6. Give five steps that can be followed in developing a biophysical analysis.
7. How can social, political, and economic elements be incorporated into a biophysical analysis?

8. What kinds of data might one want to gather in a biophysical assessment?

9. What is a simple way of translating a simple growth rate into a doubling time? For example, the United States had 300 million people growing at 1% a year in about the year 2000. If this 1% a year growth rate continues when would the United States have 600 million people? How old would you be then if you were still alive?

10. What are time series data? How do they help us to understand biophysical economics?

11. What kind of corrections need to be made for raw economic data (e.g., GDP) when examining data over time?

12. What is the Gini index? How does that help to put a more nuanced perspective on, for example, GDP data?

13. What are a few important considerations in how imports, exports, and their difference might influence our economic policies?

14. How does a prediction of land use change account for possible economic possibilities? How does that relate to land quality?

15. What are some of the pitfalls that await even the best possible plan that one might develop? How can citizen involvement assist in that process?

References

1. Hall. C.A.S. (editor, Gregoire Leclerc and Carlos Leon, associate Editors) 2000. Quantifying sustainable development: The future of topical economies. Academic Press, San Diego.

2. We refer the reader to the following web sites as good examples of other definitions of ecological economics and how ecological economics can actually be done. http://www.wordiq.com/definition/Ecological_economics; http://www.fs.fed.us/eco/s21pre.htm, http://www.anzsee.org/ANZSEE8.html.

3. World-wide Fund for Nature International (WWF). 2004. Living Planet report 2004. Global Footprint Network, UNEP World Conservation Monitoring Centre, WWF, Gland Switzerland; Mathis Wackernagel @ http://www.youtube.com/watch?v=94tYMWz_Ia4.

4. Herendeen, R. and C. Bullard. 1975. The energy costs of Goods and Services. 1963 and 1967, Energy policy 3: 268.

5. Bullard, C.W.; Hannon, B.; Herendeen, R.A. Energy Flow through the US Economy. University of Illinois Press: Urbana, 1975.

6. Hannon B. (1981). Analysis of the energy cost of economic activities: 1963 2000. Energy Research Group Doc. No. 316. Urbana: University of Illinois.

7. Hall, C.A.S., C.J. Cleveland and R. Kaufmann. 1984. Energy and Resource Quality: The ecology of the economic process. Wiley Interscience, NY. 577 pp. (Second Edition. University Press of Colorado).

8. Cleveland, C.J. 2004. The encyclopedia of energy. Elsevier.

9. Green Design, Carnegie-Mellon University: http://www.ce.cmu.edu/GreenDesign/research/lca.html.

10. Odum, H.T. 1996. Environmental accounting, emergy and environmental decision making. John Wiley & Sons. New York, N. Y.

11. Brown, M. 2004. Energy quality, emergy, transformity: The contributions of H.T. Odum to quantifying and understanding systems. Pp. 201–214. In M. Brown and C.A.S. Hall (eds). Through the Macroscope: the legacy of H.T. Odum. Ecological Modelling, special issue: Volume 178.

12. Herendeen, R. 2004. Energy analysis and emergy analysis – a comparison. Pp. 227–238. In M. Brown and C.A. Hall (eds). Through the Macroscope: the legacy of H.T. Odum. Ecological Modelling, special issue: Volume 178.

13. Brown, M. and R. Herendeen. 1996. Embodied energy analysis and emergy analysis: a review. Ecological Economics: a comparative review. 19:219–236.

14. Brown, M., M. Wackernagel and C. A. S. Hall. 2000. Comparable estimates of sustainability: Economic, resource base, ecological footprint and emergy, pp 695–714 in 1 above.

15. Foran, B., M. Lenzen, C. Dey and M. Bilek. 2005. Integrating sustainable chain management with triple bottom line accounting I Integrating sustainable chain management with triple bottom line accounting Ecological Economics 52: 143–157. 5.

16. LeClerc, G. and Charles Hall (Eds.) 2008. Making development work: Scientific alternatives to neoclassical economic theory. University of New Mexico Press. Albuquerque.

17. Pontius, R. G. Jr, J. Cornell and C.A.S. Hall. 1995. Modeling the spatial pattern of land-use change with GEOMOD2: application and validation for Costa Rica. Agriculture, Ecosystems & Environment 85: 191–203.

18. Hall, C.A.S., H. Tian, Y. Qi, G. Pontius and J. Cornell. 1995. Modeling spatial and temporal patterns of tropical land use change. Journal of Biogeography 22:753–757.

19. Detwiler, P. and C.A.S. Hall 1988. Tropical forests and the global carbon cycle. Science. 239:42–47.

20. Barreteau, O, M. Antona, P. d'Aquino, S. Aubert, S. Boissau, F. Bousquet, W. Daré, M. Etienne, C. Le Page, R. Mathevet, G. Trébuil, J. Weber, 2003. Our companion modelling approach. Journal of Artificial Societies and Social Simulation vol. 6, no. 1 (http://jasss.soc.surrey.ac.uk/6/2/1.html).

21. Swartz P., 1996. "The art of the long view: planning for the future in an uncertain world", Doubleday, New York, 272p.

22. Hallock, J., Tharakan, P., Hall, C., Jefferson, M. and Wu, W. 2004. Forecasting the availability and diversity of the geography of oil supplies. *Energy* 30:2017–201.

Part V

Understanding How Real-World Economies Work

Peak Oil, Market Crash, and the Quest for Sustainability: Economic Consequences of Declining EROI

18

Introduction

Much of what traditional economics believes "works" because of clever technology, substitutions, and intelligent investments, in fact does so only because we have had huge amounts of cheap energy to throw at the problem [1]. Our present situation is perhaps most readily captured by the phrases "the end of cheap oil" and "the second half of the age of oil," created by petroleum geologists Colin Campbell and Jean Laherrère. These concepts also apply to a very much broader suite of the basic resources and environmental conditions required to fuel our economy (Fig. 18.1). Although many people are taught and believe that technology has made natural resources increasingly irrelevant, this book contains a great deal of evidence to show the contrary. Our national and global society is becoming more, not less, dependent upon natural resources, as oil, for example, underlies essentially everything we do economically. Second, many of the things that are treated as externalities in conventional economics, that is, as supposedly secondary issues not properly included in prices, are instead what we believe to be often the main issues of economics. Depletion of highest quality fuels is one such issue. More generally, understanding and protecting the basic systems of the Earth, such as the atmosphere, far from being a luxury or an externality as is indicated in conventional economic analysis, are the critical issues for economics.

As we write these final chapters, in 2010 and the first half of 2011, the U.S. national economy continues to struggle. There has been little inflation-corrected growth of the economy for 6 years, unemployment remains stubbornly high, and the stock market, although doing better than unemployment, remains far below its peak, even when not correcting for inflation. All but a couple of our states are facing severe budgetary problems, schools and colleges everywhere are facing severe budget shortfalls, and federal and state debt is of increasing concern.

The most abrupt change in our economy began in the summer of 2008 with the highest oil prices ever (almost $150 a barrel) and historically high prices for other energy and most raw materials. The Dow Jones Industrial average was down from its historic high of 14,198 the preceding fall to 11,734 the first week of August. Then, a series of disasters struck the financial markets, with many of the largest, most prestigious, and seemingly impervious companies declaring bankruptcy. Each week the stock market lost 5% or 10% of its value until, by the end of November, the Dow Jones had dropped to as low as 8,000, barely half the peak. Many investors lost from one third to one half of the value of their stocks. Since then the market recovered somewhat but has continued to be unpredictable and very low relative to what it had been. Unemployment remains persistently in excess of 9% while budget deficits exceed one trillion dollars per year. It is not known how long these recessionary conditions will last.

C.A.S. Hall and K. Klitgaard, *Energy and the Wealth of Nations: Understanding the Biophysical Economy*,
DOI 10.1007/978-1-4419-9398-4_18, © Springer Science+Business Media, LLC 2012

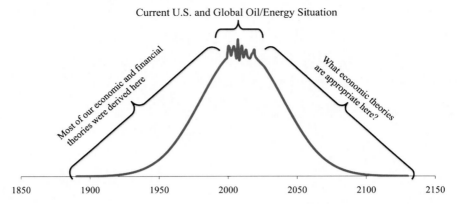

Fig. 18.1 Model of the relation of the development of economic and financial theory and concepts plotted in time along with an approximate representation of the Hubbert curve for global petroleum. In some sense it was easy to make all kinds of economic and financial models "work" when the real work done in an economy was expanding at 2% or 3% a year. Things may be more difficult now as we have reached peak oil, and will certainly be even more difficult for our persisting theories on the downside of the Hubbert curve

Few understand the underlying role of energy. The summer of 2011 also saw the sixth year in a row in which the global production of conventional oil did not rise, leading some to say that the long predicted "peak oil," the time of maximum global oil production, had indeed arrived. Total energy use in the United States had not increased for at least 3 years. The use of oil went down by about 8% from its peak in early 2008.

Who says the economy has to grow again? If the economy is as dependent upon energy as the rest of this book suggests then perhaps sustained growth in either energy or the economy can never occur again. A group to which we belong believes that the world has entered a new mode, one that was predicted in many ways in the 1960s and 1970s by some geologists, ecologists, and economists. This is a world of limits, one in which our once-trusted tools of conventional economics are no longer sufficient by themselves, if indeed they ever were, of righting economic wrongs and allowing us all to maximize our material well-being. There is no question that under the auspices of conventional economics many parts of the Western world, and increasingly Asia, have done very well in increasing human material well-being. The perspective that we raise, however, is whether our growth in wealth has been due to really understanding our economies or, as

we believe, more simply to our increased ability to pull more cheap oil, gas, and coal out of the ground to allow the increased economic work that is the basis of our wealth. To some extent any set of theories about economics in the past was bound to be at least partly correct because with more and more energy it was possible to generate more and more wealth, whatever one's theoretical premises.

What Is the Source of the Crash of 2008?

Many factors merged to cause the financial crash of 2008: the subprime mortgage crisis, high foreclosure rates, and Wall Street's sale of opaque financial products known as derivatives. Behind these are many aspects of greed, corruption, and malfeasance, not to mention the moral hazard caused by lax political oversight. It is not the intent of this book to focus on the personalities and moral shortcomings behind these issues but we believe one good and detailed summary of much of this can be found at http://www.informationclearinghouse.info/article28189.htm. Although we do not wish to downplay these moral issues we also believe that the root cause that initiated the current downturn and our difficulty in climbing out of the recession was the same one that sparked four out of the last five world recessions: the high price of oil. Why did most economists and financial analysts (and models such as the Wharton model) not see this coming? One hypothesis, advanced by Nobel laureate Paul Krugman is that the economics profession, "Went astray because economists, as a group, mistook beauty, clad in impressive-looking mathematics, for truth." We agree. As the market debacle has shown, mathematical elegance in economics is not a substitute for scientific rigor, something we have discussed in many previous papers [2, 3], and in Chap. 16. If physical quantity of energy and its effect on energy prices are crucial functions affecting the economy, and they are not in our models, then of what utility are the models?

As of this writing global oil production has been essentially flat since 2004, so that peak oil, or at least a cessation of reliable growth at the former rate of 2–4% per year, appears to have occurred, with the remaining debate only about whether there may be a subsequent peak and how soon we begin a slide down the other side. If we have passed the global peak in oil production then indeed the end of cheap oil will soon be upon us, and our ability to grow or even maintain economies is likely to decrease. Because of the critical importance of liquid and gaseous petroleum for essentially everything we do we have serious reservations as to whether conventional economics and business or governmental policies can guide us again to growth or indeed to manage an economy where growth is no longer possible (e.g., Fig. 14.1). Thus the question becomes: "Can we improve upon our ability to do economics and financial analysis by using procedures that focus more on the energy available (or not) to undertake the activity in question?" In other words, are finances beholden to the laws of physics?

We think yes. Thus the question becomes: can we supplement or improve upon our ability to do economics? Resource scientists have predicted such a financial crash, or more accurately cessation of growth, for a long time [4–7]. Any good physical or biological scientist knows that all

activity in nature – or anywhere – is associated with energy use. Consequently, many in the scientific community were not the slightest bit surprised by the financial crash or its timing. For example, Colin Campbell, a former oil geologist and co-founder of ASPO, the Association for the Study of Peak Oil, predicted serious financial responses to peak oil in his (and Jean Laherrère's) classic *Scientific American* article "The End of Cheap Oil" [7]. As mentioned earlier, he was more explicit at the ASPO meeting in Pisa, Italy, in 2006 when he said that we are likely to see an end of year after year economic growth and a movement to an "undulating plateau" in oil production, prices, and economic activity, with periodic high prices in oil generating financial stress and a cessation or even reduction of growth. These financial strains would, in turn, cause a decrease in oil use and hence a price decline, with lower oil prices then leading to new economic growth and new increases in oil use and, eventually, oil prices. In other words he foresaw very large effects of restrictions in oil availability, and consequent price increases, on the market. According to Campbell, "Every single company on the stock market is overvalued from the perspective of what the cost of running that company will be after peak oil. Value is determined by performance which has been based on cheap oil."

Many other analysts have remarked upon, and even predicted, the probable impact of peak oil, or at least oil price increases, on the financial status of the United States and the world [8]. A thoughtful, chilling, and ultimately correct view of the implications of peak oil on the American economy was presented by Gail Tverberg in January, 2008 on the energy blog site, "The Oil Drum" [9]. Her predictions, which we thought impossibly pessimistic at the time, have been vindicated in great detail. Many analysts foresaw these issues as early as the 1970s, including the authors of the famous but cavalierly dismissed *Limits to Growth* study of 1972, ecologists Garrett Hardin and Howard Odum, economists Kenneth Boulding, and others. But for those who bothered to read and think about what these authors were saying, the future was clear [10]. The reason is that all of these people understood that, of necessity, real growth is based on growth in real resources, and that there are limits to those resources. The case for peak oil was clearly laid out almost 60 years ago by Hubbert [11, 12] who predicted, in 1955 that the U.S. peak in oil production would occur in 1970, which it did.

Many economists place a great deal of faith in increasing technology. In fact technology does not operate on a static playing field but continually competes with declining resource quality. In addition, technological change can be destabilizing, as new technologies supplant old ones before they are fully amortized. This can lead to excess capacity and financial loss, and is a factor in the collapse of a social structure of accumulation as well. There is little or no evidence that technology is winning this game over time because the energy return on investment keeps falling [13–15]. It is important to understand that, at least so far, the limits to growth model is an almost perfect predictor of our current situation [16]. Resource-based analysts understand and appreciate that the recent turmoil in much of our financial structure has many plausible causes. But they also know energy underlies all of these issues. The fundamental dilemma is this: if oil, the most important energy source to fuel the economy, goes through the inevitable path of growth, plateau, and eventual decline (i.e., peak oil) and the financial market is built on the assumption of unfettered growth, then something has to give (Fig. 14.1). Eventually the aspirations and assumptions of indefinite growth in assets, production, and consumption must collide with the reality of an ever-constricted source of the energy that fuels real growth.

Part of the financial stress is attributable to cheap oil that then becomes dear. Starting in the early 1990s, relatively inexpensive oil, declining interest rates, and globalization all contributed to economic growth and to declines in risk premiums for virtually all asset classes. Capital went farther out on the risk curve to make up for reduced returns, and increased leverage (i.e., a reduction in "money in the vault" relative to what was loaned) became the new norm. As volatility seemed to disappear, even more leverage was piled onto the system. Along with the changing landscape in global credit markets came cheap financing for

U.S. home buyers. The low price of energy also greatly increased discretionary income which further encouraged people to take advantage of this cheap financing, adding to massive residential development. According to financial analyst George Soros, this created a self-reinforcing "reflexive" system, where increasing home values increased collateral, which encouraged further borrowing in the household sector and in lines of credit for consumption and so on [17]. The system had been built on the premise that large amounts of discretionary spending would always be available and the notion that everyone was entitled to a McMansion, a "lawyer-foyer," and a home theater. Because the construction of homes far outpaced population growth, most of the growth was due to the perceived demand for these larger houses. To get the area needed we had to build out from the cities. The largest growth in real estate had been in the exurban areas, which were most vulnerable to gas price spikes.

Discretionary wealth, that which is available for nonessential investments and purchases, is extremely sensitive to volatile energy prices [18]. Because most oil use is not discretionary but needed for getting to or undertaking work, it is relatively inelastic: people must continue to buy it even when prices increase substantially. Consequently discretionary income dropped substantially when gasoline and other energy prices, which had been creeping up from a very low level in 1998, increased sharply in 2007–2008. The United States reached a "tipping point" in 2006–2008 [19] as the price of oil rose temporarily to nearly $150 a barrel. The assumption that the suburban lifestyle would be sustainable became a question in many a potential owner's mind. This perception appeared to be an important initiator of a decline in aggregate demand, particularly for exurban real estate. It also may have initiated the massive deleveraging we are now experiencing globally. (There is a good summary of the various analyses by Rubin, Hamilton, and others who argue that oil price increases were behind this, and past, recessions [20, 21].) Massive household debt could not be supported when the value of the underlying collateral declined: a decline triggered by the spike in energy prices. As the collateral disappeared, huge derivative positions that had been built in the previous decade experienced margin calls. A spiral of forced selling pressured all asset classes further, and forced the banking sector essentially to freeze in September of 2008. Will this faltering of the suburban model be a preview of our ultimate response to peak oil? Perhaps. Examining the general pattern of oil price increases and probability of them continuing can help us understand these things better in the longer term.

Energy Price Shocks and the Economy

At the start of 1973, oil was cheap at $3.50 a barrel. The United States was still the world's largest producer. Peak oil had just occurred in the United States in 1970, but no one noticed. The economy kept growing fueled by increasing oil imports. As domestic oil production in the United States declined from 1970 to 1973, foreign suppliers gained leverage. In late 1973 both political events that precipitated the Arab oil embargo and an accident that had severed an export oil pipe in the Middle East caused the price of oil to jump from $3 to $12 a barrel. In a matter of months these events created the largest recession since the Great Depression. The price spike had at least four immediate effects upon and within the economy: (1) oil consumption declined, (2) a large proportion of capital stocks and existing technology became too expensive to use, (3) the marginal cost of production increased for nearly every manufactured good, and (4) the cost of transportation fuels increased.

By 1979 the price of oil had increased by a factor of 10, to $35 a barrel. The proportion of gross domestic product that went to buying energy increased from 6% to 8% to 14%, restricting discretionary spending while causing previously unseen "stagflation" (Fig. 18.2). The prices of other energies, and commodities more generally, increased at nearly the same rate, driven in part by the price increase of the oil that was behind all economic activities. Then, in the 1980s, all around the world, oil that had been found but not developed (as it had not been worth much

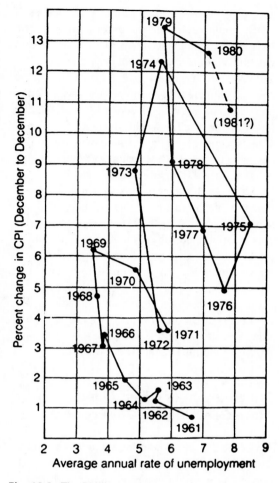

Fig. 18.2 The Phillips curve theory worked well from 1961 to 1969, however, when the price of oil increased during the 1970s the theory fell apart

previously) suddenly became profitable, and it was developed and overdeveloped. By the 1990s the world was awash in oil and the real price fell to nearly what it was in 1973. The energy portion of the GDP fell to about 6%, essentially giving everyone an extra 8% of their incomes to play with. The impact on discretionary income, perhaps a quarter of the total, was enormous. Many invested in the stock market, but then found themselves victims of the "tech bubble" of 2000. Real estate was considered a "safe" bet, so many invested in what was really surplus square footage. Speculation became rampant as real estate became valued for its financial returns rather than as a place to live. For a while it seemed as if investment in real estate was a sure path to wealth. As we now recognize, most of that increase in wealth was illusory. With energy price increases from 2008 to the summer of 2008, an extra 5–10% "tax" from increased energy prices was added to our economy as it had been in the 1970s, and much of the surplus wealth disappeared. Large-scale speculation in real estate was no longer desirable or possible as consumers tightened their belts because of higher energy costs.

Although this energy perspective is not a sufficient explanation for all that has happened, the similar economic patterns in response to the energy price increases of both the 1970s and of the last decade give the "energy trigger" considerable credibility. In systems theory language, the endogenous aspects of the economy that the economists focus on (Fed rates, money supply, etc.) became beholden to the exogenous forcing functions of oil supply and pricing that are not part of economists' usual framework.

The Relation of Oil and Energy More Generally, to Our Economy

Economics is overwhelmingly taught as a social science, in fact, our economy is completely dependent upon the physical supply and flow of resources. Specifically, our economy is overwhelmingly dependent upon oil, which supplied about 40% of U.S. energy use in the 2000s, followed by natural gas and coal at about 25% each, and nuclear at a little less than 5%. Hydropower and firewood supply no more than 4% each. Wind turbines, photovoltaics, and other "new solar" technologies together account for less than 1% (although that percentage may be increasing). Global percentages are similar. Our economy has been based on increased use of fossil fuels for most of its growth. Until 2008 we added much more new capacity with fossil fuels than with new solar, which has added a bit to the total use rather than displaced fossil fuels. Since 2008 all growth has essentially ceased, and the remaining economic activity is still based on about the same energy mix.

Because of the enormous interdependency of our economy, there is not a huge difference in the energy requirements for the various goods and services that we produce. A dollar spent for most final demand goods and services uses very roughly the same amount of energy because of the interdependencies of the economy. An exception is money spent for energy itself, which includes the chemical energy plus another 10% or so which is the energy needed to get it (i.e., the embodied energy). For 2005 an average dollar spent for final demand products required about 8 or 9 MJ (1 MJ equals 240 kcal) for that activity. Money spent in the arts might use only 2, and for chemicals such as paint, 16, but for most final demand goods and services the number is nearer to the mean. For heavy construction in the petroleum industry the estimate is about 14 MJs per dollar and for very heavy industry such as obtaining oil and gas about 20 MJs per dollar [15]. Year by year less energy is used per dollar. Most of the decline is due to inflation but there is some, and some would say substantial, increase in the efficiency with which we turn energy into goods and services. There continues to be decreasing energy return on energy invested (EROI) for our major fuels as we just go after ever more difficult resources [13–15].

Energy and the Stock Market

We include here some preliminary analyses that we think show the importance of energy to Wall Street and the economy more generally. First, Wall Street prices reflect not only something about the real operation of the economy but also a large psychological factor often called "confidence." Our hypothesis is that the energy used by the economy is in some sense a proxy for the amount of real work done. Thus over time the inflation-corrected Dow Jones Industrial Average (ICDJ), an index of industrial wealth generated, should have the same basic slope as the use of energy in society. It should also "snake" around the real amount of work done, reflecting issues of confidence, speculation, and so on. Over sufficient time, however, the ICDJ must return approximately to the real energy use line. To test

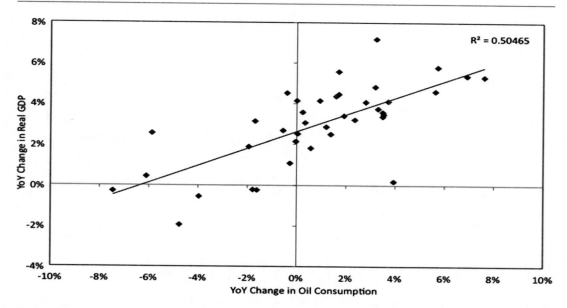

Fig. 18.3 Correlation of year-on-year (YoY) changes in oil consumption with YoY changes in real GDP, for the United States from 1970 through 2008 (*Source*: Oil consumption data from the BP Statistical Review of 2009 and real GDP data from the St. Louis Federal Reserve)

this hypothesis we plotted the ICDJ from 1915 until 2008 along with the actual use of energy by the U.S. economy. Our hypothesis would be supported if the slope of these two lines were similar over the longer time period. In fact from 1915 until 2010 the ICDJ had the same basic slope as the use of energy, and it has greater variability, consistent with our hypothesis (Fig. 1.8). We hypothesize that the Dow Jones will, over the long run, continue to snake about the total energy use in response to periods of irrational exuberance and the converse. If U.S. total energy use continues to stagnate or decrease, as it has since 2008 this hypothesis implies no sustained real growth for the Dow Jones, as in fact has been the case for that period.

In the past we also hypothesized that the amount of wealth generated by the U.S. economy should be closely related to fuel energy use. Cleveland et al. found that the gross national product of the United States was highly correlated with quality-corrected energy use from 1904 to 1984 ($R^2 = 0.94$) [22]. This high correlation appeared to be much poorer for the period 1984 until 2008, a period during which inflation-corrected GDP doubled while energy increased by only a third. It is possible that the divergence is due not to increasing efficiency but rather an increasing proclivity of governments to "cook the books" on inflation (see the online group, shadowstatistics. com). Correcting for this, if indeed that is needed, would make the relation of energy use and GDP growth much tighter through the 1990s and 2000s (Figs. 18.3 and 18.4).

A Financial Analyst Concurs

Jeff Rubin, chief economist at CIBC World Markets, wrote in a recent report that defaulting mortgages are only one symptom of the high oil prices [23]. Higher oil prices caused Japan and the European nations to enter into a recession even before the most recent financial problems hit. According to Rubin, oil shocks create global recessions by transferring billions of dollars of income from economies where consumers spend every cent they have, and then some, to econo-

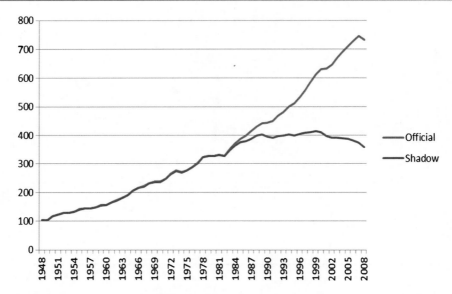

Fig. 18.4 One attempt to correct the GDP for the "deflated" inflation factor by using the inflation corrections year by year since 1984 supplied by the group shadowstatistics. If larger inflation estimates are used the economy has grown very little since 1984, and there may have been no improvement in efficiency which is how energy is changed to GDP (*Source*: Hannes Kunz; see also: http://www.leap2020.eu/the-true-us-gdp-is-30-lower-than-official-figures_a5732.html)

mies that sport the highest savings rates in the world. Although those petro-dollars may get recycled back to Wall Street by sovereign wealth fund investments, they don't all get recycled back into world demand. The leakage, as income is transferred to countries with savings rates as high as 50%, is what makes this income transfer far from demand neutral. By any benchmark the economic cost of the recent rise in oil prices is nothing short of staggering, much more so than the impact of plunging housing prices on housing starts and construction jobs, which, according to the press, has been the most obvious brake on economic growth from the housing market crash. And those energy costs, unlike the massive asset writedowns associated with the housing market crash, were borne largely by Main Street, not Wall Street, in both America and throughout the world. This big increase in oil prices has caused the annual fuel bill of OECD countries to increase by more than $700 billion a year, with $400 billion of this going to OPEC countries. Rubin asks: "Transfers a fraction of today's size caused world recessions in the past. Why shouldn't they today?" We and others believe that there is ample evidence that our economy is beholden to energy supplies and prices, and that good investors and good economists need to learn a great deal more about energy. This is one reason why we are attempting to tackle this problem head on through the development of biophysical economics. But getting the economists to rethink their intellectual training will be a tough job, no matter how much that is needed.

Is Growth Still Possible?

There was little inflation-corrected growth of the United States economy or in its use of energy from 2004 through the end of 2010. Is this just part of the normal business cycle or something new? Numerous theories have been posited over the past century that have attempted to explain business cycles. Each offers a unique explanation for the causes of, and solutions to, recessions, including: Keynesian theory, the monetarist model, the rational expectations model, real business cycle models, neoKeynesian models, and so on and on. Yet, for all the differences among these theories, they all share one implicit assumption: a return to a growing economy is both desirable and possible; that is, GDP can grow indefinitely. Historically, economists such as Baran and Sweezy and the social structure of accumulation school have analyzed the powerful internal tendencies that keep a conentrated economy from growing. In its monopoly phase a market economy stagnates due to internal forces alone. However peak oil adds another sobering dimension to the problem. But if we are entering the era of peak oil, then for the first time in history we may be asked to grow the economy while simultaneously decreasing oil consumption, something that has yet to occur in the United States for 100 years. Oil more than any other energy source is vital to today's economies because of its ubiquitous application as transportation fuel, as a portable and flexible energy carrier and as feedstocks for manufacturing and industrial production. Historically, spikes in the price of oil have been the proximate cause of most recessions. On the other hand, expansionary periods tend to be associated with the opposite oil signature: prolonged periods of relatively low oil prices that increase aggregate demand and lower marginal production costs, all leading to, or at least associated with, economic growth.

By extension, for the economy to sustain real growth there must be an increase in the flow of net energy (and materials). Quite simply economic production is a work process and work requires energy. Thus to increase production over time (i.e., to grow the economy) we must either increase the energy supply or increase the efficiency with which we use our source energy. This is called the energy-based theory of economic growth. This logic is an extension of the laws of thermodynamics, which state that: (1) energy can neither be created nor destroyed, and (2) energy is degraded during any work process so that the initial inventory of energy can do less work as time passes. As Daly and Farley [22] describe, the first law places a theoretical limit on the supply of goods and services that the economy can provide, and the second law sets a limit on the practical availability of matter and energy. In other words, to produce goods and services

energy must be used, and once this energy is used it is degraded to a point where it can no longer be reused to power the same process again.

An Energy-Based Theory of Economic Growth

This energy-based theory of economic growth is supported by data: the consumption of every major energy source has increased with GDP since the mid-1800s at essentially the same rate that the economy has expanded (Fig. 3.1). Throughout this growth period, however, there have been numerous oscillations between periods of growth and recessions. Recessions are defined by the Bureau of Economic Research as "a significant decline in economic activity spread across the economy, lasting more than a few months, normally visible in real GDP, real income, employment, industrial production, and wholesale-retail sales" [24]. From 1970 until 2007, there have been five recessions in the United States. Examining these recessions from an energy perspective elucidates a common mechanism underlying each recession: oil prices are lower and oil consumption increases during periods of economic expansion while oil consumption decreases and oil prices are higher during recessions (Fig. 18.5). Oil price increases precede essentially all recent recessions.

Plotting the year-on-year (YoY) growth rates of oil consumption and real GDP provides a more explicit illustration of the relation between economic growth and oil consumption (Fig. 18.6). But correlation is not causation, and an important question is whether increasing oil consumption causes economic growth, or conversely, whether economic growth causes increases in oil consumption [25]. Cleveland et al. [26] analyzed the impact of these two factors on the causal relation between energy consumption and economic growth. Their results indicated that increases in energy consumption caused economic growth, especially when they adjusted the data for quality and accounted for substitution. Other subsequent analyses that adjusted for energy quality support the hypothesis that energy consumption causes economic

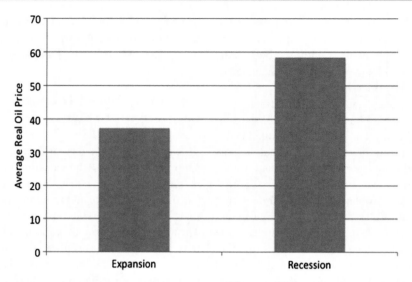

Fig. 18.5   Real oil prices averaged over expansionary and recessionary periods from 1970 through 2008

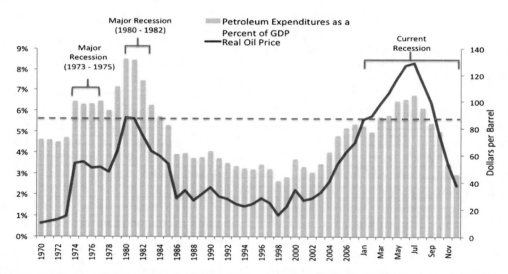

Fig. 18.6 Petroleum expenditures as a percentage of GDP and real oil price. The *dotted line* represents the threshold above which the economy moves towards reces-sions. Petroleum expenditures includes distillate fuel oil, residual fuel oil, motor gasoline, LPG, and jet fuel. The last values are for 2008

growth, not the converse [27]. In sum, our analysis indicates that about 50% of the changes in economic growth over the past 40 years are explained, at least in the statistical sense, by the changes in oil consumption alone. In addition, the work by Cleveland et al. [26]. indicates that changes in oil consumption cause changes in economic growth. These two points support the idea that energy consumption, and oil consumption in particular, is of the utmost importance for economic growth.

Yet changes in oil consumption are rarely used by neoclassical economists as a means of explaining economic growth. For example, Knoop [28] describes the 1973 recession in terms of high oil prices, high unemployment, and inflation, yet omits mentioning that oil consumption declined 4% during the first year and 2% during the second year. Later in the same description, Knoop claims that the emergence from this recession in 1975 was due to a decrease in both the price of oil and inflation, and an increase in money supply. To be sure, these factors contributed to the economic expansion in 1975, but what is omitted, again, is the simple fact that lower oil prices led to increased oil consumption and hence greater physical economic output. Oil is treated by economists as a commodity, but in fact it is a more fundamental factor of production than either capital or labor. Thus we again present the hypothesis that higher oil prices and lower oil consumption are both precursors to, and indicative of, recessions. Likewise, economic growth requires lower oil prices and simultaneously an increasing oil supply. The data support these hypotheses: the inflation-adjusted price of oil averaged across all expansionary years from 1970 to 2008 was $37 per barrel compared to $58 per barrel averaged across recessionary years, whereas oil consumption grew by 2% per year on average during expansionary years compared to decreasing by 3% per year during recessionary years (Figs. 18.5 and 18.6).

Although this analysis of recessions and expansions may seem like simple economics (i.e., high prices lead to low demand and low prices lead to high demand), the exact mechanism connecting energy, economic growth, and business cycles is rather more complicated. Hall et al. [18] and Murphy and Hall [29] report that when energy prices increase, expenditures are reallocated from areas that had previously added to GDP, mainly discretionary consumption, towards simply paying for the more expensive energy. In this way, higher energy prices lead to recessions by diverting money from the general economy towards energy only. The data show that recessions occur when oil expenditures as a percent of GDP climb above a threshold of roughly 5.5%, or, stated somewhat differently, when all energy becomes more than 12% of the economy (Fig. 18.6).

Predicting Future Economic Expansion

Each time the U.S. economy emerged from a recession over the past 40 years there was an increase in the use of oil even while a low oil price was maintained. Unfortunately oil is a finite resource. What are the implications for future economic growth if following a recession oil supplies are unable to increase with demand, or oil supplies increase but at an increased price? To undertake this inquiry we must examine first the current and probable future status of the oil supply; then we can make inferences about what the future of the oil supply and price may mean for economic growth.

Because oil consumption causes change in economic growth, understanding how both peak oil and net energy will affect oil supply and price is important to understanding the ability of our economy to grow in the future. To that end, we review both the theory and current status of peak oil and net energy as they pertain to oil supply, and then discuss how both of these may influence oil price. Optimists about future oil availability usually start with the correct observation that there is a great deal of oil left in the Earth, probably three to ten times what we have extracted, and, usually, with the assumption that future technology driven by market signals will get much of that oil out. There are at least two problems with that view. The first is that of peak oil. It is clear we have, or soon will, reach a physical limit in our ability to pump more oil out of the ground. For a long time oil production grew at 3–4% a year. Now there has been little or no growth in global oil production since 2004. The second problem is that the oil left in the ground will require an increasing quantity of energy to extract, at some point as much as is in the oil. There is a clear trend that the EROI of oil production is declining in each region for which data are available. This shows that depletion is more important than technical advances. Gagnon et al. [15]. report that the EROI for global oil extraction declined from about 36:1 in the 1990s to 18:1 in 2008. This downward trend results from at least two factors: first, increasingly supplies of

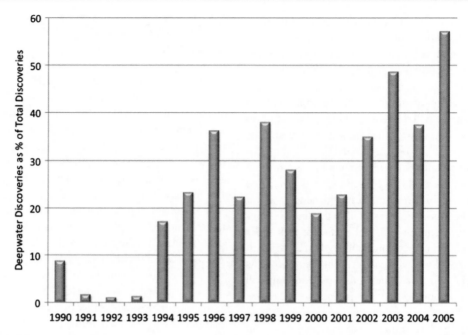

Fig. 18.7 Deepwater oil discoveries as a percent of total discoveries from 1990 through 2005 (Source: Jackson 2009)

oil must come from sources that are inherently more energy-intensive to produce, simply because firms have developed cheaper resources before expensive ones. For example, in 1990 only 2% of oil discoveries were located in ultra-deepwater locations, but by 2005 this number was 60% (Fig. 18.7). Second, enhanced oil recovery techniques, such as the injection of steam or gases are increasingly being implemented. For example, nitrogen injection was initiated in the once super-giant Cantarell field in Mexico in 2000, which boosted production for 4 years, but since 2004 production from the field has declined precipitously. Although enhanced oil recovery techniques increase production in the short term, they also significantly increase the energy input to production, offsetting much of the energy gain for society. Thus it seems that additional oil is unlikely to be available and if so it will have a low EROI and hence high price.

Forecasting the price of oil, however, is a difficult endeavor as oil price depends, in theory, on the demand as well as the supply of oil. Following the economic "crash" of 2008 most Western economies have been contracting or at least not

growing. The flat rate of oil production since 2004 did not cause a huge sustained increase in the price of oil. One thing we can do with some accuracy is to examine the cost of production of various sources of oil to calculate the price at which different types of oil resources become economical (Fig. 18.8). We can then estimate how much oil would be available at a given price. If the price of oil is below the cost of production, then most producers of that oil will cease operation. If we examine the cost of production in the areas in which we are currently discovering oil, hence the areas that will provide the future supplies of oil, we can calculate a theoretical floor price below which an increase in oil supply is unlikely.

Roughly 60% of the oil discoveries in 2005 were in deepwater locations (Fig. 18.7). Based on estimates from Cambridge Energy Research Associates [30], the cost of developing that oil is between $60 and $85 per barrel, depending on the specific deepwater province. Therefore oil prices must exceed roughly $60 to $90 per barrel to support the development of even the best deepwater resources. These data indicate that an

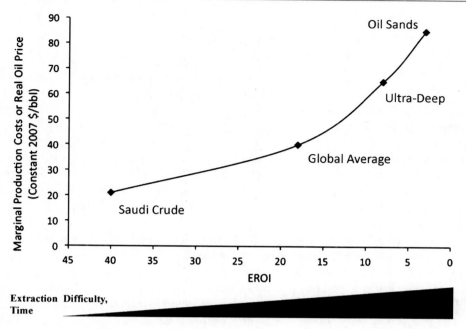

Fig. 18.8 Oil production costs from various sources as a function of the EROI of those sources. The *dotted lines* represent the real oil price averaged over both recessions and expansions during the period from 1970 through 2008

(Source: Data on EROI from Murphy and Hall [29], Gagnon et al. [32], and the data on the cost of production come from CERA [30])

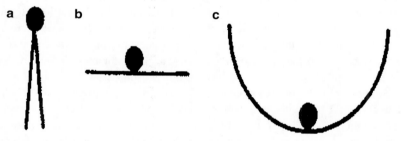

Fig. 18.9 Three types of equilibrium: unstable (**a**), neutral (**b**), and stable (**c**). The third situation seems to represent what we face in the world today

expensive oil future is necessary if we are to expand our total use of oil, that is, to grow economically. But these prices will discourage that very growth (Fig. 18.8). Indeed, it may be difficult in the future even to produce the remaining oil resources at prices the economy can afford. As a consequence, the economic growth witnessed by the United States and the globe over the past 40 years may be a thing of the past.

One way to think about this situation is to borrow a concept from systems theory. A very general concept is that many systems seek an equilibrium point because there are dynamic forces that resist change. An example is a marble in a bowl (Fig. 18.9c). The marble seeks its equilibrium position at the bottom of the bowl. One can push the marble up the side with a finger, but the marble easily slips off the finger and goes back to the equilibrium position. This might represent the situation our economy is in now, kept at a more or less constant GDP by growth being discouraged by rapidly increasing oil prices at levels of consumption barely above where we are now, but maintained from further shrinking by decreased oil prices with contraction; indeed this is a recipe for a steady-state economy (Fig. 18.10).

EROI and Prices of Fuels

Because EROI is a measure of the efficiency with which we use energy to extract energy resources from the environment, it can be used as a proxy to estimate generally whether the cost of production of a particular resource will be high or low, or perhaps even energy costs themselves [31]. For example, production from Canadian oil sands have an EROI of roughly 3:1, whereas the production of conventional crude oil has an average EROI of about 10–20:1 and Saudi crude much higher. The production costs for oil sands are roughly $85 per barrel compared to roughly $60 for average U.S. oil and $20 to 40 per barrel for Saudi Arabian conventional crude. Thus there is an inverse relation between EROI and price, indicating that high EROI resources are generally relatively inexpensive to develop and low EROI resources are generally more expensive to develop (Fig. 18.8). As oil production continues, we can expect to move farther toward the upper right of figure. We see no evidence that technology has lowered EROI even as it extends our resources. In summary, relatively low EROI appears to translate directly into higher oil prices, so that if we have to move to lower EROI oil in the future the price is likely to be higher which will, in all likelihood, be exacerbated by climate change, restricting economic activity and growth [32].

Summary

The main conclusions to draw from this discussion are: (1) over the past 40 years, economic growth has required increasing oil consumption, (2) the supply of high EROI oil cannot increase beyond current levels for any prolonged period of time, (3) the average global EROI of oil production will almost certainly continue to decline as we search for new sources of oil in the only places we have left: deepwater, arctic, and other hostile environments, (4) we have globally no more than 20–30 years of conventional oil remaining at anything like current rates of consumption and anything like current EROIs, and less if oil consumption increases and/or EROI decreases, (5) increasing oil supply in the future will require a higher oil price because mostly only low-EROI, high-cost resources remain to be discovered or exploited, (6) developing these higher cost resources is likely to cause economic contraction as oil costs exceed 5%, and total energy costs exceed 10% of GDP, and (7) using oil-based economic growth as a solution to recessions is untenable in the long-term, as both the gross and net supplies of oil have begun, or will begin, at some point, an irreversible decline. A similar assessment could be developed for other energy resources.

This growth paradox leads to a highly volatile economy that oscillates frequently between expansion and contraction periods, and as a result, there may be numerous peaks in economic activity and in oil production but little trend. In terms of business cycles, the main difference between the pre- and peak oil era is that business cycles

Peak Era Model of Economic Growth

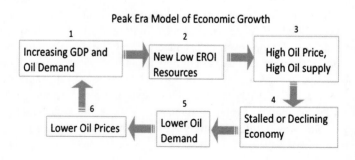

Fig. 18.10 Peak oil era model of the economy. Cycle of relation of economic growth (or recession) and oil prices

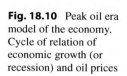

appear as oscillations around an increasing trend in the prepeak era but as oscillations around a flat trend following the peak. For the economy of the United States and most other growth-based economies the prospects for future, oil-based economic growth are bleak, and we do not have another model that would allow for growth. It seems clear that the economic growth of the past 40 years will not continue for the next 40. A resolution to these problems can occur when economic growth is no longer the primary goal. Society must begin to emphasize energy conservation over growth, and adjust our jobs, living patterns, and aspirations accordingly.

Questions

1. What events of 2008–2011 might be construed as indicating some limits to the 3–4% per year growth that the United States and much of the world had previously expected? These new limits may or may not be related to biophysical limitations. How would you assess this situation?

2. Have events since the publication of this book in late 2011 changed your answer to the previous question?

3. What was the main reason that Nobel Prize economist Paul Krugman put forth for the market crash in 2008?

4. Do you think that finances are beholden to the laws of physics? Why or why not?

5. What is the relation of an "undulating plateau" to peak oil?

6. Can you discuss financial "leverage" with respect to energy and other resources?

7. Discuss some of the financial issues that were related to the "oil crises" of the 1970s.

8. Do you think that price gives signals as to the future availability of energy? Why or why not?

9. If energy supplies are indeed restricted is economic growth still possible? What would be the requirements for that?

10. What is the relation historically between the price of energy and discretionary spending?

11. What has been the relation between the amount of oil that is consumed in a given year and the price of that oil? What might be a reason for that?

12. As the EROI for a given source of oil declines how does that relate to its price?

13. How might we best respond to a future of limited oil supplies should it occur, which seems likely?

14. Due to the depletion of high EROI oil the economic model for the peak era, that is, roughly 2000–2020, is much different when viewed as net rather than gross energy from oil. Why is that?

References

1. Derived in part from: Hall, C.A. S. and Groat, A. 2010. Energy price increases and the 2008 financial crash: A practice run for what's to come. The Corporate Examiner 37: 19–30 and Hall, C. A. S. and D. J. Murphy. Adjusting to the new energy realities in the second half of the age of oil. Ecological Modeling, in press.

2. Hall, C. A. S., Lindenberger, D., Kummel, R., Kroeger, T. and Eichhorn, W. 2001.The need to reintegrate the natural sciences with economics. BioScience 51, 663–673.

3. Gowdy, J., C. Hall, K. Klitgaard and L. Krall. 2010. The End of faith based economics. The Corporate Examiner 37: 5–11.

4. Odum, H. T. 1973. Energy, ecology, and economics. Ambio, 2: 220–227.

5. Huang J., R. Masulis, H. Stoll. 1996. Energy shocks and financial markets. Journal of Futures Markets 16: 1–27.

6. Deffeyes. 2001. Hubbert's Peak: The impending world oil Shortage. Princeton University Press.

7. Campbell and Laherrere. 1998. The end of cheap oil. Scientific American, 278: 78–83.

8. Campbell, C. 1997. The coming oil crisis. Multi-Science Publishing Company and Petroconsultants.

9. Tverberg, G. January 9, 2008. Peak oil and the financial markets. A forecast for 2008. The oil Drum. http://www.theoildrum.com/node/3382.

10. Hall, C. 2004. The myth of sustainable development: Personal reflections on energy, its relation to neoclassical economics, and Stanley Jevons. Journal of Energy Resources Technology 126:86–89.

11. Hubbert, M.K. 1969. Energy resources. In the National Academy of Sciences–National Research Council, Committee on Resources and Man: A Study and Recommendations. W. H. Freeman. San Francisco.

12. Hubbert M.K. 1974. U.S. Energy Resources: A review as of 1972. Background paper prepared for U.S. Senate Subcommittee on Interior and Insular affairs, 93d Congress, 2nd Session. Serial 93-94(92–75) . U.S. Government Printing Office, Washington.

13. Hall, C.A.S. and C.J. Cleveland. 1981. Petroleum drilling and production in the United States: Yield per effort and net energy analysis. Science 211: 576–579.

14. Cleveland, C. J. (2005). Net energy from the extraction of oil and gas in the United States. Energy: The International Journal, 30(5), 769–782.
15. Gagnon, N. C. A.S. Hall, and L. Brinker. 2009. A Preliminary Investigation of Energy Return on Energy Investment for Global Oil and Gas Production. Energies 2009, 2(3), 490–503.
16. Hall, C. A. S. and J. W. Day, Jr. Revisiting the Limits to Growth After Peak Oil. American Scientist, Volume 97: 230–237.
17. Soros, G. 1987. The Alchemy Furnace. Reading the mind of the market. John Wiley and Sons. N. Y.
18. Hall, Powers and Schoenberg. 2008. Peak oil, EROI, investments and the economy in an uncertain future. In Biofuels, solar and wind as renewable energy systems: Benefits and Risks. Pimentel Ed. Springer. The Netherlands.
19. Gladwell, M . 2000. The tipping point : how little things can make a big difference.
20. Rubin, J. Just how big is Cleveland. CIBC World markets. http://research.wibcwm.com/economic_public/download/soct08.pdf.
21. Sauter R. and S. Awerbach. 2002. Oil price volatility and economic activity: A survey and Literature Review. IEA Research paper, IEA , Paris.
22. Daley, H. and J. Farley 2004. Ecological Economics: Principles and applications. Island Press. Washington, D.C.
23. Rubin J. Just how big is Cleveland? CIBC World markets. http://research.wibcwm.com/economic_public/download/soct08.pdf
24. NBER. 2010. US Business Cycle Expansions and Contractions. National Bureau of Economic Research.
25. Karanfil. 2009. How many times again will we examine the energy-income nexus using a limited range of traditional econometric tools? Energy Policy, 37: 1191–1194.
26. Cleveland, C., R. Kaufmann and D. Stern. 2000. Aggregation and the role of energy in the economy. Ecological Economics, 32, 301–317.
27. Stern, D. 2000. A multivariate cointegration analysis of the role of energy in the US macroeconomy. Energy Economics, 22: 267–283.
28. Knoop. T.A. 2010. Recessions and Depressions: Understanding Business Cycles. Praeger. N.Y.
29. Murphy, D. J. and C.A.S. Hall. 2010. Year in Review - EROI or Energy Return on (Energy) Invested. New York Annals of Science, 1185: 102–118.
30. CERA. 2008. Ratcheting Down: Oil and the Global Credit Crisis. Cambridge Energy Research Associates.
31. King, C. and C. Hall. (in press). Relating financial and energy return on investment. Sustainability.
32. Cleveland, C.J., R. Costanza, C.A.S. Hall and R. Kaufmann. 1984. Energy and the United States economy: a biophysical perspective. Science 225: 890–897.

Environmental Considerations

19

Once, thousands of years ago, all humans were supported directly and entirely by nature. Our food, water, and everything else came directly from natural ecosystems as our ancestors, hunter-gatherers, went about their business for a million or more years. We cannot go back easily to that state because of population growth, for natural ecosystems alone could probably support no more than a few hundreds of millions of people. Today fossil-fueled systems of agriculture, water supply, and waste disposal support seven billion people on the planet. Most humans live in environments of concrete, boards, and macadam largely disconnected from the natural world. Although nature remains very popular in zoos and on television, and lucky youngsters still go camping with their parents, our population is increasingly disconnected from experiencing real nature or even rural agricultural landscapes, or from understanding our dependence upon these systems. Food comes from markets, water from faucets, entertainment from electronics encased in plastic boxes, and so on. But in fact all of these resources and toys and much more are all ultimately derived from nature, and their provision is usually associated with some degradation of nature and diminishment of natural resources. In general we do not pay for nature's goods and services but only for the energy, labor, and equipment to extract them. In fact we might argue that it is only because we do not pay for these things that we can afford to live at all, or certainly at the present level of general affluence.

No Free Lunch

Probably no single entity in contemporary life generates as much environmental impact as the use of energy. Land deformation, spills, generation of polluted water and air, emissions of greenhouse gases, and many other impacts are routinely associated with the extraction and consumption of each form of energy. More generally, our attempts to provide the American dream for our current population has resulted in the depletion of our conventional onshore oil and gas wells, and hence the Gulf oil spill and hydrofracking in Pennsylvania, mountaintop removal mining in Appalachia, as well as climate change, the continuing biodiversity crisis, and the massive depletion of fisheries and aquifers. Clearly from an environmental perspective we have reached many physical and biological limits of our nation to provide the resource-intensive "American dream" for more people, at least without extremely serious environmental disruption. More or less daily the media, local, national, and global, are filled with various outraged citizens and environmental groups who are distressed by the environmental impact of one energy technology or another. For example, many Americans are justifiably very upset about the oil spill in the Gulf of Mexico in the summer of 2010. At the same time there is increasing attention paid to the procedure known as "mountaintop coal removal in Southern Appalachia. Coal emissions are very much the source of CO_2 and of complaints about

C.A.S. Hall and K. Klitgaard, *Energy and the Wealth of Nations: Understanding the Biophysical Economy*, DOI 10.1007/978-1-4419-9398-4_19, © Springer Science+Business Media, LLC 2012

that. There are 100 large nuclear reactors that are reaching retirement during the next decade or so, with no new ones even started. Antinuclear activists were the one of the primary factors that led Wall Street to abandon nuclear financing in the late 1970s. Our electricity needs have been met instead by more coal plants, mostly. In effect are the antinuclear environmentalists responsible for the CO_2, mercury, and sulfur from these plants, as they will be for the forthcoming sudden drop off in nuclear electricity?

Biofuels are supposedly "green," but corn-based ethanol, our largest liquid biofuel by far, uses about as much fossil energy as is found in the product, so we have equal pollution from the fossil energy input plus erosion from growing corn. Biofuels made from corn have led to large environmental concerns as the effects of less exported American corn leads to increased deforestation from increased cropland in Bolivia and Brazil. The city of Syracuse, New York burns its garbage to generate electricity for 30,000 homes. This action is attacked by environmentalists because of fears of the (low levels of) dioxins and other toxins released. Upstate New Yorkers are also very much aware of the possibility of enormous acceleration of obtaining natural gas from "nontraditional" sources such as shale, where the low gas concentrations require very complex horizontal drilling and hydrofracking (pumping water, sand, and chemicals into the rock formations to loosen the gas). Energy extraction, in whatever form or aspect, tends to be an environmentally damaging process.

Of course there is nothing new about environmental impacts of obtaining energy. A large proportion, as much as 20%, of the land surface of Illinois, Wyoming, Pennsylvania, and West Virginia has been affected by surface mining. The same type of damage is occurring in many other states. Decades ago there was an enormous impact of oil and gas exploitation in the marshes of Southern Louisiana. Now that impact (including the similar effects of artificial levees) means that roughly half the land area of Southern Louisiana has been lost, and along with it its extremely important role in fish and wildlife production. These past environmental impacts, often as large and destructive as the ones seen in our

media today, were barely recognized or understood by ordinary citizens outside, and even in, the areas affected. In a sense as we exploit and deplete our most economic and highest EROI fuels we move from one "energy sacrifice area" to another: from one area whose environment is essentially "written off" to supply energy to the nation. These areas included in the past Southern Louisiana and much of Appalachia, the Rocky Mountain front in Colorado, and Wyoming in the present decade, and increasingly Pennsylvania and adjacent states such as our own New York as unconventional gas is increasingly exploited. The hydraulic fracturing debate revolves around the trade off between water quality and the local level and the provision of energy to a world market.

IPAT

What is the source of environmental impact? Probably the most general approach is what is called the IPAT equation, derived by Paul Ehrlich and John Holdren in various publications in the 1970s. The equation says that:

$$Impact = Population \ times \ affluence \ times \ technology$$

In other words the total environmental impact depends on the total number of people, each of whom is generating impacts according to their affluence, and hence resource use, and some factor of technology that indicates the impact per unit of resources used. The technological factor can increase or decrease over time. Inasmuch as in the United States there is very little enthusiasm for discussing population level, let alone restricting affluence, almost all of the discussion has focused on using technologies that decrease environmental impact. Meanwhile the population is doubling every 70 years or so and, at least until recently, affluence has increased considerably as well. The net effect is hard to evaluate but we have seen an increase in many impacts such as production of CO_2.

Internalizing Externalities

Environmental considerations, more or less completely ignored by governments (with some important exceptions) and much of the public until about 1950, are now an important factor in public debate. In general, environmental considerations have become, and are likely to be more and more, important in energy assessments as the energy resources available tend to be of lower quality (i.e., less concentrated), and no longer far from population centers or regions of less politically powerful people who were grateful for jobs.

Historically environmental impacts were dealt with, if they were, as externalities, which is an economic term. In market economies prices are established basically by the relation between costs to the firms and the consumer's willingness to pay. But there are many costs not borne by the seller or the buyer but by someone else or by society as a whole. These are called externalities and include pollution and other by-products of production, assessed in dollar terms. For example, the production of oil and gas often created the production of brines which were once (and sometimes still) dumped into a neighboring stream. This could kill fish and destroy the livelihood of commercial fishermen or sports fishing guides downstream. The idea of externalities is to assess these costs and add them in some way to the market price so that the market price would come closer to reflecting the real costs. But without governmental intervention there would be no incentive for the pollution to be assessed or cleaned up, and the price of the oil would not include the real costs to these other people and ecosystems. Many considered this sort of thing a *market failure*. It was one of the first areas in which it became clear that "free" (i.e., unregulated) markets do not resolve all issues relating to the production and sale of an item. As knowledge of, and public outrage about, pollution became stronger, laws were passed and fines levied against the polluters so that increasingly the externalities were internalized to the companies, who would then pass their new costs along to consumers. In this way the costs of the pollution,

or of its prevention or control, were passed along to those using the products.

The idea of internalizing externalities has spread to many other aspects of the price of energy in market economies. Examples include the costs when companies are forced to clean up pollutants, reduce carbon emissions, mitigate health impacts, and so on. Less clear is to what degree externalities should be internalized, or whether internalizing externalities is sufficient to fix the many problems of the market. Some environmentalists believe that the language of externalities belittles what many believe are intrinsic costs far greater than production costs. For example, as of this writing a barrel of oil costs about $90. Because this is about three times what it cost 5 years ago many people think this is a "high" price for oil. But what is the real price of oil? In 1979 Staubaugh and Yergin [1], two experts on this issue, calculated that if you included all of the externalities associated with that barrel of oil, including the environmental costs associated with transporting and burning it and the military expenditures to maintain an American military presence in the Middle East, the cost would be some $400 a barrel, or about $1,000 in 2010. In fact the future environmental cost of burning that oil in terms of climate change alone might be far greater than even that. Or one might argue that if you remove a barrel of oil from known inventories one "should" replace it, either by finding another barrel of oil or, because that seems increasingly difficult, developing a substitute. This is the idea behind a concept known as strong sustainability, which puts very serious limits on what it might be possible for us to do and call it sustainable. But meanwhile the general concept of externalities is well accepted for many issues.

How Should We Proceed? Command and Control Versus Incentives

Assuming that we have identified an environmental problem and that we wish to fix it, how should we persuade companies or individuals to change their behaviors to do so, inasmuch as it usually is costly in terms of money, time, or something else? There are two general approaches: *command and control*, where government imposes penalties, usually financial, to impose environmentally sound policies. An example is fines levied when pollutants are spilled. Another way to accomplish pollutant reductions is *incentives*, that is, making it worthwhile in some way to the companies or individuals to undertake "proper" behavior. Sometimes the incentive is to avoid more stringent approaches later.

For example, various compounds of chlorine and fluorine such as freon were found to be excellent refrigerant chemicals: stable, effective, and cheap. Unfortunately their stability meant that they were not readily broken down in nature so they escaped into the atmosphere (such as when an old refrigerator goes to a dump and the pipes corrode away). The chemicals were found to be extremely damaging to the ozone layer, a high altitude strata of O_3 which had been found to be decreasing each year. Because this layer intercepts high-energy ultraviolet light from the sun, it serves as an important protection to human skin, the eyes of wildlife, and many other things. (Ozone at the Earth's surface is another matter as it is a strong and aggressive oxidant.) The industrial nations that produced these chemicals agreed to change the formulation of refrigerants to a less aggressive form through what is known as the Montreal Protocol. This highly successful political document led to mutually agreed upon restrictions and, eventually, to a natural "reconstruction" of the damaged ozone layer. In this case no country wanted to not sign the protocol and hence be the bad guy, that was the incentive.

There has been considerable interest in how one might go about the process of cleaning up the environment efficiently, that is, with the least cost in energy and especially dollars. One such approach is "cap and trade," where a "cap" (maximum allowable for a nation or region) is set by decree and licenses to release distributed as shares. Because some factories or other entities can reduce carbon much more readily than others the idea is that the companies that reduce emissions more than their share can sell (or trade) them to other companies where reduction is more expensive. Some argue that this makes the most

sense, but others believe that it gives many companies "a license to pollute." At this point we can probably say that it worked for sulfur dioxide but not for the much more difficult problem of carbon dioxide.

Acknowledging Ecosystem Services

In the first decade of the 2000s a new concept for environmental conservation emerged that focused on understanding, appreciating, and protecting what had come to be known as "*ecosystem services*." There was a somewhat similar concept earlier called assessing and protecting "the public service functions of ecosystems" that had little impact [2]. The ecosystems services concept began to enumerate the many contributions to human welfare from intact ecosystems that had great value but did not enter markets, and hence did not, and should not, have a price associated with them. For example, all natural areas above a reservoir, such as the forested lands in the Catskill Mountain region, maintain the purity of the water that falls on their watersheds. Using ecosystems to undertake important services is not necessarily free; there are issues with the need to protect the land, clean up agricultural activities, and deal with many landowner's outrage at being told what to do, but this approach is generally considered to be a successful and cost-effective means of environmental protection.

It is easy to see the importance of these ecosystem services. Consider the situation we discussed in Chap. 11. If you lived in New York City what if you were somehow personally responsible to get your own water, which presently comes mostly from the Catskill Mountains a hundred or more kilometers to the northwest of the city. How would you do it, especially without fossil fuels? You could go to the ocean off Long Island with a couple of buckets, dip up the seawater, desalinize it with a wood fire and a still, then carry the buckets across the Hudson River and up to the top of the mountains and empty them into the watersheds that ultimately supply the city's water. This would be absurd of course, but it helps us to focus on the huge and amazing amount of work that nature does for our economy. In fact each person in New York City gets hundreds of liters of water each day from the Catskills "for free" (once the pipes were built, although maintenance and land costs are not trivial) and in addition gets extremely high-quality water without the need for treatment because the forests of the Catskills are mostly intact. At present there appears not to be a need to build a $10 billion treatment plant because of the good job the forests are doing for free, although there are some threats from the increasing area of impervious surfaces [3].

Beyond Externalities, Beyond Ecosystem Services: The Case of Climate Change

It has become clear to many environmentalists that the problem of pollution is much larger than just the more obvious and local items that sometimes were being internalized. Rather the problem included far more pervasive, insidious, and large-scale issues such as the contribution of energy and fuel burning to acid rain, development and land use change, mercury pollution effects on third-world people, and especially climate change.

Our view of the facts about climate change, given as responses to some fundamental questions, are given in Chap. 11. Such scientific thinking tends to be lost in the often acrimonious and un-useful political dialogue that has enveloped the issue. Obviously if we are to do something about reducing the release of CO_2 this would have enormous effects on our economy, indeed on just about everything humans do. Unfortunately it seems extremely difficult to extend the relatively successful environmental protection concepts used to protect ozone to climate change. With few exceptions we have not learned how to, or are unwilling to, implement programs to compensate for such things as climate change. Such limited enthusiasm as there is for restricting carbon emissions seems to be evaporating with the economic issues associated with the prolonged recession or near recession following the "crash" of 2008. The second massive international meeting to restrict carbon emissions in Copenhagen in 2009 achieved

far less than enthusiasts had anticipated, and far less than the Rio Conference in 1991. Environmental enthusiasts respond with a strong case for "green power," with the idea that a massive expenditure on solar energy (wind turbines, photovoltaics, biomass fuel, etc.) will actually create jobs while supplying us with all the power we need. In addition, in 2008 the influential Stern report undertaken in England concluded that not undertaking climate protection actions would cost more money in the long run due to climate disruption of agriculture, coastlines, hurricanes, and other impacts. But we do not know the economic implications of deliberately shifting to lower EROI or intermittent power supplies.

The United States and other industrialized countries have reduced their carbon emissions per unit of GDP (at least as officially measured, see Fig. 18.4) substantially. This sounds extremely admirable, and it is often promoted as evidence that effective technology can make a large difference. This is true. But we need to look a little further. For one thing, thanks to the federal Clean Air act and associated regulations much of U.S. heavy industry has been "outsourced," meaning moved to overseas locations. Hence today much of the steel used in the United States comes from Brazil or Korea, and much of our oil is refined in other places, such as Trinidad. Much of our GDP growth was in the financial sector which may or may not have generated much real wealth. Inflation corrections, and hence real GDP growth, remain suspect, making the United States appear more efficient than it really is. One of the principle ways that U.S. industries are meeting air quality or CO_2 reduction goals is by shifting from coal or oil to natural gas. This does indeed reduce emissions, but at a price that is rarely mentioned, which is the depletion of a very special energy and chemical feedstock resource. Will our grandchildren castigate us for using natural gas for such "trivial" purposes as making electricity when it is desperately needed to make fertilizer or bake bread? Will the costs of climate change be even greater than the enormous costs of attempting to keep it from happening? We really have no precise idea, although we do believe the costs of remediation will reduce affluence on a presumably warmer and wetter planet.

Resource Depletion and Environmental Costs

As the highest-quality energy resources are depleted and exhausted, energy extraction shifts increasingly toward lower-quality resources that tend to be less concentrated, require more infrastructure to extract, have larger environmental impacts, and are sometimes closer to population centers. For example, natural gas was once obtained from giant fields such as the Hugoton complex in Oklahoma and Texas, which had initially about 2 trillion cubic meters, of which about half has been extracted. The size of the field is roughly 2 million hectares, and there are about 10,000 wells, so there is one well per 200 hectares. In contrast the well density for shale gas is typically one per 60 hectares and the gas flow from each well is for a much shorter time. Thus whatever environmental impact is associated with a gas well will be much larger per cubic meter exploited for the lower-grade shale gas fields. In addition, our scientific understanding of environmental impacts has increased dramatically, so that now much of the public has a basic understanding of climate change, water contamination, mercury in gaseous emissions, acid rain, and so on.

How Green Is Green?

The American public is bombarded continuously with advertisements claiming that energy corporations, various cars or railroads, biomass fuels, ecotourism retreats, and even shopping patterns contribute to a "greener" Earth. Surely any reader of this book must realize that any time you are spending money you are using fuels, with all the environmental impact that entails. There may be approaches that are more or less damaging, but we know of no way to judge this without very detailed analyses. For example, there are many problems with substitutes for oil. As of this writing the most commonly used substitute for oil in the United States

is ethanol distilled from corn. This idea is very popular in corn states such as Iowa because it provides a large market for formerly surplus corn. Due to the large influence of corn state senators, lobbyists in Washington, and early Presidential caucuses, corn-based ethanol is highly subsidized at about $1 per gallon. However numerous scientific studies have shown that it takes roughly one gallon of petroleum to make one gallon of ethanol. It may even take more energy to run the tractors, make the fertilizer, pump the water for irrigation, run the harvesters, and distill the corn than you get from the alcohol that is produced! The net effect of corn-based ethanol with an EROI of close to 1:1 may be only the soil erosion generated. More generally, finding any substitute for oil on the scale that would be needed is extremely and perhaps impossibly difficult. So the truth is that we hardly know how to evaluate what the real price of a barrel of oil "should" be if all externalities were internalized, nor do we know to whom it "should" be paid, nor do we know in any detail how green, or even how viable, the alternatives are. If the money goes to taxes do governments generate any less CO_2 per dollar than general consumers?

One of your authors (Hall) is especially sensitive on the issue of displaced impacts: how resolving an impact at one place generates an impact at another location. One of his first big projects as a Post Doc at a National Laboratory was to examine the impacts of a series of nuclear and nonnuclear power plants on the fish community of the Hudson River [4]. The power plants used so much water for cooling that that they would pump towards or through the plant many millions of fish each day, especially larval fish in season. The little fish would be pulled through the cooling systems, subject to pressure and temperature shocks, and dumped back into the river where, if not dead, they were especially susceptible to predation. Probably hundreds of thousands of fish would be killed this way each day in the Hudson River, especially during the warmer months. One of the fish especially affected was the striped bass, a popular sport fish. In response to us (and others) New York City and their principal electricity suppliers decided to meet their electricity needs from another supplier, Hydro Quebec. Their electricity was generated by flooding many thousands of square kilometers of land in Northern Quebec, eliminating the hunting and fishing grounds upon which the Cree Indian tribes who lived there depended. Some of them even developed mercury poisoning as the reservoir intercepted mercury-rich rocks and the mercury entered the food chains, eventually poisoning the natives who ate the fish. So in this case wealthy Americans wishing to protect a sport fish affected the basic health and subsistence of innocent people who were affected by the consequences.

Similarly we must ask if whether, when the United States imposes a ban on drilling in the Gulf of Mexico, to protect the area from additional oil spills, this means that more oil will be developed from, for example, the coast of Nigeria where there are frequent spills and other environmental impacts.

The net effect of all this uncertainty is that we do not truly understand the environmental impact of different energy alternatives. Rather impacts are played out a piece at a time, sometimes resolving one issue at the expense of another. There has been, to our knowledge, only one attempt to generate a list of impacts of all different energy resources, and this has been an examination by Nate Hagens of the impact of the water use per MJ delivered of different energy sources. Perhaps surprising to many readers, the water used per MJ is much greater for most so-called green energy resources. One other comprehensive approach, although not aimed specifically at energy, is the ecological footprint analysis, where Mathis Wackernagel and colleagues generate more sophisticated assessments of the total resources used by each human in each country. The results are pretty scary, and indicate that we are living at least three times beyond the possible carrying capacity of the Earth.

An important question, and one that we cannot answer, is whether the impact per GJoule delivered is increasing or decreasing. There are many interacting factors. Certainly laws are more stringent and more likely to be enforced now than in the past. But the highest grade resources, those that tend to be exploited first, have a smaller "footprint" on the Earth, so that exploiting a thinner or deeper coal seam because the thicker or more shallow one is gone means that more of the Earth must be moved, dug into, or whatever. This plays out in many small oil fields, such as in Eastern Ohio. As the oil fields mature, the water cut (barrels of water per barrel of oil in the fluid pumped to the surface) increases because the initial thick pure oil field is gone. This water is often salty (brine) and is often dumped into local creeks. Because of the increasing water cut the impact tends to increase over time.

At this time we need a much more comprehensive systems-oriented environmental assessment of all energy resources to help us put all the specific environmental studies into context. Until that time we can hardly evaluate what fuels are greener than others.

Questions

1. Discuss the question "We do not pay for resources, but only for the cost of extracting them."
2. Do you think that environmentalists' opposition to nuclear energy has made our environment cleaner?
3. Which form of energy do you think is the dirtiest? The cleanest? Why?
4. What is the relation between externalities and market failures?
5. What are two general approaches used by regulatory agencies to get polluters to reduce their pollution?
6. What is an example of an effective program to reduce an important pollutant?
7. Give some examples of free services that ecosystems provide for you.
8. Why is CO_2 reduction such a difficult procedure?
9. Discuss the statement: "Any time one spends money one is using fuels and hence contributes to environmental impact."

References

1. Staubaugh, R. and D. Yergin. 1979. Energy future: The report of the energy project at the harvard business school. New York: Random House, N.Y.
2. Hall, C.A.S. 1975. The biosphere, the industriosphere and the interactions. Bull. At. Sci. 31: 11–21.
3. Hall, M.H., R. Germain, and M. Tyrrell. 2010. Predicting Future Water Quality from Land Use Change Projections in the Catskill-Delaware Watersheds. Final report to the NY State Department of Environmental Conservation, Albany, NY. http://redir.aspx?C=95f139 48884f410183203d8059421dca&URL=http%3a%2f %2fwww.esf.edu%2fes%2fhall%2fwater.pdf" http://www.esf.edu/es/hall/water.pdf and HYPERLINK "redir.aspx?C=95f13948884f410183203d8059421dca &URL=http%3a%2f%2fresearch.yale. edu%2fgisf%2fCatskill_report%2findex.htm"http:// research.yale.edu/gisf/Catskill_report/index.htm.
4. Hall, C.A.S. 1977. The Hudson River striped bass example. In Hall, C. A. S. and J. W. Day. Ecosystem modeling in theory and practice. Wiley Interscience.
5. Hagens, N., K. Mulder and N. Fisher. 2010. Burning water: Energy return on water invested. AMBIO Volume 39, Number 1/February, 2010.

Living the Good Life in a Lower EROI Future

20

We are sometimes labeled as pessimists, probably because we do believe that the future will have less oil and perhaps less energy than it does now, because we believe that the energy costs of getting whatever fuels we do use will become greater and greater, and because we think these issues will have serious energy impacts. But we do not see this automatically as a bad future, depending on how we deal with it. As boys we each had a wonderful childhood on opposite coasts in the 1950s and 1960s during a period when the U.S. energy use was only 10% or 20% of what it is now. We could go fishing and surfing (respectively) on our bicycles, and had no need for soccer moms driving us around in an SUV. We played sports all the time with neighborhood friends, and went camping and hiking to our heart's content. There was little of today's perspective that children must be driven everywhere for protection because we lived in neighborhoods and communities where everyone knew everyone else.

For the record we are neither optimists (which is our nature) nor pessimists about our energy and economic future. We really have no way to predict the future beyond some easy and very coarse extensions of present trends (for demographics, probably oil, possibly gas, conceivably coal). Declining EROI seems likely to cut into societal affluence no matter how much fuel we are able to access. The hardest things to predict would be human behavior: will we go quietly into declining affluence (as we seem to be doing now)?

Will the unemployed or never-to-be-employed cause riots? Will we be able to do things with human hands we do now with fossil fuel? Will we be able to make some kind of transition to a new energy source? If the economic pie must shrink will the rich freak out and keep their absolute quantity constant, while the poor get a smaller part of a shrinking pie? Or what? Although deeply involved in all this as professionals and modelers since the 1960s and 1970s, we can be neither optimists nor pessimists because we cannot predict these things, and do not trust anyone who says they can. We think we have to go into the future with the following model and something like the following probabilities (you can choose your own percentages): we will go off the cliff, energetically, economically, or environmentally (25%), we will make a transition to a new energy source that will benevolently replace oil (25%), or we will muddle along, gradually getting materially poorer but adjusting to that (50%). The point is that we do not think anyone knows those percentages, and so we must go into the future with a huge amount of uncertainty. That in itself might be pretty difficult. Some would trust the market to adjust, others might not, or have other mechanisms. Many people who think about these things retreat to a bunker mentality and are stocking their country houses with food and ammunition. We, on the other hand, think that is a little foolish; we will probably weather this storm all together or not at all.

C.A.S. Hall and K. Klitgaard, *Energy and the Wealth of Nations: Understanding the Biophysical Economy*, DOI 10.1007/978-1-4419-9398-4_20, © Springer Science+Business Media, LLC 2012

What Are the Main Issues for Transitioning to the Future

The main problem that we face is that we will require massive new investments in whatever might become the next energy source at a time when most citizens will be experiencing a decline in their own purchasing power. For example, if gasoline today costs $4 a gallon (and this just to extract that gallon from an aging, energy-requiring field) who will want to pay an extra $5 a gallon as an investment in whatever fuel or whatever will be needed to replace that gallon? The answer is probably few, if any, and that implies that we just continue on the path of using ever-lower grade, more expensive conventional resources, slowly grinding into ever-greater poverty.

If one accepts the importance of a biophysical basis for economics then are there some important implications of our analysis for economics and for society? The first issue pertains to the economic pie and how we will cut it. As discussed in some detail in Chap. 1, the American dream gave the hope of significant and ever-increasing prosperity to a broad swath of people through a number of generations and for an entire nation. We believe it is not clear at all that this prosperity will continue. In fact there is a great deal of evidence that we have reached the end of any increase in affluence: the GDP and the inflation-corrected stock market indices of the United States have barely budged for 5–10 years (Fig. 1.8). There is increasing evidence that such growth as took place from the mid-1990s until 2010 was based in large part on debt or speculation. As we have pointed out currently some 46 of 50 state governments are broke, unemployment remains stubbornly high, colleges and universities are having increasing difficulties balancing their budget, retirement plans have lost a great deal of their net worth, housing prices remain greatly depressed, and so on. Of course none of this is new, for the United States has gone through depressions often enough before, and many believe that if we just wait, we will come out of the present depression or that "they" will think of something. But there are at least three factors that are different: we do

not have new quantities of high EROI fuels waiting to be exploited, we have much more debt, and much of our public believes that affluence is an American birthright. What if the present depression is not part of a cycle but is the new reality? What if David Murphy's concept discussed in Chap. 18, that any increase in growth sets into motion its own demise because of the need to use much more expensive oil, is indeed the new reality? In other words, there seem to be serious energy constrictions on continuing growth. What if the national (and global) economic pie can no longer grow?

Traditionally, as we discussed in Chap. 1, the concept of the American dream, is the continually growing pie. This prospect has resolved or defused many issues in the United States for some time: labor has made more in their salaries (at least until the late 1990s) and management has made much more. Large portions of total wealth were "skimmed off" by Wall Street and other entities and it was hardly noticed largely ignored. Government could be corrupt or inefficient and still the roads got fixed and public universities expanded and so on. There were few complaints because everyone made at least a little more. But that is no longer the case. If any one group does better now it seems to be at the expense of some other group or some other use of the money. Now the question is: if the pie is no longer getting larger, indeed if because of energy constraints it can no longer get larger, how will we slice it? This may force some ugly debates back into the public vision. Indeed if total energy availability should actually shrink then we will need to ask some very hard questions about how we should spend our money, as appears to be beginning with the "Wall Street protests" of the fall of 2011.

Probably this will force individuals and our nation to focus on what is most important. One way to think about this is Maslow's hierarchy of human needs. This theory, proposed by Abraham Maslow in his 1943 paper, "A Theory of Human Motivation," [1] proposes that humans will attempt to meet their needs in more or less the following order. First they will meet their physiological needs which are the literal requirements for human survival, including breathing, nutrition, water, sleep, homeostasis, excretion, and reproductive activity. These require clean air and water, food, clothing, and shelter. Second, once physiological needs are satisfied an individual will attempt to meet safety needs in an attempt to attain a predictable orderly world in which perceived unfairness and inconsistency are under control, the familiar frequent and the unfamiliar rare. Third, once the above needs are met humans seek love and belonging, that is, emotionally based relationships in general, such as friendship, intimacy, and family. Fourth, again once the above have been met humans seek esteem, to be respected, and to have self-esteem and self-respect and also the esteem of others. Finally, according to Maslow, people seek self-actualization, the need to understand what a person's full potential is and to realize that potential, to become everything that one is capable of becoming, for example an ideal parent, athlete, painter, or inventor.

Maslow's theory has been criticized from a number of angles including the lack of evidence that humans in fact follow that hierarchy, or indeed any such hierarchy, and from the perspective that his "pyramid of needs" may be more representative of people from an individualist versus socialist society. Nevertheless his theory is broadly accepted in psychology and even marketing. Our own research on the implications of declining net energy, although not consciously based on Maslow's theories, is consistent with them. We have the sense that discretionary spending will be increasingly abandoned as humans attempt to meet their needs for food, shelter, and clothing (see Fig. 15.3). Presumably as the amount of net energy declines due to peak oil and declining EROI, humans will increasingly give up categories higher on the pyramid (fifth above) and concentrate increasingly on the more basic requirements including food, shelter, and clothing. What this may mean in modern society is that performance art, then expensive vacations, then education, then healthcare would be abandoned by the middle class if and as the economy is increasingly restricted.

Labor

During the last four decades under the pressure of profit maximization, the economies of the United States, Japan, and Germany have been substituting powerful cheap energy and increasingly automated capital for weak expensive labor. In other words labor productivity, the amount of value added per hour that the laborer works, has been greatly increased by subsidizing the efforts of a laborer with more fossil energy. This substitution has not occurred to the degree that it might for various reasons [2] but nevertheless has vastly reconfigured the role of labor while contributing to unemployment. Will an increase in the price of energy relative to labor substantially increase the amount of labor employed? If labor can again be more valuable in production, real wages would have to fall because goods and services would become more expensive relative to real purchasing power of salaries (otherwise the labor would not become relatively cheaper). Jean Laherrère has shown an uncanny relation between oil price and unemployment (Fig. 20.1) which may be something to worry about.

Debt

An enormous, perhaps overwhelming, aspect of our future will be debt. The concept and importance of debt to the American dream was presented in Chap. 1, and the connection of debt to energy in Chap. 4. Where once we could grow our way out of the importance of debt this looks more and more difficult if growth becomes a thing of the past. Federal government debt has become an enormous political football as of 2011 with some very curious political dimensions because nominally fiscal conservatives in the past generated the largest part of our debt, at least until the current situation. Given our belief that debt is a lien against future energy use then some portion of a nation's future energy use must be diverted to payment of interest and principle on debt. It is worrisome to consider

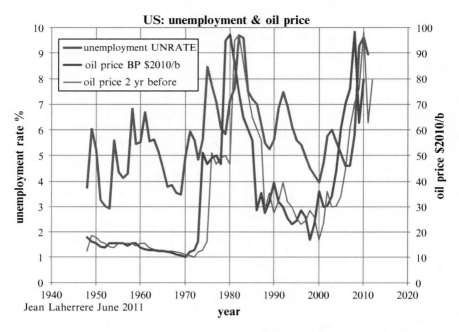

Fig. 20.1 The relation between oil price and unemployment the following year for the United States

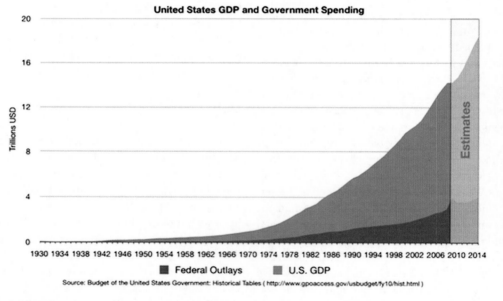

Fig. 20.2 Total American GDP and Federal expenditures (not corrected for inflation)

that in the future when we will need large amounts of our energy resource to invest in new energy technologies (including conservation) that some significant portion of whatever energy is available will be diverted to meeting debt loads. One undesirable way out of our huge debt load (Fig. 20.2), which we may never be able to pay, would be to greatly reduce the energy/dollar relation through massive or hyperinflation (Chap. 14).

International

This book focuses on the United States, and it must seem clear that we have problems enough with energy. But it is worse for many other countries. The United States imports about one third of its energy, whereas Europe and Asia import two thirds, making these countries far more vulnerable to whatever future energy situation arises. Europe had a momentary respite with the enormous North Sea oil fields, from which nearly 40 billion barrels have been extracted, and another 10–30 billion might yet be extracted but with lower EROI. This oil bonanza allowed Britain to have a few decades of affluence, and led many to believe that Margaret Thatcher's political policies had somehow saved the day. But now the oil and gas reserves of the British portion of the North Sea are nearly gone. Britain is struggling with the fact that the oil was essentially spent in a wild binge, and a new cold hard reality is upon her as civil servants and students are hit with the drastic cuts in government largess. Norway, on the other hand, has developed its oil and gas at a more measured pace and placed much of the revenues into a trust fund to help all future Norwegians, one of the relatively few examples we have of a mineral bonanza being used to help all citizens [3].

At the extreme many tropical developing countries are especially vulnerable because of their increasing reliance on oil to support increasing populations, especially through increasing use of fertilizer and other input to agriculture, and often the economic importance of tourism. The first author has a great deal of experience attempting to understand the relation of energy to what is normally called "development" in the tropics. Many tropical countries are poor and essentially all wish to become more wealthy. Hall was initially attracted to the country of Costa Rica which was promoting itself as a "laboratory for green, sustainable development." Unfortunately his experience from years of living there and studying quantitatively all major aspects of its economy (detailed in two books on the subject [4, 5]), was that Costa Rica, no matter how lovely and how well developed the ecotourism and solar

energy (mostly hydroelectric) industries, was at least as dependent upon petroleum as any other place, was far from sustainable for at least 18 reasons, and had no real plan as to how to continue its moderate standard of living without oil. This is for a country that is relatively well off with respect to its sustainability! Unfortunately, we think peak oil is likely to hit the developing world especially hard. Likewise peak oil is likely to affect fully developed but highly oil-dependent entities such as Puerto Rico especially hard. These regions, whose economy once depended almost entirely on agricultural production unsubsidized by fossil fuel, cannot possibly feed their swollen populations from indigenous agriculture. They have no contingency plans for peak oil.

Likewise agricultural production for the world more generally may be very susceptible to peak oil, peak gas (which would limit the production of nitrogen fertilizer), and peak other inputs. A website ironically called "Sustainable Phosphorus Futures" suggests global peak phosphorus by 2030, and some analyses indicate that it is here now [6, 7]. Irrigation, used on perhaps 15% of U.S. crops, is often dependent upon deep groundwater that requires more energy over time as it is pumped from deeper and deeper depths as the fossil water is depleted. More generally around the world agriculture has shifted to procedures that are energy intensive in many ways, and we expect all to be affected significantly by peak oil. Inasmuch as the growth of the global population has historically matched the growth of fossil energy, we would not be surprised to see those curves to continue to be related on the downslope of the energy curve. As the physicist Albert Bartlett states [8] there is little doubt that populations will decline; what we have a choice about is whether it will be due to procedures that we might like or dislike mildly (i.e., reproductive control) or the things we like much less, such as starvation, disease, pestilence, and war.

Choosing a Better Future

To the best of the authors' imperfect ability to predict, it appears very unlikely that there is a "supply" approach out of the circumstances we'll be left with by peak oil Every realistic analysis shows a future with either peak oil about now or at best an "undulating plateau" for no more than a decade, and then declining oil into the future. Coal and natural gas may be able to fill in part of the gap (but with enormous difficulty for liquid fuels) for some additional decades, but growth or probably even a steady-state energy economy seems unlikely after a decade. To us it seems that the die is cast because we simply are not finding oil as rapidly as we are using it (Fig. 3.2). Globally 80% of our oil comes from some 400 large oil fields discovered before 1970. Production in at least a quarter of these fields is declining and more fields will join that group soon. Whatever new oil we find will have to make up for that decline, and it is almost guaranteed that there will not be enough to add to any substantial increase in oil supplies worldwide (Fig. 3.2). There are indeed enormous quantities of low-grade fossil fuels left in the ground but the low EROI and huge investments required make it unlikely that they can replace the role of oil or offset the forthcoming shutdown of 100 U.S. nuclear plants. Low-grade types of oil such as tar sands are making a small but almost inconsequential difference on a global basis. All new oil supplies are likely to be much more expensive than existing oil production (Fig. 18.5). Natural gas may not peak for several decades but is unlikely to more than compensate for declining oil at best.

Few if any alternatives, including conservation, appear to be able to fill in for the decline of oil and eventually gas, and they would take an enormous investment in money, energy, and time to be viable. Biomass (other than traditional solid forms such as firewood) can make a certain gross contribution but unless things change considerably little net. For example presently about 10% of U.S. "gasoline" is ethanol from corn, but since the EROI of ethanol is barely different from 1:1, that fuel seems to deliver little net energy. New solar (including wind turbines and photovoltaics) are a great hope for the future but to date contribute far less than 1% of our total energy supplies, and the pace of development has slowed considerably with the recent financial collapse. All of these alternatives (including

backups) would have a much lower EROI than we are used to. Thus we do not necessarily foresee a future United States without energy, but rather substantial problems in providing the liquid and gaseous hydrocarbons that have been our lifeblood.

Coal is harder to predict. There is a lot of talk about "peak coal" (e.g., Patzek and Croft [9]) much of which is based on mining capacity, not the actual size of the resource. Peak coal is most likely to affect the world's largest coal user, China [10], but clearly in the United States, Russia, and some few other regions coal remains extremely abundant. Alaska alone has huge resources of exploitable high-quality coal. United States production in 2009 was about one billion tons, and the Powder River formation in Wyoming alone contains some 40 billion recoverable tons. Recoverable reserves at presently operating mines is about 18 billion tons, and the total recoverable coal base is estimated by the U.S. EIA as about 500 billion tons. Thus it seems that if we are willing to make the investment and deal with the environmental consequences coal can be as abundant as we wish it to be, although probably with a declining EROI, for a century at least. Curiously, the use of coal declined by about 10% from 2008 through 2010, so it is a bit hard to predict the future patterns of consumption which may have more to do with economic circumstances.

Unless we as a country decide to increase our coal use enormously, which would be difficult but certainly not impossible, given present environmental concerns and infrastructure limitations, it seems likely that the future will be one of an increasingly constricted energy supply. This implies, as discussed again and again in this book, the end of economic growth and some extremely large adjustments of our citizens to a new steady-state or declining economic condition. If we pay off our huge international debts this implies an even more constricted economic situation. Although for many this will seem like a very gloomy future, for us this is not necessarily the case. It depends upon how we do it. Others have written better or at least more comprehensively on this issue, however, we do wish to summarize some few aspects of this issue.

The Prosperous Way Down

Howard Odum was our mentor and guide, and we respected enormously his contributions to systems analysis, ecological modeling, ecological energetics, and an understanding of the relation of humans to nature and to energy. He understood how the world worked in so many fundamental ways. So it is fitting to choose the title of his last book [11], *The Prosperous Way Down*, as a guide for where we should be going.

Odum believed that a lower energy future was inevitable as fossil fuels peaked and declined. He did not write too much about the details, for to him it was just a fact. But he was interested in how humans might respond to this. He believed that a lower energy future could be a good future, even as his title indicates a prosperous time. The authors of this book agree, for we grew up in the United States, on opposite coasts, during a time when per capita U.S. energy use was only about a quarter or a third of what it is now. Can we recapture the low energy, happy childhoods of the authors with a new low energy but happy life for American citizens based on a decreasing use of energy? Can we redesign communities so that people can walk to work and to food stores, to entertainment and culture? How can we employ more people when most our production is undertaken by machines or is done overseas? Can we

take the cleared and fertile landscaper that is now suburban lawns and grow our food there? Will our present population, conditioned by advertising and corporate interests to want ever more affluence be happy with less or with other means of generating happiness? Can we be satisfied with fewer square feet per person in our houses and much less use of automobiles? Or have we cast the die with the ways that we have constructed our suburban landscapes and programmed our children to demand affluence?

So here are some aspects that might actually be better in an energy-constrained world, but only one where people had adjusted well to this new reality.

First, is wealth as measured by GDP necessarily something that leads to happiness? In fact where this has been studied (which is not easy): considerably, and the answer is yes, but that other things are more important. For example, Richard Layard of the London School of Economics found a peak in U.S. happiness in 1956, which is not too different from the results of the NGO Redefining Progress study that came up with a "Genuine Progress Indicator" and found a peak for the United States in 1977 (Fig. 20.3). Inglehart and Klingemann (and others) [12, 13] have measured subjective estimates of happiness in the world and found that after a given minimum level of income there was no correlation between either

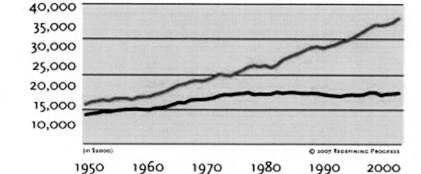

Fig. 20.3 The "genuine progress indicator" has remained constant while the official estimates of GDP has increased substantially

GROSS PRODUCTION VS. GENUINE PROGRESS, 1950-2004

— GDP Per Capita   — GPI Per Capita

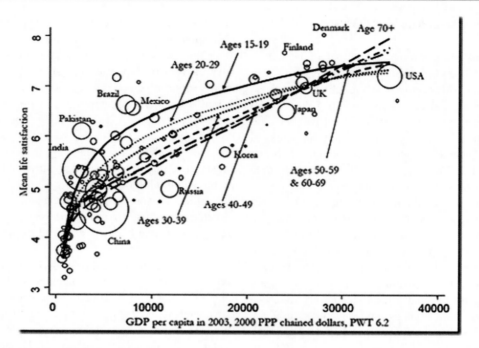

Fig. 20.4 Asymptotic relation of happiness and wealth

income or long-term growth in income and personal happiness (Fig. 20.4). The countries with the most happy people, Ireland, Nigeria, Mexico, and Venezuela were certainly not the wealthiest, and the countries with the least number of self-described happy people, Russia, Armenia, and Romania, were not among the poorest. Instead happiness seemed to depend a great deal on a sense of personal freedom and control over one's life. The Eurobarometer Ranking of the happiness index, that is, how much people enjoy their life as a whole on scale 0–10 again found little correlation with GDP. So, overall, the answer to this question appears to be that wealth, as measured by GDP, is a necessary component of personal happiness if you are poor, but has little importance above some minimum level. We can start educating our young people to this perspective now.

Second, there are many indications that a less energy-intensive lifestyle can be one of much greater community and healthier too. This is the explicit objective of various grassroots groups such as "The New Road Foundation" and "The evolution of transition" towns [14, 15], where

transition means a transition to a postpeak oil world. Surely our present success-driven, affluence-seeking, status-driven world is not one that generates the greatest happiness and respect for others.

Third, our economy is so wasteful that it should be easy to use only half as much energy and maintain something very much like the same lifestyle. For example, our railroads could be electrified, generating less energy-intensive freight transfer [16]. Sedans that deliver essentially the same services on half the gasoline already exist, and older buildings can be retrofit with insulation.

Why We Are Not Entirely Optimistic

Although we believe that a relatively smooth transition to a lower energy future with a prosperous lifestyle is quite possible we are not necessarily optimistic that it will occur. Unfortunately the American public is almost completely ignorant about peak oil, which indeed is the simplest part of the dilemma, "the energy mess," that we have inherited [16]. Quite curiously neither the press nor the national funders of science (NSF, DOE, etc.) have shown any particular interest in this issue and, if anything, have attempted to suppress any research or discussion to the subject [18]. This is quite surprising considering the enormous amount of attention given to possible climate change. We wish in no way to belittle the importance of the attention paid by both the press and the science community on climate change but we find it curious that peak oil, a situation that seems to be more immediate, more certain, and perhaps more devastating receives essentially zero press or funding, at least as of 2011. As part of this problem the public is fed a constant stream of advertisements and programs promising green clean energy when the quantitative nature of the contributions – all trivial – is never mentioned.

A second reason we are not optimistic is that Americans (and most others in the world) have been conditioned by a lifetime of television and other advertisements all indicating that happiness, sexual fulfillment, you name it, are possible only through a never-ending stream of purchases. This seems to be so engrained in our culture that it is hard to imagine it otherwise.

A third important reason that we cannot be too optimistic about making a timely transition will be the political response to this situation. This of course requires that people understand what is happening, and that political advantage can be found in adjusting to this new reality. There are many thoughtful papers that have attempted to examine the potential transition in various and often quite sophisticated ways [19, 20]. All agree that a critical first step is to question a belief in growth as the universal panacea. How this can be undertaken in the current political climate where even far less controversial legislation is stalled is beyond our comprehension. Possibly peak oil will pound some sense into the electorate's head, but more likely there will simply be a blame game inasmuch as no political party can bring back the good old days where the American dream was realized for generation after generation. If there is to be a new American dream it has to be based on something besides ever more affluence and

material consumption. But there are simple things we can start doing. Two simple things to do are simply to live near where you work and contribute to making sure your neighborhood, and neighborhoods in general, provide the necessities of life to decrease your and our dependence upon automobiles. We like the ideas of Will Allen (growingpower.inc) to bring agriculture into the central cities and think that technological optimists like Jeremy Rifkin may have some very good ideas.

Why We are Somewhat Optimistic

Human beings often respond well to crises and it is entirely possible that the self centered and materialistic behavior we witness today, that indeed is central to conventional economics, will not be the same behavior that will prevail when if and when the effects of energy constraints and, perhaps, climate change or other environmental issues make growth no longer possible nor desirable. Faced in 1988 with the cutoff of Soviet oil, Cuba emerged from their "special period" of adjustment as possibly the world's most sustainable economy. Monoculture sugar cultivation gave way to smaller scale food production. While salaries of teachers and doctors fell the national commitment to health care and education was not abandoned.

When peak oil and perhaps climate change can no longer be denied perhaps environmental sustainability will be pursued even at the expense of growth, which may not be possible anyway. In this environment it is possible that community and better health could prevail. This appears to be taking place in Japan, where after the "lost decade" – now approaching two decades – where economic growth ceased there is widespread and seemingly relatively happy adjustment of many to what we can only call a steady state economy [21].

What will labor look like in this post peak economy? Maybe we will have to work longer hours at more physical labor as the energy basis of past productivity gains decline. Perhaps work can become more meaningful with a reunification of head, hand and spirit [22]. That kind of work might produce fewer but more long lasting goods. Shipboard and truck transport may decline, generating strong reasons for local production of needed products. The concept of financial services for other than local investments may wither away. Workers may be focused on the basics.

Thus a good future and even, if needed, a prosperous way down is, we believe, quite possible for economic and political reasons, but very unlikely due to psychological and conditioning issues relating to the attitude of the American people relating to advertisement, growth, and wealth as status. We conclude that what we need most is to create a biophysically based approach and model for economics, one that would serve on at least an equal footing with the present firm–household–market based model. The actual implementation of any such project mostly remains for the future and a very different book.

Questions

1. Are you an optimist or a pessimist about the future? Why? About what?
2. What are Maslov's hierarchy of human needs? Can you list them in order?
3. What are some ways that we can make more jobs available for labor? What would be some good and some bad sides to that?
4. Name five ways that food production depends upon oil.
5. What are your views about the future of coal in the world economy? What factors might be especially important in influencing this?
6. Do you think that GDP is an adequate measure of our wealth? Why or why not?
7. What are some of the advantages that might come from a less energy-intensive lifestyle?
8. What ideas do you have to provide for a better future for all Americans and all people of the world?

References

1. Maslow, A. 1943. A theory of human motivation. Psychological Review, 50, 370–396.
2. Hall, C. A. S., Lindenberger, D., Kummel, R., Kroeger, T. and Eichhorn, W. 2001. The need to reintegrate the natural sciences with economics. BioScience 51, 663–673.
3. http://www.regjeringen.no/en/dep/fin/Selected-topics/the-government-pension-fund.html
4. Hall, C., 2000. Quantifying sustainable development: the future of tropical economies. Academic Press, San Diego.
5. LeClerc, G. and Charles Hall (Eds.) 2008. Making development work: A new role for science. University of New Mexico Press. Albuquerque
6. Vaccari, D. 2009. Phosphorus famine: The threat to our food supply. Scientific American June 3, 2009. P. 36.
7. Feiffer, D. A. 2003. Eating fossil fuels. From the Wilderness Publications. Sherman Oaks, Cal.
8. Bartlett, A. 1997. Reflections on sustainability, population growth and the environment – revisited. Renewable Resources Journal, Vol. 15, No. 4, Winter 1997–98, Pgs. 6–23
9. Patzek, T. and G. Croft. 2010. A global coal production forecast with multi-Hubbert cycle analysis. Energy 35: 3109–3122.
10. http://eclipsenow.wordpress.com/2010/05/06/peak-coal-hits-china-richard-heinbergs-article/l
11. Odum, H.T. and E. C. Odum. 2001. The prosperous way down. Univ. Press of Colorado, Boulder
12. Inglehart R, Klingemann H-D. Genes, culture, democracy, and happiness. In: Diener E, Suh EM, eds. Culture and subjective well-being. Cambridge, MA: MIT Press, 2000:165–83.
13. Sitglitz, J.E., A. Sen, J-P Fitoussi. 2010. Mismeasuring our lives. Why GDP doesn't add up. The New Press, New York
14. Brownlee, Michael The evolution of transition in the U.S. http://transition-times.com/blog/2010/11/26/the-evolution-of-transition-in-the-u-s/
15. Drake, A. Electrify trains http://www.theoildrum.com/node/4301
16. Hirsch, R. R. Bezdek and R. Wending 2010. The impending world energy mess. Apogee Prime.
17. Interview with Robert Hirsch on oil drum
18. Sorrell, S. 2010. Energy, growth and sustainability: five propositions. SPRU Sussex University, Working paper 85.
19. Beddoe, R., R. Costanza, J. Farley, E. Garza, J. Kent, I. Kuniszewski, L. Martinez, T. McCowen, K. Murphy, N. Myers, Z. Ogden, K. Stapleton and J. Woodward. Overcoming systematic roadblocks to sustainability: The evolutionary design of worldviews, institutions and technologies. Proceedings of the National Academy of Sciences. Vol. 106: 2483–2489.
20. Hirsch, R., R. Bezdec, and R. Wending. 2005. Peaking of world oil production: impacts, mitigation and risk management. U.S. Department of Energy. National Energy Technology Laboratory. Unpublished Report.
21. Justin Klitgaard, personal communication.
22. Kunstler, J. H. 2008. World made by hand. Grove/Atlantic. N.Y.

Index

A

Acid rain, 24, 215, 283, 391, 392
Adirondack mountains, 283
African origin of humans, 45–47
Agriculture, 12, 13, 19, 42, 47–54, 57–59, 61, 62, 66,
 67, 71, 74, 99, 102–104, 107, 111–113, 169, 197,
 207, 214, 226, 239, 247, 249, 260, 267, 268, 274,
 275, 281, 291, 294, 301, 317, 324, 338, 342, 355,
 357, 364, 387, 391, 399, 403
Al-Quaddafi, Muammar, 177
Aluminum, 7, 260, 266, 268
American dream, 3, 11–12, 16, 21, 22, 35–38, 189, 387,
 396, 397, 403
Aquifer 75, 274
Arab, 60, 86, 177, 326, 375
Arabian-American Oil Company (Aramco), 173
Arab oil boycott, 173, 177, 178
Arsenal of democracy, 171

B

Baran, Paul, 156
Bill, C., 29, 38, 183, 190, 198
Biomass energy, 59, 88, 99, 103, 234, 282, 336, 337,
 358, 391, 392
Biomass, energy potential, 400
Biophysical, basic concept vii, 1
Biophysical economics, 8, 106, 137, 200, 205–207, 257,
 277, 296, 353–366, 378
Bituminous coal, 315
Boiling water reactor, 265
Bretton Woods Institutions
 General Agreement on Tariffs and Trade (GATT), 172
 International Monetary Fund (IMF), 172, 197, 199, 200
 World Bank, 172, 197, 199–201, 294, 361
Budget deficit, 3, 21, 165, 178, 183, 188
Bush, George H.W., 29, 166, 167, 181, 183, 184, 198

C

Capacity utilization, 151, 185, 189
Capital, 5, 97, 134, 148, 164, 194, 217, 248, 302, 317,
 325, 355, 374, 397
 cost, 335
 labor accord, 18, 164, 171, 178, 179

Carbon, 62, 71, 72, 87, 155, 166, 230, 232, 233, 235,
 236, 239, 240, 243, 244, 263, 265–267, 270, 271,
 338, 388, 389, 391
Carrying capacity, 24, 99, 100, 191, 214, 217, 356,
 359, 393
Carter, Jimmy, 24, 179, 189
Chemical defense, 269
Classical model, 137, 305
Climate, 12, 21, 22, 44, 46, 47, 62, 65, 71, 74, 96–98,
 100, 101, 119, 128, 133, 142, 143, 149, 163, 186,
 190, 205, 209, 216, 244, 255, 263, 265, 271–272,
 275–280, 282, 284, 325, 330, 338, 343, 346, 347,
 359, 387, 389, 391, 392, 403
Clinton, Bill, 29, 38, 183, 190, 198
Coal, 8, 9, 13–15, 17–19, 22, 23, 26, 51, 67, 71, 72, 87,
 89, 96–98, 100, 102, 111, 112, 126, 127, 129,
 154, 157, 159, 160, 171, 183, 195, 200, 215, 216,
 225, 231, 232, 235, 236, 248, 258, 260, 263,
 265–267, 270, 277, 309, 314, 315, 317, 323, 324,
 328, 336, 337, 355, 358, 372, 376, 387, 388, 391,
 393, 395, 400, 401
Clark, Colin, 30, 37, 77, 81, 82, 97, 219, 232, 263, 325,
 371, 373
Collateralized debt obligation (CDO), 186
Commonwealth Club speech (1932), 168
Competition, 18, 35, 47, 60, 101, 105, 109, 110,
 117–122, 148–153, 156–158, 160, 164, 170, 174,
 176, 178, 180–182, 305, 348
Copper, 50, 51, 58, 62, 63, 124, 209, 232, 260–262, 265,
 266, 268, 328
Corn, 8, 49, 85, 104, 114, 119, 120, 126, 195, 231,
 233, 254, 255, 311, 312, 314, 319, 336, 338, 388,
 392, 400
Cost-push inflation, 182, 183
Credit default swap (CDS), 186
Cycle, 16, 30, 66, 75, 87, 134–135, 175, 228, 233, 234,
 239, 271–275, 378, 379, 384, 396

D

Dams, 14, 66, 169, 337
Deforestation, 50, 61, 169, 388
Deindustrialization, 166
Demand–pull inflation, 179
Diminishing returns, 101, 121, 135, 262, 365

C.A.S. Hall and K. Klitgaard, *Energy and the Wealth of Nations: Understanding the Biophysical Economy*,   403
DOI 10.1007/978-1-4419-9398-4, © Springer Science+Business Media, LLC 2012

Discounting, 303
Dow Jones, 30, 31, 371, 376, 377

E

Ecology, 24, 209, 214, 230, 243, 281–283, 288, 299,
    350, 354
Economics
    biophysical, 1, 6, 8, 34, 95, 96, 98, 99, 102, 106,
        116, 125, 129, 133–135, 137, 200–207,
        257, 277, 296, 301–306, 346, 348, 353–366,
        378, 396
    classical, 4–6, 98, 101, 102, 104–118, 121, 125–126,
        128, 167, 193, 271, 301
    growth, 7, 12, 16, 18, 21–24, 33–35, 38, 72, 74, 87,
        93, 96, 100, 102, 106, 115, 125, 127, 128, 135,
        137, 160, 161, 163–166, 168, 171, 174–176,
        182, 183, 185, 187, 197, 203, 205, 206, 215, 219,
        241, 294, 295, 326, 348, 359, 373, 374, 378–381,
        383–385, 401
    neoclassical, 3, 5, 6, 98, 101, 102, 104–107,
        110–112, 116–118, 120–122, 124–126, 128, 133,
        135–137, 140, 142, 143, 158, 165, 196, 197, 199,
        200, 205, 206, 215, 302–305, 348, 353, 358, 365,
        366, 380
    physiocrats, 4, 6, 21, 96, 99, 101, 102, 104, 107, 108,
        112, 118, 301
Ecosystems, 1, 12, 66, 74, 95, 141, 214, 236, 239,
    245, 246, 268, 269, 273, 274, 281–284, 354, 387,
    389, 390
Efficiency, 16, 23, 25, 29, 33, 74, 105, 111, 121,
    135, 136, 143, 148, 158, 197–205, 207,
    215–217, 226, 228, 229, 232, 245, 247, 270, 283,
    310, 311, 315, 318, 323, 328, 337, 349, 363,
    376–379, 384
Einstein, Albert, 227, 230, 236, 255, 299, 350
Electric power plant, 200
Ellsberg, Daniel, 165
Empire, 9, 53–67, 86, 152–153, 166, 167, 180, 194,
    248, 249
Energy definition, 225
Energy and
    agriculture, 12, 13, 19, 42, 47–54, 57–59, 61,
        62, 66, 67, 71, 74, 99, 102–104, 107, 111–113,
        169, 197, 207, 214, 226, 239, 247, 249, 260,
        267, 268, 274, 275, 281, 291, 294, 301, 317,
        324, 338, 342, 355, 357, 364, 387, 391, 399, 403
    fisheries, 62, 275, 343, 365, 387
    history, 1, 2, 5, 16, 18, 21, 23, 27, 34, 41, 42, 50,
        51, 54, 55, 58, 63–66, 71, 85, 97, 99, 101, 113,
        118–126, 148, 150, 158, 163, 164, 167, 171, 173,
        174, 190, 193, 196, 206, 225, 226, 240, 258, 282,
        294, 311–312, 337, 342, 361, 379
    mining, 6, 13, 19, 25, 54, 107, 110, 164, 248, 261,
        268, 269, 301, 311, 355, 388, 400
Energy budget, 48
Energy cost, 7, 33, 43, 49, 55, 59, 66, 75, 85, 86, 89, 95,
    137, 179, 180, 215, 218, 233, 245, 262, 279, 283,
    309, 311–319, 327–330, 337, 358, 366, 376, 378,
    384, 395

Energy crisis, 23, 24, 26, 63, 176,
    178, 215
Energy/GNP ratio, 183
Energy gradient, 233
Energy opportunity cost, 241
Energy prices, 23, 28, 74, 89, 183, 217, 309, 310, 318,
    373, 374, 376, 381
Energy price shocks, 375–376
Energy quality, 379
Energy return on investment (EROI)
    biomass energy, 88, 336
    definition, 43
    of different fuels, 312
    energy quality, 311
    food capture, 43
    hunter-gatherer, 44
    insulation, 402
    nuclear power, 314
    oil shale, 314, 325, 337
    solar, 88, 400
    wood plantations, 336
Energy storage, 241–242
Enhanced oil recovery (EOR), 75, 77, 382
Entropy, 41, 135, 233–236, 241, 243, 260–261,
    304, 328
Entropy, economic systems, 260
Environment, 9, 43, 86–87, 97, 135, 158, 194, 229, 253,
    304, 337, 349, 354, 384, 388
Erosion, 58, 61, 62, 134, 169, 207, 226, 311, 330, 344,
    357, 359, 364, 388, 392
Evolution
    cultural, 53, 72, 247, 253
    organic 277–281, 285
Exploration, 75, 77, 86, 107, 148, 219, 327,
    329, 330
Exponential growth, population, 294
Externalities, 89, 96, 270, 347, 349, 371, 389,
    391, 392

F

Falkowski, Paul, 243
Famine, 210, 214
Federal Reserve System, 178
Fertilizer, 7, 8, 19, 87, 89, 95, 100, 112, 169, 201, 205,
    207, 214, 218, 255, 267–269, 283, 294, 311, 338,
    362, 363, 391, 392, 399
Financial return on investment, 365
Fiscal policy, 102, 175, 176, 179
Fisheries, 62, 275, 343, 365, 387
Fossil fuels, 6, 14, 17, 22, 35, 65–67, 71, 72, 88, 89,
    96–100, 102, 106, 108–111, 115, 125–127,
    135, 147, 150, 156, 157, 163, 207, 210, 214,
    218, 219, 221, 236, 247, 249, 260–266, 283, 314,
    317, 323–325, 327–329, 337, 347, 358, 376, 390,
    400, 401
Free market, 29, 108, 110, 119, 128, 133, 137, 194, 196,
    198, 200, 202, 346
Fuels, 233, 384
Fuel wood, 4, 227

G

GDP. *See* Gross domestic product (GDP)
*General Theory of Employment, Interest, and
        Money*, 101, 105, 122, 160, 161, 173, 189
Glass–Steagall Act, 184
Gold standard, 20, 32, 123, 159, 160, 164, 167, 170, 172
Gordon, David, 175, 181
Greece, 48, 51, 55–57
Gross domestic product (GDP), 3, 10, 22, 27, 29–31, 72,
        74, 124, 137, 184, 187–189, 199–203, 205, 206,
        233, 289, 295, 309, 310, 316, 318, 323, 330, 331,
        335, 338, 341, 354, 362, 363, 365, 375, 377–381,
        383, 384, 391, 396, 400–403
Gross national product, 17, 157, 173, 176, 182, 183, 377
    inadequacy, 24
Ground water, 260, 359, 399
Growth, 1, 41, 72, 96, 134, 149, 163–191, 193, 209–221,
        226, 254, 289, 302, 348, 359, 371–385, 387, 396

H

Hadley cells, 259
Hansen, Alvin, 167
Hawley–Smoot Tariff, 166
Homeostasis, 397
Hoover, Herbert, 16, 159, 166–168
Hubbert curve, 37, 79, 81, 102, 372
Hubbert, M. King, 24, 27, 30, 37, 77, 79, 81, 82, 87, 89, 102,
        164, 165, 176, 213, 214, 217, 325, 346, 372, 374
Hubbert production cycle, 87
Hugoton-field, 390
Hydrogen, 19, 71–73, 85, 87–89, 190, 230, 232, 233, 239,
        240, 242–244, 258, 263, 266, 267, 270, 345

I

Ickes, Harold, 170
Imports, 30, 103, 114, 123, 178, 195, 197, 203, 204, 207,
        330, 363, 367, 375, 399
Industrial concentration, 149, 155–157
Industrialization, 13–14, 17, 19, 34, 65, 97–98, 111, 115,
        140, 156, 157, 196
Inflation
    biophysical economic model, 24
    neoclassical model, 165
    stagflation, 165
Insulation, 163, 402
Investments, 9, 20, 29, 53–55, 59, 61, 62, 88, 89, 95, 96,
        122–124, 126, 129, 157, 183, 198, 241, 243, 271,
        277, 279, 289, 290, 294, 323–338, 371, 374, 378,
        396, 400
Irrigation, 14, 49, 54, 100, 254, 274, 392, 399

J

Joule, 147, 228–232, 358

K

Keeling, Charles, 164
Kemp–Roth Tax Cut, 181, 183

Kennedy, John F., 22, 175
Keynes, John Maynard, 16, 29, 101, 112, 117, 122, 173,
        188, 189
Keyserling, Leon, 175
Kung, 42–44, 247

L

Labor, 4, 48, 74, 101, 134, 150, 164, 195, 217, 261, 301,
        311, 316, 336, 356, 381, 387, 396
    productivity, 14, 18, 20, 21, 23, 38, 108, 109, 115,
        120, 173, 183, 360, 397
Laherrere, Jean, 30, 77, 78, 81, 83, 324, 325, 338, 371, 397
Lavoissier, Antoine, 229
Liebig law of minimum, 99
Libya, 76, 176, 177
Limits to growth, 24, 33, 34, 96, 102, 125, 149, 209–221,
        373, 374
Liquefied natural gas, 73
Louisiana, 9, 328, 336, 388

M

Maintenance respiration, 230
Management, 18, 21, 122, 128, 150–152, 159, 160, 179,
        184, 188, 281, 294, 341, 350, 357, 396
Manhattan project, 337
Markets, 3, 48, 72, 96, 133, 148, 164, 194, 209, 239, 268,
        290, 302, 310, 323, 346, 355, 371, 387, 395
Marshall plan, 20, 172, 173
Mathematics, 48, 55, 60, 254, 287–290, 296, 298, 299,
        304, 341–343, 346, 350, 373
Mattei, Enrico,
Maxwell, James Clerk., 227
Maximum power, 99, 245, 249
Mercury, 229, 277, 343, 388, 391–393
Michelis-Menten, 292
Model, 24, 77, 105, 133, 148, 174, 196, 213, 234, 257,
        298, 302, 330, 341, 353, 372, 395
Mortgage-backed securities, 186

N

National debt, 188, 346
Natural energies, 138
Natural gas, 19, 26, 71–73, 75, 76, 81, 85, 87, 88, 95, 97,
        100, 183, 215, 216, 234, 236, 248, 260, 265–267,
        309, 314, 315, 323–328, 338, 376, 388, 391,
        392, 400
    liquids, 87, 323, 326
Natural resources, 4, 5, 26, 58, 62, 110, 116, 136–138,
        168, 205, 211, 271, 331, 358, 371, 387
Natural selection, 67, 234, 241, 243–245, 248, 253, 254,
        257, 267, 268, 270, 275, 277–281, 283, 305, 344,
        345, 350
Nature, 4, 42, 71, 95, 133, 196, 209, 225, 253, 288, 302,
        331, 342, 353, 373, 387, 395, 401
Negentropy, 234, 235, 260
Neoclassical economics
    failures, 116
    growth model, 214, 215, 217, 218

natural resources, 110, 116, 136
pareto efficiency, 105, 111, 349
Neoliberalism, 182, 193–207
New deal agencies
  Civilian Conservation Corps (CCC), 16, 168
  Federal Housing Administration (FHA), 169
  Federal National Mortgage Association
    (FNMA), 169
  Home Owners Loan Corporation (HOLC), 169
  National Labor Relations Board (NLRB), 170, 181
  National Recovery Administration (NRA), 169
  Public Works Administration (PWA), 169
  Tennessee Valley Authority (TVA), 169
  Works Project Administration (WPA), 16, 170, 175
New economists, 16, 174, 175, 179
Newton, Isaac, 226, 227, 254, 256, 288, 296
  energy cost, 7, 239, 244, 267–269
  fertilizer, 87, 267, 268, 399
NSC–68, 175
Nutrients, 41, 96, 241, 249, 281, 282

O
Odum Howard, 25, 41, 99, 116, 214, 241, 245, 249, 309,
    311, 338, 355, 356, 373, 401
Oil, 7, 61, 71, 95, 133, 148, 163, 197, 209, 225, 258, 288,
    303, 309, 323, 341, 353, 371, 387, 395
OPEC. *See* Organization of petroleum exporting countries
    (OPEC)
Organic matter, 265
Organization of Petroleum Exporting Countries (OPEC),
    23, 76, 165, 177, 179, 180, 182, 183, 378
  orographic, 273

P
Pax Americana, 18, 164, 178, 182
Peak everything, 209, 219
Peak oil, 8, 30, 34, 37, 74–75, 81, 82, 85, 96, 114, 149,
    163, 176–177, 189, 191, 209, 210, 219, 303, 319,
    323–338, 371–385, 397, 399, 400, 403
Pennsylvania, 149–151, 153, 154, 170, 180, 195, 263,
    346, 387, 388
Perlin, John, 41, 49, 51, 248
Petroleum, 2, 9, 15–18, 23, 24, 36, 37, 71–90, 147–161,
    164, 165, 171, 173, 177, 186, 218, 219, 232, 234,
    237, 248, 249, 265, 266, 294, 310–312, 314, 319,
    323–330, 336–338, 371–373, 376, 380, 392, 399
Phosphorus, peak phosphorus, 399
Polanyi, Karl, 8
Population, exponential growth, 214, 294
Prices, 5, 55, 74, 100, 134, 148, 164, 196, 209,
    225, 275, 288, 302, 309, 324, 347, 354, 371,
    389, 396
Priestly, Joseph, 229
Primary productivity, 38, 284
Production, 5, 41–68, 72, 96, 133, 148, 164, 194, 209,
    225–250, 253, 290, 301, 310, 323, 341, 362, 372,
    388, 397
Productivity, 6, 14, 18, 20–23, 25, 26, 34, 35, 38, 53,

58, 61, 64, 85, 99, 108, 109, 111–113, 115–118,
    120–122, 126, 150, 155, 157, 158, 160, 163, 164,
    166, 173–176, 178, 180, 182, 183, 189, 214, 284,
    359, 360, 397
  growth, 22, 25, 158, 173, 174, 178, 182

R
Radiation, 229, 238, 259, 276
Rain shadow, 274
Rankine, William, 228
Reagan, Ronald, 26, 28, 38, 165, 180, 181
Reconstruction Finance Corporation (RFC), 167, 168
Recycling, 261, 269
Reparations, 32, 159, 160, 166
Reserves, 7, 14, 15, 18, 32, 56, 64, 75–77, 85, 89, 90, 166,
    167, 170, 176, 178–181, 183, 185, 186, 232, 277,
    279, 318, 327, 338, 347, 358, 377, 399, 401
Resources, quality, 75, 136, 216, 217, 328,
    389, 392, 400
Revelle, Roger, 164
Ricardo, David, 5, 99, 101, 104, 108, 109, 113, 114, 126,
    193, 233, 271, 301, 328, 365
Rigor
  mathematical, 257, 288, 299, 342, 346
  scientific, 257, 288, 342, 373
Romans, 53, 56–61, 64, 248, 254
Roosevelt, Franklin, 16, 17, 29, 34, 38, 167, 168, 189

S
Sadi, C., 228, 229
Samuel, B., 175, 181
Science
  biophysical, 3, 8, 160
  natural, 3, 7, 34, 129, 255, 257, 298, 302, 303, 305,
    323, 348
Scientific method, 2, 129, 198, 206, 253–257, 281, 299,
    303, 305, 341, 345, 348, 350, 366
Secular stagnation, 191
Smith, Adam, 4, 6, 8, 21, 98, 99, 101, 104, 107, 109,
    112, 113, 115, 118, 125, 129, 148, 157, 200,
    215, 301
Soil, 1, 11–13, 31, 34, 36, 49, 51, 53, 54, 56–59, 67, 74,
    95, 96, 99, 100, 113, 148, 149, 169, 205, 209, 214,
    233, 236, 237, 241, 246, 248, 249, 253–256, 260,
    271, 273–276, 281, 284, 295, 301, 311, 314, 328,
    338, 356, 357, 359, 365, 392
Solar energy, 4, 6, 13, 14, 58–60, 64–67, 71, 72, 75, 99,
    102, 229, 237, 238, 241, 246–248, 258, 259, 272,
    273, 277, 282, 328, 356, 391, 399
Solar radiation, 229, 259
Spindletop, 9–11, 14, 154, 170
Steel, 7, 12, 13, 48, 67, 96, 148, 159, 200, 234, 247,
    311, 391
Supply-side economics, 26, 180–184
Sustainability, 35, 84, 88, 141, 143, 161, 191, 207, 311,
    357, 362, 371–385, 389, 399
Sustainable development, 207, 356, 399
Sweezy, Paul, 156, 167, 191

Systems, 1, 41, 74, 95, 134, 155, 165, 196, 210, 230, 253, 287, 303, 311, 328, 341, 355, 371, 387, 401
Systems approach, viii, 255, 269
Systems thinking, 391

T
Tainter, Joseph, 41, 58–59, 66, 67, 248
Technology, resource quality, 374
Texas railroad commission, 170, 177
Thatcher, Margaret, 29, 30, 165, 399
Thermodynamics, 134–137, 140, 200, 228, 232, 235–236, 239, 243, 260, 276, 304–306, 348, 379
Thomas, M., 101, 104, 114, 126, 210, 301
Thomas,W., 175, 181
Thompson, Benjamin (Count Rumford), 228
Thorium, 88
Three-Mile Island, 180
Thucydides, 51, 55, 56
Transportation, 4, 14, 15, 25, 72, 103, 107, 150–152, 154–158, 176, 177, 225, 234, 260, 315–317, 360, 375, 379
Treaty of Detroit, 164, 173
Trophic efficiency, fisheries, 62, 275, 343, 365, 387

Tropics, 47, 199, 237, 247, 259, 263, 265, 271, 276, 356, 399
Troubled Assets Relief Program (TARP), 167, 185

U
Unemployment, 16, 18, 22, 23, 26, 35, 74, 101, 105, 120, 122, 123, 135, 159, 165, 167, 168, 170, 171, 173, 174, 176, 178–181, 184, 185, 188, 189, 210, 215, 336, 371, 382, 396–398
Uranium, 17, 88, 216

W
Water, 6, 41, 71, 97, 155, 203, 209, 225, 253, 295, 323, 355, 382, 387, 397
Watt James, 6, 248
Work, 3, 48, 72, 95, 133, 147, 163, 196, 215, 226, 254, 287, 302, 311, 325, 343, 353, 372, 391, 403

Y
Yergin, Daniel, 17, 151, 165